Shlomo Libeskind • Isa S. Jubran

Euclidean, Non-Euclidean, and Transformational Geometry

A Deductive Inquiry

Shlomo Libeskind
Dept of Mathematics
University of Oregon
Eugene, OR, USA

Isa S. Jubran
Dept of Mathematics
SUNY Cortland
Cortland, NY, USA

ISBN 978-3-031-74152-4   ISBN 978-3-031-74153-1 (eBook)
https://doi.org/10.1007/978-3-031-74153-1

Mathematics Subject Classification (2020): 51-01

© Springer Nature Switzerland AG 2024

This work is subject to copyright. All rights are solely and exclusively licensed by the Publisher, whether the whole or part of the material is concerned, specifically the rights of translation, reprinting, reuse of illustrations, recitation, broadcasting, reproduction on microfilms or in any other physical way, and transmission or information storage and retrieval, electronic adaptation, computer software, or by similar or dissimilar methodology now known or hereafter developed.

The use of general descriptive names, registered names, trademarks, service marks, etc. in this publication does not imply, even in the absence of a specific statement, that such names are exempt from the relevant protective laws and regulations and therefore free for general use.

The publisher, the authors and the editors are safe to assume that the advice and information in this book are believed to be true and accurate at the date of publication. Neither the publisher nor the authors or the editors give a warranty, expressed or implied, with respect to the material contained herein or for any errors or omissions that may have been made. The publisher remains neutral with regard to jurisdictional claims in published maps and institutional affiliations.

This book is published under the imprint Birkhäuser, www.birkhauser-science.com by the registered company Springer Nature Switzerland AG
The registered company address is: Gewerbestrasse 11, 6330 Cham, Switzerland

If disposing of this product, please recycle the paper.

Contents

	Preface	ix
1	**Surprising Results and Basic Notions**	**1**
	1.1 Surprising Results and Unexpected Answers	1
	1.1.1 Problem Set	6
	1.2 Basic Notions	7
	1.2.1 Problem Set	20
2	**Congruence, Constructions, and the Parallel Postulate**	**21**
	2.1 Angles and Their Measurement	24
	2.1.1 Problem Set	27
	2.2 Triangles and Congruence of Triangles	29
	2.2.1 Problem Set	55
	2.3 The Parallel Postulate and Its Consequences	58
	2.3.1 Problem Set	65
	2.4 Parallel Projection and the Midsegment Theorem	69
	2.4.1 Problem Set	80
	2.5 More on Constructions	85
	2.5.1 Problem Set	92
3	**Circles**	**97**
	3.1 Central and Inscribed Angles	97
	3.1.1 Problem Set	110
	3.2 Inscribed Circles	115
	3.2.1 Problem Set	124
	3.3 More on Constructions	130
	3.3.1 Problem Set	139
4	**Area and the Pythagorean Theorem**	**141**
	4.1 Areas of Polygons	141
	4.1.1 Problem Set	154
	4.2 The Pythagorean Theorem	159
	4.2.1 Problem Set	170

5 Similarity — 177
- 5.1 Ratio, Proportion, and Similar Polygons — 177
  - 5.1.1 Problem Set — 194
- 5.2 Further Applications of the Side-Splitting Theorem and Similarity — 200
  - 5.2.1 Problem Set — 209
- 5.3 Areas of Similar Figures — 216
  - 5.3.1 Problem Set — 222
- 5.4 The Golden Ratio and the Construction of a Regular Pentagon — 223
  - 5.4.1 Problem Set — 229
- 5.5 Circumference and Area of a Circle — 233
  - 5.5.1 Problem Set — 241

6 Isometries and Size Transformations — 245
- 6.1 Reflections, Translations, and Rotations — 245
  - 6.1.1 Problem Set — 259
- 6.2 Congruence and Euclidean Constructions — 267
  - 6.2.1 Problem Set — 274
- 6.3 More on Extremal Problems — 277
  - 6.3.1 Problem Set — 286
- 6.4 Similarity Transformation with Applications to Constructions — 290
  - 6.4.1 Construction of a Common Tangent — 292
  - 6.4.2 Problem Set — 295

7 Composition of Transformations — 297
- 7.1 Introduction — 297
- 7.2 In Search of New Isometries — 298
  - 7.2.1 Problem Set — 315
- 7.3 Composition of Rotations, the Treasure Island Problem, and Other Treasures — 319
  - 7.3.1 Problem Set — 330

8 More Recent Discoveries — 335
- 8.1 The Nine-Point Circle and Other Results — 335
  - 8.1.1 Problem Set — 340
- 8.2 Complex Numbers and Geometry — 342
  - 8.2.1 Problem Set — 356

9 Inversion — 363
- 9.1 Introduction — 363
  - 9.1.1 Problem Set — 368
- 9.2 Properties of Inversions — 372
  - 9.2.1 Problem Set — 383
- 9.3 Applications of Inversions — 388
  - 9.3.1 Shoemaker's Knife Problem — 389

	9.3.2 Apollonius' Problem	393
	9.3.3 Peaucellier–Lipkin Linkage	394
	9.3.4 Ptolemy's Theorem	396
	9.3.5 Miquel's Theorem	398
	9.3.6 Problem Set	402
9.4	The Nine-Point Circle and Feuerbach's Theorem	409
	9.4.1 Problem Set	416
9.5	Stereographic Projection and Inversion	419
	9.5.1 Problem Set	429

10 Hyperbolic Geometry — 431

- 10.1 Introduction … 431
- 10.2 Hyperbolic Geometry … 437
  - 10.2.1 Problem Set … 454
- 10.3 Models of Hyperbolic Geometry … 458
  - 10.3.1 Problem Set … 475
- 10.4 Compass and Straightedge Constructions in the Poincaré Disc Model $\mathbb{D}^2$ … 476
  - 10.4.1 Problem Set … 491
- 10.5 Hyperbolic Tessellations … 494
  - 10.5.1 Problem Set … 496

11 Elliptic Geometries — 499

- 11.1 Introduction and Basic Results … 499
- 11.2 Models of Elliptic Geometry … 508
  - 11.2.1 Problem Set … 519

12 Projective Geometry — 525

- 12.1 Introduction and Early Results … 525
  - 12.1.1 Problem Set … 539
- 12.2 Projective Planes … 541
  - 12.2.1 Problem Set … 543
- 12.3 The Real Projective Plane … 543
  - 12.3.1 Problem Set … 550
- 12.4 Homogeneous Coordinates … 551
  - 12.4.1 Problem Set … 554
- 12.5 Duality: Poles, Polars, and Reciprocation … 555
  - 12.5.1 Problem Set … 561
- 12.6 Polar Circles and Self-Polar Triangles … 564
  - 12.6.1 Problem Set … 566

13 Taxicab Geometry — 569
- 13.1 Introduction — 569
- 13.2 Taxicab Versus Euclidean — 571
  - 13.2.1 Problem Set — 578
- 13.3 Distance from a Point to a Line — 582
- 13.4 Taxicab Midsets — 584
  - 13.4.1 Problem Set — 588
- 13.5 Circle(s) Through Three Points — 589
- 13.6 Conics in Taxicab Geometry — 593
- 13.7 Taxicab Incircles, Circumcircles, Excircles, and Apollonius' Circle — 595
- 13.8 Inversion in Taxicab Geometry — 599
  - 13.8.1 Problem Set — 601

14 Fractal Geometry — 607
- 14.1 Introduction — 607
- 14.2 Fractal Dimension — 611
  - 14.2.1 Problem Set — 615
- 14.3 Affine Transformations — 618
- 14.4 Iterated Function Systems — 619
  - 14.4.1 Problem Set — 623
- 14.5 The Julia and Mandelbrot Sets — 626
  - 14.5.1 Problem Set — 628
- 14.6 Linear Transformations and 2×2 Matrices: A Brief Summary — 629

15 Solid Geometry — 631
- 15.1 Objectives — 631
- 15.2 Fundamental Concepts — 633
  - 15.2.1 Problem Set — 642
- 15.3 Polyhedra — 644
- 15.4 Descartes' Lost Theorem — 646
- 15.5 Euler's Formula and Its Consequences — 653
  - 15.5.1 Problem Set — 659

Bibliography — 663

Index — 669

Preface

Our goal in writing this text is to spark in college geometry students a passion for problem solving and a love for mathematics. The first eight chapters of the text evolved out of the text *Euclidean and Transformational Geometry: A Deductive Inquiry* (Jones and Bartlett 2008). The subsequent seven chapters mainly deal with various non-Euclidean geometries. The proof and problem-solving strategies used in this book guide students toward successfully solving unfamiliar problems and proving results on their own. Some of the questions students will ask and often find answers to in the text include:

- How does one know where to begin and how to proceed?
- Which approach is more promising, and why?
- Are different solutions possible, and how do they compare?

Our experience shows that many students, at first frustrated with the challenge of non-routine, proof-oriented problems, ultimately develop into passionate problem solvers. They especially appreciate realizing that proofs and constructions do not come "out of thin air," but are based on logical thinking processes. Students can apply such deductive reasoning skills to other areas of learning/teaching. Therefore, in order to improve the teaching/learning of mathematics in general and geometry in particular, we must stress that teachers possess a deep level of understanding of the subject matter, enjoy solving nonroutine problems, and be passionate about the subject. This will lead to learners' having a better understanding of the concepts being taught. We believe that teaching/learning from this text will help achieve these goals.

Experience at most universities shows that the majority of students remember very little geometry from courses taken in high school. Furthermore, most do not have experience with constructing proofs or with solving nonroutine problems. Because of these gaps in student experience, Chapters 1–8 of this text do not assume previous knowledge of geometry. Definitions of basic terms used in geometry are included for those who lack previous knowledge, and should help those with some previous knowledge to validate the accuracy of their knowledge.

In Chapters 1–8, we discuss, at a higher level and in much greater depth, topics that are normally taught in a high school geometry courses. These topics include definitions and axioms, congruence, circles and related concepts, area and the Pythagorean theorem, similarity, isometries and size transformations, and composition of transformations. Constructions and the use of transformations to carry out constructions are emphasized. Problem sets appearing at the end of each section include nonroutine problems that students, based on our experience, will enjoy exploring, and some challenging problems whose solutions are surprisingly not too difficult once discovered.

For students and pre-service teachers to truly understand and appreciate some of the concepts in Chapters 1–8, they need to examine what happens when one or more of the building blocks of Euclidean geometry (points, lines, axioms, etc.) are altered. For instance, altering the parallel postulate to state that a line has more than one parallel or that a line has no parallel lines results in hyperbolic geometry or elliptic geometry, respectively. In Chapters 9–15, we discuss in some detail a number of topics and geometries that are normally not taught at the high school level. The treatment of these topics is not meant to be comprehensive; it is rather intended to supplement the material in Chapters 1–8 by introducing interesting topics mainly dealing with non-Euclidean geometry. Including some of these ideas in a high school honors geometry course, a sophomore/junor-level college geometry course, or a second college geometry course will help learners realize that Euclidean geometry is not the only valid geometry. As Henri Poincaré is quoted to have said, "One geometry cannot be more true than another; it can only be more convenient" and "Geometry is not true, it is advantageous."

In Chapter 9, we consider circle inversion both as a unique transformation and as an important problem-solving tool. We use it to solve the shoemake's knife problem, Apollonius's problem, and the Peaucellier–Lipkin linkage apparatus. We also use it to prove Ptolemy's theorem, Miquel's theorem, nine-point circle theorem, and Feuerbach's theorem. Finally, we relate circle inversion to stereographic projection. Inversion becomes a recurring theme used as a useful tool for exploring concepts from various modern geometries.

In Chapter 10, we consider hyperbolic geometry. We review Saccheri's work that led to the discovery of this geometry. Special attention is given to models of the hyperbolic plane, especially the Poincaré disc model $\mathbb{D}^2$. Inversion in a hyperbolic line is shown to be an isometry on $\mathbb{D}^2$ which leads to the conclusion that the angle of parallelism associated with a point not on a line is uniquely determined by the distance between the point and the line. Given the availability of free dynamic geometry software such as *GeoGebra*, constructions in hyperbolic geometry using inversion are emphasized, and hyperbolic tessellations are briefly explored.

In Chapter 11, we consider single and double elliptic geometries, and describe models of these geometries including the Beltrami–Riemann sphere model and the Klein half-sphere and disc models. We also review some of Saccheri's work that led to the discovery of these geometries and prove Girard's theorem. Inversion is used for constructions in the Klein disc model.

In Chapter 12, we consider projective geometry. We introduce central projection and use it to prove classical results such as Desargue's theorem, the invariance of the cross ratio under projection, Pascal's theorem, and Pappus' theorem. Models of the projective plane are briefly described and homogeneous coordinates are introduced. Poles, polars, and reciprocation with respect to a fixed circle are studied to emphasize the planar principle of duality. Finally, we show that for an obtuse triangle, the circumcircle and the nine-point circle are inverted onto each other under inversion in the polar circle of the triangle.

In Chapter 13, we consider taxicab geometry. We define the taxicab metric and describe a taxicab angle measure analogous to the radian measure of Euclidean angles, which allows us to define taxicab trigonometric functions. The distance between a point and a line is defined and used to define taxicab midsets and discuss some of their applications. Taxicab circles through three points, conics, incircles, circumcircles, and excircles are also discussed. Finally, inversion in taxicab geometry is explored.

In Chapter 14, we consider fractal geometry. We define the similarity dimension and box dimension and apply these to a number of fractal objects. Iterated function systems are then used to generate complex fractal images similar to objects found in nature. The availability of free fractal-generating software such as the *IFS Construction Kit* makes hands-on dynamic exploration of fractals possible. Finally, the Julia and Mandelbrot sets are defined and briefly explored.

In Chapter 15, we consider solid geometry. We briefly explore various fundamental concepts of three-dimensional space and give a concise definition of polyhedra. We then investigate Descartes' lost theorem, prove Euler's formula, and prove that there are only five regular convex polyhedra. Finally, some important consequences of Euler's formula are discussed.

1 Surprising Results and Basic Notions

> *It may well be doubted whether, in all the range of science, there is any field so fascinating for the explorer–so rich in hidden treasures–so fruitful in delightful surprises –as Pure Mathematics.*
>
> Lewis Carroll (1832–1898), best known as author of *Alice's Adventures in Wonderland* and *Through the Looking Glass*

Introduction

Frequently a course in high school geometry starts with a long list of definitions and proofs of theorems that are intuitively obvious. It rarely includes theorems and proofs that would make a lasting impression on the student.

Pause for a moment and recall some geometric theorems. Are there any on your list that contain an unexpected or surprising result? Which are especially beautiful? Are any of them especially important and, if so, why?

The first section of this chapter introduces some problems and theorems from Euclidean geometry whose solutions or statements you will very likely find surprising. In several cases, you will be asked to conjecture a solution or a theorem through a sequence of suggested experiments. Later in this text, you will learn how to prove some of these conjectures and statements. Several of the proofs are particularly beautiful owing to their unexpected simplicity and applicability to new problems.

To complete the problems in this introductory chapter (and in other chapters), you will need a compass, a ruler, and some blank sheets of paper. Dynamic geometry software such as *GeoGebra, Geometer's Sketchpad* and *Geometry Expressions* is especially helpful in investigating geometric properties and making conjectures. Throughout the text we will, when appropriate, suggest optional activities using the software. It should be noted, however, that the text can be read and studied without geometry software.

We begin by taking the terms *point*, *line*, *plane*, and *space* as undefined (not in the sense that we don't know what they are, but in the sense that they have not been given formal definitions). We also take as undefined the concepts of *set*, *belongs to*, or *is an element of* a set. In the formal axiomatic approach that we will be taking, in this introductory chapter we use only the properties of the undefined terms that we state in the axioms—we are not allowed to make conclusions based on drawings.

1.1 Surprising Results and Unexpected Answers

The Treasure Island Problem

Among her great-grandfather's papers, Ansley found a parchment describing the location of a pirate treasure buried on a deserted island. The island contained a coconut tree, a banana tree, and a gallows where traitors were hanged. It was accompanied by the following directions:

> *Walk from the gallows to the coconut tree, counting the number of steps. At the coconut tree, turn 90° to the right. Walk the same distance and put a spike in the ground. Return to the gallows and walk to the banana tree, counting your steps. At the banana tree, turn 90° to the left, walk the same number of steps, and put another spike in the ground. The treasure is halfway between the spikes.*

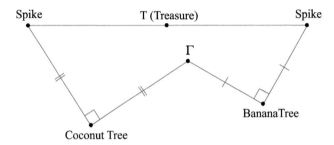

Figure 1.1

Ansley found the island and the two trees but could find no trace of the gallows or the spikes, as both had probably rotted. Nevertheless, because she knew geometry she was able to locate the treasure. Your quest is to devise a plan to find the exact location of the treasure.

If you have spent enough time pondering a solution but could not find one, try the following: Fix the positions of the two trees. Choose an arbitrary position for the gallows and mark it Γ_1. Follow the directions to find the corresponding spikes and the midpoint between them, T_1. Next choose another position for the gallows Γ_2 and follow the directions to find the corresponding location of the treasure, T_2. Repeat the procedure for at least two more positions of the gallows (geometry software is especially convenient here). What do you notice about T_1, T_2, T_3, and T_4? Now conjecture how to locate the treasure.

The Pythagorean Theorem

One of the best-known and most useful theorems in geometry and perhaps all of mathematics is the Pythagorean Theorem, which was discovered in the sixth century B.C.E. (Before the Common Era). There are numerous known proofs of the theorem (*The Pythagorean Proposition* by E. Loomis contains hundreds of them). Do you recall any proof of the Pythagorean Theorem? Can you prove it on your own?

The following is one way to state the theorem: If squares are constructed on the sides of any right triangle (a triangle with a 90° angle), then the area of the largest square equals the sum of the areas of the other squares. In Figure 1.2, if the areas of the squares are A, B, and C, then $A + B = C$ or, equivalently, $a^2 + b^2 = c^2$, where a and b are the lengths of the legs of the triangle and c is the length of the hypotenuse (the side opposite and right angle).

1.1 Surprising Results and Unexpected Answers

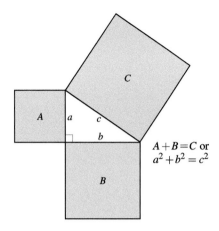

Figure 1.2

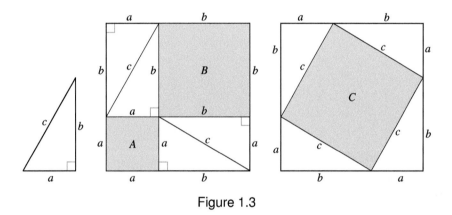

Figure 1.3

Figures 1.2 and 1.3 can be used to justify the Pythagorean Theorem. Can you see how?

In Chapters 4 and 5, we will give several proofs of the Pythagorean Theorem. In Chapter 5 we will also explore what other figures can be constructed on the sides of a right triangle so that the area of the figure constructed on the hypotenuse is equal to the sum of the areas of the figures constructed on the other two sides. We will also generalize the theorem for triangles that are not necessarily right triangles.

The Hiker's Path

A hiker H in Figure 1.4 needs to get first to the river r and then to her tent T. Find the point X on the bank of the river so that the hiker's total trip $HX + XT$ is as short as possible. This problem will be investigated in Chapters 2 and 6.

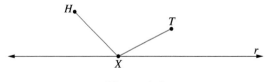

Figure 1.4

The Shortest Highway

A highway connecting two cities A and B as in Figure 1.5 needs to be built so that part of the highway is on a bridge perpendicular to the parallel banks b_1 and b_2 of the river. Where should the bridge be built so that the path $AXYB$ is as short as possible? This problem will be investigated in Chapter 6.

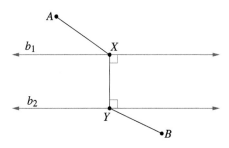

Figure 1.5

Steiner's Minimum Distance Problem

One of the greatest geometers of all time, and certainly of the 19th century, was Jacob Steiner (1796–1863). Born in Switzerland but educated in Germany, Steiner discovered and proved new theorems and introduced new geometrical concepts. In particular, he was interested in the solutions of maximum and minimum problems using purely geometric methods—that is, without using calculus or algebra. Among the many results, he proved the theorem illustrated in Figure 1.6.

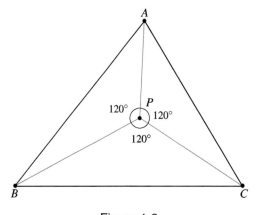

Figure 1.6

1.1 Surprising Results and Unexpected Answers

If A, B, and C are three points forming a triangle such that each of the angles of $\triangle ABC$ is less than $120°$, then the point for which the sum of the distances from P to the vertices of the triangle (that, is $PA + PB + PC$) is at its minimum has the property that each of the angles at P is $120°$. Steiner also dealt with the case when one of the angles is $120°$ or greater than $120°$. In addition, he generalized the problem to n points.

Many maximum and minimum problems can be solved more efficiently using purely geometric methods rather than calculus. Such problems will be investigated using transformational geometry in Chapter 6.

The Nine-Point Circle

The 19th century experienced a renewed interest in classical Euclidean geometry. Probably the most spectacular discovery was the **Nine-Point Circle**, which was investigated simultaneously by the French mathematicians Charles Jules Brianchion (1785–1864) and Jean-Victor Poncelet (1788–1867), who published their work jointly in 1821. The theorem is, however, commonly attributed to the German mathematician and high school teacher Karl Wilhem Feurbach (1800–1834), who independently discovered the theorem and published it with some related results in 1822.

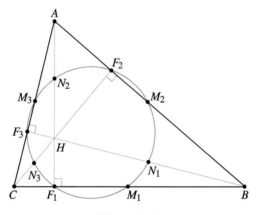

Figure 1.7

Given any triangle ABC, nine particular points can be associated with it, as shown in Fig. 1.7. The first three points M_1, M_2, and M_3 are the midpoints of the three sides of the triangle. The next three points N_1, N_2, and N_3 are the midpoints of the segments joining the vertices A, B, and C with the point H where H is the point of intersection of the three altitudes of the triangle (we will prove in Chapter 2 that the three altitudes of any triangle intersect in a single point). The final three points F_1, F_2, and F_3 are the points of intersection of each altitude with each corresponding side (these points are known as the "feet" of the altitudes).

The theorem states that all nine points lie on one circle called the *Nine-Point Circle*. For an acute triangle (all its angles are less than $90°$), the midpoints and the feet of the altitudes, lie on the triangle itself; for an obtuse triangle (one angle is greater than $90°$) two of the feet are outside the triangle but they still lie on the circle. The *Nine-Point Circle Theorem* and some related results will be proved in Chapter 8.

Morley's Theorem

In 1899, Frank Morley, professor of mathematics at Haverford College and later at Johns Hopkins University, discovered an unusual property of the trisectors of the three angles of any triangle: If the angle trisectors are drawn for each angle of any triangle, then the adjacent trisectors of the angles meet at vertices of an equilateral triangle. In Figure 1.8, the adjacent trisectors of angles A and B meet at D, the adjacent trisectors of angels A and C meet at E, and the adjacent trisectors of angles C and B meet at F. Morley's theorem states that triangle FDE is equilateral (i.e., its three sides are all the same length). Since 1899, many different proofs of this theorem have been published, including several since 2000. A proof of Morley's Theorem will be presented in Chapter 8.

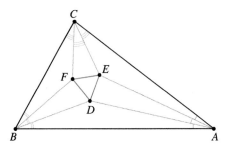

Figure 1.8

1.1.1 Problem Set

In each of the following problems, you may use any tools to perform the experiments (geometry software such as the GeoGebra, Geometer's Sketchpad or Geometry Expressions is especially convenient but not necessary).

1. **a.** Conjecture the solution to the *Treasure Island problem* by choosing at least four different positions for the gallows as described in the text.
   b. Does your conjecture hold for some Γ positioned below the line connecting the coconut and banana tree? (Make an appropriate construction.)
   c. Place Γ at one of the trees and find the corresponding treasure.
   d. Place Γ on the line connecting the trees halfway between them and find the corresponding treasure.
   e. Based on your experiments in (a) through (d), what seems to be the simplest way to find the treasure?

2. **a.** Draw a circle on transparent (or see-through) paper. How would you find the center of the circle by folding the circle onto itself?
   b. Draw an arc of a circle. How could you find the center now?

1.2 Basic Notions

3. Let $ABCD$ be any convex quadrilateral. On each side of the quadrilateral, construct a square as shown in Figure 1.9. Find the centers C_1, C_2, C_3, and C_4 of the squares, where C_1 and C_3 are centers of opposite squares. How are the segments C_1C_3 and C_2C_4 related? Repeat the experiment starting with a different quadrilateral.

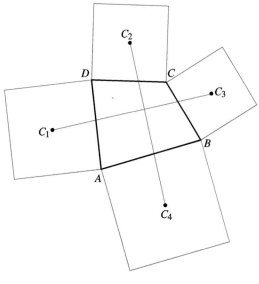

Figure 1.9

4. Construct the *Nine-Point Circle* for an arbitrary triangle.

5. Check Morley's Theorem experimentally for arbitrary triangles.

6. Use Figure 1.3 to prove the Pythagorean Theorem.

1.2 Basic Notions

In this section we set out the basic definitions and axioms that form the foundation for the work we do in the rest of the book. Many of the notions presented there are intuitive, and we merely formalize them by stating them as axioms or theorems that can be proved.

Axiom 1.1

Line, planes, and space are set of points. Space contains all points.

In geometry, it is common to use the terms *on*, *in*, *passes through*, and *lies on*. We say that a point is *on* a line rather than *belongs* to a line. Synonymously, we say that a line *passes through* a given point. We also say that a point is *on* a plane or *in* a plane. A line *lies* in a plane if it is a subset of the plane—that is, if every point on the line is also in the plane. Points that are on the same line are called **collinear**. Points that are on the same plane are **coplanar**.

The following axioms describe the fundamental relationships among points, lines, and planes. The accompanying figures are merely models for the relationships; they do not represent the only possible configuration.

Axiom 1.2

Any two distinct points are on exactly one line. Every line contains at least two points.

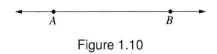

Figure 1.10

Remark 1.1

It may seem that Axiom 1.2 implies that all lines are straight; otherwise, more than one line could be drawn through two points. First, notice that "straightness" has not yet been defined. Second, it is possible to show an example of a geometry in which lines are objects that satisfy Axiom 1.2 and other axioms in this section but are not "straight."

Axiom 1.3

Any three noncollinear points are on exactly one plane. Each plane contains at least three noncollinear points.

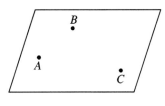

Figure 1.11

Axiom 1.4

If two points of a line are in a plane, then the entire line is in the plane.

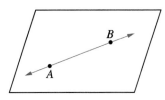

Figure 1.12

1.2 Basic Notions

Axiom 1.5

In space, if two planes have a point in common, then the planes have an entire line in common.

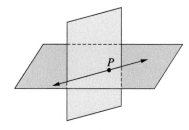

Figure 1.13

Axiom 1.6

In space, there exist at least four points that are noncoplanar.

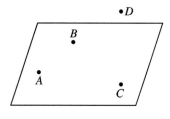

Figure 1.14

Remarks

- Axiom 1.2 is often encountered in an equivalent form: "Two points determine a unique line" or "There is one and only one line passing through two distinct points."
- An equivalent form of Axiom 1.3 is "Three noncollinear points determine a unique plane."
- Notice the second sentence in Axioms 1.1 and 1.2. We know intuitively that lines and planes contain infinitely many points, but this fact does not follow from the preceding axioms. Additional axioms will be needed to assure infinitude of points on a line.

Notation for Points, Lines, and Planes

It is customary to denote points by capital letters of the Latin alphabet. Axiom 1.2 assures that a line can be named by any two points on it. The line containing points A and B, as in Figure 1.10, will be denoted by $\overleftrightarrow{AB}$ or line AB. Whenever convenient, we also may name a line by a single letter. In this text, only lowercase letters from the Latin alphabet are used to name lines.

Axiom 1.3 assures that a plane can be named by any three noncollinear points in the plane. For example, the plane containing the upper face $ABCD$ of the box in Figure 1.15 can be named in each of the following ways: plane ABC, plane BCD, plane ADC, and plane ABD. (Of course, the plane can also be named by any three other noncollinear points not labeled in the figure.) Whenever convenient, we may also name a plane by a single letter.

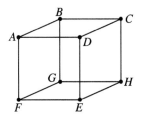

Figure 1.15

Recall that the intersection of two sets is the set of all elements that are common to both sets. We know intuitively that two distinct lines either do not intersect (have no points in common) or intersect in exactly one point. This understanding leads to our first theorem.

Theorem 1.1

If two distinct lines intersect, they intersect in exactly one point.

Proof. Let the lines be k and ℓ. It is given that the lines intersect. Therefore, there exists a point P that is on both lines. We want to show that k and ℓ have no other points in common. For that purpose, we use an indirect proof. Suppose there is another point Q on k and ℓ, as in Figure 1.16.

Figure 1.16

By Axiom 1.2, there is a unique line through P and Q. Hence, $k = \ell$, which contradicts the hypothesis that the lines are different. Consequently, the existence of another intersection point Q must be rejected. ∎

Axiom 1.3 assured us that three noncollinear points determine a plane. Can you come up with another way of uniquely determining a plane? You may have come up with another way already—namely, "two parallel lines uniquely determine a plane." However, before the concept of parallel lines has any meaning in our system, we must formally define parallel lines.

1.2 Basic Notions

Definition of Parallel and Skew Lines

Two lines are parallel if they lie in the same plane and do not intersect. Lines that do not intersect and are not contained in any single plane are called skew lines.

If ℓ and m are parallel, we write $\ell \| m$. In Figure 1.15, for example, $\overleftrightarrow{AB} \| \overleftrightarrow{DC}$, $\overleftrightarrow{AB} \| \overleftrightarrow{EH}$, and $\overleftrightarrow{AC} \| \overleftrightarrow{FH}$, but $\overleftrightarrow{AB}$ and $\overleftrightarrow{DE}$ are skew lines.

Now Solve This 1.1

1. What is the maximum number of intersection points determined by n lines in the same plane?
2. What is the maximum number of lines determined by n points? Does it matter if the points are in the same plane?

Coordinate Systems and Distance

Following G. D. Birkhoff's (1884–1944) axioms, we now introduce the concept of distance, assuming the existence and properties of real numbers. We start with an intuitive background that will motivate the axioms and definitions that follow.

Historical Note: *George David Birkhoff*
George David Birkhoff was one of the most distinguished leaders in American mathematics and the preeminent U.S. mathematician of his time. He taught at Harvard from 1912 until his death in 1944. Birkhoff proposed an axiom system for Euclidean geometry in his 1941 text *Basic Geometry*.

Given any line, a **coordinate system** on the line can be created by choosing an arbitrary point O on the line and having that point correspond to 0. This point O is called the **origin**. Next to the right of point O, another point P is chosen (see Figure 1.17) to which corresponds the number 1. The segment $\overline{OP}$ is called **unit** segment.

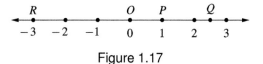

Figure 1.17

By marking off segments equal to the length of $\overline{OP}$ repeatedly to the left and right of O, we find points corresponding to the integers. By dividing segments into an appropriate number of equal parts, we find points that correspond to all rational numbers. You most likely know that any real number (and not only rational numbers) corresponds to some point on the line and, conversely, that every point of the line corresponds to some real number. (In Chapter 4, we show how to find points that correspond to real numbers such as $\sqrt{2}$ and $\sqrt{5}$.) Thus, there is one-to-one correspondence between the points on a line and the real numbers. Such a correspondence is called a **coordinate system** for a line. The number corresponding to a given point P is called the **coordinate** of P. Thus the coordinate of Q in Figure 1.17 is 2.5.

If we denote the line in figure 1.17 by x, we write the coordinate of Q as x_Q. Thus $x_Q = 2.5$ and $x_R = -3$. We can find the distance between two points by using the coordinates of the points. For example, in Figure 1.17, we have $PQ = OQ - OP = x_Q - x_P = 2.5 - 1 = 1.5$ We can also find RP by finding the difference between the coordinates of the points: $RP = x_P - x_R = 1 - (-3) = 4$. Because distance is a non-negative number and it is cumbersome to indicate which point has the greater coordinate, we use the absolute value function. Thus $AB = |x_A - x_B|$.

Based on this discussion we introduce the following axiom and definitions.

Axiom 1.7: The Ruler Postulate

The points on a line can be put in one-to-one correspondence with the real numbers.

Definition of a Coordinate System for a line

The correspondence in Axiom 1.7 is called a coordinate system for a line. A line with a coordinate system is called a number line.

Definition of a Distance Between Two Points

In a coordinate system for $\overleftrightarrow{AB}$, the distance between points A and B, denoted by AB, is the real number $|x_A - x_B|$, where x_A and x_B are the coordinates of A and B, respectively.

You may have already observed that the distance between two points depends on the unit chosen for the coordinate system. If Q and R stay in the same plane but we change the position of the point that corresponds to the number 1, the QR will change as well. Also, because the distance from a point A to the origin is the real number $|x_A|$, and there exist real numbers as large as we wish, we can conclude that there are points on a line as far from the origin as we wish. Therefore, we can say that a line is infinite in length.

We can use the concept of distance to define what we mean when we say that a point is between two other points.

Definition of Betweenness

B is between A and C if and only if A, B, and C are collinear and $AB + BC = AC$ (see Figure 1.18). In this case we write $A - B - C$.

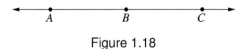

Figure 1.18

1.2 Basic Notions

Definition of a Segment

The segment $\overline{AB}$ consists of the points A and B and all points between A and B.

The length of segment $\overline{AB}$ is the distance between A and B, denoted by AB. A point M between A and B such that $AM = MB$ is a **midpoint** of $\overline{AB}$. Our intuition tells us that every segment has exactly one midpoint. We can prove this fact by using Axiom 1.7 and our definitions. You will be guided in the process of finding a proof in the problem set at the end of this chapter. Most of us probably find it more satisfying to prove statements that are not intuitively obvious. Nevertheless, you may find the challenge of proving a statement rewarding in itself even if the statement seems obvious. We also want you to realize that even intuitively obvious statements can be logically deduced from the axioms and definitions.

Example 1.1

Given two points A and B on a number line with coordinates x_A and x_B, respectively, find the coordinate x_M of M, the midpoint of $\overline{AB}$. See Figure 1.19.

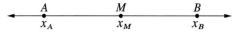

Figure 1.19

Solution. Without loss of generality assume $x_B > x_A$. The definition of the midpoint implies that $AM = MB$. This equation implies

$$x_M - x_A = x_B - x_M,$$
$$2x_M = x_A + x_B,$$
$$x_M = \frac{x_A + x_B}{2}.$$

Now Solve This 1.2

A student approaches the solution of Example 1.1 as follows: Because $AB = x_B - x_A$ and the midpoint of AB is halfway between A and B, the coordinate of the midpoint should be $\frac{1}{2}(x_B - x_A)$. The student realizes the answer is wrong but would like to know why and how to use her approach to obtain the correct answer. How would you respond?

Definition of a Ray

The ray $\overrightarrow{AB}$ [shown in Figure 1.20(a)] is the union of $\overline{AB}$ and the set of all points C such that B is between A and C. The point A is called the endpoint of the ray. The rays having a common endpoint and whose union is a line are opposite rays. In Figure 1.20(b), $\overrightarrow{AC}$ and $\overrightarrow{AB}$ are opposite rays.

(a)   (b)

Figure 1.20

> **Definition of an Angle**
>
> An angle is a union of two rays with a common endpoint. The common endpoint is the vertex of the angle, and the two rays are the sides of the angle.

In Figure 1.21(a), the angle shown is the union of $\overrightarrow{AB}$ and $\overrightarrow{AC}$ and its vertex is A. (Using set notation, the angle is $\overrightarrow{AB} \cup \overrightarrow{AC}$.) The angle in Figure 1.21(a) is denoted by $\angle BAC$ or $\angle CAB$. When there is no danger of ambiguity, it is common practice to name an angle by its vertex. Thus the angle in Figure 1.21(a) can also be denoted by $\angle A$. If the rays $\overrightarrow{AB}$ and $\overrightarrow{AC}$ are on the same line–that is, if A, B, and C are collinear, as in Figure 1.21(b)—then $\angle ABC$ is called a **straight angle**.

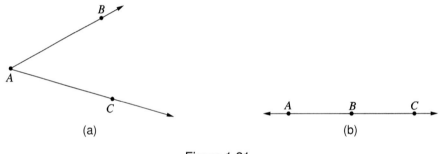

(a)   (b)

Figure 1.21

If A, B, and C are three noncollinear points, then the union of the three segments $\overline{AB}$, $\overline{BC}$, and $\overline{AC}$ is a **triangle** and is denoted by $\triangle ABC$ (shown in Figure 1.22). The three segments are the **sides** of the triangle, and the three points are the **vertices**.

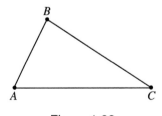

Figure 1.22

At this juncture, we could prove several statements that are intuitively obvious. For example, if $A - B - C$ (B is between A and C), then $C - B - A$ (B is between C and A). Also, $\overline{AB} = \overline{BA}$ and $\triangle ABC = \triangle CBA$. By "equal," we mean "exactly" equal in the set theory sense: Two sets are equal if they contain the same elements. To give a rigorous

1.2 Basic Notions

treatment of the next topics we will discuss, we need the concept and properties of half-planes. Intuitively, we know that any line divides the plane into two parts separated by the line and that each part is referred to as the **half-plane**. This fact will be introduced in Axiom 1.8, the Plane-Separation Axiom. Before we get to this axiom, however, it will be useful to define what we mean by a convex set.

> **Definition of a Convex Set**
>
> A set is convex is for every two points P and Q belonging to the set, the entire segment $\overline{PQ}$ is in the set.

Notice that the interiors of the triangle as well as the circle in Figure 1.23 are convex sets. (We are making this statement on an intuitive basis; the interior of a triangle has not been defined yet.) Also, the segment $\overline{AB}$ is a convex set. However, the interiors of the figures in Figure 1.24 are not convex, as the segment $\overline{PQ}$ in each figure does not lie entirely in the interior of the figure.

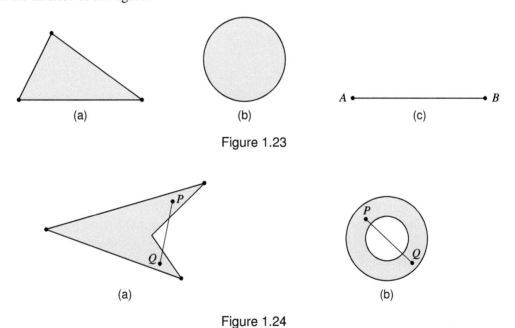

Figure 1.23

Figure 1.24

Properties of convex sets have been extensively investigated, and the concept of convex sets has important applications in mathematics. For example, in the interior of any closed curve (see Figure 1.25) there exist a point P and three chords through P (a chord of a closed curve is a segment connecting any two points of the curve) with the following property: The six angles formed at P are 60° each, and P is the midpoint of each of the chords. (The proof of this statement requires more extensive study of a convex sets. Figure 1.25 illustrates the statement for a circle.)

Some questions involving convex sets have not been answered yet. One such question involves equichordal points. A point P in the interior of a region is an **equichordal** point

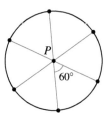

Figure 1.25

if all chords through P are of the same length. For example, the center of a circle is an equichordal point. However, there exist noncircular regions that have equichordal points. In 1916, Fujiwara raised the question of whether there exists in a plane a convex region that has two equichordal points. No one yet has been able to give a complete answer to this question. In 1984, Spaltenstein described a construction of a convex region on a sphere that has two equichordal points. (This does not answer the question posed in 1916, as the region is on a sphere and hence is not planar.)

Axiom 1.8: The Plane-Separation Axiom

Each line in a plane separates all the points of the plane that are not on the line into two nonempty sets, called half planes, with the following properties:

1. The half-planes are disjoint (have no points in common) convex sets.
2. If P is in one half-plane and Q is in the other half-plane, the segment $\overline{PQ}$ intersects the line that separates the plane.

Notice that neither of the half-planes in Axiom 1.8 includes the line. Thus a line divides the plane into three mutually disjoint subsets: the two half-planes and the line. Also, it follows from Axiom 1.8 that a half-plane is determined by a line and a point not on the line. Thus we can refer to the two half-planes in Figure 1.26 as "the half-plane of ℓ containing P" and "the half-plane of ℓ containing Q."

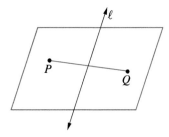

Figure 1.26

Using Axiom 1.8, it is possible to prove a theorem named after the German mathematician Moritz Pasch (1843–1930), which states that if a line intersects one side of a triangle and does not go through any of its vertices, it must also intersect another side of the triangle (see

1.2 Basic Notions

Figure 1.27). In 1882, Pasch published one of the first rigorous treatises on geometry where he stated this theorem as an axiom (he did not use Axiom 1.8 as an axiom). Pasch realized that Euclid often relied on assumptions made visually from diagrams and contributed to filling the gaps in Euclid's reasoning.

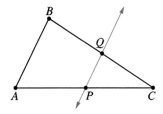

Figure 1.27

We state Pasch's axiom as a theorem and leave its proof for you to explore in the problem set at the end of this chapter.

> **Pasch's Axiom**
>
> If a line intersects a side of a triangle and does not intersect any of the vertices, it also intersects another side of the triangle.

Using Axiom 1.8, it becomes possible to precisely define the interior of an angle and hence the interior of a triangle.

> **Definition of the Interior of an Angle**
>
> If A, B, and C are not collinear, then the interior of $\angle BAC$ is the intersection of the half-plane of $\overleftrightarrow{AB}$ containing C with the half-plane of $\overleftrightarrow{AC}$ containing B. (See Figure 1.28.)

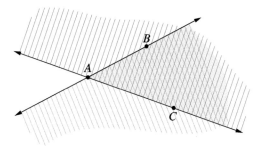

Figure 1.28

> **Remark 1.2**
>
> This definition is valid (or, as it is commonly referred to in mathematics, **well-defined**) if it is independent of the choices of B and C on the sides of the angle. It can be proved that this indeed is the case.

Using the preceding definition, we can define the interior of a triangle. Coming up with an appropriate definition is left to the reader as an exercise. The definition of the interior of an angle can also be used to define **betweenness** for rays.

> **Definition of Betweenness of Rays**
>
> $\overrightarrow{AD}$ is between $\overrightarrow{AB}$ and $\overrightarrow{AC}$ if and only if $\overrightarrow{AB}$ and $\overrightarrow{AC}$ are not opposite rays, and D is in the interior of $\angle BAC$. (See Figure 1.29.)

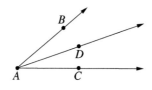

Figure 1.29

We can use this definition to prove the following visually obvious theorem. Its proof is also left as an exercise.

> **Theorem 1.2**
>
> $\overrightarrow{CD}$ intersects side $\overline{AB}$ of $\triangle ABC$ between A and B if and only if $\overrightarrow{CD}$ is between $\overrightarrow{CA}$ and $\overrightarrow{CB}$. (See Figure 1.30).
>
>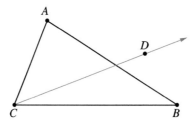
>
> Figure 1.30

Angle Measurement

Angles are commonly measured in degrees with a protractor. (Another unit of measurement for angles is the **radian**.) To measure $\angle EAB$, we place the protractor as shown in Figure 1.31 and read off the measure of the angle as 140°.

1.2 Basic Notions

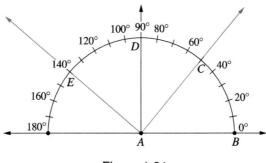

Figure 1.31

The following axioms and definitions formalize our intuitive knowledge of the protractor.

Axiom 1.9: The Angle Measurement Axiom

There is a real number between 0 and 180 that corresponds to every angle. The number 180 corresponds to a straight angle.

The number in Axiom 1.9 is called the **degree measure of the angle** and is written $m(\angle BAC)$ (m stands for measure). In Figure 1.31, $m(\angle BAC) = 50°$ and we say that $\angle BAC$ is a 50-degree angle, written as $50°$.

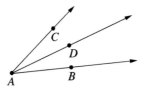

Figure 1.32

Axiom 1.10: The Angle Construction Postulate

Let $\overrightarrow{AB}$ be a ray on the edge of a half-plane. For every real number r, $0 < r < 180$, there is exactly one ray, with C in the half plane, such that $m(\angle CAB) = r$ degrees.

Axiom 1.11: The Angle Addition Postulate

If D is a point in the interior of $\angle BAC$ (see Figure 1.32), then $m(\angle BAD) + m(\angle DAC) = m(\angle BAC)$.

Notice that Axiom 1.11 implies that measures of angles can also be computed by subtraction. For example, $m(\angle BAD) = m(\angle BAC) - m(\angle DAC)$.

1.2.1 Problem Set

In the following problems, use the axioms, definitions, and theorems presented in this chapter to prove what is required.

1. Prove that:

   a. If two lines intersect, they lie in exactly one plane.

   b. Two parallel lines determine a unique plane.

2. Consider the undefined terms: *ball*, *player*, and *belongs to*. Also consider the following axioms:

   Axiom 1: There is at least one player.

   Axiom 2: To every player belong two balls.

   Axiom 3: Every ball belongs to three players.

   Prove or disprove each of the following:

   a. There are at least two balls.

   b. There are at least three players.

   c. There is always an even number of balls.

   d. The number of players is always odd.

   (Hint: You may want to model the balls by points and players by segments.)

3. Consider the undefined terms: *line*, *point*, and *belongs to*. Also consider the following axioms:

   Axiom 1: Any two lines have exactly one point in common.

   Axiom 2: Every point is on (belongs to) exactly two lines.

   Axiom 3: There are exactly four lines.

   Answer each of the following:

   a. How many points are there? Prove your answer.

   b. Prove that there are exactly three points on each line.

   Hint: You may substitute the term point by person and line by committee.

4. Using the terminology presented in this chapter, come up with a precise definition of the interior of a triangle.

5. Which of the following is true: *always*, *sometimes* but not *always*, or *never*? Justify your answers.

   a. The intersection of two convex sets is convex.

   b. The intersection of two nonconvex sets is a nonconvex set.

   c. The union of two convex sets is convex.

   d. The union of two nonconvex sets is nonconvex.

6. Prove Pasch's Axiom.

2 Congruence, Constructions, and the Parallel Postulate

> *At the age of eleven, I began Euclid, with my brother as my tutor. This was one of the great events of my life, as dazzling as first love. I have not imagined there was anything so delicious in the world... From that moment until... I was thirty-eight, mathematics was my chief interest and my chief source of happiness.*
>
> Bertrand Russell (1872–1970) *The Autobiography of Bertrand Russell* (London: G. Allen and Unwin, 1968)

Historical Note: *Bertrand Russell*
Lord Bertrand Arthur William Russell was a British philosopher, logician, and mathematician who made important contributions to the foundations of mathematics. In 1910, he became a lecturer at Cambridge University but was dismissed and later jailed for making pacifist speeches during World War I. He abandoned pacifism during World War II in the face of the Nazi atrocities in Europe and the Nazi threat to Great Britain. After the war he reverted back to pacifism, becoming a leader in the anti-nuclear and anti-Vietnam War movements. Later Russell taught at several U.S. universities, including Harvard University and the University of Chicago. He won the Nobel Prize for Literature in 1950. Russell died in 1970 at the age of 98.

Introduction

In this chapter we lay the foundation for Euclidean and non-Euclidean geometry. About 250 B.C.E., Euclid systematically collected and organized the geometrical knowledge of his time in a treatise composed of 13 books, called *The Elements*. (This treatise also included number theory and topics in algebra.) Euclid started out with a list of statements called *axioms* or *postulates* that he assumed to be true and then showed that geometric statements followed logically from his assumptions. However, he did not realize that some geometric terms could not be defined. Some terms must be left undefined to avoid "circular" definitions. (For example, in one dictionary a *line* is defined as "the path traced by a moving point" and then a *path* is defined as "a line obtained by a moving point.") Euclid "defined" a *point* as that which has no part and a *line* as breadthless length, when neither *part* nor *breadth* nor *length* had been defined.

In his proofs Euclid used unstated assumptions and relied on diagrams to make what seemed to him to be obvious conclusions. Despite these shortcomings, Euclid's achievements were monumental. In presenting a vast amount of mathematics by starting with a few basic assumptions and then logically deducing other mathematical statements, Euclid laid the foundation for a deductive approach to mathematics. Only toward the end of the 19th century was a rigorous foundation for Euclidean geometry established. In 1899, David Hilbert in his book *Foundations of Geometry* established a set of axioms along with undefined terms for Euclidean geometry and succeeded in proving Euclid's theorems relying solely on logic. Hilbert's success had its roots in a revolution that had taken place in geometry some years earlier. In 1829, the Russian mathematician Nikolai Lobachevsky—and independently two years later the Hungarian mathematician Janos Bolyai—established a new geometry referred to as non-Euclidean geometry. At the same time, the great German mathematician Karl Friedrich Gauss was very likely aware of the new results, but did not publish them as he was worried about the controversy that they might arouse.

Non-Euclidean geometry is based on an axiom that denies Euclid's Fifth Postulate. (Euclid used the term *postulate* for geometric assumptions and the term *axiom* for general mathematical non-geometrical assumptions. For example, one of Euclid's axioms is "The part is smaller than the whole" and one of his postulates is "A straight line segment can be drawn joining any two points.") Euclid's *Fifth Postulate* or the *Parallel Postulate* is equivalent to the following:

Axiom 2.1: Euclidean Parallel Postulate

Through a given point P not on a line ℓ there is exactly one line parallel to ℓ. (See Figure 2.1.)

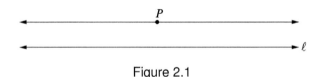

Figure 2.1

For generations, mathematicians believed it was possible to prove the Parallel Postulate using Euclid's other postulates and axioms, and many tried to do so. Only Lobachevsky, Bolyai, and Gauss were bold enough to replace Euclid's Parallel Postulate by another one that denied Euclid's postulate. The postulate they used is referred to as the Hyperbolic Parallel Postulate.

Axiom 2.2: Hyperbolic Parallel Postulate

Given a line ℓ and a point P not on ℓ, there exist at least two lines through P that are parallel to ℓ.

Euclid's approach to geometry was directed by physical reality. Thus, points and lines were idealized mathematical terms for what we perceive as points and straight lines, Postulates were "self-evident" truths. To the originators of hyperbolic geometry, points and lines were undefined terms satisfying certain axioms; as such they did not necessarily represent any idealized physical objects. It is possible to find familiar objects to represent points, lines, and planes that satisfy the Hyperbolic Parallel Postulate. This will be done in a later chapter.

In Sections 1.2 and 1.3, we develop neutral geometry (also called absolute geometry) in which the Parallel Postulate is not used. Consequently, theorems in neutral geometry are true in Euclidean geometry as well as hyperbolic geometry. We adopt a set of axioms that constitute a slight modification of the axioms introduced by the American mathematician George David Birkhoff (1884–1944). The axioms are *consistent* and *independent*. A set of axioms is said to be consistent if no contradictions can be derived from the set; and it is said to be independent if no axiom of the set is implied by the other axioms in the set.

In 1899, David Hilbert introduced axioms that extended the original set of axioms put forth by Euclid. Hilbert's goal was to make Euclid's system complete and introduce a level of rigor into geometry that did not depend upon figures. Hilbert's system included 16 axioms and proofs, though using it can make proofs long and cumbersome. A student of the well-known mathematician Hugo Steinhaus wrote a formal proof of the Pythagorean Theorem from Hilbert's axioms, a treatise which ran 80 pages long. Hilbert's purely geometric axiom system differs from Birkhoff's in that it does not assume any properties of the real numbers. Because Birkhoff assumed properties of real numbers, his set of axioms is remarkably concise.

Nevertheless, Hilbert's approach to geometry is a monumental contribution to mathematics as a whole! His book, *Foundations of Geometry*, heralded the beginning of the modern axiomatic method in mathematics. Hilbert emphasized that it is not necessary to assign a specific meaning to an undefined concept. In an oft-quoted piece of conversation, Hilbert asserted that the undefined terms of point, line, and plane could be substituted respectively with tables, chairs, and beer glasses. What matters is the way in which the axioms relate.

Because the intention of this textbook is to obtain results quickly and as simply as possible, we base our axioms on a modification of Birkhoff's axioms, introduced in Chapter 1.

> **Historical Note:** *Euclid (Third Century B.C.E.)*
> The place and exact year of Euclid's birth are not known. Historians believe he was the first mathematics professor at the University of Alexandria. Euclid's *The Elements* was the first treatise on mathematics as a deductive system and is considered to be one of the greatest achievements of humankind. It has been more widely studied than any other book except for the Bible. More than 1000 editions of *The Elements* have appeared since its first printing in 1482. However, no Greek copy from Euclid's time has been found; instead, *The Elements* reached the West from an Arabic translation. The teaching of geometry has been dominated by *The Elements* for more than 20 centuries, and Euclid's work has had a profound influence on scientific thinking.

2.1 Angles and Their Measurement

Pairs of Angles

Certain pairs of angles occur often enough that it is convenient to give them special names.

Two angles are **adjacent** if they lie in the same plane, they share a common side, the interiors of the angles have no point in common, and the vertices are the same. ∠CAD and ∠DAB in Figure 2.2 are adjacent angles. The common side of two adjacent angles of equal measure is called an **angle bisector**.

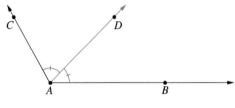

Figure 2.2

Two angles form a **linear pair** if they are adjacent and the noncommon sides are opposite rays (see Figure 2.3(a). Two angles are **vertical** if their sides form two pairs of opposite rays (see Figure 2.3(b)).

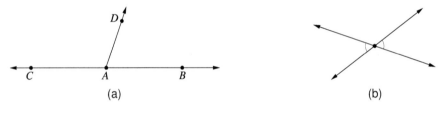

Figure 2.3

It seems obvious and can be readily deduced that the sum of the measures of two angles in a linear pair is 180°. We often encounter pairs of angles whose measures add up to 180° or 90°. Such angles have special names. Two angles are **supplementary** if the sum of their measures is 180°; each angle is called the **supplement** of the other. Two angles are **complementary** if the sum of their measures is 90°; each angle is called the **complement** of the other.

Notice that two angles in a linear pair are supplementary but not every two supplementary angles form a linear pair. In fact, it follows that if two angles are supplementary and adjacent, then they form a linear pair. An angle measuring less than 90° is termed **acute**, and one measuring more than 90° and less than 180° is termed **obtuse**. (For now all our angles have non-negative measure.)

It is common to call a 90° angle a **right angle**. An angle measuring 180° is a **straight angle**. **Congruent angles** are angles that have the same measure (we denote the measure of an angle by m). Thus, if $m(\angle A) = m(\angle B)$, then ∠A and ∠B are **congruent**. We use the symbol ≅ for congruent. Hence ∠A ≅ ∠B. Notice that the equality symbol (=) is

2.1 Angles and Their Measurement

reserved for real numbers or for sets that have the same elements (the degree measure is a real number). Because two angles that have the same measure are not necessarily the same set of points, we do not use the equality symbol between congruent angles. Similarly, we do not use the equality symbol between congruent segments.

Notation for Angles and Their Measures

It is often cumbersome to use the notation $m(\angle A)$ or $m(\angle BAC)$ for the measure of an angle. For this reason, we commonly use lower case Latin or Greek alphabet letters for measures of angles.

In Figure 2.4 we denoted the measures of the three angles of $\triangle ABC$ by α, β, and γ (notice the correspondence to A, B, and C). Thus $m(\angle A) = \alpha$, $m(\angle B) = \beta$, and $m(\angle C) = \gamma$. Because α is already the measure of $\angle A$, we do not write $m(\alpha)$ or $m(\angle \alpha)$.

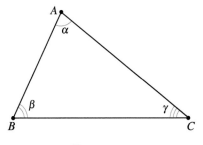

Figure 2.4

The following theorems follow immediately from the preceding definitions. We prove Theorem 10.12.

Theorem 2.1

1. *Supplements of congruent angles are congruent.*
2. *Complements of congruent angles are congruent.*
3. *The sum of the measures of two angles in a linear pair is* $180°$.

Theorem 2.2

Vertical angles are congruent.

Proof. Denoting the measures of the angles in Figure 2.5 by α, β, γ, and δ, we need to show that $\alpha = \beta$ and $\gamma = \delta$. Notice that $\alpha + \gamma = 180°$ and $\beta + \gamma = 180°$, which implies that $\alpha = \beta$. (Alternatively, we could say that α is a supplement of γ and that β is a supplement of γ and, therefore, by Theorem 10.11, $\alpha = \beta$.)

26 Chapter 2. Congruence, Constructions, and the Parallel Postulate

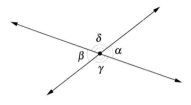

Figure 2.5

When two lines intersect, they form four angles as shown in Figure 2.5. If one of the angles formed is a right angle, the lines are **perpendicular**. Previous theorems imply that in this case all four angles are right angles. The fact that lines are perpendicular or equivalently that an angle formed is a right angle is denoted in drawings by the symbol ⌐ as shown in Figure 2.6. We also write $a \perp b$ to denote that lines a and b are perpendicular.

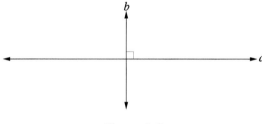

Figure 2.6

When a line intersects a segment at its midpoint and is perpendicular to the segment, it is called the **perpendicular bisector** of the segment. The line ℓ in Figure 2.7 is the perpendicular bisector of $\overline{AB}$. Notice that M is the midpoint of $\overline{AB}$ and that the congruent segments $\overline{AM}$ and $\overline{MB}$ are marked by the same symbol.

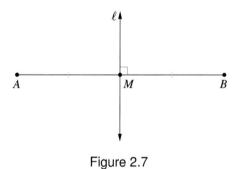

Figure 2.7

Example 2.1

In Figure 2.8, $\angle AOB$ and $\angle BOC$ form a linear pair. The rays $\overrightarrow{OD}$ and $\overrightarrow{OE}$ are their angle bisectors, respectively. Prove that $\overleftrightarrow{OE} \perp \overleftrightarrow{OD}$.

2.1 Angles and Their Measurement

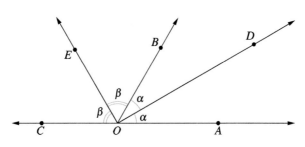

Figure 2.8

Solution. Because $\overrightarrow{OD}$ and $\overrightarrow{OE}$ are angle bisectors, two pairs of congruent angles are formed. We denote the measures of the congruent angles in each pair by α and β, respectively, as shown in Figure 2.8. Because $\angle AOB$ and $\angle BOC$ are a linear pair, we have $2\alpha + 2\beta = 180°$. Hence $\alpha + \beta = 90°$.
Because $m(\angle EOD) = \alpha + \beta$, it follows that $m(\angle EOD) = 90°$ and therefore $\overleftrightarrow{OE} \perp \overleftrightarrow{OD}$.

2.1.1 Problem Set

1. Consider the four angles α, α_1, β, and β_1 created by two intersecting lines as shown. Prove that if one of the angles is a right angle, then the other three are also right angles.

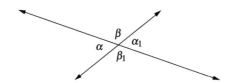

2. Two lines AB and CD intersect at O. If $\overrightarrow{OP}$ is the bisector of $\angle BOD$, prove that when $\overrightarrow{OP}$ is extended, the extension bisects $\angle AOC$.

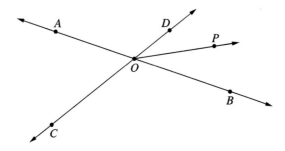

3. Points A, O, and B are on line ℓ with O between A and B. The line is drawn on a sheet of paper and then the paper is folded about O so that $\overrightarrow{OA}$ "falls" on $\overrightarrow{OB}$. Explain why when the paper is unfolded the crease is perpendicular to $\overleftrightarrow{AB}$.

4. $\overline{OA} \perp \overline{OC}$ and $\overline{OB} \perp \overline{OD}$. Which of the non-right angles formed in the diagram are congruent? Justify your answer.

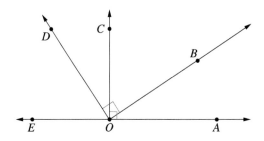

5. Examine the following argument showing that the sum of the measures of the angles in any triangle is 180°. Is the argument valid? Justify your answer.

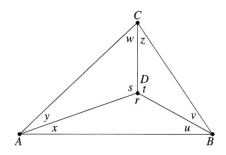

In $\triangle ABC$ we choose an arbitrary point D in the interior of the triangle and connect it to the three vertices of the triangle. We then mark the measures of the nine angles as shown in the figure and let the sum of the measures of the three angles in a triangle be k. We show that $k = 180°$.

Adding up all the measures of the angles in the three triangles $\triangle ABD$, $\triangle ACD$ and $\triangle CBD$, we get $3k$. Hence $(x+u+r)+(v+z+t)+(w+y+s) = 3k$. By the commutative and associative properties of addition, we get $(x + u + v + z + w + y) + (r + t + s) = 3k$. Notice that the quantity in the first parentheses represents the sum of the measures of the angles in $\triangle ABC$ and for that reason equals k. This and the fact that $r + t + s = 360$ imply that $k + 360 = 3k$ or $k = 180°$. (Notice that we put the degree mark only in the final answer.)

6. Jaimee announced that she has her own somewhat different proof of Example 2.1. Examine Jaimee's proof and compare it with the one given in the text. Answer the following questions:

   a. What are the strengths and weaknesses of Jaimee's exposition? In the case of shortcomings, suggest improvements.

   b. Which proof do you like better? Why?

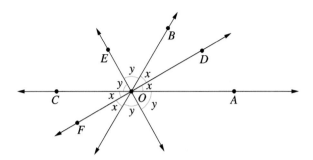

Jaimee:
I know that if a ray divides a straight angle into two congruent angles, each must be 90°. Thus, to show that $EO \perp OD$, I need to prove only that $\angle FOE = \angle EOD$. I extend the rays in Figure 2.8 and use the fact that the bisector of an angle divides it into two congruent angles and the fact that vertical angles are congruent. I mark the angles by x or by y as shown in my figure above. $\angle FOD$ is a straight angle because I extended the ray OD. I see that ray OE divides $\angle FOD$ into two angles, each equal to $x + y$. Hence each angle is 90°.

7. Find the exact time(s) between noon and 2PM when the hour and minute hands of a clock are on a straight line. Explain your reasoning.

2.2 Triangles and Congruence of Triangles

Informally, when two figures have the same size and shape, they are congruent. In Section 2.1, we defined congruent segments as segments that have the same length, and congruent angles as angles that have the same measure. When two figures are congruent, it is always possible to fit one figure onto the other so that matching sides and angles are congruent. This is the basis for defining congruent triangles.

> **Definition of Congruent Triangles**
>
> Two triangles are congruent if there is a one-to-one correspondence between their vertices so that corresponding sides are congruent and corresponding angles are congruent.

In Figure 2.9, $\triangle ABC$ and $\triangle DEF$ are congruent. Corresponding sides and angles are marked as shown. Notice that if we fit one triangle onto the other, vertex A will correspond to vertex D, B to E, and C to F. (You should be able to superimpose one triangle onto the other by tracing $\triangle DEF$ on a sheet of paper. In some cases you will need to flip the sheet and trace the triangle on the other side of the paper before superimposing it on a congruent triangle.) The fact that the triangles are congruent is written as $\triangle ABC \cong \triangle DEF$. Whenever the symbol $\cong$ is used, corresponding vertices must be written in the same position. Thus,

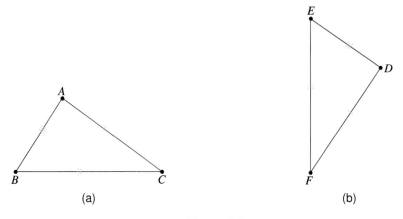

Figure 2.9

without reference to a figure, the congruence $\triangle ABC \cong \triangle DEF$ implies $\angle A \cong \angle D$, $\angle B \cong \angle D$, $\angle C \cong \angle F$, $\overline{AB} \cong \overline{DE}$, $\overline{BC} \cong \overline{EF}$, and $\overline{AC} \cong \overline{DF}$.

If the three sides and three angles of one triangle are congruent to the corresponding sides and angles of another triangle, by definition the two triangles are congruent. However, congruence of fewer corresponding parts is sufficient to determine that two triangles are congruent.

Figure 2.10 shows $\triangle ABC$ and $\triangle DEF$ in which $\overline{AB} \cong \overline{DE}$, $\angle A \cong \angle D$, and $\overline{AC} \cong \overline{DF}$.

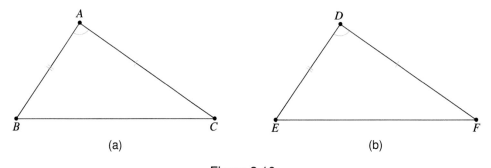

Figure 2.10

These conditions assure that the triangles are congruent, which can be seen as follows. Because $\angle A \cong \angle D$, we could superimpose $\angle D$ on top of $\angle A$ so that ray $\overrightarrow{DE}$ falls on ray $\overrightarrow{AB}$ and $\overrightarrow{DF}$ falls on $\overrightarrow{AC}$. Because $\overline{AB} \cong \overline{DF}$ and $\overline{AC} \cong \overline{DF}$, the vertex E will fall on B and the vertex F on C. Thus the vertices D, E, and F will fall on the corresponding vertices A, B, and C and, therefore, the triangles are congruent.

Notice, however, that this argument is an intuitive justification and not a formal proof. We do not have a rigorous definition of what it means to superimpose one figure on top of another. In a later chapter, we will define transformations, which will enable us to give a precise definition of what it means to *move* a figure and hence a definition of congruent figures. Meanwhile we state the congruence condition discussed above as an axiom.

2.2 Triangles and Congruence of Triangles

Axiom 2.3: The Side – Angle – Side (SAS) Triangle Congruence Condition

If two sides of a triangle and the angle included between these sides, are congruent to the corresponding sides and angle of a second triangle, then the triangles are congruent.

We can use the SAS condition to prove other congruence conditions, but first we need a theorem concerning the angles of an **isosceles triangle**. A triangle is isosceles if at least two of its sides are congruent. If all the sides of a triangle are congruent, the triangle is **equilateral**. (Notice that the definitions imply that an equilateral triangle is isosceles but an isosceles triangle is not necessarily equilateral.)

The two congruent sides of an isosceles triangle are called the **sides** of the triangle and the third side is the **base** of the triangle. The angle of an isosceles triangle opposite the base is the **vertex angle**, and the other two angles are the **base angles**. In an equilateral triangle, any side can be a base. A triangle with no congruent sides is called **scalene**. A triangle with all acute angles is called an **acute** triangle. A triangle with an obtuse angle is called an **obtuse** triangle.

Theorem 2.3: The Isosceles Triangle Theorem

If two sides of a triangle are congruent, then the angles opposite these sides are congruent.

Theorem 10.13 is frequently stated as: *The base angles of an isosceles triangle are congruent.*

Figure 2.11(a) gives a quick pictorial representation of this theorem. The marks on the sides indicate that the sides are congruent and the marks on the base angles indicate the conclusion that the angles are congruent.

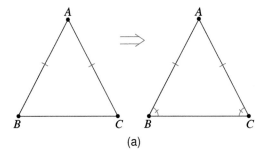 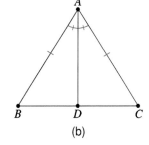

(a) (b)

Figure 2.11

We formally state the hypothesis (what is given) and the conclusion (what we need to prove) of the theorem:

Given: In $\triangle BAC$ (see Figure 2.11(a)), $\overline{AB} \cong \overline{AC}$.

Prove: $\angle B \cong \angle C$.

Plan: One approach is to "divide" $\triangle ABC$ into two triangles that seem to be congruent and in which $\angle B$ and $\angle C$ will be corresponding angles. Anticipating the use of the SAS condition, we bisect $\angle A$ in Figure 2.11(b). If we could show that the resulting two triangles are congruent, it would follow that the corresponding angles $\angle B$ and $\angle C$ are congruent.

Proof. Bisect $\angle A$. The angle bisector intersects $\overline{BC}$ at D as shown in Figure 2.11(b).

We have now $\triangle ABD \cong \triangle ACD$ by SAS ($\overline{AB} \cong \overline{AC}$, $\angle BAD \cong \angle CAD$, and $\overline{AD} \cong \overline{AD}$). Because the corresponding parts of the triangles are congruent, $\angle ABD \cong \angle ACD$. ∎

Alternative Proof. $\triangle BAC \cong \triangle CAB$ by SAS because $\overline{BA} \cong \overline{CA}$, $\angle BAC \cong \angle CAB$, and $\overline{AC} \cong \overline{AB}$. Because the corresponding angles in these triangles are congruent, we have $\angle B \cong \angle C$. ∎

> **Remark 2.1**
>
> You may feel uneasy about the second proof of Theorem 10.13, perhaps because you are used to seeing two separate triangles when you are proving that triangles are congruent. Note that the definition of congruence of triangles does not preclude the possibility that $\triangle BAC \cong \triangle CAB$ as indicated in the second proof. Notice that if $\triangle BAC$ is not isosceles, then $\triangle BAC \not\cong \triangle CAB$.

Theorem 10.13 implies the following:

> **Corollary 2.1**
>
> All the angles of an equilateral triangle are congruent; that is, an equilateral triangle is equiangular.

We are now ready to prove two other congruence conditions for triangles.

> **Theorem 2.4: The Angle – Side – Angle (ASA) Condition**
>
> *Given a one-to-one correspondence between the vertices of two triangles, if two angles and the included side of one triangle are congruent to the corresponding parts of the second triangle, the two triangles are congruent.*

Given: $\triangle ABC$ and $\triangle A_1B_1C_1$, the correspondence $A \leftrightarrow A_1$, $B \leftrightarrow B_1$, $C \leftrightarrow C_1$ and $\angle A \cong \angle A_1$, $\overline{AB} \cong \overline{A_1B_1}$, $\angle B \cong \angle B_1$.

Prove: $\triangle ABC \cong \triangle A_1B_1C_1$.

Plan: Because the only congruence condition we can use is the SAS condition, we try in Figure 2.12(b) to construct an additional side so that SAS can be applied. One approach is to construct a point C_2 so that $\overline{A_1C_2} \cong \overline{AC}$. Then, by SAS, $\triangle ABC \cong \triangle A_1B_1C_2$. We then prove that $C_2 = C_1$.

2.2 Triangles and Congruence of Triangles

Proof. Construct on $\overrightarrow{A_1C_1}$ in Figure 2.12 a point C_2 such that $A_1C_2 = AC$. By SAS regardless of whether $C_2 \neq C_1$, $C_2 = C_1$ or $A_1C_2 < A_1C_1$, we have

1. $\triangle ABC \cong \triangle A_1B_1C_2$.
   To show that $C_2 = C_1$, notice that (1) implies
2. $\angle A_1B_1C_2 \cong \angle ABC$.
   In addition, the hypothesis tells us that
3. $\angle ABC \cong \angle A_1B_1C_1$.

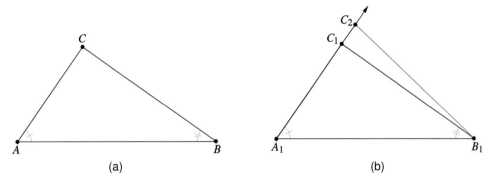

(a)   (b)

Figure 2.12

Statements 2 and 3, along with the transitive property of congruence, imply that $\angle A_1B_1C_1 \cong \angle A_1B_1C_2$. We show now that this congruence of angles implies $C_2 = C_1$. Indeed, if $C_1 \neq C_2$, then either C_1 is between A_1 and C_2 or C_2 is between A_1 and C_1. In the first case, C_1 would be in the interior of $\angle A_1B_1C_2$, which would imply that $m(\angle A_1B_1C_1) < m(\angle A_1B_1C_2)$ and would contradict the fact that the angles are congruent. Similarly, we can show that C_2 between A_1 and C_1 contradicts the equality of the angles. Thus $C_2 = C_1$, and from (1) we obtain $\triangle ABC \cong \triangle A_1B_1C_1$. ∎

> **Now Solve This 2.1**
>
> Use the ASA congruence condition to prove the following statements:
> 1. If two angles of a triangle are congruent, then the sides opposite these angles are congruent. [Notice that this statement is the converse of Theorem 10.13. In Figure 2.12(a) show that if $\angle B \cong \angle C$ then $\triangle BAC \cong \triangle ABC$.]
> 2. An equiangular triangle is equilateral.

Medians, Perpendicular Bisectors, and Additional Properties of Isosceles Triangles

The segment connecting the vertex of a triangle to the midpoint of the opposite side is called a **median**. If M is the midpoint of $\overline{BC}$ in Figure 2.13, then $\overline{AM}$ is a median. Every triangle has three medians.

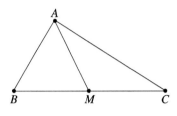

Figure 2.13

Now Solve This 2.2

To investigate the relationship among a median, angle bisector, and perpendicular bisector of a base in an isosceles triangle, fold or draw an isosceles triangle ABC as shown in Figure 2.14(a), where $AB = AC$ and $\overline{BC}$ is the base. Fold the paper so that C falls on top of B and then unfold it. The unfolded triangle and the crease are shown in Figure 2.14(b).

(a)

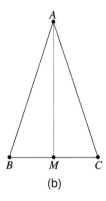
(b)

Figure 2.14

While performing the paper-folding activity, you likely noticed that $\triangle ACM$ was folded onto $\triangle ABM$ [see Fig. 2.14(b)]. Thus it seems that $\angle CAM \cong \angle BAM$, $\overline{BM} \cong \overline{MC}$, and $\angle AMB \cong \angle AMC$. The last congruence of angles means that $\overline{AM} \perp \overline{BC}$. These observations tell us that in an isosceles triangle, the angle bisector of the vertex angle (opposite the base) is also the perpendicular bisector of the base. We state these observations in the following theorem:

Theorem 2.5

The median to the base of an isosceles triangle is the perpendicular bisector of the base as well as the angle bisector of the vertex angle.

Proof. Assume that in Figure 2.14(b), $\overrightarrow{AM}$ is the angle bisector of $\angle A$. It is sufficient to prove that $\triangle ABM \cong \triangle ACM$. The congruency of these triangles follows from SAS. ∎

2.2 Triangles and Congruence of Triangles

> **Remark 2.2**
>
> In the proof of Theorem 10.14 and from now on we assume that an angle has a unique angle bisector. For a proof of this fact see Moise [65, p. 109].

Theorem 10.14 can be viewed as follows: If vertex A in Figure 2.14(b) is equidistant from the endpoints of the segment $\overline{BC}$ (that is, $AC = AB$), then A lies on the perpendicular bisector of $\overline{BC}$. It seems that the converse, which we state in the next theorem, is also true.

> **Theorem 2.6**
>
> *Every point on the perpendicular bisector of a segment is equidistant from the endpoints of the segment.*

Given: Line m is the perpendicular bisector of $\overline{AB}$ (see Figure 2.15) and P is any point on m.

Prove: $\overline{AP} \cong \overline{BP}$.

Proof. There are two cases. If P is on $\overline{AB}$ (not shown), then because it is on the perpendicular bisector of $\overline{AB}$, it must be the midpoint of $\overline{AB}$ and hence equidistant from A and B. If P is not on $\overline{AB}$ as in Figure 2.15, we have $\triangle AMP \cong \triangle BMP$ by SAS because $\overline{AM} \cong \overline{BM}$, $\overline{MP} \cong \overline{MP}$, and the included angles at M are congruent as each is a right angle. From the congruence of the triangles, we have $\overline{AP} \cong \overline{BP}$. ∎

Figure 2.15

Theorem 2.6 tells us that every point on the perpendicular bisector of a segment is equidistant from the endpoints of the segment. Are there any other such points? That is, are there points not on the perpendicular bisector that are equidistant from the endpoints of the segment? By Theorem 10.14, if P is equidistant from A and B, it must be on the perpendicular bisector of $\overline{AB}$. Consequently, the perpendicular bisector of a segment is the set of all points equidistant from the endpoints of the segment. In geometry, a set of points satisfying a certain property is often called the locus. Thus we have the following corollary to Theorem 2.6:

> **Corollary 2.2**
>
> A point is equidistant from the endpoints of a segment if and only if it is on the perpendicular bisector of the segment. Equivalently, the locus of all points equidistant from the endpoints of a segment is the perpendicular bisector of the segment.

Figure 2.16 shows two points P and Q that are equidistant from A and B. From Corollary 2.2 we know that P and Q are on the perpendicular bisector of $\overline{AB}$. Because a line is determined by two points, any two points on the perpendicular bisector of a segment determine the perpendicular bisector. Hence $\overleftrightarrow{PQ}$ is the perpendicular bisector of $\overline{AB}$. This is stated in the following corollary to Theorems 10.14 and 2.6:

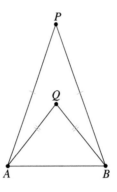

Figure 2.16

> **Corollary 2.3**
>
> If each of two points is equidistant from the endpoints of a segment, then the line through these points is the perpendicular bisector of the segment.

Corollary 2.3 can be represented pictorially as shown in Figure 2.17.

> **Now Solve This 2.3**
>
> 1. Use any tools to construct an acute scalene triangle and the three perpendicular bisectors, three angle bisectors, three medians, and three altitudes. What do you notice?
> 2. Repeat part 1 for an obtuse triangle.

Euclidean Constructions

Euclid and other Greek mathematicians required that construction of geometric figures be done using only a compass and a straightedge (an unmarked ruler). The Greek compass

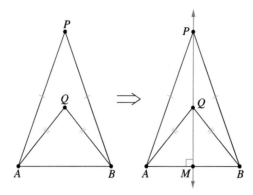

Figure 2.17

was a collapsible compass, which loses its radius once it is picked up. As a consequence, once a circle was constructed one could not move the compass to construct another circle having the same radius at a new center. Thus the Greek compass could not be used to mark off distances. Although today it is common practice to use a noncollapsible compass, any construction using a modern compass can also be accomplished using a collapsible compass.

The following rules apply to Euclidean constructions (constructions with a compass and straightedge in a plane).

1. Given two points, a unique straight line can be drawn containing the points as well as the unique segment connecting the points. (This is accomplished by aligning the straightedge across the points.)
2. It is possible to extend any part of a line.
3. A circle can be drawn given its center and radius.
4. Any finite number of points can be chosen on a given line, segment, or circle.
5. Points of intersection of two lines, two circles, or a line and a circle can be used to construct segments, lines, or circles.
6. No other instruments (such as a marked ruler, triangle, or protractor) or procedures can be used to perform constructions.

In reality, compass and ruler constructions are subject to error. For example, a geometrical line is an ideal line with zero width. However, a drawing of a line, no matter how sharp the pencil and how good the ruler and compass are, has a nonzero width.

From now on when we ask for a **construction** of a geometric object, we will mean a compass and straightedge construction that follows the rules given above. For convenience, we will frequently substitute "straightedge" with the one word: **ruler**. To construct a line or a line segment, the ruler must be aligned across two fixed points. Students sometimes ignore this requirement, leading to invalid constructions.

Example 2.2: Invalid Construction of a Tangent to a Circle

Given a circle with center O and point P as shown in Figure 2.18, construct through P a tangent to the circle (a line that touches the circle in a single point).

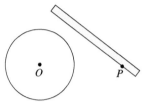

Figure 2.18

Invalid Procedure. Rotate the ruler about point P counterclockwise until it just touches the circle as shown in Figure 2.18.
This approach, which is sometimes called "eyeballing," involves looking at the figure and guessing. (A valid construction will be explored in Chapter 3.)

Example 2.3: Invalid Construction of a Perpendicular to a Line

Given a line ℓ and a point P on the line as in Figure 2.19, construct the perpendicular to the line through the point.
Invalid Procedure. Use a drafting triangle to align one of the legs of the triangle with the line as shown in Figure 2.19. Then move the triangle along the leg until the other leg passes through the given point. Notice that this is a practical way to construct the perpendicular, but because it does not follow the Euclidean construction rules it is an invalid construction.

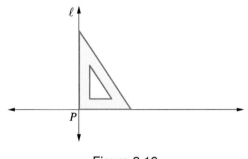

Figure 2.19

In what follows, we assume that the reader can construct a circle (the locus of all points at a given distance from a given point—the center) given its center and radius and can construct a segment congruent to a given segment. Let's proceed to our first construction.

2.2 Triangles and Congruence of Triangles

Construction 2.2.1 Constructing an Equilateral Triangle

To construct an equilateral triangle on a given segment AB as in Figure 2.20, we want to find point C whose distance from point A as well as from point B is AB. We know how to construct all the points whose distance from A is AB: the circle whose center is A and whose radius is AB. Our desired point C lies somewhere on this circle. To determine its location, we find all the points whose distance from B is also AB. Such points are on the circle with center B and radius AB. Because vertex C lies on both circles, it can only be where the two circles intersect—that is, at point C or C' as shown in Figure 2.20. Thus $\triangle ABC$ or $\triangle ABC'$ is the required equilateral triangle.

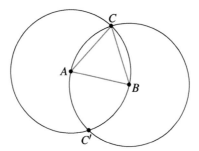

Figure 2.20

Construction 2.2.2 Construction of a Perpendicular Bisector of a Segment

Corollary 2.3 tells us that if we find any two points equidistant from the endpoints of a segment, the line through these points is the perpendicular bisector of the segment. In Figure 2.21 we use an approach similar to the one used in construction 2.2.1 to find two such points P and Q. $\overleftrightarrow{PQ}$ is the perpendicular bisector of $\overline{AB}$. Figure 2.20 suggests a somewhat different way to construct the perpendicular bisector; $\overleftrightarrow{CC'}$ is the perpendicular bisector of $\overline{AB}$.

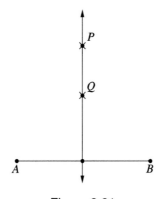

Figure 2.21

Construction 2.2.3 Construction of a Perpendicular to a Line Through a Point on the Line

In Figure 2.22(a), M is a point on ℓ and we want to construct the perpendicular to ℓ through M. We wish to use the previous construction; to that end, we construct an arbitrary segment $\overline{AB}$ so that M is the midpoint. This is done by drawing a circle (or semicircle) centered at M as pictured in Figure 2.22(b). The circle intersects ℓ at points A and B, both of which are equidistant from M. We now need only to find another point P equidistant from A and B. The line $\overleftrightarrow{PM}$ is the perpendicular bisector of $\overline{AB}$ and, in particular, is perpendicular to ℓ through M, as required.

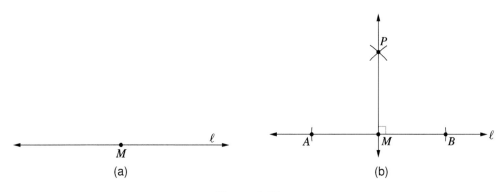

Figure 2.22

Properties of a Kite

A quadrilateral that has two pairs of congruent adjacent sides is called a **kite**. It can also be described as the quadrilateral created when two isosceles triangles share a common base (the base becomes then a diagonal of the kite). In Figure 2.23(a), $ABCD$ is a convex kite; in Figure 2.23(b), $ABCD$ is a non-convex kite.

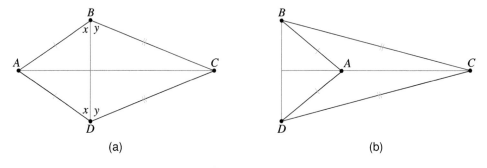

Figure 2.23

If we fold each of the kites in Figure 2.23 along the diagonal $\overline{AC}$ or the line containing $\overline{AC}$, we observe that B falls on D. Hence the diagonals of a kite are perpendicular to each

2.2 Triangles and Congruence of Triangles

other, and $\overline{AC}$ bisects both $\overline{BD}$ and the angles at A and C. We state these properties in the following theorem:

> **Theorem 2.7**
>
> *The diagonal of a kite connecting the vertices where the congruent sides intersect bisects the angles at these vertices and is the perpendicular bisector of the other diagonal.*

Proof. In Figure 2.23, the vertices A and C are equidistant from the endpoints of $\overline{BD}$. Hence $\overleftrightarrow{AC}$ is the perpendicular bisector of $\overline{BD}$ (by Corollary 2.3). Because the perpendicular bisector of the base of an isosceles triangle is also the angle bisector of the vertex angle, $\overleftrightarrow{AC}$ bisects each of the angles at A and C. In Figure 2.23(b), $\overleftrightarrow{AC}$ bisects $\angle BAD$, which implies that $\angle BAC \cong \angle DAC$. ∎

A quadrilateral in which all the sides are congruent is called a **rhombus**. Because a rhombus is a kite, we have the following corollary to Theorem 2.7:

> **Corollary 2.4**
>
> The diagonals of a rhombus are perpendicular bisectors of each other, and each bisects a pair of opposite angles.

> **Now Solve This 2.4**
>
> 1. Prove the converse of the statement in Corollary 2.4: A quadrilateral in which the diagonals are perpendicular bisectors of each other is a rhombus.
> 2. State and prove a statement similar to that given in Part 1 for a kite.
> 3. Classify each of the following quadrilaterals using angle bisectors. Your statement should start like the one in Part 1, using (i) a rhombus, (ii) a kite. Justify your answers.

Basic Constructions Using the Properties of a Kite

The properties of a kite or a rhombus can be used to perform some basic constructions. To **bisect a given segment** $\overline{AB}$, we construct a kite or a rhombus so that $\overline{AB}$ is the common base of the two isosceles triangles, as shown in Figure 2.24(a). The diagonal $\overline{DC}$ will be the perpendicular bisector of $\overline{AB}$ (actually $\overleftrightarrow{DC}$ and not $\overline{DC}$ is the perpendicular bisector of $\overline{AB}$).

Construction 2.2.4 Bisect a Given Angle

One way to bisect a given angle such as $\angle A$ in Figure 2.24(b) is to create a kite (or a rhombus) in which A is the vertex where the two congruent sides of the kite intersect. To do so, we

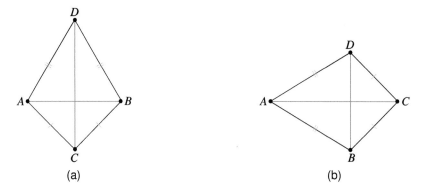

Figure 2.24

construct two isosceles triangles with a common base. The diagonal $\overline{AC}$ will bisect the angles at A and C. Notice that constructing a rhombus will also accomplish the task of bisecting a given angle.

> **Remark 2.3**
>
> Notice that the constructions taking advantage of the properties of a kite involve basically the same steps as the corresponding ones done earlier using the properties of a perpendicular bisector and an isosceles triangle. Because a kite is a concrete and appealing object, some students find it easier to recall how to perform these constructions by referring to the properties of a kite.

The properties of a kite can be used to prove the Side – Side – Side congruency condition stated in the following theorem:

> **Theorem 2.8: Side – Side – Side (SSS) Triangle Congruence Condition**
>
> *Given a one-to-one correspondence among the vertices of two triangles, if the three sides of one triangle are congruent to the corresponding sides of the second triangle, then the triangles are congruent.*

Given: $\triangle ABC$, $\triangle A_1 B_1 C_1$, and the correspondence $A \leftrightarrow A_1$, $B \leftrightarrow B_1$, $C \leftrightarrow C_1$ and $\overline{AB} \cong \overline{A_1 B_1}$, $\overline{AC} \cong \overline{A_1 C_1}$, $\overline{BC} \cong \overline{B_1 C_1}$.

Prove: $\triangle ABC \cong \triangle A_1 B_1 C_1$.

Plan: The idea is to place a flipped-over copy $\triangle A_1 B_1 C_1$, so that the copy and $\triangle ABC$ share a side [side $\overline{AC}$ in Figure 2.25(b). In this way a kite is formed]. We can then use what we know about kites to show that the triangles are congruent.

2.2 Triangles and Congruence of Triangles

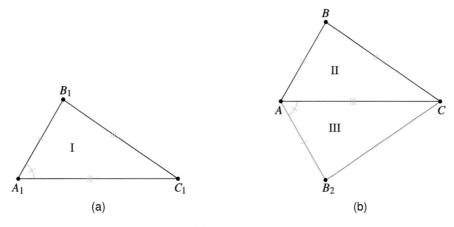

Figure 2.25

Proof. In Figure 2.25(b), construct triangle III congruent to triangle I using the SAS condition. This can be accomplished by constructing $\angle CAB_2$ congruent to $\angle C_1 A_1 B_1$ and $AB_2 = A_1 B_1$. It follows that $ABCB_2$ is a kite. ($AB_2 = A_1 B_1$ by construction, $A_1 B_1 = AB$, and, therefore, $AB_2 = AB$.) Also by **CPCT** (corresponding parts of congruent triangles), $B_2 C = B_1 C_1$ and $B_1 C_1 = BC$, which implies $B_2 C = BC$. From the properties of a kite, $\overline{CA}$ bisects $\angle BCB_2$ and, consequently, triangles III and II are congruent by ASA. From the triangle congruences $\Delta \text{I} \cong \Delta \text{III}$ and $\Delta \text{III} \cong \Delta \text{II}$, it follows (by transitivity) that $\Delta \text{I} \cong \Delta \text{II}$, i.e., that $\Delta A_1 B_1 C_1 \cong \Delta ABC$. ∎

An approach similar to the one used in the proof of the SSS theorem will soon be used to determine a useful condition for two right triangles to be congruent (Theorem 2.11). A triangle in which one of the angles is a 90° angle is called a **right triangle**. The side opposite the right angle is the **hypotenuse** and the other two sides are **legs**. (We will soon prove that a triangle cannot have two right angles.)

The Relative Size of an Exterior Angle of a Triangle

When a side of a triangle is extended as in Figure 2.26, an angle supplementary to an angle of the triangle is formed. Such an angle is called an **exterior angle** of the triangle. In Figure 2.26, $\angle BAC$ is one of the angles of the triangle and both $\angle BAD$ and $\angle CAE$ are exterior angles. A triangle has six exterior angles. Notice, however, that pairs of exterior angles at the same vertex are vertical angles and therefore congruent. In Figure 2.26, $\angle BAD$ and $\angle CAE$ are such a pair and hence congruent. Notice that $\angle BAD$ and $\angle CAE$ are vertical angles and hence congruent.

Any angle of the triangle that is not adjacent to an exterior angle is called a **remote interior** angle of the exterior angle. Thus $\angle B$ and $\angle C$ are remote interior angles of $\angle BAD$ (or $\angle CAE$). The next theorem relates the size of an exterior angle to the size of a remote interior angle. (We say that one angle is greater than another if the measure of that angle is greater than the measure of the other angle.) This major theorem will be used to prove several other theorems.

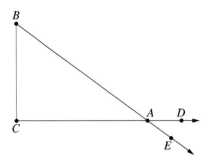

Figure 2.26

Theorem 2.9: The Exterior Angle Theorem

An exterior angle of a triangle is greater than either of the remote interior angles.

Given: $\triangle ABC$ in which $\overline{BA}$ has been extended as in Figure 2.27.

Prove: $m(\angle CAD) > m(\angle ACB)$.

Plan: To prove that $\angle CAD$ is greater than $\angle C$, we try to "fit" the latter angle inside the exterior angle. This can be done by creating congruent triangles.

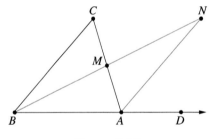

Figure 2.27

Proof. Let M be the midpoint of $\overline{AC}$. Extend $\overline{BM}$ such that $\overline{BM} \cong \overline{MN}$. Now $\triangle MCB \cong \triangle MAN$ by SAS (why?) and, therefore, $\angle CAN \cong \angle C$. Because N is in the interior of $\angle CAD$, $\angle CAN$ is smaller than $\angle CAD$. Because $\angle CAN \cong \angle C$, it follows that $\angle C$ is smaller than $\angle CAD$. To prove that $\angle CAD$ is greater than $\angle B$, extend $\overrightarrow{CA}$ to create the ray $\overrightarrow{CA}$. Then an exterior angle congruent to $\angle CAD$ is formed. Now "fit" $\angle B$ inside that angle by a proper construction. The details are left as an exercise. ∎

Notice that we have not justified the intuitively obvious fact that N is in the interior of $\angle CAD$. This can be proved using the Plane Separation Axiom 1.8. We leave the details to the interested reader.

Corollary 2.5 follows immediately from the Exterior Angle Theorem:

2.2 Triangles and Congruence of Triangles 45

> **Corollary 2.5**
>
> Through a point not on a line, there is a unique (one and only one) perpendicular to the line.

Given: A line ℓ as in Figure 2.28(a) and P a point not on ℓ.

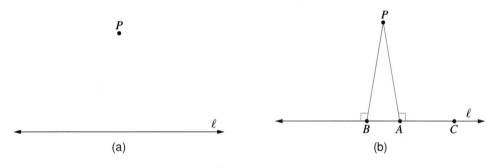

Figure 2.28

Prove: There exists one and only one line through P perpendicular to ℓ.

Proof. The existence of a line through P perpendicular to ℓ can be established by constructing a segment on ℓ such that P is equidistant from the endpoints of the segment and then constructing another point equidistant from the same endpoints. This construction is explored in Now Solve This 2.5.

To prove the uniqueness, assume that multiple perpendiculars through P exist. Let $\overline{PA}$ and $\overline{PB}$ be two perpendiculars to ℓ as shown in Figure 2.28(b). Then in $\triangle PAB$ the exterior angle PAC and its remote interior angle PBA are right angles. This contradicts the Exterior Angle Theorem. Consequently, more than one perpendicular to ℓ through P cannot exist. ∎

> **Theorem 2.10: Hypotenuse-Leg (H-L) Congruence Condition**
>
> *If the hypotenuse and a leg of a right triangle are congruent to the hypotenuse and a leg of another right triangle, then the triangles are congruent.*

Given: $\triangle ABC$ and $\triangle A_1 B_1 C_1$ with right angles at C and C_1, $\overline{BC} \cong \overline{B_1 C_1}$ and $\overline{AB} \cong \overline{A_1 B_1}$.

Prove: $\triangle ABC \cong \triangle A_1 B_1 C_1$.

Proof. We extend $\overline{AC}$ and find A_2 so that $CA_2 = C_1 A_1$. Then $\triangle \text{II}$ and $\triangle \text{III}$ are congruent by SAS. Consequently, $BA_2 = B_1 A_1$, and therefore $\triangle A_2 BA$ is isosceles. We know (Theorem 10.14) that the angle bisector of the vertex angle in an isosceles triangle is also the perpendicular bisector of the base $\overline{AA_2}$. Because the perpendicular from a point to a line is unique (Corollary 2.5) it must be the angle bisector of $\angle B$. Consequently, $\angle ABC \cong \angle A_2 BC$ and therefore by SAS triangles I and III are congruent. Because triangles II and III were shown to be congruent, it follows that I and II are congruent. ∎

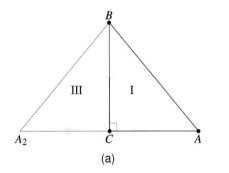
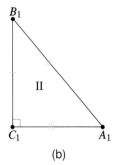

Figure 2.29

> **Now Solve This 2.5**
>
> In Figure 2.28(a), construct a circle with center at P and intersecting ℓ in two points. Next construct a point Q equidistant from these two points. Why is $\overleftrightarrow{PQ}$ perpendicular to ℓ?

The Exterior Angle Theorem or the uniqueness of the perpendicular to a line through a point can be used to prove the following theorem.

Theorem 2.11: Hypotenuse-Acute Angle Triangle Congruence Condition

If the hypotenuse and an acute angle of one right triangle are congruent to the hypotenuse and an acute angle of another right triangle, then the triangles are congruent.

Given: $\triangle ABC$ and $\triangle A_1 B_1 C_1$ in which $\overline{AB} \cong \overline{A_1 B_1}$, $\angle A \cong \angle A_1$, and $\angle C$ and $\angle C_1$ are right angles.

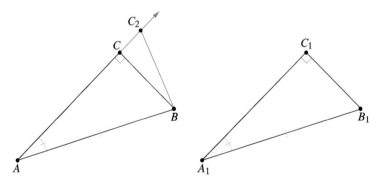

Figure 2.30

Prove: $\triangle ABC \cong \triangle A_1B_1C_1$.

Proof. In Figure 2.30, the marked parts of the right triangles $\triangle ABC$ and $\triangle A_1B_1C_1$ are congruent. If $\overline{AC} \cong \overline{A_1C_1}$, then the triangles are congruent by SAS. If $AC \neq A_1C_1$, then $AC < A_1C_1$ or $AC > A_1C_1$. In either case, we can find a point C_2 on $\overrightarrow{AC}$ such that $\overline{AC_2} \cong \overline{A_1C_1}$. Then $\triangle ABC_2 \cong \triangle A_1B_1C_1$ by SAS and, consequently, $\angle AC_2B$ is a right angle. Then $\overline{BC_2}$ and $\overline{BC}$ are two different perpendiculars from B to $\overleftrightarrow{AC}$. This contradicts the uniqueness of a perpendicular from a point to a line. Thus $AC \neq A_1C_1$ must be rejected and, therefore, $\overline{AC} \cong \overline{A_1C_1}$ and $\triangle ABC \cong \triangle A_1B_1C_1$ by SAS. ∎

The existence of a unique line perpendicular to a given line through a given point enables us to define the concept of a height, which is important in finding the areas of triangles and other figures.

> **Definition of the Distance from a Point to a Line**
>
> The distance from a point P to a line ℓ is the length of the segment connecting P with the **foot** of the perpendicular to ℓ through P. (The foot of the perpendicular is the point of intersection of the perpendicular and ℓ.)

Notice that the distance from P to ℓ is well defined because through a point not on a line the perpendicular is unique (Corollary 2.5.).

In Figure 2.31, the length of $\overline{PQ}$ is the distance from P to ℓ.

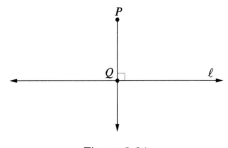

Figure 2.31

The segment from a vertex of a triangle perpendicular to the opposite side or the line containing the side is an **altitude** to that side. The distance from a vertex of a triangle to the line containing the opposite side is called a **height**. Thus a height is the length of an altitude. A triangle has three heights corresponding to its three sides. Figure 2.32 shows an obtuse $\triangle ABC$ and the three altitudes $\overline{CD}$, $\overline{BF}$, and $\overline{AE}$, and the corresponding heights h_1, h_2, and h_3. Notice that the lines containing the three altitudes intersect at a single point P in Figure 2.32. When three or more lines intersect in a single point, we say that they are **concurrent**. The proof that the lines containing the altitudes of a triangle are concurrent follows from the fact that the perpendicular bisectors of the sides of a triangle are concurrent. These and other concurrency theorems will be explored in the problem sets.

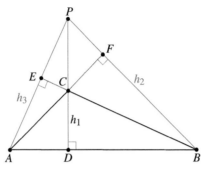

Figure 2.32

We saw earlier that a perpendicular bisector of a segment can be characterized as follows: A point is on the perpendicular bisector of a segment if and only if it is equidistant from the endpoints of the segment. Can an angle bisector be characterized in a similar way? The definition of the distance between a point and a line enables us to do so, as stated in the following theorem:

Theorem 2.12

A point is on the angle bisector of an angle if and only if it is equidistant from the sides of the angle.

1. **Given:** P on the angle bisector of $\angle A$ as in Figure 2.33(b).

   Prove: $\overline{PB} \cong \overline{PC}$ ($\overline{PB}$ and $\overline{PC}$ are perpendiculars to the sides of the angle).

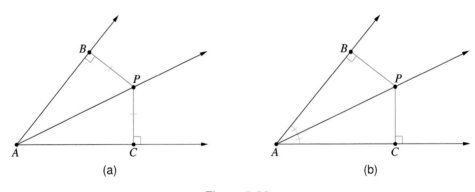

Figure 2.33

2. **Given:** The distances from P to the sides of the angle are equal as in Figure 2.33(a).

   Prove: P is on the angle bisector of $\angle A$.

2.2 Triangles and Congruence of Triangles

Proof of part 1. Because P is on the angle bisector of $\angle A$, $\triangle ABP \cong \triangle ACP$ by the Hypotenuse–Acute Angle Theorem. Consequently, $\overline{PB} \cong \overline{PC}$ (by CPCT). (See Figure 2.33(b).)

Proof of part 2. Since $PB = PC$ (given), $\overline{PB} \cong \overline{PC}$ and hence $\triangle ABP \cong \triangle ACP$ by the Hypotenuse–Leg Condition. Consequently, $\angle PAB \cong \angle PAC$ (by CPCT). (See Figure 2.33(a).)

We can use the Exterior Angle Theorem to prove the following theorem concerning the relative sizes of the angles opposite non-congruent sides of a triangle, and conversely, the relative sizes of the sides opposite non-congruent angles. The proofs of these theorems will be explored in the problem set at the end of this section.

Theorem 2.13

Given two non-congruent sides in a triangle, the angle opposite the longer side is greater than the angle opposite the shorter side.

Theorem 2.14: The Converse of Theorem 2.13

Given two non-congruent angles in a triangle, the side opposite the greater angle is longer than the side opposite the smaller angle.

Theorem 2.14 will be used to prove the intuitively well-known fact that the shortest path connecting any two points A and B is the segment $\overline{AB}$. This understanding is formalized in the following theorem, which is fundamental in most branches of mathematics.

Theorem 2.15: The Triangle Inequality

The sum of the lengths of any two sides of a triangle is greater than the length of the third side.

Given: $\triangle ABC$ with sides length a, b, and c (shown in Figure 2.34(a)).

Prove: $a + b > c$, $a + c > b$, and $b + c > a$.

Plan: We prove the first inequality $a + b > c$. (The proofs of the other two inequalities are analogous.) Our plan is to construct a triangle with one side of length $a + b$ and another side of length c, and then use Theorem 2.14 to show that the angle opposite side $a + b$ is greater than the angle opposite side of length c and hence conclude from Theorem 2.14 that $a + b > c$.

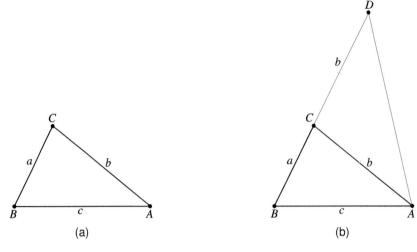

Figure 2.34

Proof. We extend side $\overline{BC}$ in Figure 2.34(b) so that $CD = b$. Next we show that $\angle BAD$ is greater than $\angle D$. Notice that $\angle D \cong \angle CAD$ because $\triangle CDA$ is an isosceles triangle. Because C is in the interior of $\angle BAD$, we have $m(\angle BAD) > m(\angle CAD)$. Consequently, $m(\angle BAD) > m(\angle D)$. Applying Theorem 2.14 to $\triangle BAD$, we have $BD > BA$ or $a + b > c$. ∎

> **Now Solve This 2.6**
>
> 1. Given the lengths of three segments a, b, and c such that $a + b > c$, show that it is not always possible to construct a triangle whose sides have lengths a, b, and c.
>
> 2. Construct three segments of lengths a, b, and c so that a triangle with these segments as sides will exist.
>
> 3. What conditions on a, b and c are necessary and sufficient for a triangle with these sides to exist?

> **Example 2.4: Hiker's Shortest Path**
>
> Given line ℓ and points A and B not on the line, find the shortest path from A to a point on the line and then to B. (See also Chapter 1, "The Hiker's Path.")
>
> **Solution**
>
> **Case 1.** In Figure 2.35(a), the points A and B are on opposite sides of the line. The segment $\overline{AB}$ intersects ℓ at P, and the "straight" path $A - P - B$ is the shortest path. This is the case because if Q is any other point on ℓ, then by the triangle inequality $AQ + QB > AB$.
>
> **Case 2.** Suppose that A and B are on the same side of ℓ, as in Figure 2.35(b). We could reduce this case to the previous one by imagining that ℓ is a mirror and finding the shortest path connecting A with a point P on ℓ and the reflection B' of B in the

mirror. The reflection of B in ℓ is obtained by dropping a perpendicular from B to ℓ and finding B' so that ℓ is the perpendicular bisector of $\overline{BB'}$. (A detailed study of reflections will be pursued in Chapter 6 which deals with transformations.)

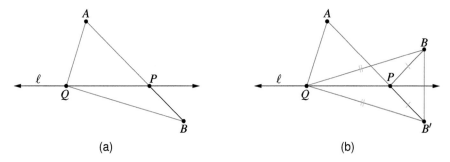

Figure 2.35

Notice that a property of a perpendicular bisector implies that if Q is any point on ℓ, then the length of the path $A - Q - B$ is the same as the length of the path $A - Q - B'$. Consequently, this case reduces to finding the shortest path connecting A to some point on ℓ and B'. It can be solved using Case 1 by connecting A with B' and finding the point P where $\overline{AB'}$ intersects ℓ. The path $A - P - B$ is the required path.

In the solution of Example 2.4, we used the concept of reflection in a line. For easy reference, here is the definition.

Definition of Reflection in a Line

A **reflection in a line** ℓ assigns to each point P in the plane not on ℓ the point P', the *image of* P, in such a way that ℓ is the perpendicular bisector of $\overline{PP'}$. If P is on ℓ, then $P' = P$. (See Figure 2.36.)

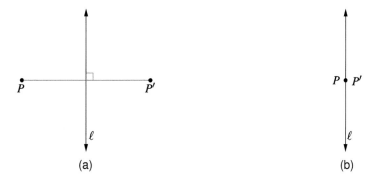

Figure 2.36

> **Remark 2.4**
> 1. When P' is the image of P under reflection in a line ℓ, we also say that P' is the *reflection of P in ℓ*.
> 2. Reflection in a line is a function from the plane to the plane.

Properties of Parallel Lines

Through a point P not on a line ℓ, there exists a line parallel to ℓ. This fact follows from the next theorem.

> **Theorem 2.16**
>
> *If two lines in the same plane are each perpendicular to a third line in that plane, then they are parallel.*

Given: k, ℓ, and m are three lines in the same plane as in Figure 2.37 such that $k \perp m$ and $\ell \perp m$.

Prove: $k \| \ell$.

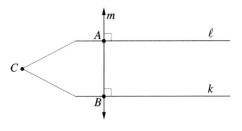

Figure 2.37

Proof. If k and ℓ are not parallel, then they must intersect at some point C. In that case, there will be two perpendiculars to m through C, which contradicts Corollary 2.5. Consequently, $k \| \ell$. ∎

We can also prove Theorem 2.16 by direct use of the Exterior Angle Theorem. The angles marked in Figure 2.37 are 90° each; hence $\angle CAB$ is also a right angle (property of vertical angles). Consequently, an exterior angle of $\triangle ABC$ is equal to a remote interior angle, which contradicts the Exterior Angle Theorem. Notice that if the marked angles in Figure 2.37 were only congruent but not necessarily 90° each, we could use the same argument to prove $k \| \ell$. This is stated in the next theorem (Theorem 2.17). Before we proceed, however, we need to define some common terminology for the angles formed when a line intersects two given lines.

Any line that intersects a pair of lines in two points is called a **transversal** of those lines. In Figure 2.38, line m is a transversal of lines k and ℓ. Angles are formed by these lines and are named according to their position relative to the transversal.

2.2 Triangles and Congruence of Triangles

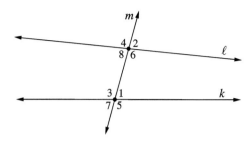

Figure 2.38

Interior angles: ∠1, ∠3, ∠6, ∠8
Corresponding angles: ∠1 and ∠2, ∠3 and ∠4, ∠5 and ∠6, ∠7 and ∠8
Alternate interior angles: ∠1 and ∠8, ∠3 and ∠6
Alternate exterior angles: ∠5 and ∠4, ∠7 and ∠2
Interior angles on the same side of a transversal: ∠1 and ∠6, ∠3 and ∠8

Remark 2.5

Two corresponding angles are congruent if and only if two alternate interior angles are congruent. (Why?)

Theorem 2.17

If two lines are cut by a transversal so that a pair of corresponding angles are congruent (or a pair of alternate interior angles are congruent), then the lines are parallel.

The proof of Theorem 2.17 is explored in Now Solve This 2.7.

Now Solve This 2.7

1. Prove Theorem 2.17 by contradiction. That is, show that if the lines were not parallel, a triangle would be formed with an exterior angle congruent to a remote interior angle, which will contradict the Exterior Angle Theorem.

2. Draw a line ℓ and a point P not on the line. Then construct a line m through the point P parallel to ℓ. Describe the construction and prove that $m \| \ell$.

3. Recall that we have defined a rhombus as a quadrilateral with all sides congruent. Prove that opposite sides of a rhombus are parallel.

4. Carry out Part 2 above by constructing a rhombus.

Remark 2.6

Part 3 of Now Solve This 2.7 implies that a rhombus is a **parallelogram**, a quadrilateral with parallel opposite sides.

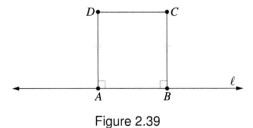

Figure 2.39

You may wonder about the converse of Theorem 2.17: *If two parallel lines are cut by a transversal, then a pair of corresponding angles is congruent.* This statement is needed to prove that the sum of measures of the angles of any triangle is 180° and, therefore, many other theorems of Euclidean geometry. In fact, the statement is equivalent to the famous Parallel Postulate (discussed in the next section) and hence distinguishes Euclidean geometry from non-Euclidean geometries. (See the Historical Note on Girolamo Saccheri.) Even the existence of rectangles cannot be established using the axioms we have stated and the theorems we have proved so far. (A **rectangle** is a quadrilateral with four right angles.) If we try to construct a rectangle as shown in Figure 2.39 by constructing a line ℓ, marking two points A and B on the line, and then constructing two congruent segments $\overline{AD}$ and $\overline{BC}$ each perpendicular to ℓ, we obtain what is called a **Saccheri quadrilateral**. The segment $\overline{AB}$ is called the *lower base* (at A and B, the angles are 90°) and the opposite side is called the *upper base*.

It is easy to prove that the diagonals of a Saccheri quadrilateral are congruent and that the upper base angles are congruent. Because we have not proved yet that the sum of the measures of the angles of a triangle is 180° (and hence that the sum of the measures in any quadrilateral is 360°), we cannot conclude that the Saccheri quadrilateral is a rectangle. This can be done with the Parallel Postulate, as will be shown in the next section.

> **Historical Note:** *Girolamo Saccheri*
> Girolamo Saccheri (1667–1733), an Italian mathematician, was dissatisfied with the *Parallel Postulate*. Like most mathematicians at the time, he believed that it should be possible to prove the postulate as a theorem using previous axioms and theorems. He wrote a book entitled *Euclid Freed of All Blemish*, in which he tried to prove the Parallel Postulate. Although Saccheri's proof was wrong, in the attempt to find a proof he did prove many new theorems that are now considered a part of *absolute* or *neutral geometry*, a geometry independent of the Parallel Postulate. The theorems we have proved so far belong to the realm of absolute or neutral geometry.

> **Now Solve This 2.8**
> Is it possible to prove that the diagonals of a parallelogram bisect each other using only the material covered so far? Explain.

2.2 Triangles and Congruence of Triangles

2.2.1 Problem Set

Complete Problems 1 through 22 without using the Parallel Postulate.

1. The congruence $\triangle ABC \cong \triangle DEF$ can also be written $\triangle BAC \cong \triangle EDF$. How many such symbolic representations exist for two congruent triangles? Do different representations give different information about the triangles?

2. State and prove a congruence condition (other than the definition) for two quadrilaterals to be congruent.

3. State and prove a theorem analogous to Theorem 2.7 for non-convex kites.

4. Write a careful proof of Theorem 2.8 (SSS condition).

5. Write a proof of Theorem 2.9 (hypotenuse-leg condition).

6. **a.** Draw an acute triangle and an obtuse triangle. In each case construct the three perpendicular bisectors of the sides of the triangle. Write a summary of your findings.

   b. In part (a) you must have noticed that the perpendicular bisectors of the sides of a triangle are concurrent—they intersect in a single point. Examine the following proof that the perpendicular bisectors of the sides of a triangle are concurrent and answer the question in part (c).
   In $\triangle ABC$, let O be the point where the perpendicular bisectors of $\overline{AB}$ and $\overline{BC}$ intersect. We need to prove only that O is also on the perpendicular bisector of $\overline{AC}$. We have proved that a point is on the perpendicular bisector of a segment if and only if it is equidistant from the endpoints of the segment. Thus, it suffices to prove that $AO = CO$. We have
   $AO = BO$ (since O is on the perpendicular bisector of $\overline{AB}$),
   $BO = CO$ (since O is on the perpendicular bisector of $\overline{BC}$).
   Consequently, $AO = CO$ and, therefore, O must be on the perpendicular bisector of $\overline{AC}$.

   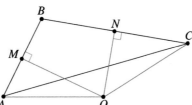

   c. The proof in part (b) is not completely rigorous; an assumption was made without proving it. What was the assumption?

   d. Prove that three angle bisectors in a triangle are concurrent.

7. Construct a line ℓ and a point P not on ℓ. Construct the perpendicular to ℓ through P. Describe the steps in the construction and prove that it is valid (i.e., that the line you constructed through P is perpendicular to ℓ).

8. A circle that passes through each vertex of a triangle is called a **circumscribed circle** or **circumcircle** as shown in figure (a). A circle that is tangent to each side of a triangle

is called an **inscribed circle** or **incircle** as shown in figure (b).

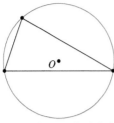
(a) Circumscribed circle

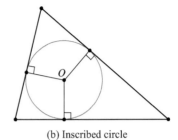
(b) Inscribed circle

    a. Assume that a tangent to a circle is perpendicular to the radius at the point of contact. Explain how to find the circumscribed circle and the inscribed circle for a given triangle. Justify your answers.

    b. Draw a triangle in which (i) all the angles are *acute*; (ii) one angle is *obtuse*; and (iii) one angle is 90°. In each case construct the circumscribed and inscribed circles.

    c. What seems to be true about the centers of the circumscribed circles?

9. **a.** Prove that the angle bisectors of a rhombus are concurrent.

    b. Prove that the point where the angle bisectors intersect is equidistant from all sides of the rhombus.

    c. Construct a rhombus and the circle inscribed in the rhombus. Clearly identify the radius of the circle.

10. **a.** Prove that the angle bisectors of a convex kite are concurrent as follows. Consider the kite $ABCD$ in the figure below and its diagonal BD. Prove that $\overline{BD}$ is on the angle bisectors of $\angle B$ and of $\angle D$. Then construct the angle bisector of $\angle A$ and point O where that angle bisector intersects $\overline{BD}$. Next let d_1 be the distance from O to $\overline{AD}$, d_2 be the distance from O to $\overline{AB}$, d_3 be the distance from O to $\overline{BC}$, and d_4 be the distance from O to $\overline{DC}$. Argue that $d_1 = d_2$, $d_2 = d_3$, and $d_4 = d_1$. Conclude that $d_3 = d_4$ and hence that O is on the angle bisector of $\angle C$.

    b. Construct any kite and the circle inscribed in the kite. Clearly identify the radius of the circle. Describe the construction and prove that it is valid.

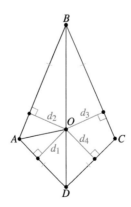

2.2 Triangles and Congruence of Triangles

11. Use the Exterior Angle Theorem to prove that in any $\triangle ABC$, $m(\angle A)+m(\angle B)+m(\angle C) < 270°$.

12. Prove that in any quadrilateral, the sum of the lengths of any three sides is greater than the length of the fourth side.

13. Prove Theorem 2.13, which states that in a triangle if $a > b$ then $\alpha > \beta$, where α is the measure of the angle opposite a and β opposite b.

14. Prove Theorem 2.14 (which is the converse of Theorem 2.13), which states that $\alpha > \beta \Rightarrow a > b$, as follows. Assume that $a > b$ is false (i.e., that $a \leq b$) and show that the cases $a = b$ and $a < b$ cause a contradiction (the latter condition contradicts Theorem 2.13).

15. a. Construct a scalene triangle and another triangle congruent to it using the three sides of your triangle. Describe your construction and explain why it is valid.

    b. Draw an angle and a ray that does not intersect the angle. Then use your answer to part (a) to construct an angle congruent to the original angle that has the ray as one of its sides. Briefly explain the idea behind the construction.

    c. Repeat part (b), but this time use an isosceles triangle to "duplicate" the angle.

16. Construct a scalene triangle and a triangle congruent to it using only two sides and the included angle of the original triangle.

17. Construct a rhombus given its diagonals (that is, given two segments congruent to the diagonals). Describe and justify your construction.

18. Given three segments of length a, b, and c, such that $b - c < a$, is it always possible to construct a triangle whose side lengths are a, b and c? Justify your answer.

19. Prove that the sum of the distances from any point in the interior of a triangle to the three vertices is greater than half the **perimeter** (the sum of the lengths of the sides) of the triangle.

20. Prove that the perimeter of a quadrilateral is greater than the sum of the lengths of the diagonals.

21. In $\triangle ABC$, $\overline{OB}$ and $\overline{OC}$ bisect $\angle ABC$ and $\angle ACB$, respectively. Prove that if $AB > AC$, then $OB > OC$.

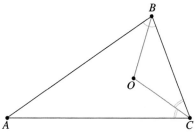

22. Prove that if a, b, and c are the lengths of the sides of a triangle, then there exists a triangle with sides of lengths $\sqrt{a}$, $\sqrt{b}$, and $\sqrt{c}$.

23. ∠AOB has vertex O, which is not on the paper. Construct the bisector of ∠AOB (without extending the sides on additional paper). Describe your construction and prove that it produces the required angle bisector.

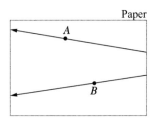

2.3 The Parallel Postulate and Its Consequences

Circa 300 B.C.E., Euclid arranged plane geometry based on five postulates. The fifth postulate, known as the Parallel Postulate, was stated by Euclid as follows:

> *If a straight line falling on two straight lines make the interior angles on the same side less than two right angles, the two straight lines, if produced indefinitely, meet on that side on which the angles are less than the two right angles.*

From ancient times this postulate was criticized for several reasons. Euclid proved 28 theorems before using the fifth postulate in a proof. The converse of the Parallel Postulate, equivalent to "The sum of two angles of a triangle is less than two right angles," was proved in Euclid's *Elements* as a theorem. In addition, mathematicians felt that because the postulate is so intuitively obvious, there must be a way to prove it. Indeed, for generations, mathematicians believed that the Parallel Postulate could be proved based on Euclid's previous postulates. As a result of the many attempts to prove this postulate, several equivalent statements have been discovered. The most common substitute for Euclid's Parallel Postulate is known as **Playfair's axiom** after the Scottish mathematician John Playfair (1748–1819), although it appears earlier in the writing of Proclus (410–485 C.E.[1]) in his *Commentary on the First Book of Euclid's Elements*. Because of its simplicity, Playfair's axiom is used today as the Parallel Postulate.

Axiom 2.4: The Parallel Postulate (Playfair's Axiom)

Given a line and a point not on the line, there exists a unique line through the point parallel to the given line.

This axiom is illustrated in Figure 2.40, where given ℓ_1 and P not on ℓ_1, there exists one and only one line ℓ_2 through P such that $\ell_2 \| \ell_1$.

---
[1]C.E. stands for "Common Era."

2.3 The Parallel Postulate and Its Consequences

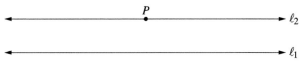

Figure 2.40

> **Historical Note:** *John Playfair*
> John Playfair (1748–1819), a Scottish mathematician and physicist, was educated at home until the age of 14, when he started his studies at the University of St. Andrews. He was ordained as a minister but continued his studies. In 1785, Playfair was awarded the chair of mathematics at the University of Edinburgh. In 1795, he wrote *Elements of Geometry*, which became a standard text in geometry and went through many editions. In that text, Playfair introduced his version of the Parallel Postulate, which remains the standard today.

The Parallel Postulate is the turning point at which different geometries branch out. By accepting the Parallel Postulate, **Euclidean geometry** is developed. If we deny the postulate, there are two possibilities:

1. There is more than one parallel to a line through a given point not on the line.
2. There is no parallel to a line through a point not on the line.

In each case we obtain a **non-Euclidean geometry**. Statement 1 is the basis for **hyperbolic geometry**, while statement 2 with some modifications to previous axioms results in **elliptic geometry**.

Using the Parallel Postulate, we can prove Theorem 2.18, the converse of Theorem 2.17 (which we state again for the reader's convenience), as well as the subsequent theorems of Euclidean geometry. Theorem 2.17 stated that: "If two lines are cut by a transversal so that a pair of corresponding angles are congruent, then the lines are parallel."

> **Theorem 2.18**
>
> *If two parallel lines are cut by a transversal, then corresponding angles are congruent.*

Given: $\ell_1 \| \ell_2$ (as in Figure 2.41) and t (a transversal).

Prove: $\angle 1 \cong \angle 2$.

Plan: We assume that the angles are not congruent and obtain a contradiction of the Parallel Postulate.

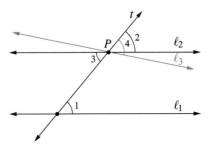

Figure 2.41

Proof. If ∠1 and ∠2 are not congruent, then there exists a line ℓ_3 through P such that $\ell_3 \neq \ell_2$ and so that ∠4, which ℓ_3 makes with t, is congruent to ∠1. Then by Theorem 2.17, $\ell_3 \| \ell_1$. Thus, through P, there are two lines ℓ_2 and ℓ_3, parallel to ℓ_1, which contradicts the Parallel Postulate. Consequently, ∠1 ≅ ∠2. ∎

Theorem 2.18, and its converse, and the fact that a pair of corresponding angles is congruent if and only if a pair of alternate interior angles is congruent imply Theorems 2.19 and 2.20.

Theorem 2.19

Two lines in a plane are parallel if and only if a pair of corresponding angles formed by a transversal is congruent.

Theorem 2.20

Two lines in a plane are parallel if and only if a pair of alternate interior angles formed by a transversal is congruent.

In Figure 2.42, $\ell_1 \| \ell_2$ and the angles are as marked. By Theorem 2.19, $\alpha = \gamma$. Because $\beta + \gamma = 180°$, we have $\alpha + \beta = 180°$. Conversely, if $\alpha + \beta = 180°$ (and it is not given that $\ell_1 \| \ell_2$), we can show that $\alpha = \gamma$ and, therefore, by Theorem 2.19 that $\ell_1 \| \ell_2$. These observations are stated in the following theorem.

Theorem 2.21

Two lines are parallel if and only if a pair of interior angles on the same side of a transversal is supplementary.

Referring to Figure 2.42, Theorem 2.21 states that: $\ell_1 \| \ell_2$ if and only if $\alpha + \beta = 180°$. We can now prove a key theorem of Euclidean geometry:

Theorem 2.22

The sum of the measures of the interior angles of a triangle is $180°$.

2.3 The Parallel Postulate and Its Consequences

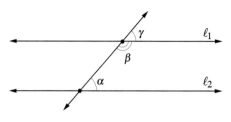

Figure 2.42

Proof. Let the measures of the interior angles of a triangle be α, β, γ; we need to prove that $\alpha + \beta + \gamma = 180°$.

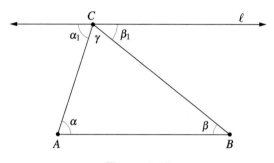

Figure 2.43

We could prove this result by showing that the sum of the measures of the angles of a triangle equals the measure of a straight angle. A straight line will create angles congruent to the angles of a triangle if it is parallel to one of the sides of the triangle. Thus we construct in Figure 2.43 line ℓ through C parallel to $\overline{AB}$. Because $\overleftrightarrow{AC}$ is a transversal for the parallel lines ℓ and $\overleftrightarrow{AB}$, $\alpha = \alpha_1$, as they are alternate interior angles. Similarly, because $\overleftrightarrow{CB}$ is a transversal for the same parallel lines, $\beta = \beta_1$. Consequently, $\alpha + \beta + \gamma = \alpha_1 + \beta_1 + \gamma = 180°$. ∎

Example 2.5

In Figure 2.44, $\ell_1 \| \ell_2$ and k is a transversal. If $\overrightarrow{AB}$ and $\overrightarrow{CB}$ are angle bisectors of the interior angles as shown, find $m(\angle CBA) = \beta$.

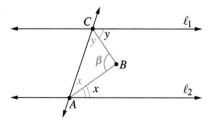

Figure 2.44

Solution.
If we mark the measures of the angles as shown in Figure 2.44, we have

$$2x + 2y = 180° \quad \text{(Theorem 2.21)}$$
$$x + y = 90°.$$

Now in $\triangle ABC$:
$$x + y + \beta = 180°.$$

Hence

$$\beta = 180° - (x + y)$$
$$= 180° - 90°$$
$$= 90°.$$

Consequently, $\angle CBA$ is a right angle.

Now Solve This 2.9

1. Find an expression for the sum of the measures of the interior angles of a convex n-gon in terms of n.
2. What is the sum of the measures of the exterior angles of a convex n-gon? Prove your answer.

We are now able to prove the following.

Theorem 2.23

In Euclidean geometry a Saccheri quadrilateral is a rectangle.

Proof. In Figure 2.45 $\angle A$ and $\angle B$ are right angles, and $\overline{AD} \cong \overline{BC}$. We need to prove that $\angle D$ and $\angle C$ are right angles.

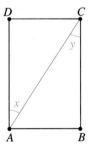

Figure 2.45

2.3 The Parallel Postulate and Its Consequences

Construct the diagonal $\overline{AC}$. To prove that ∠D is a right angle, we show that the diagonal divides ABCD into congruent triangles. Notice that $x = y$ as alternate interior angles formed by the parallel lines AD and BC. Thus $\triangle ABC \cong \triangle CDA$ by SAS; consequently, $\angle D \cong \angle B$ and hence ∠D is a right angle. Because the sum of the measures of the angles in any quadrilateral is 360° (see Now Solve This 2.9 or simply draw a diagonal and consider the sum of the angles in the two triangles), it follows that $m(\angle C) = 90°$. ∎

We define now some common terms of Euclidean geometry. We will investigate the properties of several quadrilaterals, such as the **trapezoid**, a quadrilateral with at least one pair of parallel sides; the **parallelogram**, a quadrilateral in which each pair of opposite sides is parallel; the **rhombus**, a parallelogram with two adjacent sides congruent (this definition is equivalent to the one given earlier, which stated that a rhombus is a quadrilateral with four congruent sides); and the **square**, a rhombus with a right angle. A **rectangle** can be defined as a parallelogram with a right angle. (It is easy to show that a quadrilateral in which all the angles are right angles is a rectangle, a property used in Theorem 2.23.)

The proofs of the following theorems are straightforward. You should prove them on your own.

Theorem 2.24

The measure of an exterior angle in a triangle is equal to the sum of the measures of its two remote angles. (In Figure 2.46 $\varepsilon = \alpha + \beta$.)

Figure 2.46

Theorem 2.25

If a transversal is perpendicular to one of two parallel lines, it is also perpendicular to the other line.

Theorem 2.26

In a parallelogram:
1. *Each diagonal divides the parallelogram into two congruent triangles.*
2. *Each pair of opposite sides is congruent.*
3. *The diagonals bisect each other.*

It is useful to know which minimal properties characterize the various types of quadrilaterals. For example, which properties characterize a parallelogram? Before going any further, list all conditions that are necessary and sufficient for a quadrilateral to be a parallelogram—without listing any more properties than needed. Then read on to compare your list of some of the useful conditions, and do the proofs as an exercise in Now Solve This 2.10.

> **Theorem 2.27**
>
> 1. A quadrilateral in which each pair of opposite sides is congruent is a parallelogram.
> 2. A quadrilateral in which the diagonals bisect each other is a parallelogram.
> 3. A quadrilateral in which each pair of opposite angles is congruent is a parallelogram.
> 4. A quadrilateral in which a pair of opposite sides is parallel and congruent is a parallelogram.

The last condition in Theorem 2.27 may be the least well-known but is often very useful. We prove part (3) of Theorem 2.27.

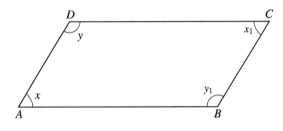

Figure 2.47

Proof of Theorem 2.27. In Figure 2.47, $ABCD$ is a quadrilateral in which the measures of the angles satisfy $x = x_1$ and $y = y_1$.

We need to prove that: (a) $\overline{AB} \| \overline{DC}$, (b) $\overline{AD} \| \overline{BC}$.

If we could show that $x + y = 180°$, statement (a) would follow (why?), and similarly if we can show that $x + y_1 = 180°$ statement (b) would follow.

Because $ABCD$ is a quadrilateral, $x + y + x_1 + y_1 = 360°$. Substituting $x_1 = x$ and $y_1 = y$, we get $(x + y) + (x + y) = 360°$ or $x + y = 180°$. Substituting $y = y_1$ in the last equation, we get $x + y_1 = 180°$. Consequently, statements (a) and (b) are true and $ABCD$ is a parallelogram. ∎

2.3 The Parallel Postulate and Its Consequences

> **Now Solve This 2.10**
> 1. Prove Theorems 2.24 through 2.27.
> 2. Part (1) of Theorem 2.27 and its converse can be written as follows: A necessary and sufficient condition for a quadrilateral to be a parallelogram is that each pair of opposite sides is congruent. Write the other parts of Theorem 2.27 and corresponding converses using the phrase "necessary and sufficient."

Construction of a Line Through P (not on ℓ) Parallel to ℓ. In Figure 2.48(a), a line ℓ and a point P are given. To construct through P a line parallel to ℓ we make P a vertex of any parallelogram so that one of the sides of the parallelogram is on ℓ. By part 1 of Theorem 2.27, to achieve this goal we need simply to make each pair of opposite sides congruent. We can do so by constructing all sides of the quadrilateral congruent as shown in Figure 2.48(b). Notice that this construction yields a rhombus and, therefore, has opposite parallel sides.

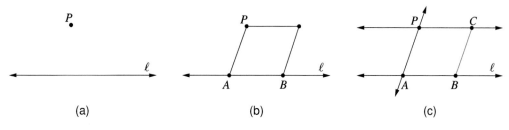

Figure 2.48

To accomplish the construction (see Figure 2.48(c)), draw any line through P that intersects ℓ and label the point of intersection A. (The labeling is, of course, not necessary for the actual construction.) Next take P and A to be the first two vertices of a rhombus. Locate the next vertex B on ℓ so that $\overline{AB} \cong \overline{AP}$. Now find the fourth vertex C so that $\overline{PC} \cong \overline{AB} \cong \overline{AP}$. (For that purpose we use a compass to construct an arc with center B and radius AB and an arc with center P and radius AP. The intersection of the two arcs is the point C. Line PC is the required line.)

2.3.1 Problem Set

You may use the Parallel Postulate in the following problems unless instructed otherwise.

1. Which part of the following statement requires the use of the Parallel Postulate and which does not? Justify your answer.
   Two lines cut by a transversal are parallel if and only if a pair of corresponding angles is congruent.
2. **a.** In Now Solve This 2.9 you found the sum of the measures of the interior angles in a convex n-gon. Does your answer apply to concave quadrilaterals as well?
   b. Find the sum of the measures of the interior angles of any convex n-gon in terms of n by drawing all the diagonals from one vertex as shown below. (You may have

used this approach in Now Solve This 2.9.)

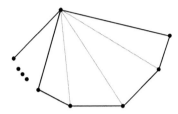

c. Prove your answer to part (b) by mathematical induction.

3. Justify your conjecture in Now Solve This 2.9 about the sum of the measures of the angles of any convex *n*-gon by choosing any point *P* in the interior and connecting *P* with each of the vertices as shown below. (You may have used this approach in New Solve This 2.9.)

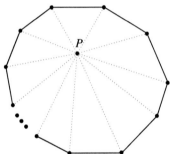

4. If $m(\angle BAC) = \alpha$, $m(\angle B'A'C') = \beta$, $\overrightarrow{AB} \| \overrightarrow{A'B'}$, and $\overrightarrow{AC} \| \overrightarrow{A'C'}$ as in the following figure, how are α and β related? Prove your answer.

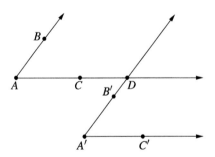

5. a. Let *m*, *n*, *a*, and *b* be lines such that $m \| n$, $m \perp a$, and $n \perp b$. Prove that $a \| b$.
    b. Use your work from part (a) to prove that any two perpendicular bisectors of two sides in a triangle must intersect.
    c. Did the proof of part (a) [and therefore of part (b)] depend on the Parallel Postulate? Justify your answer.

6. a. In the accompanying figure, the measures of the indicated angles are as shown. Complete the following and prove the completed statement:

$a \| b$ if and only if x, y, and z satisfy the condition _____.

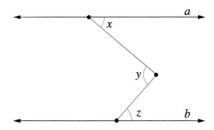

b. Write and prove a statement similar to the one in part (a) that corresponds to the figure below.

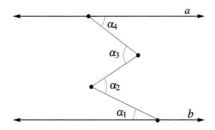

c. Write and prove a statement similar to the one in part (b) for
   (i) Five angles
   (ii) Six angles

d. Generalize the statement in part (c) for n angles $\alpha_1, \alpha_2, \ldots, \alpha_n$.

e. Prove your generalization in part (d) by mathematical induction.

7. In the accompanying figure, $a \| b$. The angles at P and Q have been trisected. The trisection rays form a quadrilateral $ABCD$. What is the most that can be said about the angles of $ABCD$? Prove your answer.

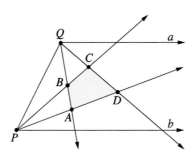

8. In △ABC, the angle bisectors at B and C form △BDC. If $m(\angle A) = \alpha$, can $m(\angle D)$ be found in terms of α? If so, find it. If not, prove that it cannot be found.

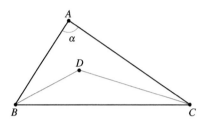

9. In △ABC the interior angles at B and C have been bisected to form △CDB as shown. Given $m(\angle A) = \alpha$, can $m(\angle D)$ be found in terms of α? If so, find it. If not, show that $m(\angle D)$ cannot be determined.

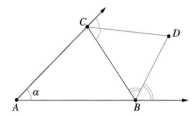

10. Complete each of the following statements and then prove the completed statements.
    a. A quadrilateral is a *rhombus* if and only if its diagonals _____.
    b. A quadrilateral is a *square* if and only if its diagonals _____.
    c. A quadrilateral is a *rectangle* if and only if its diagonals _____.
    d. A trapezoid that is not a parallelogram is an *isosceles trapezoid* (a trapezoid having a pair of congruent non-parallel sides) if and only if its diagonals _____.

11. Prove that in a 30°–60°–90° triangle, the side opposite the 30° angle is half as long as the hypotenuse. (*Hint*: Such a triangle is "half" an equilateral triangle.)

12. Devise three different methods for constructing a line through a given point parallel to a given line. Justify each method.

13. The following diagram suggests a proof of Theorem 2.13 that $a > b \Rightarrow \alpha > \beta$. Based on these diagrams, write a proof of Theorem 2.13.

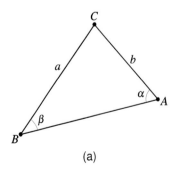
(a)

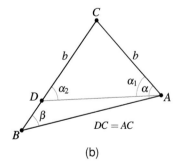
(b)

2.4 Parallel Projection and the Midsegment Theorem

14. What kind of quadrilateral is obtained by connecting the intersection points of the angle bisectors of a rectangle? Provide the most information you can about the obtained quadrilateral and prove your answer.

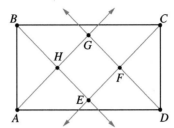

15. For the following figures, what kind of quadrilateral is always obtained by connecting the intersection points of the angle bisectors? Provide the most information you can about the quadrilaterals obtained, and prove your answers.
    a. A parallelogram
    b. An isosceles trapezoid

16. In the figure below, $ABCD$ is a square and all marked angles are congruent. What is the most that can be said about quadrilateral $EFHG$? Prove your answer.

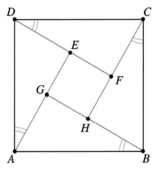

2.4 Parallel Projection and the Midsegment Theorem

If P is a point not on a line ℓ, as in Figure 2.50(a) then the point P', where the perpendicular through P to ℓ intersects ℓ, is called the **perpendicular projection** of P onto ℓ. We can actually project a point in any direction, not just the perpendicular one. In Figure 2.50(b), k and ℓ are two lines and m is a transversal. The line through P parallel to m intersects ℓ at P'. The point P' is the **projection of P on ℓ parallel to m**.

If we consider a line as a set of points, then a parallel projection is a function from k to ℓ that assigns to each point of k a corresponding point of ℓ. If we denote the function by f, then $f(P) = P'$ and $f(Q) = Q'$, where P, Q are on k and P', Q' on ℓ. Notice that the domain of this function is k and the range is ℓ. This function is called the **projection**

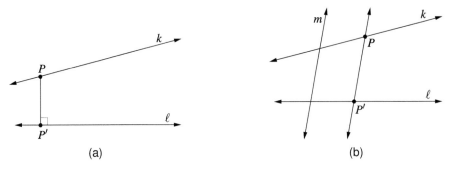

Figure 2.50

of k on ℓ in the direction of m and is denoted by $f : k \to \ell$. We refer to such functions as *parallel projections*. It is straightforward to show that a parallel projection function is one-to-one and onto and, therefore, a one-to-one correspondence. The next two theorems describe two useful properties of parallel projections.

> **Theorem 2.28**
>
> *Parallel projection preserves betweenness.*

We need to show that if $f : k \to \ell$, is a parallel projection; and A, B, C are points on k; $f(A) = A'$, $f(B) = B'$, $f(C) = C'$, and A–B–C (i.e., B is between A and C), then A'–B'–C'.

This theorem is illustrated in Figure 2.51. Because the result is so plausible, we will omit the proof here.

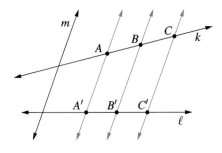

Figure 2.51

> **Theorem 2.29: Property of Parallel Projection**
>
> *A parallel projection preserves congruence of segments belonging to the same line.*

Proof. Figure 2.52 shows lines k, ℓ, and m; points A, B, C, and D on k such that $AB = CD$, as well as points A', B', C', and D' on ℓ so that $\overleftrightarrow{AA'}$, $\overleftrightarrow{BB'}$, $\overleftrightarrow{CC'}$, and $\overleftrightarrow{DD'}$ are parallel to m. We need to show that $A'B' = C'D'$.

2.4 Parallel Projection and the Midsegment Theorem

We distinguish two cases:

Case 1. If $k \parallel \ell$, then $\overline{AB}$ and $\overline{A'B'}$ are opposite sides of a parallelogram and hence $AB = A'B'$. Similarly, $CD = C'D'$. Since $AB = CD$, it follows that $A'B' = C'D'$.

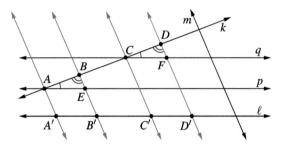

Figure 2.52

Case 2. If k is not parallel to ℓ, to prove that $A'B' = C'D'$, we try to construct congruent triangles with corresponding sides AB and CD as well as sides congruent to $A'B'$ and $C'D'$. Keeping in mind the proved Case 1, this can be accomplished by constructing through A and C lines p and q parallel to ℓ, as shown in Figure 2.52. From Case 1, it follows that $A'B' = AE$ and $C'D' = CF$. Therefore, it will suffice to prove that $AE = CF$. For that purpose we prove that $\triangle ABE \cong \triangle CDF$. Because each pair of similarly marked angles comprises corresponding angles between parallel lines, the similarly marked angles are congruent. Hence by ASA, $\triangle ABE \cong \triangle CDF$. Thus $AE = CF$ and, therefore, $A'B' = C'D'$. ∎

In Chapter 5, we will prove that parallel projections also preserve the ratio of the lengths of the segments. Functions preserving certain geometric properties will play a crucial role in later chapters.

An immediate and useful consequence of Theorem 2.29 arises when the congruent segments are adjacent—that is, when $B = C$ in Figure 2.52. We obtain the following corollary.

> **Corollary 2.6**
>
> If three or more parallel lines intercept congruent segments on one transversal, then they intercept congruent segments on any other transversal.

Corollary 2.6 is illustrated in Figure 2.53. If $A_1A_2 = A_2A_3 = A_3A_4 = \cdots = A_{n-1}A_n$ and $\overline{A_1A'_1} \parallel \overline{A_2A'_2} \parallel \overline{A_3A'_3} \parallel \cdots \parallel \overline{A_nA'_n}$ then $A'_1A'_2 = A'_2A'_3 = A'_3A'_4 = \cdots = A'_{n-1}A'_n$.

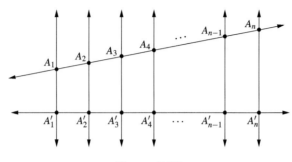

Figure 2.53

Corollary 2.6 can be used for the following construction.

Construction 2.4.1 Division of a Segment into Any Number n of Congruent Parts

Given $\overline{AB}$ as in Figure 2.54, we illustrate the construction for $n = 3$; i.e., we need to divide $\overline{AB}$ into three congruent segments.

Using Corollary 2.6, we can achieve the construction as follows:

Through A, construct any ray $\overrightarrow{AC}$ and mark on $\overrightarrow{AC}$ any point P. Construct Q and R so that $\overline{AP} \cong \overline{PQ} \cong \overline{QR}$. Connect R with B. Through Q and P, draw lines parallel to $\overline{BR}$. The points of intersection P' and Q' accomplish the required construction since by Corollary 2.6, $\overline{AP'} \cong \overline{P'Q'} \cong \overline{Q'B}$.

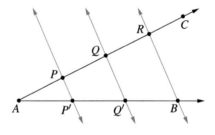

Figure 2.54

Figure 2.55 shows a special case of Corollary 2.6 for a triangle. In $\triangle ABC$, M is the midpoint of $\overline{AC}$, and a line is drawn through M parallel to $\overline{AB}$ that intersects $\overline{BC}$ at N. From Corollary 2.6, we can conclude that N is the midpoint of $\overline{BC}$. Thus we have the following theorem:

Theorem 2.30

A line passing through the midpoint of one side of a triangle and parallel to another side bisects the third side.

2.4 Parallel Projection and the Midsegment Theorem

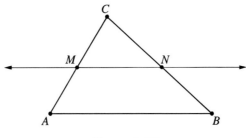

Figure 2.55

The segment connecting the midpoints of two sides of a triangle is called the **midsegment**. Each triangle has three midsegments. Because a segment has a unique midpoint, it follows from the last corollary that a midsegment of two sides of a triangle is parallel to the third side.

How does the length of the midsegment in a triangle compare to the length of the parallel side? Informally given that any triangle is "half" of a parallelogram (a diagonal divides a parallelogram into two congruent triangles), we could turn $\triangle ABC$ in Figure 2.56(a) into a parallelogram by tracing $\triangle A_1 B_1 C_1$ over $\triangle ABC$ and then turning it upside-down as shown in Figure 2.56(b). (We will see later that this operation amounts to rotating $\triangle ABC$ about N by 180°.) We obtain a parallelogram if we make $B_1 = C$ and $C_1 = B$. It seems now that $MN = \frac{1}{2}MM_1$ and $MM_1 = AB$ and consequently that $MN = \frac{1}{2}AB$. We state this result in the following theorem. A more rigorous argument based on the ideas presented above is explored in Now Solve This 2.11. Here we prove the theorem in a somewhat different way.

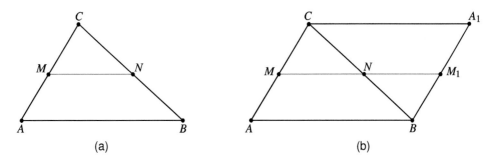

Figure 2.56

Theorem 2.31: The Midsegment Theorem

The segment connecting the midpoints of two sides of a triangle is parallel to the third side and half as long as that side.

Proof. In $\triangle ABC$ in Figure 2.56, M and N are midpoints of $\overline{AC}$ and $\overline{BC}$.

We need to prove: (1) $\overline{MN} \| \overline{AB}$; (2) $MN = \frac{1}{2}AB$.

Part (1) follows from Theorem 2.30. Let's see how. In Figure 2.57 M and N are midpoints.

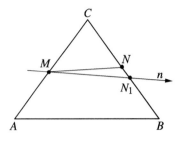

Figure 2.57

We construct through M a line n parallel to $\overline{AB}$ intersecting $\overline{CB}$ at N_1. Theorem 2.30 tells us that N_1 is the midpoint of $\overline{CB}$. Because the midpoint of a segment is unique $N_1 = N$, and since we constructed $\overline{MN_1} \| \overline{AB}$, it follows that $\overline{MN} \| \overline{AB}$.

Part (2) can be proved in a variety of ways including the approach in the next "Now Solve This." We use here a somewhat different approach. In Figure 2.58(a), we mark point D on $\overline{AB}$ so that $\overline{AD} \cong \overline{MN}$ and prove that $\overline{DB}$, the remaining part of $\overline{AB}$, is also congruent to $\overline{MN}$. It would then follow that $AB = 2MN$.

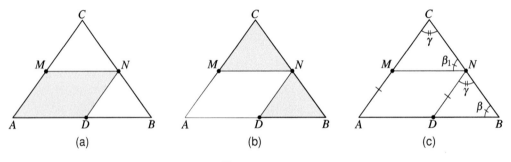

Figure 2.58

Because in quadrilateral $MNDA$ opposite sides $\overline{MN}$ and $\overline{AD}$ are parallel and congruent, $MNDA$ is a parallelogram. To prove that $\overline{DB} \cong \overline{MN}$, we search for two congruent triangles in which $\overline{DB}$ and $\overline{MN}$ are corresponding sides. The shaded triangles $\triangle MCN$ and $\triangle DNB$ seem to be such triangles. Indeed $\overline{ND} \cong \overline{MA}$ (why?) and $\overline{MA} \cong \overline{MC}$ (given) and hence $\overline{ND} \cong \overline{CM}$.

We next focus on finding congruent angles. In Figure 2.58(c), $\beta = \beta_1$ (corresponding angles, $\overleftrightarrow{AB} \| \overleftrightarrow{MN}$ and $\overleftrightarrow{CB}$ as transversal) and $\gamma = \gamma_1$ (corresponding angles, $\overleftrightarrow{ND} \| \overleftrightarrow{AC}$ and $\overleftrightarrow{CB}$ as transversal).

Consequently, $\angle M \cong \angle D$ and by ASA, $\triangle DNB \cong \triangle MCN$ and hence $\overline{DB} \cong \overline{MN}$. ∎

2.4 Parallel Projection and the Midsegment Theorem

> **Now Solve This 2.11**
>
> To prove part (2) of Theorem 2.31, we want to obtain a parallelogram and therefore draw through C a line parallel to $\overline{AB}$ and through B a line parallel to $\overline{AC}$ as shown in Figure 2.56(b). The two lines intersect at A_1. It follows that ABA_1C is a parallelogram. We extend $\overline{MN}$ until it intersects $\overline{BA}$ at M_1. At this point it seems obvious that $NM_1 = MN$ and that $MM_1 = AB$. Thus, if $MN = x$, then $2x = MM_1 = AB$ and hence $x = \frac{1}{2}AB$. Prove that $NM_1 = MN$.

> **Remark 2.7**
>
> We could have postponed the Midsegment Theorem until Chapter 5 and deduced it from theorems concerning similar triangles. Instead, we chose to prove the theorem here because the methods used give us valuable practice and an opportunity to see different approaches.

> **Now Solve This 2.12**
>
> We have seen that the diagonals of a parallelogram bisect each other. Suppose they bisect each other at O. If we draw a line through O and it intersects two parallel sides of the parallelogram at points P and Q, respectively, prove that $OP = OQ$.

Medians of a Triangle

A **median** of a triangle is a segment connecting a vertex to the midpoint of the opposite side. The Midsegment Theorem can be used to prove a surprising property of the medians of a triangle. If you construct the three medians of an arbitrary triangle, you will notice that the medians are concurrent. In search of other properties of medians, we construct two medians $\overline{AM}$ and $\overline{CN}$ in $\triangle ABC$ as shown in Figure 2.59. As in Figure 2.56(b), we trace $\triangle ACB$ to obtain $\triangle A_1C_1B_1$ and place it "upside down" so that $B_1 = C$ and $C_1 = B$. Rather than relying on this kind of informal tracing, we could also proceed formally as follows: We extend $\overline{AM}$ to A_1 so that $\overline{MA_1} \cong \overline{MA}$. We obtain the quadrilateral ABA_1C whose diagonals bisect each other. By Theorem 2.27, ABA_1C is a parallelogram. It seems that $AO = OO_1 = O_1A_1$ and $OM = MO_1$. Therefore if $OM = x$, then $OO_1 = 2x$ and $AO = 2x$. Hence $AO = 2 \cdot OM$. We ask you to justify these statements on your own. Meanwhile we state what we have observed in the following theorem and prove it in a different way.

> **Theorem 2.32: Properties of Medians**
>
> 1. Any two medians of a triangle intersect at a point that divides each median into segments whose lengths are in the ratio $2 : 1$.
> 2. The three medians in a triangle are concurrent.

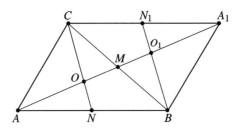

Figure 2.59

Proof. In Figure 2.60(a) $\overline{AM}$ and $\overline{CN}$ are two medians, intersecting at point O.

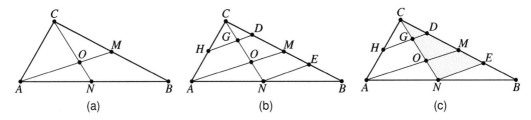

Figure 2.60

We need to prove that $CO = 2 \cdot ON$ and $AO = 2 \cdot OM$.

Let D and E in Figure 2.60(b) be the midpoints of $\overline{CM}$ and $\overline{MB}$, respectively. Because M is the midpoint of $\overline{CB}$, we have $CD = DM = ME = EB$. Through D construct $\overline{DH}$ parallel to $\overline{AM}$ intersecting $\overline{CO}$ at G.

Because $CD = DM$ it follows that $CG = GO$. Next connect E and N. Then $\overline{EN}$ is a midsegment in $\triangle AMB$. Hence $\overline{EN} \| \overline{AM}$. Now the parallel lines DG, MO, and EN project the congruent segments CD, DM, and ME onto congruent segments CG, GO, and ON.

Thus if $ON = d$, then $CO = 2d = 2 \cdot ON$. Hence any median divides a second median into segments whose length are in the ratio 2 : 1, and it follows that $AO = 2 \cdot OM$. ∎

To prove that the medians are concurrent, notice that if $ON = d$ then $CO = 2d$ and $CN = 3d$. Thus, $CO = \frac{2}{3}CN$. Because this is true for any two medians, for the third median from B intersecting $\overline{CN}$ at O_1 (not shown) we have $CO_1 = \frac{2}{3}CN$. Consequently $CO_1 = CO$ and therefore $O_1 = O$. Thus the three medians intersect at the same point O, that is, are concurrent.

> **Now Solve This 2.13**
>
> Another proof that medians of a triangle are concurrent.
> In Figure 2.60(c), extend $\overline{CN}$ to point I so that $NI = d$ as shown in Figure 2.61. Then show that $AOBI$ is a parallelogram. Connect B with O and I with H and show that H, O, B are collinear by showing that $\overline{OH} \| \overline{AI}$ and $\overline{OB} \| \overline{AI}$.

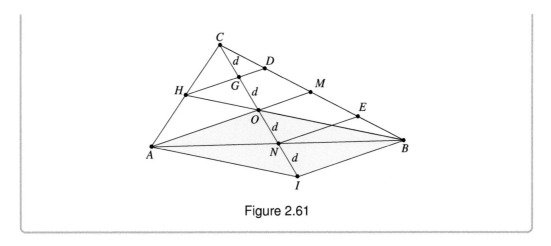

Figure 2.61

The point of concurrency of the medians of a triangle is called the **centroid**. The centroid is the center of gravity of the triangle when a triangle is considered to be a thin plate of uniform density. (The center of gravity can be defined mathematically; the definition can be found in most calculus texts.)

Example 2.6

What kind of quadrilateral is obtained when the midpoints of consecutive sides of a convex quadrilateral are connected?

Solution. In Figure 2.62, the midpoints M, N, P, and Q of the sides of quadrilateral $ABCD$ are connected to form the quadrilateral $MNPQ$. Starting with different quadrilaterals, accurate constructions reveal that in each case $MNPQ$ is a parallelogram. This fact can actually be discovered without any experimentation and proved for all quadrilaterals as follows.

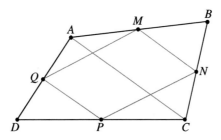

Figure 2.62

The fact that M and N are midpoints suggests the use of the Midsegment Theorem. The side $\overline{MN}$ is a midsegment in $\triangle ABC$. Thus $\overline{MN} \| \overline{AC}$ and $MN = \frac{1}{2}AC$. Similarly, in $\triangle ADC$, $\overline{QP} \| \overline{AC}$ and $QP = \frac{1}{2}AC$. Both imply that $\overline{MN} \| \overline{PQ}$ and $\overline{MN} \cong \overline{PQ}$. This happens if and only if $MNPQ$ is a parallelogram.

Example 2.7

Complete and prove each of the following statements assuming that the quadrilateral $MNPQ$ is the quadrilateral defined in Example 2.6.

1. $MNPQ$ is a rhombus if and only if $ABCD$ is a quadrilateral such that _____.
2. $MNPQ$ is a rectangle if and only if $ABCD$ is a quadrilateral such that _____.

Solution.

1. Using the result of Example 2.6, $MNPQ$ in Figure 2.63 is a rhombus if and only if $\overline{MN} \cong \overline{MQ}$. We need to write an equivalent condition involving characteristics of $ABCD$. We have $MN = \frac{1}{2}AC$ and $MQ = \frac{1}{2}BD$. Thus $MN = MQ$ if and only if $\frac{1}{2}AC = \frac{1}{2}BD$; that is, $\overline{AC} \cong \overline{BD}$. Consequently, $MNQP$ is a rhombus if and only if the diagonals of $ABCD$ are congruent.

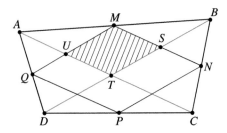

Figure 2.63

2. It is straightforward to show that if $ABCD$ is a rhombus or even a kite, then $MNPQ$ is a rectangle. However, neither condition is necessary: $MNPQ$ can be a rectangle if $ABCD$ is not a kite (recall that a rhombus is a special case of a kite). To find an "if and only if" condition, we can proceed as follows. Since $MNPQ$ is always a parallelogram, it will be a rectangle if and only if one of its angles is 90°. Notice that $MSTU$ is also a parallelogram ($\overline{MS}\|\overline{UT}$ and $\overline{MU}\|\overline{ST}$) and therefore $\angle UMS \cong \angle UTS$. Thus $\angle UMS$ is a right angle if and only if $\angle UTS$ is a right angle—that is, if and only if the diagonals of $ABCD$ are perpendicular. Consequently, Statement 2 could be completed by "its diagonals are perpendicular."

Example 2.8

1. State and prove the midsegment theorem for trapezoids.
2. Prove that the segment connecting the two midpoints of the diagonals of a trapezoid is parallel to the bases of the trapezoid and find the length of this segment if the lengths of the bases are a and b.

2.4 Parallel Projection and the Midsegment Theorem

Solution.

1. We use the Midsegment Theorem for triangles to deduce that the midsegment of a trapezoid (the segment connecting the midpoints of the sides) is parallel to the bases and its length is one half the sum of the lengths of the bases.

In Figure 2.64, $\overline{BC} \| \overline{AD}$, and M and N are the midpoints of $\overline{AB}$ and $\overline{CD}$, respectively. We need to show that $\overline{MN} \| \overline{AD}$ and $MN = \frac{1}{2}(AD + BC)$.

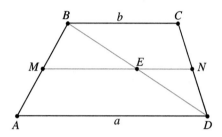

Figure 2.64

Through M, construct line ℓ (not shown in Figure 2.64) parallel to the bases. A property of parallel projections implies that ℓ intersects $\overline{CD}$ at its midpoint. Because the midpoint of a segment is unique, it follows that ℓ intersects $\overline{CD}$ at N. Consequently, $\overrightarrow{MN}$ is the line ℓ and, therefore, is parallel to the bases.

To find the length of $\overline{MN}$ given the lengths of the bases a and b, consider the two triangles created by the diagonal $\overline{BD}$ (the other diagonal could be used equally well). We have $ME = \frac{a}{2}$ and $EN = \frac{b}{2}$. Thus $MN = \frac{a+b}{2}$.

2. In Figure 2.65, P and Q are the midpoints of the diagonals.

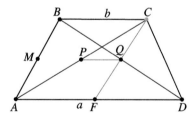

Figure 2.65

We need to show that $\overline{PQ}$ is parallel to the bases of the trapezoid. Also, express PQ in terms of a and b.

We try to make $\overline{PQ}$ a midsegment of a triangle. One such triangle can be obtained by connecting C with Q. The line CQ intersects $\overline{AD}$ at F. We show that $\overline{PQ}$ is a midsegment in $\triangle ACF$. Because P is the midpoint of $\overline{AC}$ (given), it remains to be shown that $\overline{CQ} \cong \overline{QF}$. For that purpose, notice that by ASA (how?) we have $\triangle BCQ \cong \triangle DFQ$. Thus $\overline{CQ} \cong \overline{QF}$ and, therefore, $\overline{PQ}$ is a midsegment in $\triangle ACF$. Consequently, we have $\overline{PQ} \| \overline{AF}$ and $PQ = \frac{1}{2}AF$. We now express $\overline{AF}$ in terms of a and b. $PQ = \frac{1}{2}AF = \frac{1}{2}(AD - FD)$. We have $AD = a$ and $FD = BC = b$. Therefore, $PQ = \frac{1}{2}(a - b)$.

Now Solve This 2.14

You can now prove your conjecture about the location of the treasure described in "The Treasure Island Problem" in Chapter 2 (see also Problem 1 in Problem Set 1.1.1). You should have conjectured that the treasure is independent of the location of the gallows Γ in Figure 2.66.

1. Show that if the conjecture is true, then by choosing the gallows Γ at T_2 (tree number 2), it follows that the treasure M is on the perpendicular bisector of the segment $T_1 T_2$ connecting the trees T_1 and T_2, half the distance between the trees, and above the lines connecting the trees.

2. Use the ideas in Figure 2.66 to prove the assertion in part (1). (We labeled $\Gamma C = h$, $S_1 A = a$, and $S_2 B = b$. You need to show that $T_1 A = T_2 B = h$, $CT_1 = a$, and $CT_2 = b$. Also show that $MN = \frac{1}{2}T_1 T_2$ and $T_2 N = T_1 N$.)

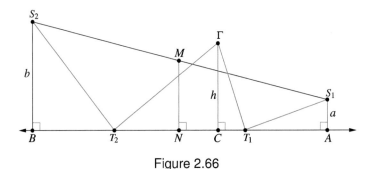

Figure 2.66

2.4.1 Problem Set

1. If k and ℓ are lines with transversal m and $f : k \to \ell$ is a parallel projection in the direction of m, explain why f is a one-to-one and onto function.

2. Divide a segment into five congruent parts. Justify your construction.

3. Give an alternative proof of the Midsegment Theorem by extending $\overline{MN}$, drawing through C line k parallel to $\overline{AB}$, labeling P as the intersection of $\overleftrightarrow{MN}$ with k, and

proving that $MPCB$ is a parallelogram.

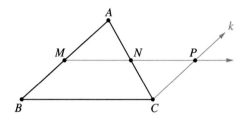

4. Give an alternative proof of the fact that any two medians divide each other in the ratio $2:1$ by proving first that if P and Q are the midpoints of $\overline{BO}$ and $\overline{CO}$, respectively, then $MNQP$ is a parallelogram.

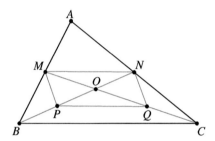

5. **a.** How does the length of the median to the hypotenuse in a right triangle compare to the length of the hypotenuse? Prove your answer.

   b. State and prove the converse of your statement in part (a).

6. Prove that the bisector of the right angle in a right triangle bisects the angle between the median and the altitude drawn to the hypotenuse.

7. Let $ABCD$ be a quadrilateral and $MNQP$ be the quadrilateral obtained by connecting the consecutive midpoints of the sides of $ABCD$. Complete the following statement and prove the completed statement:

   $MNQP$ is a square if and only if $ABCD$ has the property _____.

8. Give an alternative proof of the Midsegment Theorem for trapezoids (in Example 2.8) by considering a proof suggested by tracing the trapezoid $ABCD$, turning the traced trapezoid "upside down," and attaching the two trapezoids so that their bases are collinear.

9. What, if anything, is missing (or wrong) in the following proof of Example 2.8, part 2? Through P, draw a line parallel to $\overline{BC}$ and intersecting $\overline{AB}$ at S. A property of parallel projection implies that S is the midpoint of $\overline{AB}$. Applying the Midsegment Theorem

in $\triangle ABD$ and $\triangle ABC$, we have $PQ = QS - PS = \frac{1}{2}AD - \frac{1}{2}BC = \frac{1}{2}(a - b)$.

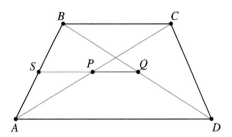

10. **a.** Let P and Q be the midpoints of the diagonals in the trapezoid $ABCD$. Draw $\overline{BP}$ and $\overline{CO}$ as shown, and express UV in terms of a and b where $AD = a$ and $BC = b$. Justify your answer.
    b. Under what condition will $U = V$?

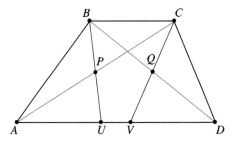

11. Prove the converse of the Midsegment Theorem: If a segment connects two points on two sides of a triangle in such a way that it is parallel to the third side and is half the length of that side, it is a midsegment.

12. $ABCD$ is a parallelogram whose diagonals intersect at E. The distances of the points $A, B, C, D,$ and E from ℓ are $a, b, c, d,$ and e, respectively,
    a. Prove that $b + d = a + c$.
    b. Find e in terms of $a, b, c,$ and d.

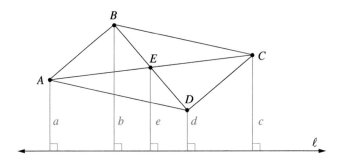

2.4 Parallel Projection and the Midsegment Theorem

13. The segment $\overline{AB}$ (of fixed length) is moving so that the endpoints A and B are on the sides of a right angle. What is the locus (set of points) of all the midpoints of $\overline{AB}$? Prove your answer.

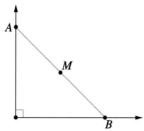

14. $ABCD$ is a parallelogram. Its diagonals divide the parallelogram into four triangles. Points F, G, H, and I are the incenters (the centers of the inscribed circles) of the corresponding triangles. What is the most that can be concluded about which type of quadrilateral $FGHI$ is? Prove your answer.

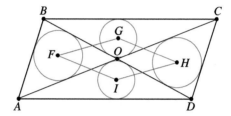

15. **a.** Prove that the three midsegments of a triangle divide it into four congruent triangles.
    b. Construct an acute angle triangle and its three altitudes.
    c. Construct an obtuse angle triangle and its three altitudes. Extend some of the altitudes to show that the lines containing the altitudes are concurrent.
    d. Prove that the altitudes of a triangle or the lines containing the altitudes are concurrent. (*Hint*: Through each vertex of the triangle, construct a line parallel to the opposite side. The three lines intersect in three points, which determine a new triangle. Show that the perpendicular bisectors of the sides of the new triangle contain the altitudes of the original triangle.)

16. In the following figure, three adjacent squares form a rectangle. Prove that $x+y+z = 90°$.

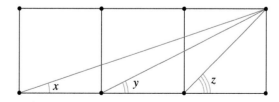

17. Prove that the sum of the distances from any point on the base of an isosceles triangle to its sides equals the height drawn to one of the congruent sides.

18. For which point P in the interior of an equilateral triangle $\triangle ABC$ will the sum of the distances from the three sides be at its minimum? (The sum of the distance is $x+y+z$.) *Hint*: Try various points and measure the corresponding values of $x+y+z$. Make a conjecture and prove the conjecture first for points on a side of the triangle. (Use only the concepts and theorems in this chapter.)

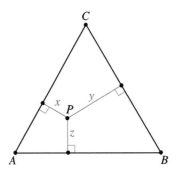

19. Describe how to perform each of the following "constructions with obstructions."

   a. Construct the perpendicular bisector of a segment that is near the bottom of a page such that no drawing below the segment is allowed.

   b. Construct the perpendicular to segment AB at A if an extension of the segment is not allowed.

20. Is the Parallel Postulate (Axiom 2.4) equivalent to the following? (Justify your answer.) *Two intersecting straight lines cannot both be parallel to the same straight line.*

21. The following is a sketch of a proof showing that the Parallel Postulate implies Euclid's Fifth Postulate. Complete the details of the proof.

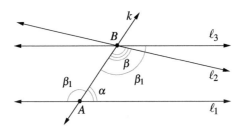

Lines ℓ_1 and ℓ_2 intersect a transversal k at points A and B, respectively, as shown in the figure. Assume that $\alpha + \beta < 180°$. If β_1 is supplementary to α, then we can conclude that $\beta_1 > \beta$. Thus there exists line ℓ_3 through B that forms with k an angle whose measure is β_1. Hence $\ell_3 \| \ell_1$. Because $\beta_1 > \beta$, $\ell_2 \neq \ell_3$. Consequently, ℓ_1 and ℓ_3 are not parallel and must meet. This will contradict the exterior angle theorem.

22. Prove that if $ABCD$ is a quadrilateral in which $MN = \frac{1}{2}(a+b)$, where M is the midpoint of $\overline{AB}$, N is the midpoint of $\overline{CD}$, $AD = a$, and $BC = b$, then $ABCD$ is a trapezoid.

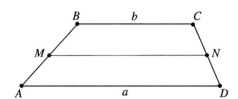

23. In the Treasure Island Problem, Kepri proved that the location of the treasure is independent of the location of the gallows by considering two positions for the gallows: one at one of the trees Γ^* and the other Γ in an arbitrary position. Can you prove it? (*Hint*: Prove that the segments $S_1 S_2$ and $S_1^* S_2^*$ bisect each other.)

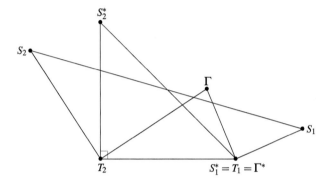

2.5 More on Constructions

Three famous construction problems (using only a straightedge and a compass) have occupied mathematicians and amateurs since antiquity:

1. **Trisecting an Angle:** Is it possible to trisect any given angle—that is, to divide the angle into three congruent angles?

2. **Doubling the Cube:** Given a side of a cube, is it possible to construct a side of a new cube whose volume is twice the volume of the given cube?

3. **Squaring the Circle:** Is it possible to construct a square whose area is the same as the area of a given circle?

In the preceding sections, we saw that many constructions can be performed with only a compass and a straightedge. These three constructions, however, eluded the best mathematical minds for generations. In the nineteenth century, several mathematicians suggested that the famous three constructions couldn't be performed with straightedge and compass alone.

The first two were proved impossible by the French mathematician Pierre Wantzel (1814–1848) in 1837. The key to Wantzel's proof was the conversion of the geometric construction problems to their equivalent algebraic equations. Wantzel obtained criteria for solutions of polynomial equations with integer coefficients to be constructible with only straightedge and compass and showed that the first two problems involve equations whose solutions are not constructible. A more modern version of the proofs can be found in many abstract algebra textbooks (e.g., Beachy and Blair, pp. 283–288). For a more elementary approach, see the beautifully written *What Is Mathematics* (Courant and Robbins, pp. 134–138).

The third problem was proved impossible in 1882 by the German mathematician Ferdinand Lindmann (1852–1939). He noted that squaring the circle of radius 1 is equivalent to solving the equation $x^2 = \pi$ and hence constructing $\sqrt{\pi}$. Lindmann proved that π is transcendental—that is, it is not a solution of any polynomial equation with integer coefficients. His proof requires a considerable knowledge of advanced mathematics. The impossibility of squaring the circle or trisecting an angle has not stopped well-meaning amateurs from coming up with their own "constructions," which, of course, are wrong.

In contrast to these impossible constructions, we will focus in this section on some possible constructions using the theorems introduced in this chapter. More constructions will follow in forthcoming chapters after new theorems will be introduced. From time to time—and especially when a construction is not immediately apparent—we will discuss how a construction can be discovered and its correctness proved by using the following structure:

1. **Investigation** (sometimes also referred to as *analysis*), in which we imagine the given problem as solved and search for properties of the figure that will enable us to accomplish its construction. This powerful problem-solving technique should enable you to discover the construction on your own.

2. **Construction**, where we describe the steps in the construction and actually perform the construction using only a compass and a straightedge.

3. **Proof**, where we prove that our construction does what was asked for.

Construction 2.5.1 Equal Distances

Given two lines k and m and the point A on line k, construct points X and Y on k such that the distance from X to A equals the distance from X to m and similarly such that the distance from Y to A equals the distance from Y to m.

Investigation: To discover the location of the desired points X and Y, we first assume that the points have been found and then search for properties of figure 2.67(a) that will help us construct these points. As $\overline{XX_1}$ and $\overline{YY_1}$ are perpendicular to m, it seems reasonable to construct the perpendicular from A to m. That perpendicular intersects line m in A_1, as shown in Figure 2.67(b). (Notice that $\overline{AA_1}$ can be easily constructed using the basic construction of drawing the perpendicular from a point to a line.)

2.5 More on Constructions

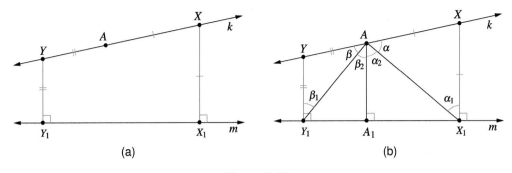

Figure 2.67

Next, since $AX = XX_1$ and $AY = YY_1$ we connect A with X_1 and with Y_1, to obtain the isosceles triangles $\triangle AXX_1$ and $\triangle AYY_1$. Because base angles in an isosceles triangle are congruent, we denote the pairs of congruent angles by α and α_1 and by β and β_1. Anticipating additional angles to be congruent, we name two more angles with vertices at A as α_2 and β_2, as shown in Figure 2.67(b). Since lines YY_1, AA_1, and XX_1 are perpendicular to m, they are parallel to one another. Consequently, $\alpha_1 = \alpha_2$ (alternate interior angles created by the parallel lines AA_1 and XX_1 and the transversal AX_1) and $\beta_1 = \beta_2$. It follows that $\alpha = \alpha_1 = \alpha_2$ and $\beta = \beta_1 = \beta_2$; therefore $\overrightarrow{AX_1}$ and $\overrightarrow{AY_1}$ are angle bisectors of $\angle XAA_1$ and $\angle YAA_1$, respectively. Notice that these angle bisectors can be constructed because the angles formed by line k and $\overline{AA_1}$ are known. The intersections of the angle bisectors and line m are the points X_1 and Y_1. The intersections of the perpendiculars to m through X_1 and Y_1 with line k determine the required points X and Y.

Construction: The steps of the construction were described toward the end of the investigation. For the sake of completeness, we repeat them here:

1. Construct the perpendicular from A to m. The foot of the perpendicular is A_1.
2. Construct the angle bisectors of the angles formed by lines k and $\overrightarrow{AA_1}$ intersecting line m at X_1 and Y_1, respectively.
3. At X_1 and Y_1, erect perpendiculars to m. These perpendiculars intersect line k at the required points X and Y.

We need to prove that the construction is valid. We refer to Figure 2.67(b). By step 2, $\beta = \beta_2$ and $\alpha = \alpha_2$. Also, because $\overline{XX_1} \parallel \overline{AA_1} \parallel \overline{YY_1}$, $\beta_1 = \beta_2$ and $\alpha_1 = \alpha_2$. Thus $\alpha = \alpha_1$ and $\beta = \beta_1$. Consequently, $\triangle AXX_1$ and $\triangle AYY_1$ are isosceles triangles and, therefore, $AX = XX_1$ and $AY = YY_1$, as required.

Construction 2.5.2 Extending a Line Beyond an Obstruction

Extend line AB beyond the small obstruction shown in Figure 2.68 without the straightedge's touching the obstruction.

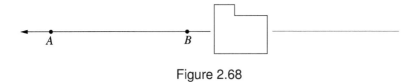

Figure 2.68

Investigation: It seems that unless we construct additional lines and points, we cannot accomplish much. Recall that in an isosceles triangle the angle bisector at a vertex where the congruent sides meet is also perpendicular to the base (the opposite side).

Thus one way to solve this problem is to construct an isosceles triangle ACD with vertex at A and to make $\vec{AB}$ the angle bisector, as in Figure 2.69. Then the perpendicular bisector of the base CD will be on the other side of the obstruction provided that the sides of the isosceles triangle are long enough. You are asked to complete the construction in the next Now Solve This.

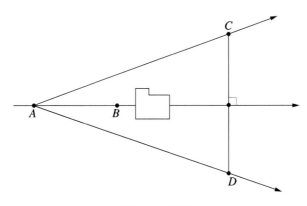

Figure 2.69

Now Solve This 2.15

1. Describe the construction steps suggested in Construction 2.5.2 and actually construct the extension of $\overline{AB}$.

2. Prove that the construction accomplishes the extension of $\overline{AB}$ (this is very short and simple).

Construction 2.5.3 A Line Through a Point in the Interior of an Angle

Point P is in the interior of $\angle A$. It is required to construct a line through P so that P is the midpoint of the line segment in the interior of the angle. In Figure 2.70(a), we need to construct a line XY so that $PX = PY$.

2.5 More on Constructions

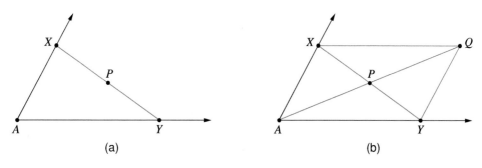

Figure 2.70

Investigation: We imagine that the points X and Y have been found. We can connect A with P. If we extend $\overline{AP}$ to point Q so that $AP = PQ$, the segments $\overline{AQ}$ and $\overline{XY}$ in Figure 2.70(b) will bisect each other at P and, therefore, $AXQY$ will be a parallelogram. In a parallelogram, opposite sides are parallel; thus we can readily find the required points X and Y. Through Q we need merely to construct lines parallel to the sides of the angle. These lines intersect the sides of the angle at the points X and Y.

Construction: The construction is embedded in the above investigation. The steps are as follows:

1. Connect A with P, and extend $\overline{AP}$ so that $AP = PQ$.
2. Through Q, construct parallels to the sides of the angle. These parallels intersect the sides in X and Y.
3. XY is the required segment.

We need to prove that the construction is valid. By our construction, $AXQY$ is a parallelogram. In a parallelogram diagonals bisect each other, so P must be the midpoint of $\overline{XY}$.

Construction 2.5.4 An Inaccessible Vertex of an Angle

In Figure 2.71(a), construct a line through a point P in the interior of an angle to the inaccessible vertex A of the angle.

Investigation: Let k and ℓ be the sides of the angle in Figure 2.71(a). Imagine that vertex A is known and that line $\overrightarrow{AP}$ is drawn. To solve the construction problem, it is sufficient to find one more point on $\overrightarrow{AP}$.

By constructing through P lines parallel to the sides k and ℓ of the angle, a parallelogram is created with $\overline{AP}$ as a diagonal. The segment joining B with D forms the second diagonal. We know that in a parallelogram the diagonals bisect each other. This means that the midpoint M of $\overline{BD}$, which we can construct, lies on $\overrightarrow{AP}$. We have found a second point on $\overrightarrow{AP}$. The line through M and P is the required line.

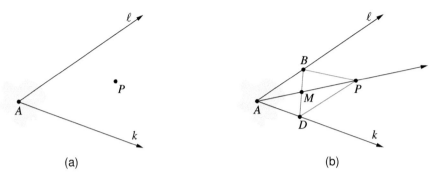

Figure 2.71

Construction: Through P, construct lines parallel to k and ℓ. Let the intersection points with k and ℓ be D and B, respectively. Construct $\overline{BD}$ and its midpoint M. Line PM is the required line.

We now show that the construction is valid. Because $ABPD$ is a parallelogram, its diagonals bisect each other. Hence $\overleftrightarrow{MP} = \overleftrightarrow{AP}$.

> **Now Solve This 2.16**
>
> In Problem 23 of Problem Set 2.2.1, we were asked to construct the angle bisector of an angle whose vertex is inaccessible (actually part of the angle bisector). Here we ask you to explore the following approaches that might differ from yours.
>
> 1. Construct an isosceles triangle ABC and then construct the perpendicular bisector of $\overline{BC}$. You will need to construct the angle first (α in Figure 2.72). The construction of the angle is suggested in Figure 2.72.
>
>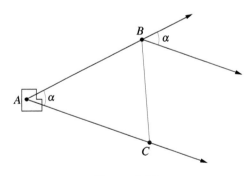
>
> Figure 2.72
>
> 2. Bisect the angle α whose vertex is B and construct the perpendicular to that angle bisector.
> 3. Use the fact that the three angle bisectors of a triangle are concurrent. Then construct any triangle with one vertex A and arbitrary points B and C on the sides of the angle (like in Figure 2.72 except that $\triangle ABC$ does not need to be isosceles).

2.5 More on Constructions

Construct the point where the angle bisectors intersect. Repeat this process for a different triangle.

Construction 2.5.5 SSA Construct a Triangle Given Two Sides and an Angle Opposite One of the Sides

Investigation: In Figure 2.73(a), we are given angle B and segments b and c congruent to sides AC and AB of $\triangle ABC$, respectively. We need to construct $\triangle ABC$.

In Figure 2.73(b), we first construct $\angle B$ and then place point A on one of the sides such that $BA = c$. To find the third vertex C, notice that C is on a ray, on the other side of $\angle B$, and at the distance b from A, that is, on the circle with center A and radius b. That circle intersects the ray in two points, C and C_1, and hence the problem has two solutions: $\triangle ABC$ and $\triangle ABC_1$.

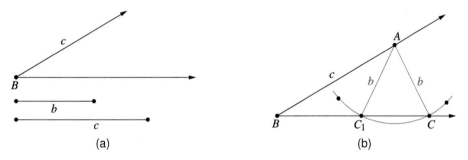

Figure 2.73

Because the construction and its proof are embedded in the investigation, they are not provided here separately. However, you are invited to further explore the construction by finding when the construction has two solutions, one solution, or no solution.

Now Solve This 2.17

Referring to Construction 2.5.4 and Figure 2.73(a), construct figures and explore a number of solutions for each of the following problems:

1. When the side b opposite the given angle is shorter than shorter than side c [Figure 2.73(b)]
2. When $b > c$
3. When $b = c$
4. When b is equal to the distance from A to the opposite side
5. When b is less than the distance from A to the opposite side.

2.5.1 Problem Set

In the following construction problems you need to include an investigation, steps of the construction, the actual construction, and finally a proof that the construction is valid.

1. The figure below suggests how to construct a trapezoid given its sides. Describe this construction.

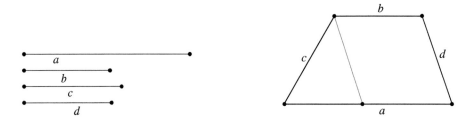

2. Construct a right triangle given its hypotenuse c and a leg a.

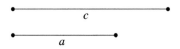

3. Construct an isosceles triangle ABC with $AB = AC$, given $\angle A$ and its perimeter (the sum of its three sides).

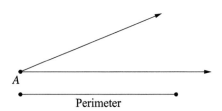

4. For each of the following, construct all the points that are equidistant from three given lines. In each case draw the lines.
   a. The lines do not intersect in a single point (that is, the lines are not concurrent) and no two lines are parallel.
   b. Exactly two lines are parallel.
   c. The three lines are concurrent.

5. Construct a triangle, given one side a, the altitude h to that side, and the median m to that side.

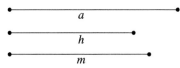

2.5 More on Constructions

6. Construct your own arbitrary $\triangle ABC$ and inscribe in it a rhombus such that one of the angles of the rhombus is an angle of the triangle.

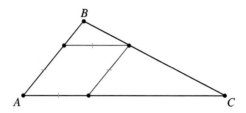

7. Construct an equilateral triangle given its height h.

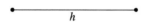

8. P is on one side of a given $\angle A$. Find point X on the other side of the angle such that $AX + PX = s$, the length of a given segment.

9. Divide a given segment into three congruent parts using a theorem about medians of a triangle.

In some of the following construction problems the given segments or angles are not drawn. When this is the case (like in the next problem) construct your own figure and extract the given data from the figure and then use only the extracted data to construct the figure.

10. Construct a quadrilateral given three of its sides and the two diagonals.

11. Construct a triangle given two of its sides and the median to the third side.

12. Construct a triangle given its medians.

13. **a.** If $\overline{AB}$ is a diameter of a semicircle and C is any point on the semicircle, prove that $\angle ACB$ is $90°$.

b. Use your result from part (a) to construct a right triangle given its hypotenuse c and one side (labeled *a* in the figure below).

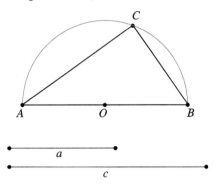

c. Solve the problem in part (b) without using the theorem in part (a).

14. Construct a right triangle given the hypotenuse c and the sum of the remaining sides a and b.

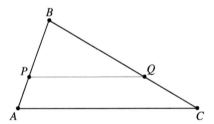

15. Construct a triangle given a side a, angle α opposite a, and the altitude h to one of the other sides.

16. Construct a square inscribed in a given rhombus.

17. In $\triangle ABC$, construct points P and Q on sides $\overline{AB}$ and $\overline{BC}$, respectively, such that $\overline{PQ} \| \overline{AC}$ and $PQ = AP + QC$.

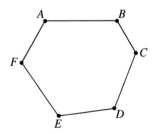

18. The point of intersection of an altitude with a side of a triangle is called the **foot of the altitude**. Given the three feet of the altitudes of a triangle, construct the triangle.

19. a. Construct a hexagon $ABCDEF$ (see the figure below), which is not regular but for which each of the interior angles is 120°.

b. If two polygons have corresponding angles that are congruent and corresponding sides that are proportional (the ratio of corresponding sides is a constant), the polygons are called *similar*. How many nonsimilar hexagons with all interior angles 120° are there? Justify your answer.

c. Prove that in the hexagon in part (a),

$$AF + AB = ED + DC.$$

20. a. Prove that if $\overline{HG}$ and $\overline{EF}$ are two perpendicular segments connecting arbitrary points on pairs of opposite sides of the square, then $HG = EF$.

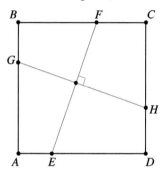

b. Suppose X, Y and Z, W are pairs of points on opposite sides of a square that remain after the sides of the squares have been erased. Reconstruct the square. Is the solution unique?

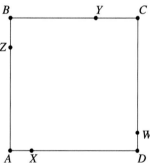

21. Given points A and B on the same side of line ℓ, find (construct) point X on ℓ such that $\angle AXC$ is twice the size of $\angle BXD$. (Points C and D are arbitrary points on ℓ such that X is between C and D. These points appear only for the purpose of naming the angles.)

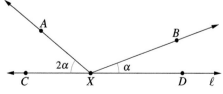

3 Circles

It is impossible to be a mathematician without being a poet in soul.

Sonia Kovalevskaya (1850–1891), a famous Russian mathematician

Introduction

A **circle** is the set of all points in the plane, at a given distance (the **radius**) from a fixed point in the plane (the **center**). Two circles are **congruent** if and only if their radii are equal. The circle is one of the most commonly encountered figures in daily life. Wheels, cross sections of trees, coins, lids, the shape of the full moon, and the equator—all suggest circles. Some of the most important concepts in mathematics, such as π and the circular or trigonometric functions, are related to the circle. Circles have many fascinating properties in their own right. We will explore some of these properties in this chapter.

3.1 Central and Inscribed Angles

A line that intersects a circle in two points is called a secant; the segment connecting those two points is called a **chord**. In Figure 3.1(a), $\overleftrightarrow{PQ}$ is a secant and $\overline{PQ}$ is a chord. A chord containing the center of the circle is a **diameter**, whose length is twice the radius. Circles that share the same center are **concentric**. The **interior of a circle** is the set of all points in the plane whose distance from the center of the circle is less than the radius. The **exterior of a circle** is the set of all points in the plane whose distance from the center is greater than the radius. When there is only one circle with a given center we will name that circle by its center. Thus the circle in Figure 3.1(b) is **circle** O.

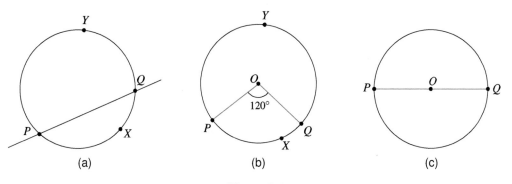

Figure 3.1

> **Historical Note:** *Euclid's definition of a circle*
> Euclid defined a circle as:
>
> > *A plane figure contained by a line so that all straight lines falling on the line from a point, called the center, are equal.*
>
> Notice how cumbersome Euclid's "definition" is. Because he did not have the concept of distance, he could not use the word *radius* either. For Euclid a circle is a *line* albeit not a *straight line*.

> **Remark 3.1**
>
> 1. Circles that are not concentric are commonly named after their centers.
> 2. The terms "radius" and "diameter" are used both for the segments and their lengths. When we refer to length, we will use the prefix "the." Thus *a radius* is a segment from the center to a point on the circle, while *the radius* is the length of any such segment.

The secant $\overleftrightarrow{PQ}$ divides the circle into two arcs. The larger of the two arcs is called the **major arc** and the smaller is called the **minor arc**. In Figure 3.1(a), $\widehat{PXQ}$ (read "arc PXQ") is the minor arc, and $\widehat{PYQ}$ is the major arc. We will denote the minor arc connecting P and Q by $\widehat{PQ}$ and the major arc by three letters corresponding to the endpoints of the arc and a point on the arc, as in $\widehat{PYQ}$ in Figure 3.1(a).

An angle whose vertex is the center of the circle is a **central angle**. In Figure 3.1(b), $\angle POQ$ is a central angle whose sides intersect the circle at P and Q. The arc $\widehat{PQ}$ is in the interior of that angle. We say that $\angle POQ$ **intercepts** $\widehat{PQ}$. The chord determined by the endpoints of an arc **subtends** the arc.

It is convenient to associate to a minor arc the measure of its central angle. If $m(\angle POQ) = 120°$, we may write $m(\widehat{PQ}) = 120°$. Thus the **measure of a minor arc** is defined as the measure of its central angle. Another measure of an arc is its length, as discussed in more detail in Chapter 5. The **measure of a major arc** is defined to be 360° minus the measure of the minor arc. If $\overline{PQ}$ is a diameter as in Figure 3.1(c), then the two arcs formed are called **semicircles** and the measure of each is 180°.

Adjacent arcs in a circle are arcs that have exactly one point in common. In Figure 3.1(a), the arcs $\widehat{PX}$ and $\widehat{XQ}$ are adjacent. It follows from the angle addition postulate that the measure of an arc which is the union of two adjacent arcs is the sum of the measures of those two arcs. **Congruent arcs** are arcs in the same circle or in congruent circles that have the same measure.

The following statements are intuitive. Their proofs are straightforward and, therefore, are omitted here. (For that reason we call them "propositions.")

3.1 Central and Inscribed Angles

> **Proposition 3.1**
>
> *Two minor arcs in the same circle or in congruent circles are congruent if and only if their central angles are congruent.*

> **Proposition 3.2**
>
> *In a circle, or in congruent circles, two arcs are congruent if and only if the corresponding chords are congruent.*

> **Proposition 3.3**
>
> *(a) A diameter is perpendicular to a chord if and only if it bisects the chord.*
> *(b) A diameter bisects each of the arcs determined by the chord if and only if it is perpendicular to the chord.*

We illustrate Proposition 3.3 in Figure 3.2, where $\overline{CD} \perp \overline{AB}$ and hence $\overline{CD}$ bisects $\overparen{BA}$ and $\overparen{BCA}$ at D and C, respectively.

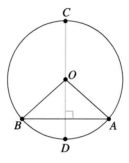

Figure 3.2

In Figure 3.3, $\overline{CD}$ is the common chord of the circles with centers at A and B. The line $\overleftrightarrow{AB}$ is the perpendicular bisector of $\overline{CD}$. We state this fact in the following proposition:

> **Proposition 3.4**
>
> *If two circles intersect in two points, the line through their centers is the perpendicular bisector of their common chord.*

Proof. Because $AC = AD$, A is equidistant from the endpoints of $\overline{CD}$ and, therefore, lies on the perpendicular bisector of $\overline{CD}$. Similarly, B is on that perpendicular bisector. Since a line is determined by any two points on it, $\overleftrightarrow{AB}$ is the perpendicular bisector of $\overline{CD}$. ∎

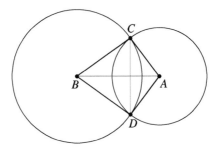

Figure 3.3

Tangent to a Circle

A **tangent** at a point P to a curve can be roughly described as a line through P that just touches the curve. More precisely, if Q is any point on the curve, $\overleftrightarrow{QP}$ is a secant. If Q gets closer and closer to P (from either side of P), the secant often approaches a fixed line t (as in Figure 3.4). That fixed line (if it exists) is called the tangent to the curve at P. A precise definition of a tangent, which involves the concept of a limit, can be found in most calculus books.

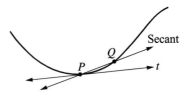

Figure 3.4

In the case of a circle, it seems that the secant in Figure 3.5 approaches the line t, which is perpendicular to the radius $\overline{PO}$. This observation is the basis of the following definition, which applies only to circles:

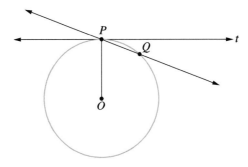

Figure 3.5

3.1 Central and Inscribed Angles

Definition of a Tangent to a Circle

The **tangent to a circle** O at point P on the circle is the line through P that is perpendicular to the radius $\overline{OP}$.

Now Solve This 3.1

A tangent to a curve may intersect the curve at more than one point. In the case of a circle, however, the tangent intersects the circle at exactly one point. (a) Prove this claim by choosing in Figure 3.5 any point T (not P) on the tangent and showing that the distance from the center to T is greater than the radius. (b) Prove that a line that intersects a circle in only one point is tangent to the circle.

An Angle Formed by a Tangent and a Secant

At any given moment, an object moving along a circular path is "facing" in the direction of the motion along the tangent to the circle. To see why, consider an object moving from P to another point on the circle very close to P. The line connecting the two points is very close to the tangent. Suppose a particle moves along the circle from point P to point Q in Figure 3.6.

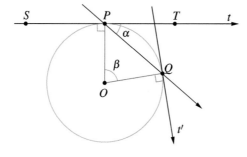

Figure 3.6

Notice that at Q the particle is "facing" in the direction of the motion along the tangent t' at that point. The particle moved along the arc $\stackrel{\frown}{PQ}$, whose measure is β, the measure of the central angle POQ. If we connect P with Q, an angle between the tangent t and the chord $\overline{PQ}$ is formed. We denote the measure of $\angle TPQ$ by α. It seems that the size of α depends on the size of β. Notice that if $\beta = 0$, then $\alpha = 0$; if $\beta = 180°$, then P and Q are diametrically opposite and $\alpha = 90°$. In fact, we can prove that $\alpha = \frac{1}{2}\beta$ in all cases as stated in the following theorem:

Theorem 3.1

The measure of the angle formed by a tangent and a chord equals half the measure of the intercepted arc.

Proof. Notice that both α and $\angle SPQ$ in Figure 3.6 are angles formed by tangent t and chord $\overline{PQ}$. These angles are supplements of each other and, therefore, if the chord is not a diameter, one of them is acute and the other is obtuse. If α is acute, we need to prove the theorem for α and for its supplement $\angle SPQ$. If t is tangent to the circle at P, $\overline{PQ}$ is a chord, α and β are the corresponding measures of $\angle QPT$ and $\angle POQ$, we need to show that

1. $\alpha = \frac{1}{2}\beta$, and
2. $m(\angle SPQ) = \frac{1}{2}$ the measure of the major arc PQ.

Notice that the first statement is for the acute angle α between a tangent and a chord, whereas the second is the theorem for an obtuse angle between a tangent and a chord.

To find a relationship between α and β in Figure 3.6, we use the fact that $t \perp \overline{OP}$ and the fact that the sum of the measures of the angles in any triangle is 180°. Because $\triangle POQ$ is isosceles and $m(\angle OPQ) = 90° - \alpha$, we have $(90° - \alpha) + (90° - \alpha) + \beta = 180°$. Consequently, $\alpha = \frac{1}{2}\beta$. To prove the second part of the theorem, we express $m(\angle SPQ)$ in terms of α and use $\alpha = \frac{1}{2}\beta$. Hence $m(\angle SPQ) = 180° - \alpha = 180° - \frac{1}{2}\beta = \frac{1}{2}(360° - \beta)$. Because $\frac{1}{2}(360° - \beta)$ is half the measure of the major arc PQ, we have proved the second part of the theorem. ∎

> **Now Solve This 3.2**
>
> Figure 3.7 suggests a quick way to prove Theorem 3.1 if the angle formed by the chord and tangent is acute. Supply the details of the proof to show that $\alpha = \frac{1}{2}\beta$.
>
>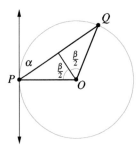
>
> Figure 3.7

The Inscribed Angle Theorem

We can use Theorem 3.1 to deduce a similar relationship between the measure of an angle formed by two chords intersecting at a point P on a circle (see Figure 3.8(a)) and the intercepted arc. Such an angle is called an **inscribed angle** and is defined more formally as an angle whose vertex is on the circle and whose sides are chords of the circle. Recall that sides of an angle are rays and not segments; however, it is common practice to refer to the segments $\overline{PA}$ and $\overline{PB}$ in Figure 3.8(b) as sides of $\angle APB$. Based on this discussion, we pose the following problem:

3.1 Central and Inscribed Angles

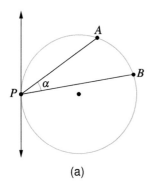
(a)

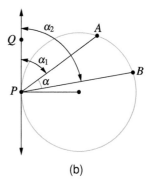
(b)

Figure 3.8

Problem 3.1

Find a relationship between the measure of an inscribed angle and the measure of its intercepted arc.

Solution. In Figure 3.8(a) we need to find a relationship between α and $m(\widehat{AB})$. To use Theorem 3.1, we draw the tangent $\overleftrightarrow{PQ}$ in Figure 3.8(b). Each of the angles that the tangent makes with $\overline{PA}$ and $\overline{PB}$ can be expressed now in terms of the measure of the arcs they intercept. In addition, α is the difference of these angles. Referring to Figure 3.8(b), we have:

$$\alpha_2 = \frac{1}{2}m(\widehat{PB}),$$

$$\alpha_1 = \frac{1}{2}m(\widehat{PA}),$$

$$\alpha = \alpha_2 - \alpha_1 = \frac{1}{2}m(\widehat{PB}) - \frac{1}{2}m(\widehat{PA}),$$

$$\alpha = \frac{1}{2}\left[m(\widehat{PB}) - m(\widehat{PA})\right],$$

$$\alpha = \frac{1}{2}m(\widehat{AB}).$$

In Figure 3.8, $\widehat{PAB}$ is a minor arc, so it can also be written as $\widehat{PB}$. Notice that the solution to Problem 3.1 does not depend on the size of α and hence is completely general. We summarize the result of Problem 3.1 in the following theorem:

Theorem 3.2: The Inscribed Angle Theorem

The measure of an inscribed angle equals half the measure of the intercepted arc.

The proof of Theorem 3.2 was given in the solution to Problem 3.1. Figure 3.9 illustrates the theorem in the case where the center of the circle lies in the interior of the inscribed angle. A different proof of Theorem 3.2 (commonly found in textbooks) that does not use Theorem 3.1 will be explored in the problem set.

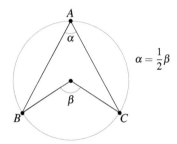

Figure 3.9

The Inscribed Angle Theorem is fundamental in the study of circles, so we shall use it frequently. Its full significance is perhaps more evident in the following corollary:

Corollary 3.1

In any circle, all the inscribed angles intercepting the same arc are congruent.

Proof. By Theorem 3.2, the measure of each of the inscribed angles in Figure 3.10 is half the measure of $\overarc{BC}$. ∎

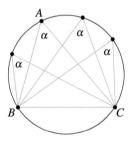

Figure 3.10

We will frequently describe the chord of the intercepted arc as being subtended by the inscribed angle. In Figure 3.10, this would mean that $\overline{BC}$ is subtended from A by $\angle BAC$ whose measure is α. Another way to say this is "α is the angle by which the chord $\overline{BC}$ is seen from A." Using this terminology, we have the following useful theorem.

Theorem 3.3

From all the points on $\overarc{BAC}$, the chord $\overline{BC}$ is subtended (or seen) by congruent angles each measuring half the measure of the intercepted arc.

3.1 Central and Inscribed Angles

In Figure 3.10, the measure of ∠BAC is denoted by α. We have shown that from every point on $\overset{\frown}{BAC}$, the chord $\overline{BC}$ is subtended by angle α. Nevertheless, we have not proved that the points on $\overset{\frown}{BAC}$ are the only points in the half-plane determined by $\overleftrightarrow{BC}$ and A (the half-plane "above" $\overleftrightarrow{BC}$) from which $\overleftrightarrow{BC}$ is subtended by α. In the next theorem, we show that this is the case—that is, there are no other points in the half-plane from which $\overline{BC}$ is seen by α.

> **Theorem 3.4**
>
> *If α is the measure of an inscribed angle subtending $\overline{BC}$ of a circle, then the locus[1] of all the points (in the half-plane determined by $\overleftrightarrow{BC}$ and point A on the circle) from which $\overline{BC}$ is subtended by α is the arc $\overset{\frown}{BAC}$.*

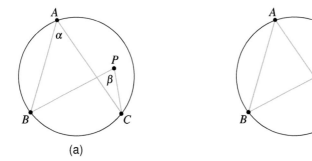

Figure 3.11

Proof. In Figure 3.11(a), let $m(\angle A) = \alpha$. We need to show that if P is not on $\overset{\frown}{BAC}$ but rather in the mentioned half-plane, then $m(\angle BPC) \neq \alpha$. We distinguish two cases: one where P is in the interior of the circle, and one where P is in the exterior of the circle. Figure 3.11(a) shows a point P in the interior of a circle and in the half-plane determined by $\overleftrightarrow{BC}$ and A. Let $m(\angle BPC) = \beta$. It seems that $\beta > \alpha$. To prove this, we extend $\overline{BP}$ until $\overrightarrow{BP}$ intersects the circle at Q as shown in Figure 3.11(b). In this way we create ∠BQC whose measure is also α (by Corollary 3.1). In $\triangle PQC$, β is an exterior angle and hence $\beta > \alpha$. The case when P is in the exterior of the circle is similar and is left for the reader to prove in Now Solve This 3.3. ∎

> **Now Solve This 3.3**
>
> 1. Prove Theorem 3.3 for the case when point P is in the exterior of the circle.
> 2. Given a segment, construct the locus of all the points in the plane from which the segment is subtended by an angle of (a) 45° and (b) 30°.

---

[1] A **locus** is the set of all points satisfying a given condition.

In Figure 3.12 each of the marked angles is inscribed in a semicircle. What can be said about such inscribed angles? Because the measure of a semicircle (as an arc) is 180°, each of the inscribed angles measures $\frac{1}{2} \cdot 180°$ or 90°. This is stated in the following corollary:

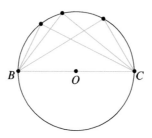

Figure 3.12

Corollary 3.2

Any angle inscribed in a semicircle is a right angle.

Now Solve This 3.4

We have proved one part of the following statement. State and prove the other part. An angle inscribed in a circle is a right angle if and only if it subtends a diameter.

Example 3.1

In Figure 3.13, $\overline{AO} \perp \overline{OC}$, O is the center of the circle, and $\overline{DE}$ is the perpendicular bisector of $\overline{OC}$. Find α and β, where α and β are measures of angles inscribed in circle O as shown in Figure 3.13.

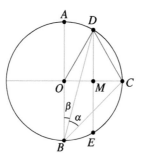

Figure 3.13

Solution. α, β, and $\alpha + \beta$ are inscribed angles. Hence each equals half the measure of the corresponding intercepted arc. Because $\overline{AO} \perp \overline{OC}$, we have $m(\widehat{AC}) = 90°$ and

$$\alpha + \beta = \frac{1}{2} m(\widehat{AC}), \qquad \alpha + \beta = \frac{1}{2} \cdot 90° = 45°.$$

3.1 Central and Inscribed Angles

Consequently, it is sufficient to find either α or β. To find α, we need to know only the measure of $\overset{\frown}{DC}$, that is, $\angle DOC$. The piece of information that we have not used yet is the fact that $\overline{DE}$ is the perpendicular bisector of $\overline{OC}$. This information implies that D is equidistant from O and C, that is, $OD = DC$. Because OD and OC are radii, $\triangle ODC$ is equilateral and thus $m(\angle DOC) = 60°$. Because α and $\angle DOC$ intercept the same arc $\overset{\frown}{DC}$, we have $\alpha = \frac{1}{2}m(\angle DOC) = \frac{1}{2} \cdot 60° = 30°$. Thus $\alpha = 30°$. Because $\alpha + \beta = 45°$, $\beta = 15°$.

Now Solve This 3.5

1. Construct a 45° angle and use the information from Example 3.1 to trisect the angle.
2. Trisect a 45° angle in a different way.

Relationship Among the Angles of a Quadrilateral Inscribed in a Circle

When each vertex of a polygon is on a circle as shown in Figure 3.14, the polygon is said to be **inscribed in a circle** or *cyclic*. The circle is said to be **circumscribing the polygon**.

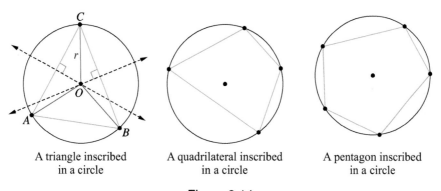

A triangle inscribed in a circle   A quadrilateral inscribed in a circle   A pentagon inscribed in a circle

Figure 3.14

In Problem 8 of Problem Set 2.2.1, we saw that for every triangle there exists a circle that circumscribes it; hence every triangle is cyclic. Because this property is so important, we look at it again here. If a triangle is inscribed in a circle with center O and radius r as in Figure 3.14 then the distance from each vertex of the triangle to the center equals r. Therefore O is equidistant from the endpoints of each side of the triangle and consequently O is on the perpendicular bisector of each of these sides. Conversely, given any triangle we can find the intersection O of two perpendicular bisectors of any two sides of the triangle. (The three perpendicular bisectors are concurrent.) O will be the center of the circumscribing circle. Discovering this property of triangles naturally leads us to wonder whether quadrilaterals have a similar property. In other words, given a quadrilateral, does there always exist a circle that circumscribes it? As there is a unique circle through three noncollinear points, given three such points *A*, *B*, and *C* as shown in Figure 3.15(a), we could choose a point *D* not on the circle that goes through *A*, *B*, and *C* and quadrilateral *ABCD* would not be cyclic. It should also be visually evident that a quadrilateral like the one in Figure 3.15(b), in which two nonconsecutive vertices are relatively close to each other, is not cyclic.

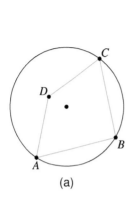 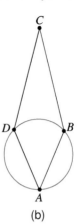

(a)   (b)

Figure 3.15

What, then, are the necessary and sufficient conditions for a quadrilateral to be cyclic? If a quadrilateral such as the one in Figure 3.16 is cyclic, the center of the circumscribing circle is equidistant from each vertex and hence is on the perpendicular bisector of each side. Conversely, if for a given quadrilateral the four perpendicular bisectors of its sides are concurrent (that is, they intersect in a single point), then that point is the center of the circumscribing circle. Thus we have the following theorem.

Theorem 3.5

A necessary and sufficient condition for a quadrilateral to be cyclic is for the perpendicular bisectors of its sides to be concurrent.

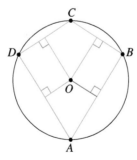

Figure 3.16

In reality, this condition is seldom useful, as there is usually no simple way to recognize whether the perpendicular bisectors of the four sides are concurrent. (A rectangle is an

3.1 Central and Inscribed Angles

exception. Why?) Because the four angles of a cyclic quadrilateral are inscribed angles, and inscribed angles have special properties, it makes sense to look for a relationship among the angles of a cyclic quadrilateral. In Figure 3.16, we have

$$m(\angle A) = \frac{1}{2}m(\widehat{DB}),$$
$$m(\angle C) = \frac{1}{2}m(\widehat{DAB}).$$

Because the union of $\widehat{DB}$ and $\widehat{DAB}$ is the entire circle, we get

$$m(\angle A) + m(\angle C) = \frac{1}{2}\text{ (entire circle) } = \frac{1}{2} \cdot 360°.$$

Our investigation implies that the sum of the measures of any pair of opposite angles in a cyclic quadrilateral is 180°. Thus a necessary condition for a quadrilateral to be cyclic is that the sum of the measures of a pair of opposite angles is 180°. It turns out that this condition is also sufficient for a quadrilateral to be cyclic, as stated in the following theorem:

Theorem 3.6

A necessary and sufficient condition for a quadrilateral to be cyclic is that the sum of the measures of a pair of opposite angles equals 180°.

Proof. Earlier, we showed that the condition is necessary; we need to show that the condition is sufficient. Suppose that $ABCD$ in Figure 3.17 is a quadrilateral in which

$$m(\angle A) + m(\angle C) = 180°. \tag{3.1}$$

We assume that $ABCD$ is not cyclic and obtain a contradiction. We construct a circle with center O that contains any three of the vertices. In Figure 3.17, O is the circle through $B, C,$ and D.

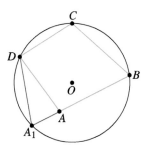

Figure 3.17

Because we assumed that $ABCD$ is not cyclic, A is not on the circle. We distinguish two cases and for each obtain a contradiction. If A is in the interior of the circle, we extend

$\overline{AB}$ so that it intersects the circle at A_1. In this way, we get quadrilateral A_1BCD inscribed in a circle. By the already proved first part of the theorem, we have

$$m(\angle A_1) + m(\angle C) = 180°. \tag{3.2}$$

Equations (3.1) and (3.2) imply that $m(\angle A) = m(\angle A_1)$, which is impossible because $\angle A$ is an exterior angle for ΔDAA_1 and, therefore, greater than any remote interior angle of the triangle. A similar contradiction is obtained for the case when A is in the exterior of the circle and is left as an exercise. ∎

> **Now Solve This 3.6**
>
> Use Geogebra or a compass and straightedge to perform the following constructions.
>
> 1. Construct a circle and a quadrilateral $ABCD$ with perpendicular diagonals inscribed in the circle as shown in Figure 3.18(a). Through P (the point of intersection of the diagonals), draw a line perpendicular to one of the sides. The line divides the opposite side into two segments of length a and b. Measure the segments, and find the ratio between their lengths. Repeat the experiment for a few other such quadrilaterals. Make a conjecture about your results and prove it.
>
> 2. Draw any quadrilateral $ABCD$ and its angle bisectors as shown in Figure 3.18(b). The points of intersection of the four angle bisectors form a new quadrilateral $WXYZ$. Construct a circle through any three of the vertices of the new quadrilateral. What do you observe about the circle? Repeat the experiment for two other quadrilaterals. Make a conjecture about your results and prove it.
>
>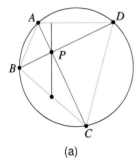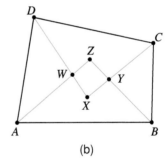
>
> (a)   (b)
>
> Figure 3.18

3.1.1 Problem Set

In the following problems, if a point is labeled by the letter O, it is the center of the circle.

1. **a.** State and prove an "if and only if" condition relating two congruent chords in a circle to their distances from the center of the circle.

   b. Given two chords in a circle, prove that the longer chord is closer to the center of the circle.

3.1 Central and Inscribed Angles

    c. State and prove the converse of the statement in part (b).

2. Prove part (a) of Proposition 3.3, that a diameter is perpendicular to a chord if and only if it bisects the chord.

3. Give an alternative proof of the *Inscribed Angle Theorem*, by first proving a special case stated in part (a) below, and then using your result from part (a) to prove the cases in parts (b) and (c). In each case show that $\alpha = \frac{1}{2}\beta$.

    a. One side of the inscribed angle contains the diameter of the circle.

    b. The center of the circle is in the interior of the angle.

    c. The center of the circle is in the exterior of the angle.

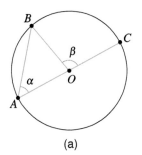 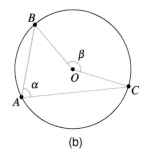 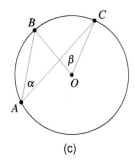

               (a)   (b)   (c)

    d. Which proof of the Inscribed Angle Theorem do you like better—the one in the text or the one outlined in Problem 4? Explain why.

4. Recall that in the text the *Inscribed Angle Theorem* followed from Theorem 3.1. As suggested in Problem 3, Theorem 3.2 can also be proved without the use of Theorem 3.1. In fact, it can even be shown that Theorem 3.1 follows from the *Inscribed Angle Theorem* by viewing a tangent as a limit of secants. Do so now by using the following figure and assuming that if P moves toward A, then the secant $\overleftrightarrow{AP}$ approaches the tangent $\overleftrightarrow{AB}$ and $\angle PAQ$ approaches $\angle BAQ$.

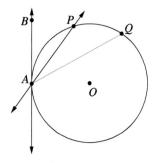

5. In each of the following cases find unique values for x and y, if possible. Otherwise prove that there are infinitely many solutions.

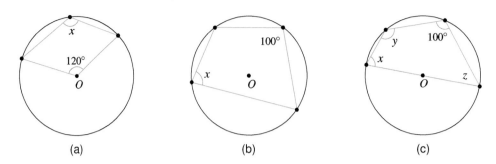

6. △ACB is inscribed in a circle. H is the orthocenter (the point where the altitudes intersect) of △ACB. Point K is the intersection of $\overrightarrow{AD}$ and the circle. Prove that $\overline{HD} \cong \overline{DK}$.

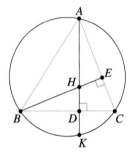

7. **a.** In the accompanying figures, two secants intersect in the interior or the exterior of the circle. In each case, find the angle α formed by the secants in terms of the measures of the intercepted arcs. State and prove the corresponding results. (*Hint*: Connect some of the points in the figures to create inscribed angles.)

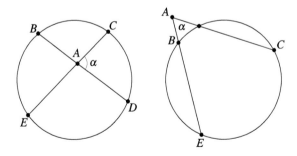

b. Explain why the Inscribed Angle Theorem is a special case of each of the results in part (a).

3.1 Central and Inscribed Angles

8. Prove that two parallel chords in a circle create congruent arcs and hence congruent chords. In the figure below prove that if $\overline{BC} \parallel \overline{AD}$, then $\widehat{AB}$ and $\widehat{CD}$ are congruent.

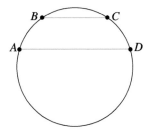

9. Problem 7(a) can now be answered in a different way using the result of Problem 8. Use the figure below to answer the first part in Problem 7(a) and sketch an analogous figure to answer the second part of Problem 7(a).

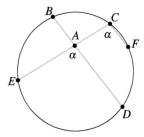

10. Answer each of the following questions with an "if and only if" statement. Justify your answers.
   a. What kind of parallelograms are cyclic?
   b. What kind of trapezoids are cyclic?

11. $\triangle ABC$ is inscribed in a circle. Line m is tangent to the circle at C, and $n \parallel m$. Line n intersects $\triangle ABC$ at D and E. Is $ABED$ cyclic? Prove your answer.

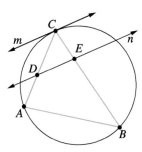

12. Acute $\triangle ABC$ is inscribed in circle O as shown below. The altitude $\overline{BH}$ intersects the diameters through A and C at D and E, respectively.
   a. Prove that $\triangle DOE$ is isosceles.

b. Is the assertion in part (a) still true if △ABC is an obtuse triangle? Why or why not?

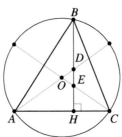

13. $\overline{AB}$ is a diameter of a semicircle with center O. Let S be on $\overrightarrow{AB}$, and let $\overleftrightarrow{SC}$ be a tangent to the circle. If $\overrightarrow{ST}$ bisects $\angle ASC$ and intersects $\overline{AC}$ at D, find the measure of $\angle SDC$. Justify your answer.

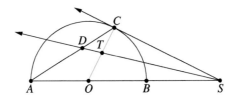

14. Let A be a point in the exterior of a circle O. (Choose A close to the circle.) Let B and C be the points of tangency of the tangents drawn from A to the circle, and let $\overline{CP}$ be a diameter. The line $\overleftrightarrow{PB}$ intersects $\overleftrightarrow{CA}$ at point E. Prove that $m(\angle BAC) = 2m(\angle BEA)$. (*Note:* You need to draw your own figure.)

15. $ABCD$ is inscribed in circle O. $\overline{AB}$ is a diameter, and E is the intersection of $\overrightarrow{AD}$ and $\overrightarrow{BC}$.
    a. Prove that $\angle ODC \cong \angle AEB$.
    b. Prove that $\overline{OD}$ is the tangent at D to the circle that circumscribes △DEC.
    c. Which of the preceding statements, if any, are true if $\overline{AB}$ is not a diameter? Justify your answer.

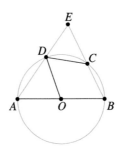

3.2 Inscribed Circles

When each side of a polygon is tangent to a circle, the circle is said to be **inscribed in the polygon**; we also say that the **polygon circumscribes the circle**. Such a polygon is called **circumscribable** and the circle is an **incircle**. The center of the incircle is the **incenter**. Figures 3.21(a) and 3.21(b) show a circumscribable triangle and a circumscribable quadrilateral, respectively.

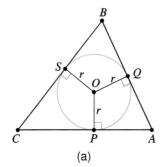
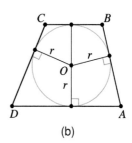

Figure 3.21

In Problem 8 of Problem Set 2.2.1, we learned that it is possible to inscribe a circle in every triangle. Let's reexamine why, by looking at the procedure for constructing such a circle, if it exists. Notice that each of the distances OP, OQ, and OS from the center O to the corresponding sides must be equal to r, the radius of the circle. In particular, O is equidistant from the sides of $\angle A$ and hence must be on the angle bisector of $\angle A$. Similarly, O is on the angle bisectors of $\angle B$ and of $\angle C$. This suggests the following theorem.

> **Theorem 3.7**
>
> *The angle bisectors of any triangle intersect at a single point O called the incenter of the triangle. The distance from O to any of the sides of the triangle is the radius of the incircle.*

Notice in Figure 3.21(a) that P, Q, and S are the points of tangency. (Why?) It seems that $AP = AQ$. This follows from the fact that $\triangle AQO \cong \triangle APO$. Similarly, $BQ = BS$ and $CS = CP$. This suggests the following useful theorem:

> **Theorem 3.8**
>
> *From a point in the exterior of a circle, the two tangent segments are congruent.*

Proof. Figure 3.22 shows the two tangent segments $\overline{AP}$ and $\overline{AQ}$ from A. We need to show that $\overline{AP} \cong \overline{AQ}$. Notice that $\triangle AQO \cong \triangle APO$ by the hypotenuse–leg condition because the angles at P and Q are right angles (definition of a tangent), and the triangles share the same hypotenuse $\overline{AO}$ and $\overline{PO} \cong \overline{QO}$. Hence $\overline{AQ} \cong \overline{AP}$. ∎

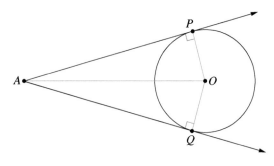

Figure 3.22

Circles Inscribed in Quadrilaterals

Figures 3.23(a) and 3.23(b) show that a circle cannot be inscribed in every quadrilateral. What makes the quadrilateral in Figure 3.23(c) circumscribable whereas the others are not?

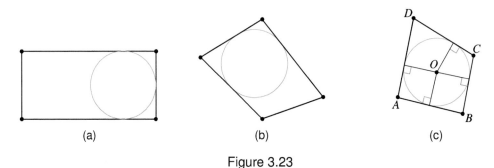

Figure 3.23

Notice that in Figure 3.23(c), the distance from the center O to each of the sides of the quadrilateral is the radius of the circle. Thus O is equidistant from each side of any of the angles. Consequently, O must be on each of the angle bisectors—that is, the angle bisectors intersect at O. Conversely, suppose that the angle bisectors of the four angles of a quadrilateral are concurrent and intersect at point O. Can we conclude, then, that the quadrilateral is circumscribable? Because a point is on the angle bisector of an angle if and only if it is equidistant from the sides of the angle, it follows that O is equidistant from all sides of the quadrilateral. We have proved the following theorem:

Theorem 3.9

A circle can be inscribed in a quadrilateral if and only if the angle bisectors of the four angles of the quadrilateral are concurrent.

Notice that a circle can be inscribed in every square because the diagonals of a square are also its angle bisectors. Can you think about other familiar quadrilaterals in which a circle can be inscribed? (Try to find one before reading on.) Such a quadrilateral is given in the following example.

3.2 Inscribed Circles

Example 3.2

Prove that a circle can be inscribed in every kite.

Proof. By Theorem 3.9, it is sufficient to prove that the angle bisectors of a kite are concurrent. Consider the kite $ABCD$ in Figure 3.24, in which $AB = BC$ and $AD = DC$.

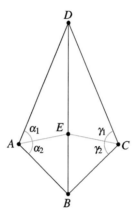

Figure 3.24

Diagonal $\overline{BD}$ bisects $\angle B$ and $\angle D$ (why?). We construct the angle bisector of $\angle A$, which will divide $\angle A$ into two congruent angles α_1 and α_2, and will intersect $\overline{BD}$ in some point E. We will show that the angle bisector of $\angle C$ also intersects $\overline{BD}$ at E. One way to achieve this goal is to connect E with C and prove that $\overline{CE}$ (which makes angles γ_1 and γ_2 with the sides $\overline{DC}$ and $\overline{BC}$, respectively) is on the angle bisector of $\angle C$. That is, we want to show that $\gamma_1 = \gamma_2$. For that purpose it suffices to show that $\alpha_1 = \gamma_1$ and $\alpha_2 = \gamma_2$ (because $\alpha_1 = \alpha_2$ this would imply that $\gamma_1 = \gamma_2$). Notice that $\triangle ABE \cong \triangle CBE$ by SAS because $AB = BC$ (definition of a kite), $BE = BE$, and $\angle ABE \cong \angle CBE$ (property of a kite). Similarly $\triangle ADE \cong \triangle CDE$. By $CPCT$, $\angle BCE \cong \angle BAE$ and $\angle DCE \cong \angle DAE$, that is, $\alpha_1 = \gamma_1$ and $\alpha_2 = \gamma_2$. Consequently, $\overrightarrow{CE}$ is the angle bisector of $\angle C$ and, therefore, all the angle bisectors of a kite are concurrent. By Theorem 3.9 a circle can be inscribed in the kite. ∎

A Relationship Among the Sides of a Circumscribable Quadrilateral

In Example 3.2, we saw that a circle can be inscribed in any kite. We also saw, in Figure 3.23, that some quadrilaterals are not circumscribable. We can, however, always find a unique circle that is tangent to any three sides of a quadrilateral (Why?), though it may not be tangent to the fourth side. If a circle can be inscribed in a quadrilateral, as in Figure 3.25, then all the sides are tangent to the circle. Therefore, by Theorem 3.8, all tangent segments from the same vertex are congruent. This fact is indicated in Figure 3.25 by denoting the lengths of the congruent tangent segments by the same letters.

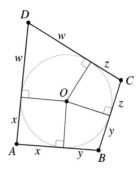

Figure 3.25

Consequently, we have

$$AB = x + y, \quad BC = z + y, \quad CD = w + z, \quad DA = w + x.$$

These equations suggest a relationship among the lengths of the four sides of the quadrilateral. Notice that

$$AB + DC = (x + y) + (w + z) = x + y + z + w,$$
$$BC + DA = (z + y) + (w + x) = x + y + z + w.$$

Consequently,

$$AB + DC = BC + DA. \tag{3.3}$$

Equation (3.3) was proved under the assumption that the quadrilateral is circumscribable. Thus we have proved the following theorem:

Theorem 3.10

If a quadrilateral is circumscribable, then the sum of the lengths of two opposite sides equals the sum of the lengths of the other two sides.

Remark 3.2

Theorem 3.10 can be stated in each of the following equivalent forms:

1. A necessary condition for a quadrilateral to be circumscribable is that the sum of the measures of two opposite sides equals the sum of the measures of the remaining opposite sides.

2. If the sum of the measures of two opposite sides in a quadrilateral does not equal the sum of the measures of the remaining two opposite sides, then the quadrilateral is not circumscribable. (This is the contrapositive of the statement in Theorem 3.10 and hence equivalent to it.)

3.2 Inscribed Circles

> **Example 3.3**
>
> 1. Prove that if a parallelogram is not a rhombus then it is not circumscribable.
> 2. On the basis of the theorems proved so far, can you conclude that the quadrilateral shown in Figure 3.26(b) is circumscribable? (The measures of the sides are as shown.)
>
> **Solution.**
>
> 1. If the parallelogram in Figure 3.26(a) is circumscribable, then $2p = 2q$ or $p = q$. The contrapositive of this statement says that if $p \neq q$, then the parallelogram is not circumscribable. Thus, if the parallelogram is not a rhombus, it is not circumscribable.
>
> 2. For the quadrilateral in Figure 3.26(b), we know that the sum of the measures of opposite sides for each pair of the sides is the same: $2 + 6 = 3 + 5$. Theorem 3.10 does *not* tell us that if Equation (3.3) is satisfied, then the quadrilateral is circumscribable; if a statement is true, its converse is not necessarily true. Consequently, we cannot conclude yet that the quadrilateral in Figure 3.26(b) is circumscribable. For that purpose, we would need to establish the converse of Theorem 3.10. This is our next goal.
>
>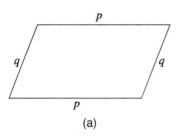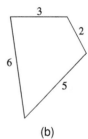
>
> (a)   (b)
>
> Figure 3.26

> **Theorem 3.11: Converse of Theorem 3.10**
>
> *If the sum of the lengths of two opposite sides of a quadrilateral equals the sum of the lengths of the two remaining opposite sides, then the quadrilateral is circumscribable.*

Proof. Suppose that the condition of the theorem is satisfied. Referring to Figure 3.27, we have

$$AB + DC = AD + BC. \tag{3.4}$$

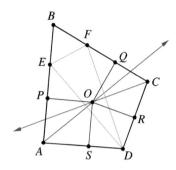

Figure 3.27

We will show that it is possible to inscribe a circle in the quadrilateral by showing that the angle bisectors of the four angles are concurrent. We first consider a special case, in which two adjacent sides of the quadrilateral are congruent. Then Equation (3.4) implies that $ABCD$ is a kite. (Why?) In this case, the proof is straightforward and was established in Example 3.2. Next, we suppose that two adjacent sides are not congruent. Without loss of generality, let $AB > AD$. Then Equation (3.4) implies

$$AB - AD = BC - DC \tag{3.5}$$

and hence $BC > DC$.

We mark E and F on $\overline{AB}$ and $\overline{BC}$, respectively, so that $\overline{AE} \cong \overline{AD}$ and $\overline{CF} \cong \overline{DC}$. Thus $\triangle DCF$ and $\triangle DAE$ are isosceles. Because $AE = AD$ and $DC = CF$, substituting AE and CF for AD and DC, respectively, in Equation (3.5), we get

$$AB - AE = BC - CF. \tag{3.6}$$

Equation (3.6) implies

$$BE = BF. \tag{3.7}$$

From Equation (3.7), we know that $\triangle BEF$ is isosceles. Our objective is to show that the angle bisectors of $ABCD$ are concurrent. Notice that because $\triangle ADE$ is isosceles, the angle bisector of $\angle A$ is also the perpendicular bisector of the opposite side $\overline{ED}$. Similarly, the angle bisector of $\angle B$ is the perpendicular bisector of $\overline{EF}$, and the angle bisector of $\angle C$ is the perpendicular bisector of $\overline{DF}$. Thus the angle bisectors of $\angle A$, $\angle B$, and $\angle C$ are the perpendicular bisectors of the sides of $\triangle DEF$, which we know are concurrent. Let O be the intersection of these three perpendicular bisectors, which are also the angle bisectors of $\angle A$, $\angle B$, and $\angle C$. From O, we construct $\overline{OP}$, $\overline{OQ}$, $\overline{OR}$, and $\overline{OS}$ (the perpendiculars to the four sides of $ABCD$). It remains to be shown that O is on the angle bisector of $\angle D$. This will be the case if and only if $OS = OR$.

Because O is on the angle bisector of $\angle A$, it is equidistant from the sides AB and AD; hence $OS = OP$. Similarly, because O is on the angle bisector of $\angle B$, $OP = OQ$. Because O is on the angle bisector of $\angle C$, $OQ = OR$. These three equations imply that $OS = OR$ and hence $OS = OP = OQ = OR$. Because these are the lengths of segments from O perpendicular to the sides of $ABCD$, O is the center of the circle inscribed in $ABCD$. ∎

3.2 Inscribed Circles

> **Now Solve This 3.7**
>
> Why in the proof of Theorem 3.11 was it important to prove first the special case when the quadrilateral is a kite?

Notice that the last two theorems can be combined into one theorem:

> **Theorem 3.12**
>
> A quadrilateral is circumscribable if and only if the sum of the lengths of two of its opposite sides equals the sum of the lengths of the other two opposite sides.

> **Example 3.4**
>
> In Figure 3.28, $\overline{AB}$ is a diameter of circle O, and $\overleftrightarrow{BN}$ and $\overleftrightarrow{EF}$ are tangents (E is an arbitrary point on the circle different from B). The tangent $\overleftrightarrow{EF}$ intersects $\overleftrightarrow{BN}$ at C, and $\overleftrightarrow{AE}$ intersects $\overleftrightarrow{BN}$ at D. Prove that $BC = CD$.

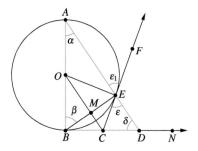

Figure 3.28

Investigation: What can we conclude about C? Because $\overline{CB}$ and $\overline{CE}$ are tangent segments from point C, by Theorem 3.8 we have

$$BC = CE. \tag{3.8}$$

If $BC = CD$ were true, then Equation (3.8) would imply that $CE = CD$. Conversely, if we could show that $CE = CD$, then by Equation (3.8) we would have $CD = BC$. It seems easier to prove that $CE = CD$, as these are sides in $\triangle ECD$ and we are more familiar with investigating triangles. In fact, it would suffice to show that the measures ε and δ of the angles opposite these sides in Figure 3.28 are equal. This is equivalent to what we need to prove and seems to be easier because we have substantial information about angles related to the circle. Thus our new subgoal is to show that $\varepsilon = \delta$.

Notice that $\varepsilon_1 = \varepsilon$ (vertical angles). Thus it suffices to show that $\varepsilon_1 = \delta$. As δ is neither an angle between a chord and a tangent nor an inscribed angle, we try to express δ in terms of such angles. In $\triangle ABD$, $\angle B$ is a right angle. (Why?) Therefore

$$\delta = 90° - \alpha. \tag{3.9}$$

We will show that $\varepsilon_1 = 90° - \alpha$. For that purpose, we connect B with E to create a right angle BEA (which intercepts the diameter $\overline{AB}$). Now we have $\varepsilon_1 = \beta$ (both angles intercept the same arc AE). In the right triangle ABE, we have $\beta = 90° - \alpha$. Thus $\varepsilon_1 = 90° - \alpha$. This result, along with Equation (3.9), implies that $\varepsilon_1 = \delta$ and hence that $\varepsilon = \delta$. Consequently, the subgoal and the required results are proved.

We next give a proof based on the above investigation but omit the motivation for each step. The proof is therefore shorter.

Proof. Because $\overleftrightarrow{BN}$ in Figure 3.28 is tangent to the circle at B, $\angle B$ is a right angle. In $\triangle ABD$, we have $\delta = 90° - \alpha$. Also $\varepsilon_1 = \varepsilon$ (vertical angles). Connect B with E. We have $\varepsilon_1 = \beta$ since both angles intercept the same arc AE (Corollary 3.2). Because $\angle AEB$ is an angle inscribed in a semicircle, it is a right angle. Hence in $\triangle ABE$, we have $\beta = 90° - \alpha$, which implies that $\varepsilon_1 = 90° - \alpha$.

Consequently, $\varepsilon_1 = \delta$ and $\varepsilon = \delta$. This result implies $CE = CD$. Because $CE = BC$ (tangent segments from C), we have $CD = BC$. ∎

In what follows, we give a somewhat different proof of the result in Example 3.4. (The plan and the proof are combined.)

An Alternate Proof. Notice that O is the midpoint of $\overline{AB}$. We need to show that C is the midpoint of $\overline{BD}$. This reminds us of the Midsegment Theorem. To apply that theorem, we need to create a triangle and its midsegment. For that purpose, we connect O with E and with C. Because $\overline{BC} \cong \overline{CE}$ (tangent segments) and $\overline{OB} \cong \overline{OE}$ (radii), it follows that $OECB$ is a kite. We know that the diagonals of a kite are perpendicular to each other, that $\overline{OC}$ bisects the angles at O and C, and that M is the midpoint of $\overline{BE}$. We focus now on $\triangle ABD$ and try to show that $\overline{OC} \| \overline{AD}$. This last relation will follow if a pair of corresponding angles created by $\overline{OC}$, $\overline{AD}$, and a transversal are congruent. Because $\angle BMO$ is a right angle, it suffices to show that $m(\angle BEA) = 90°$. Indeed, $\angle BEA$ is a right angle because it intercepts the diameter AB. Consequently, $\overline{OC} \| \overline{AD}$. Because O is the midpoint of $\overline{AB}$ and $\overline{OC} \| \overline{AD}$, it follows that C is the midpoint of $\overline{BD}$ (property of parallel projections). ∎

> **Now Solve This 3.8**
>
> 1. The following is yet another idea for a proof of Example 3.4. First notice that in Figure 3.28, C is the midpoint of $\overline{BD}$. Then prove that $\overline{OC} \| \overline{AD}$ by showing that $\alpha = m(\angle BOC)$, and complete the proof.
> 2. Prove that $ABCE$ in Figure 3.28 is not circumscribable.

3.2 Inscribed Circles

Example 3.5

If $\triangle ABC$ is an equilateral triangle inscribed in a circle and P is any point on $\overset{\frown}{AC}$, prove that $AP + PC = PB$. (See Figure 3.29)

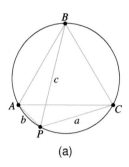 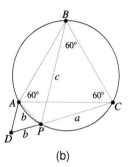

(a)   (b)

Figure 3.29

Plan: Referring to Figure 3.29(a), we need to show that $a + b = c$. Notice that c is a side in $\triangle ABP$. Our experience in working with triangles tells us to extend a by b or b by a (see the proof of triangle inequality, Theorem 2.15). Then $a + b$ becomes a side in a triangle. We extend $\overline{PC}$ so that $DP = b$, as shown in Figure 3.29(b). Then $a + b$ is a side in $\triangle ACD$. We want to prove that $\overline{DC} \cong \overline{BP}$, so it makes sense to try to show that $\triangle ABP$ and $\triangle ACD$ are congruent. This will be investigated next in Now Solve This 3.9.

Now Solve This 3.9

1. Complete the proof that $AP + PC = PB$ (in Figure 3.29(b)) by showing that
   i. $m(\angle APD) = 60°$
   ii. $\triangle APD$ is equilateral
   iii. $\angle BAP = \angle DAC$
2. Alternative transformational approach: Show that $m(\angle APC) = 120°$. Then let D be the image of A under rotation by $60°$ counterclockwise about P. Finally, rotate $\triangle ADC$ about A by $60°$ counterclockwise, show that its image is $\triangle APB$, and complete the proof.

Remark 3.3

Figure 3.29(b) is a vivid example of the fact that congruence of two corresponding sides and an angle does not ensure congruence of triangles. In $\triangle ABP$ and $\triangle ACP$, $\overline{AP}$ is a common side, $\overline{AC} \cong \overline{AB}$, and $\angle ABP \cong \angle ACP$, yet the triangles are not congruent because $m(\angle APB) = 60°$ and none of the angles in $\triangle ACP$ is $60°$. (Why?) Because this condition is true for any P on $\overset{\frown}{AC}$, we have exhibited infinitely many such pairs of triangles.

3.2.1 Problem Set

1. $\triangle ABC$ is inscribed in a circle. Point D is the incenter. Prove that $\triangle DAE \cong \triangle ADE$.

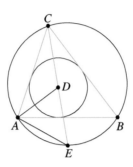

2. If a circle with center O is inscribed in a trapezoid whose bases are tangent to the circle at points P and Q (see figure below), prove each of the following statements:
   a. O, P, and Q are collinear.
   b. The diameter of the circle equals the height of the trapezoid.
   c. $m(\angle COB) = 90°$.

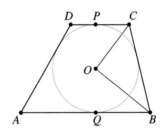

3. Two congruent circles intersect at P and Q as shown. A line is drawn through Q that intersects the circles at A and B. Another line is drawn through P that intersects the circles at A and C.
   a. Prove that $PB = PA$.
   b. If $\overline{AQ}$ is a diameter, prove that $CP = AP$ and that $m(\angle CBA) = 90°$.

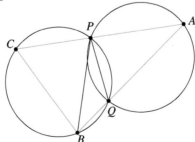

4. The two circles shown below share a common center (they are *concentric circles*). If $\overline{AB}$ and $\overline{CD}$ are two chords in the larger circle tangent to the smaller circle, prove that

$AB = CD$.

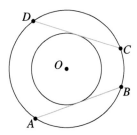

5. In the accompanying figure, $\overline{PA}$ and $\overline{PB}$ are tangent to circle O. If Q is on $\widehat{AB}$ and $\overline{CD}$ is tangent at Q, prove that:
   a. $m(\angle COD)$ is the same regardless of the position of Q on $\widehat{AB}$.
   b. The perimeter of $\triangle PCD$ is the same regardless of the position of Q on $\widehat{AB}$.

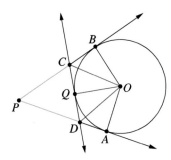

6. Suppose $\overleftrightarrow{PA}$ and $\overleftrightarrow{PB}$ are tangents to a circle at A and B, respectively, and $m(\angle BPA) = \alpha$. Answer the following:
   a. Express the marked angles γ and δ at C and D, respectively, in terms of α.
   b. Prove that $\alpha = \gamma - \delta$.

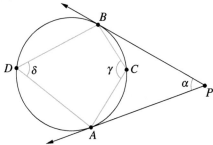

7. $ABCD$ is a quadrilateral with right angles at A and at D. Circle O is tangent to the sides $\overline{AB}$ and $\overline{AD}$ at B and E, respectively. The diagonal $\overline{AC}$ contains O. The side $\overline{CD}$ intersects the circle at P, and $\overline{PB}$ intersects $\overline{AC}$ at Q.
   a. Find the angles of $\triangle CQP$.

b. Prove that $AQ = r$, where r is the radius of the circle. That is, prove that $\overline{AQ} \cong \overline{CO}$.

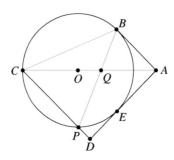

8. A circle is tangent at P to the side $\overline{BC}$ of the square $ABCD$. The vertices A and D are on the circle as shown. The side $\overline{DC}$ intersects the circle at Q. Prove that

  a. $\angle QPA$ is a right angle.

  b. $\overline{AP}$ bisects $\angle QAB$.

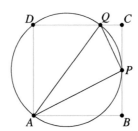

9. Determine whether it is possible to find a circumscribable, isosceles trapezoid that is not a rectangle or a parallelogram and which has each of the following properties. Justify your answers.

  a. A diagonal bisecting an angle of the trapezoid.

  b. The trapezoid is cyclic.

  c. The diagonals are perpendicular to each other.

10. a. Circle O is inscribed in a rhombus. $ABCD$ is a quadrilateral whose vertices are the four points of tangency. What kind of quadrilateral does $ABCD$ seem to be? Prove your answer.

  b. State and prove the converse of the theorem suggested in part (a).

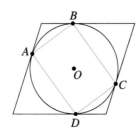

3.2 Inscribed Circles

11. Circle O is inscribed in trapezoid $ABCD$, and $\overline{MN}$ is the midsegment of the trapezoid. Prove the following:
   a. $\overline{MN}$ contains O.
   b. $ON = NB$ and $OM = MA$.

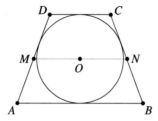

12. In a trapezoid $ABCD$, $\overline{MN}$ is the midsegment and P is a point on $\overline{MN}$ such that $PN = NB$ and $PM = MA$.
   a. Explain how to inscribe a circle in $ABCD$? Justify your answer.
   b. Prove that P is the center of the inscribed circle.

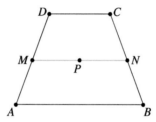

13. Use the accompanying figure to prove that the centers O_1 and O_2 of two **disjoint circles** (circles that have no common points) and the point P, where the two tangents intersect are collinear. (Assume that neither circle is contained in the other.)

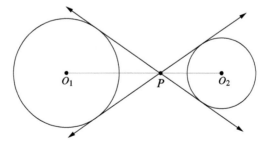

14. Two circles that have only one point in common are called **tangent circles**. Assume that neither circle is contained in the other and prove the following:
   a. The common point of two tangent circles and the centers of the circles are collinear. (*Hint*: Assume the contrary and use the triangle inequality.)

b. The tangent to one of the circles at the point of contact is also tangent to the other circle.

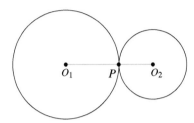

15. Circles O_1 and O_2 are tangent to each other at B. A is any point on the common tangent through B, and $\overrightarrow{AP}$ and $\overrightarrow{AQ}$ are tangents to the circles O_1 and O_2, respectively. Prove that $AP = AQ$.

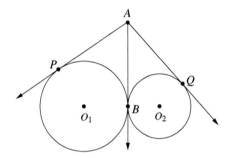

16. The circles O_1 and O_2 are tangent at P. The line AB is tangent to the circles at A and B, respectively. If C is on $\overrightarrow{AB}$ and $\overrightarrow{CP}$ is the tangent through P, prove that
   a. $AC = CB$.
   b. $m(\angle APB) = 90°$.

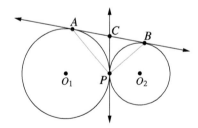

17. The circles O_1 and O_2 are tangent at P. A line through P intersects the first circle at A and the second circle at B. Prove that the tangent at A to the first circle is parallel to the

3.2 Inscribed Circles 129

tangent at B to the second circle.

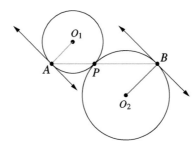

18. Two congruent circles O_1 and O_2 intersect at A and B. A third circle with center at B intersects O_1 at C and O_2 at D. (These points are in the same half-plane determined by $\overline{AB}$, and the radius of circle B is less then AB.) Prove that A, C, and D are collinear. (*Hint:* Let E be the intersection of $\overrightarrow{AC}$ with circle O_2. Then prove that $BC = BE$.)

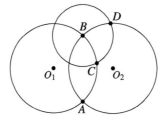

19. The vertices of $\triangle EFD$ are on the sides of $\triangle ABC$ as shown. Construct the three circles that circumscribe $\triangle ADE$, $\triangle DFC$, and $\triangle EBF$. Repeat the construction for differently positioned triangles EFD.
   a. Based on what you observed, make a conjecture concerning the three circles.
   b. Prove your conjecture in part (a).

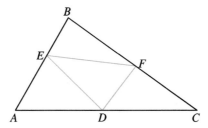

20. a. Prove that whenever three circles (as shown below) are tangent to each other, the three tangents at the points of contact are concurrent.

b. Use the result in part (a) to construct three noncongruent circles tangent to each other. Describe your construction.

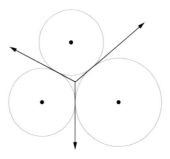

3.3 More on Constructions

In this section we will investigate several constructions based on the Inscribed Angle Theorem (Theorem 3.2), its corollary, and Theorem 3.4 restated here for convenience.

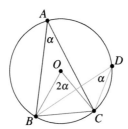

Figure 3.30

From all the points of $\overset{\frown}{BAC}$ (see Figure 3.30), $\overline{BC}$ is subtended by congruent angles whose measure is half of the measure of the intercepted arc. Moreover, the arc BAC consists of all the points (in a half plane determined by $\overleftrightarrow{BC}$ and point A) from which $\overline{BC}$ is subtended by that angle.

These facts will be used in many of the construction problems in this section. Each problem will include three parts:

1. **Investigation** (sometimes referred to as analysis), in which we imagine the given problem as solved and search for properties of the figure that will enable us to accomplish the construction. This is a powerful problem-solving technique that should enable you to discover the construction on your own.

2. **Construction**, in which we describe the steps in the construction and actually perform the construction using only a compass and a straightedge.

3. **Proof**, in which we prove that our construction does what was asked for.

3.3 More on Constructions

Construction 3.3.1

Construct the locus from which a given segment a is subtended by a given acute angle α (see Figure 3.31).

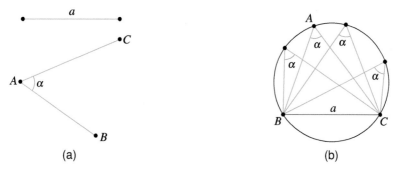

Figure 3.31

Investigation: We are imagining that the problem has been solved and, therefore, that $\overline{BC}$ in Figure 3.31(b) is subtended by α from any point on $\widehat{BAC}$. Theorem 3.4 tell us that $\widehat{BAC}$ is the only set of points in the half-plane determined by $\overleftrightarrow{BC}$ and A, for which $\overline{BC}$ is subtended by α. (The reflection of $\widehat{BAC}$ in $\overleftrightarrow{BC}$ is the set of points in the other half-plane from which $\overline{BC}$ is subtended by α.)

To construct the required locus (that is, $\widehat{BAC}$) we need to find the center O of the circle to which the arc belongs. Thus we connect in Figure 3.32, O with B and C. Notice that O is equidistant from B and C and, therefore, lies on the perpendicular bisector n of $\overline{BC}$. Line n includes $\overline{OM}$ the angle bisector of $\angle BOC$ (because $\triangle BOC$ is isosceles). If we could construct line k in Figure 3.32, then we would be able to find O as the intersection point of lines n and k. In $\triangle MBO$, $\angle MBO$ measures $90° - \alpha$, which suggests drawing line p perpendicular to $\overrightarrow{BM}$ at B. Let D be any point on p "above" $\overleftrightarrow{BC}$. Then we would have $m(\angle DBO) = \alpha$ (which also follows from the fact that $\overleftrightarrow{BD} \parallel \overleftrightarrow{OM}$). Since α is given, we now know how to construct line k, which contains a side of α.

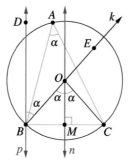

Figure 3.32

Construction:

1. On any line, construct $\overline{BC}$ congruent to the given segment a (see Figure 3.33).
2. Construct the perpendicular bisector n of $\overline{BC}$.
3. Through B, construct line p perpendicular to $\overline{BC}$ (or, equivalently, parallel to n).
4. Let D be any point on line p above $\overleftrightarrow{BC}$. Construct $\angle DBE$ congruent to the given angle α.
5. Let O be the intersection of $\overleftrightarrow{BE}$ and n.
6. Construct the circle with center O and radius $\overline{BO}$. The major arc BC in the half-plane determined by $\overleftrightarrow{BC}$ and O is the solution. That arc, along with its reflection in $\overleftrightarrow{BC}$, is the solution in the entire plane.

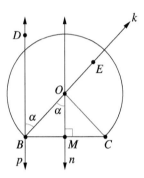

Figure 3.33

To show that the major arc $\overset{\frown}{BC}$ is the solution, it will suffice to prove that $m(\angle BOC) = 2\alpha$. Notice that $p \| n$ implies that $m(\angle BOM) = \alpha$. Also $m(\angle COM) = \alpha$, because $\overrightarrow{OM}$ is the angle bisector of $\angle BOC$. Consequently, $m(\angle BOC) = 2\alpha$, and hence the major arc BC is the solution.

Now Solve This 3.10

1. Modify if necessary, steps 1 through 6 above and Figure 3.32 for the case when α is obtuse.
2. Construction 3.3.1 can also be approached in the following way: Construct any triangle with side a and opposite angle α. If α is acute, it is convenient to construct a right triangle as suggested in Figure 3.34. Next, construct the circle that circumscribes the triangle. The major arc CB is the solution.
   a. Perform this construction for a given segment a and a given angle α. (Draw the segment and the angle before performing the construction.) Justify your construction.

3.3 More on Constructions

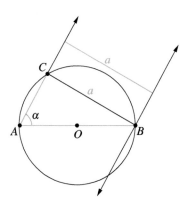

Figure 3.34

b. Investigate how to use this approach for an obtuse angle α. (*Hint:* Consider the supplementary angle $180° - \alpha$.)

Notation for Construction Problems

A, B, C	vertices
a, b, c	sides
α, β, γ	angles
$\angle A, \angle B, \angle C$	
h_a, h_b, h_c	altitudes
m_a, m_b, m_c	medians
d_A, d_B, d_C	angle bisectors
R	radius of circumscribed circle
r	radius of inscribed circle

It is understood that angle α is at vertex A and is opposite side a. It is common for a to stand for both line segment and the length of that segment. It is from the context that we know which meaning is intended. For example, the problem "Construct a triangle from a, b, and c" means "Construct a triangle given three of its sides." (Of course the existence of such a triangle is not guaranteed. If $a + b \leq c$, the problem has no solution.)

Construction 3.3.2

Construct a triangle from a, α, h_a.

Investigation: Figure 3.35 shows a triangle with the given data. We need to construct $\triangle ABC$ given the data to the left of the figure.

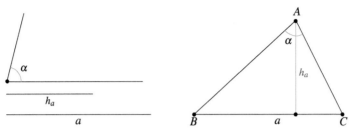

Figure 3.35

We can duplicate segment a anywhere, and mark its endpoints B and C. Now we need to find only the third vertex A. What do we know about A? Since h_a is given, we know that A is at the distance h_a from $\overleftrightarrow{BC}$. There are infinitely many such points. In fact, the locus of all points A at the distance h_a from $\overleftrightarrow{BC}$ is a line m parallel to $\overleftrightarrow{BC}$ (line m is shown in Figure 3.36). We also know that from A the side $\overline{BC}$ is subtended by the given angle α. We know how to construct the locus of all such points A from which $\overline{BC}$ is subtended by α (Construction 3.3.1). If α is acute, that locus is the major arc BC, as shown in Figure 3.36. The intersection of the two loci—the arc and line m—comprises the possible points A.

Construction: The steps are as follows (Figure 3.36):

1. Construct $\overline{BC}$ congruent to the given segment a.
2. Construct the locus m of all points (in a half-plane) at the distance h_a from $\overline{BC}$. One way to accomplish this goal is to construct a rectangle with sides $\overline{BC}$ and h_a. A similar but somewhat different approach is suggested in Figure 3.36.

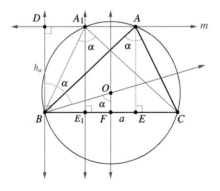

Figure 3.36

3. Construct the locus of all points (in the half-plane of step 2) from which $\overline{BC}$ is subtended by α.
4. Determine the points of intersection (if any) of the loci of steps 2 and 3. These points are A and A_1 in Figure 3.36.
5. $\triangle ABC$ and $\triangle A_1 BC$ are the only solutions.

3.3 More on Constructions

Proof. $\triangle ABC$ and $\triangle A_1BC$ each have side $\overline{BC}$ congruent to the given side a. The angles at A and A_1 are α by virtue of our construction (Construction 3.3.1). AE and A_1E_1 are the heights of the triangles; each equals h_a because the distance from every point on line m to $\overleftrightarrow{BC}$ is h_a. ∎

> **Remark 3.4**
>
> The location of the vertex A in Construction 3.3.2 was found as the intersection of two loci: a line and an arc. A similar approach is useful in most construction problems. The number of points of intersection determines the number of solutions.

> **Now Solve This 3.11**
>
> Prove that in Figure 3.36, $\triangle BCA \cong \triangle CBA_1$.

Construction 3.3.3

Construct a triangle from a, α, m_b as shown in Figure 3.37. Assume that α is acute.

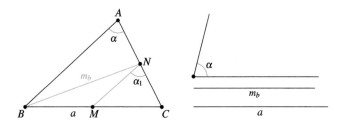

Figure 3.37

Investigation: In $\triangle ABC$ in Figure 3.37, note that from A, the side BC is seen by α. We need to construct the locus from which $\overline{BC}$ is seen by α, i.e., that A is on, but we do not see an obvious way to find it. We note that point N is on a circle with center B and radius m_b, but this information does not seem to help us in finding A or N. To find the location of N, we need another locus to which N belongs. For that purpose, we construct through N a line parallel to $\overleftrightarrow{AB}$ intersecting $\overline{BC}$ at M. It follows from the Midsegment Theorem that M is the midpoint of $\overline{BC}$. As $\overleftrightarrow{NM} \parallel \overleftrightarrow{AB}$ in Figure 3.37, $\alpha_1 = \alpha$. Now we know how to construct $\triangle BNC$ as N is on two loci: the circle with center B and radius m_b, and the locus of all points from which $\overline{MC}$ is seen by α_1 that equals α. We can now extend $\overline{NC}$ by its length to obtain A.

Construction:
1. Construct $\overline{B_1C_1}$ congruent to a (Figure 3.38).
2. Construct the midpoint M_1 of $\overline{B_1C_1}$.
3. Construct the locus of all points from which $\overline{M_1C_1}$ is seen by α (the major arc M_1C_1).

4. Construct the locus of all points at distance m_b from B_1 (the circle with center at B_1 and radius m_b).
5. Find N_1, the point of intersection of the loci in steps 3 and 4.
6. Extend $\overline{C_1 N_1}$ by its length to locate A_1.
7. Connect A_1 with B_1. Now $\triangle A_1 B_1 C_1$ is the required triangle.

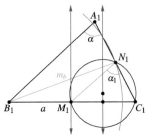

Figure 3.38

To show that the construction creates the required triangle, notice that $\overline{B_1 C_1}$ is congruent to a by construction. Because M_1 and N_1 were constructed to be midpoints of $\overline{B_1 C_1}$ and $\overline{A_1 C_1}$, respectively, $\overline{M_1 N_1}$ is a midsegment in $\triangle A_1 B_1 C_1$ and hence $\overline{M_1 N_1} \| \overline{A_1 B_1}$. Thus $\angle B_1 A_1 C_1 \cong \angle M_1 N_1 C_1$. Notice that $\angle M_1 N_1 C_1$ was constructed to be congruent to α and hence $m(\angle B_1 A_1 C_1) = \alpha$. Also, $\overline{B_1 N_1}$ was constructed to be congruent to m_b. By construction, N_1 is the midpoint of $\overline{A_1 C_1}$, so $\overline{B_1 N_1}$ is the required median.

> **Now Solve This 3.12**
>
> Figure 3.39 suggests a somewhat different approach to Construction 3.3.3. Extend $\overline{BN}$ by its length to obtain point D and hence the parallelogram $ABCD$. Complete the investigation, describe the construction, and prove that it satisfies the given requirements.
>
>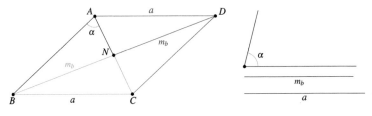
>
> Figure 3.39

Construction 3.3.4

From a point P outside a circle, construct tangents to the circle with center O.

Investigation: In Figure 3.40, imagine that $\overleftrightarrow{PT_1}$ is one of the required tangents. Connect O with T_1. Since $\overleftrightarrow{PT_1}$ is a tangent, it is perpendicular to the radius $\overline{OT_1}$. In addition to T_1 being on the given circle with center O, because $\angle OT_1 P$ is $90°$, we know that T_1 is also on the

3.3 More on Constructions   137

locus of all points from which $\overline{OP}$ is seen at a right angle. This locus (in one half-plane) is a semicircle with center M, which is the midpoint of $\overline{OP}$ and radius $\overline{OM}$. The intersection of this semicircle with the given circle is T_1. The other semicircle intersects the given circle at T_2.

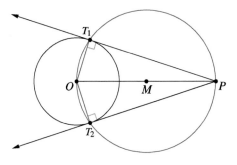

Figure 3.40

The construction and its proof follow directly from the investigation and are left to the reader.

Construction 3.3.5 Common External Tangents to Two Circles

Given circles with centers O and O_1 and different radii, construct their common external tangents, as shown in Figure 3.41.

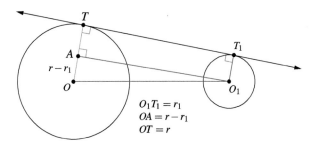

Figure 3.41

Investigation: The construction of interior common tangents will be explored in Now Solve This 3.13 at the end of this section. In Figure 3.41 we imagine one of the tangents touching the circles at T and T_1. We know that the radii OT and O_1T_1 are perpendicular to the common tangent. We also know the locations O and O_1 and hence $\overline{OO_1}$. To aid in constructing part of Figure 3.41, and because the angles at T and T_1 are right angles, we draw through O_1 a line that is perpendicular to $\overline{OT}$ and intersects $\overline{OT}$ at A. Because $OA = r - r_1$, and $\triangle OAO_1$ is a right triangle, we know how to construct that triangle. After $\triangle OAO_1$ is constructed, we extend $\overline{OA}$ to the point T, where the line intersects the circle. We notice that ATT_1O_1 is a rectangle.

Now we have several choices regarding how we construct the common tangent. We could construct through T the line perpendicular to $\overline{OT}$ and claim that it is the common tangent, or we could construct through O_1 the line perpendicular to $\overline{AO_1}$; where that line intersects the smaller circle is T_1. We would then claim that $\overline{TT_1}$ is a common tangent.

Construction: In Figure 3.42, we construct the right triangle OAO_1 using $r - r_1$ as one of the sides and $\overline{OO_1}$ as the hypotenuse. The steps of the construction are suggested in the preceding investigation and in Figure 3.42. We choose the second option for constructing the point of tangency T_1; that is, we construct through O_1 a line parallel to $\overline{OT}$. That line intersects the smaller circle at T_1, which is the other point of tangency. $\overleftrightarrow{TT}_1$ is one of the common tangents.

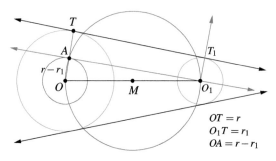

Figure 3.42

We now prove that $\overleftrightarrow{TT'}$ is the required common tangent. By construction the angle at A is 90°. At O_1, we constructed the line parallel to $\overleftrightarrow{OA}$ and intersecting the smaller circle at T_1. We need to prove that $\overleftrightarrow{TT}_1$ is a common tangent—that is, that the angles at T and T_1 are 90°. For that purpose, we prove that ATT_1O_1 is a rectangle. By our construction, the angles of ATT_1O_1 at A and O_1 are 90°. First we show that ATT_1O_1 is a parallelogram. Notice that $AT = OT - OA = r - (r - r_1) = r_1$. Because $O_1T_1 = r_1$, we have $AT = O_1T_1$. Because $\overline{AT}$ and $\overline{O_1T_1}$ are parallel and congruent, ATT_1O_1 is a parallelogram, but a parallelogram with a right angle is a rectangle, and the proof is completed.

> **Now Solve This 3.13**
>
> 1. As mentioned in the investigation of Construction 3.3.5, the common external tangent to the circles can be constructed by first finding the point of tangency T to the larger circle and then at T constructing the perpendicular to $\overline{OT}$. Prove that this perpendicular is also tangent to the smaller circle. (*Note:* You can't assume that the perpendicular intersects the smaller circle.)
>
> 2. Given two circles like the ones in Figure 3.43, construct their interior tangents. Write an investigation, describe the construction, and write a proof showing that the construction is valid.

3.3 More on Constructions

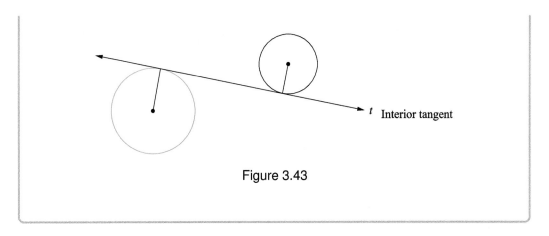

Figure 3.43

3.3.1 Problem Set

For each of the problems, except for Problem 1, provide an investigation, a construction, and a proof.

1. Construct the locus of all points (in one half-plane) from which a given segment (that you draw) is seen by
   a. 90°
   b. 60°
   c. 120°
   d. 105°
2. In $\triangle ABC$, α is an obtuse angle. Use only a, α, and m_b to reconstruct the triangle.
3. Draw a right triangle and then construct a triangle congruent to it using only a leg, the hypotenuse, and the fact that it is a right triangle.
4. Which of the following (taken from an actual triangle) determines a unique triangle (up to congruence; i.e., two congruent triangles are considered to be the "same")? Justify your answers.
   a. a, α, R
   b. a, h_a, h_b
   c. a, b, R
   d. a, α, r
5. Construct a rhombus given a side and the radius of the inscribed circle.
6. Construct a triangle given a, r, R.
7. Construct a square and four random points $P, Q, S,$ and T on its sides, with one point on each side. Trace the four points on a semitransparent page without tracing the square, The problem is to construct the original square using only the four points $P, Q, S,$ and T. Problems 7 and 8 appear in Problem Set 1.10. Here you are asked to solve them

using properties of circles.

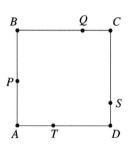

8. Given a line ℓ and points A and B on the same half plane determined by ℓ, construct point X on ℓ such that $\angle PXA$ is twice the size of $\angle QXB$ (the angles are marked by α and 2α, but neither is given). Is the answer unique?

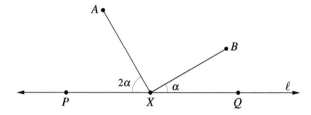

4 Area and the Pythagorean Theorem

> *I think it is said that Gauss had ten different proofs for the law of quadratic reciprocity. Any good theorem should have several proofs, the more the better. For two reasons: usually, different proofs have different strengths and weaknesses, and they generalise in different directions—they are not just repetitions of each other.*
>
> Sir Michael Atiyah, interview in *European Mathematical Society Newsletter*, September 2004

Introduction

In this chapter we explore the concept of area, develop the formulas for the areas of various polygons, and solve related problems. We give several proofs of the Pythagorean Theorem using area, and apply the theorem to construction problems. *The Pythagorean Proposition* by Elisha Loomis contains a collection of 367 proofs of the Pythagorean Theorem. The excellent website www.cut-the-knot.org/pythagoras contains 98 proofs and various generalizations of the theorem.

4.1 Areas of Polygons

To measure the length of a segment, we choose a unit segment and find how many of those unit segments cover the segment we want to measure. Similarly, to measure the area of a region, we choose a unit area and find how many of these units cover the region. Given a unit of length, the corresponding unit of area is the area of a *square* with sides of unit length. Thus area is measured in square units. In Figure 4.1(a), the rectangle $ABCD$ has sides 3 units and 4 units long. In Figure 4.1(b), 3 rows of 4 squares each have been drawn. Because the rectangle contains $3 \cdot 4 = 12$ squares, the area of the rectangle is 12 square units. If the lengths of the sides of a rectangle are a units and b units, respectively, where a and b are whole numbers, we could divide the rectangle into a rows of b unit squares each, that is, into ab unit squares. The rectangle would have an area of ab square units.

If the lengths of the sides of a rectangle are rational numbers, we can show that its area is also the product of the lengths of its sides. For example, consider the rectangle $ABCD$ in Figure 4.2(a), whose two adjacent sides are $\frac{3}{2}$ and $\frac{5}{3}$ units long, respectively. We know how to find the area of a rectangle if the rectangle is divided into a whole number of congruent squares. This can be accomplished if we could find a common measure for the sides. In other words, we could divide each of the sides AB and BC into a whole number of parts of the same length (using the same unit of length for each side). To find a common measure, we write the fractions with a common denominator:

$$\frac{3}{2} = \frac{3 \cdot 3}{2 \cdot 3} = \frac{9}{6} = 9 \cdot \frac{1}{6} \quad \text{and} \quad \frac{5}{3} = \frac{5 \cdot 2}{3 \cdot 2} = \frac{10}{6} = 10 \cdot \frac{1}{6}.$$

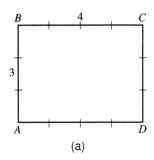
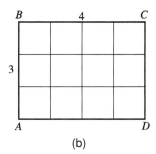

Figure 4.1

The common measure for the two sides is $\frac{1}{6}$ of the unit. Thus we can divide side AB, whose length is $\frac{3}{2}$ units, into 9 equal parts, each $\frac{1}{6}$ of the unit long, and side BC, whose length is $\frac{5}{3}$ units, into 10 equal parts, each also $\frac{1}{6}$ of the unit long.

In Figure 4.2(b), there are $9 \cdot 10$ squares created in this way. Notice that square $AGFE$ has sides that are one unit long and contains $6 \cdot 6$ of the smaller squares (each $\frac{1}{6}$ of a unit on a side). Because $6 \cdot 6$ of these small squares make one unit of area, one small square is $\frac{1}{6 \cdot 6}$ units of area. Thus $9 \cdot 10$ small squares (which cover rectangle $ABCD$) are $9 \cdot 10 \cdot \frac{1}{6 \cdot 6}$ units of area. Because $9 \cdot 10 \cdot \frac{1}{6 \cdot 6} = \frac{9}{6} \cdot \frac{10}{6} = \frac{3}{2} \cdot \frac{5}{3}$, we see that the area of the rectangle $ABCD$ in Figure 4.2 is the product of the lengths of its two adjacent sides. In an analogous way, it can be shown that if the lengths of adjacent sides of a rectangle are any rational numbers, then the area of the rectangle is the product of their lengths (see Problem Set 4.1.1).

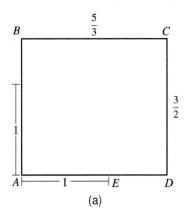
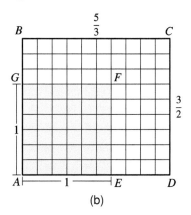

Figure 4.2

What happens if the lengths of one or both of the sides are irrational numbers? In that case a common measure for the adjacent sides cannot always be found (see Problem Set 4.1.1). We can, however, approximate the lengths using rational numbers. For example, suppose that one of the sides is $\sqrt{2}$ units long and the other is $\sqrt{3}$ units long. Suppose a calculator displays the first eight digits of the decimal expansion as:

$$\sqrt{2} \approx 1.4142136 \quad \text{and} \quad \sqrt{3} \approx 1.7320508$$

4.1 Areas of Polygons

In Figure 4.3, $\varepsilon_1 = \sqrt{2} - 1.4142136$ and $\varepsilon_2 = \sqrt{3} - 1.7320508$. This suggests that the area of the rectangle whose sides are the rational numbers 1.4142136 and 1.7320508 should be very close to the area of the rectangle whose sides are $\sqrt{2}$ and $\sqrt{3}$. The area of the unshaded rectangle whose side lengths are rational numbers is $(1.4142136) \cdot (1.7320508)$, which is very close to $\sqrt{2} \cdot \sqrt{3}$. Because we can approximate $\sqrt{2}$ and $\sqrt{3}$ to any degree of accuracy using rational numbers, it seems that the exact area of the original rectangle is $\sqrt{2} \cdot \sqrt{3}$ square units. A rigorous proof can be accomplished using the concept of a limit. The preceding discussion suggests that the area of any rectangle is the product of the lengths of its two adjacent sides.

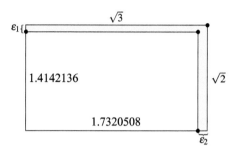

Figure 4.3

SPOTLIGHT on Teaching

When teaching the concept of area, intuitive "hands-on" approaches should precede the development of formulas. Such activities can be accomplished via a geoboard, which is constructed out of a piece of wood with equally spaced nails. Using rubber bands, one can create various polygons using the nails as vertices. Figure 4.4(a) shows a geoboard and a polygon created using a rubber band.

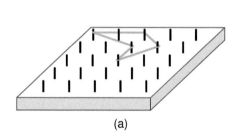 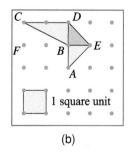

Figure 4.4

If the distance between two (horizontally or vertically) adjacent nails or dots is 1 unit, then the area of the smallest square is 1 square unit as shown in Figure 4.4(b). That square unit can be used to find the area of the polygonal region $ABCDE$ (defined by the rubber band) in Figure 4.4(a). In Figure 4.4(b), we have divided that polygon into three triangles CDB, DBE, and ABE. The area of $\triangle CDB$ is half the area of the rectangle $CDBF$. Because the area of the rectangle is 2 square units, the area of $\triangle CDB$ is $\frac{1}{2} \cdot 2$, or 1 square unit. Similarly, the areas of $\triangle DBE$ and $\triangle ABE$ are $\frac{1}{2}$ square unit each. Thus the area of the polygonal region is $1 + \left(\frac{1}{2}\right) + \left(\frac{1}{2}\right)$, or 2 square units.

Figure 4.5 and the corresponding computation suggest how to find the area of the polygon $ABCD$ shown.

$$\text{Area}(ABCD) = \text{Area}(EFGD) - \Big[\text{Area}(\Delta I) + \text{Area}(\Delta II) + \text{Area}(\Delta III)\Big]$$
$$= 4 \cdot 3 - \left(\frac{1}{2} \cdot 2 \cdot 1 + \frac{1}{2} \cdot 1 \cdot 3 + \frac{1}{2} \cdot 4 \cdot 1\right) = 7\frac{1}{2} \text{ square units.}$$

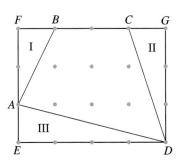

Figure 4.5

In Figure 4.6, the region is not a polygonal region and hence cannot be divided so easily. It is possible, however, to estimate the area of the region by finding the areas of two polygonal regions: one contained in the region and another such that the curved region is containing the region. Can you show that the area of the curved region is between 25 and 68 square units? How would you go about finding better bounds? This approach of "squeezing" the area of a region between two areas of polygons will be used in the next chapter to find the area of a circle.

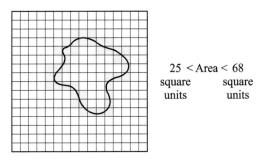

Figure 4.6

An Axiomatic Approach to Area

So far our approach to the concept of area has been intuitive. We have shown that the formula for the area of a rectangle is plausible, but we have not proved it. In what follows we show how area can be treated rigorously by introducing the following axioms.

Axiom 4.1

The area of any polygonal region (a polygon and its interior) is a unique positive real number.

Axiom 4.2

The area of any point or line segment is 0.

Axiom 4.3

Congruent polygonal regions have the same area.

Axiom 4.4

Area is additive. That is, if we can divide a figure into nonoverlapping parts, or parts that share lines or line segments, then its area is the sum of the areas of the parts.

Axiom 4.5

A square of length a units has area of a^2 square units.

Using some of these axioms, we will justify the well-known formulas for the areas of a rectangle, a parallelogram, a triangle, and a trapezoid.

Theorem 4.1: Area of a Rectangle

The area of a rectangle[1] with sides of length a and b is ab.

Proof. Consider a rectangle with sides of length a and b as in Figure 4.7(a). Denote its area by A. To use Axiom 4.5, we construct a square as in Figure 4.7(b). By Axiom 4.5, the area of the square with side $a + b$ is $(a + b)^2$. This square can also be regarded as the union of two rectangles with area A, and two squares with area a^2 and b^2, respectively. This is equivalent to each of the following:

$$(a + b)^2 = 2A + a^2 + b^2,$$
$$a^2 + 2ab + b^2 = 2A + a^2 + b^2,$$
$$2ab = 2A.$$

---

[1] Strictly speaking, we should talk about the area of a rectangular region. For brevity's sake, however, we will adopt the convention of calling the area of any polygonal region the "area of the polygon."

Therefore, $A = ab$.

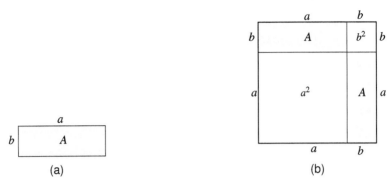

Figure 4.7

> **SPOTLIGHT on Teaching : Area and Perimeter**
>
> Many students in middle schools and high schools incorrectly believe that figures having equal perimeters necessarily have equal areas. This belief is even common among teachers (Liping Ma, pp. 84–98). Consider a square and a rectangle with equal perimeters. A counterexample will show that they can have the same perimeter but different areas, e.g., a square with side 6 units long and a rectangle with sides 10 and 2 units long. Each has a perimeter of 24 units but the areas are 36 and 20 square units, respectively. One can ask several related questions: How large and how small can the area of the rectangle with perimeter 24 units get? Is there a rectangle with half the area of the square but the same perimeter?

Areas of a Parallelogram, a Triangle, and a Trapezoid

Knowing the formula for the area of a rectangle, we can use it to derive formulas for the area of a parallelogram, a triangle, and a trapezoid. We will use the following terminology. Any side of a triangle or a parallelogram will be referred to as a **base**. The **height of a triangle** to a given base is the distance from the line containing the base to the vertex opposite the base. A height is the length of an altitude. A triangle has three heights. A **height of a parallelogram** is the distance between two parallel sides. A parallelogram has two heights. The parallel sides of a trapezoid are the **bases of the trapezoid**. The distance between the bases is the **height of the trapezoid**.

There are many ways to derive formulas for the areas of a parallelogram, a triangle, and a trapezoid. Some approaches will be explored in the problem set at the end of this section. One important strategy is known as **dissection**. In dissection, we cut a figure with unknown area into a number of nonoverlapping pieces. By reassembling these pieces, we can then obtain a figure whose area we know how to find.

For example, to find the area of the parallelogram, in Figure 4.8(a), we can cut out a triangular piece of the parallelogram and move it to obtain a rectangle, see Figure 4.8(a).

4.1 Areas of Polygons

Because the shaded triangles are congruent, the area of the parallelogram in Figure 4.8(a) is the same as the area of the rectangle in Figure 4.8(b). The area of that rectangle is bh; therefore, the parallelogram also has area bh.

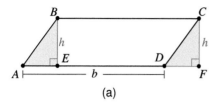 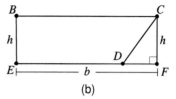

(a)   (b)

Figure 4.8

One way to make the preceding argument more rigorous is shown in the incomplete proof or the following theorem.

> **Theorem 4.2: Area of a Parallelogram**
>
> *The area of a parallelogram equals the product of the length of a base and the corresponding height.*

Proof. In the parallelogram $ABCD$ in Figure 4.8(a), we drop the perpendicular from B to the side AD. From the Hypotenuse – Leg congruency condition, $\triangle DCF \cong \triangle ABE$. Consequently, since $EBCF$ is a rectangle we have

$$\begin{aligned}\text{Area}(\square ABCD) &= \text{Area}(\triangle ABE) + \text{Area}(EBCD) \\ &= \text{Area}(\triangle DCF) + \text{Area}(EBCD) \\ &= \text{Area}(EBCF) \\ &= bh.\end{aligned}$$

∎

> **Remark 4.1**
>
> The preceding proof of Theorem 4.2 is incomplete because it does not apply to a parallelogram like the one in Figure 4.9(a), where the perpendicular from B to $\overleftrightarrow{AD}$ does not intersect the side AD of the parallelogram. Try to prove the theorem for this parallelogram before reading on. The proof is explored in Now Solve This 4.1.
>
>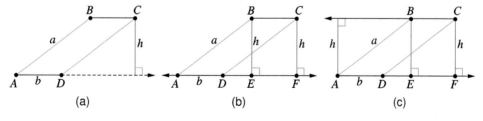
>
>   (a)   (b)   (c)
>
> Figure 4.9

Now Solve This 4.1

Prove the formula given for the area of a parallelogram like the one in Figure 4.9 in two different ways:

1. Notice that

$$\text{Area}(ABCD) = \text{Area}(ABCF) - \text{Area}(\Delta DCF),$$
$$\text{Area}(BCFE) = \text{Area}(ABCF) - \text{Area}(\Delta ABE).$$

2. Enclose the parallelogram $ABCD$ in a rectangle with sides AF and CF as shown in Figure 4.9(c), and subtract from the area of that rectangle the area of two triangles.

Remark 4.2

Axiom 4.1 tells us that the area of the parallelogram $ABCD$ in Figure 4.9(a) is unique. A proof analogous to the one given earlier, however, shows that the area of the parallelogram $ABCD$ is also ag, where a is the other base of the parallelogram and g is the corresponding height. Therefore, $bh = ag$.

To find a formula for the area of a triangle, we could also dissect the triangle and reassemble the pieces into a rectangle. This approach will be explored in the problem set at the end of this section. Here we use a somewhat different approach. We can trace or cut out triangle ABC in Figure 4.10 and mark the traced triangle $A'B'C'$ on a separate patty paper (or any other paper). We retrace $\Delta A'F'C'$ next to the original triangle so that A falls on A', B falls on B', and C' is as shown. The resulting figure $ACBC'$ turns out to be a parallelogram. Consequently, the area of ΔABC is half the area of $ACBC'$—that is, $\frac{1}{2}bh$. We state this result in the following theorem and outline a proof.

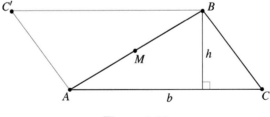

Figure 4.10

Theorem 4.3: Area of a Triangle

The area of a triangle is half the product of the length of a base and the corresponding height.

4.1 Areas of Polygons

Proof. Let M be the midpoint of a side of the triangle (side AB in Figure 4.10) and C' the point collinear with M and C such that $C'M = MC$. The quadrilateral $ACBC'$ is a parallelogram (why?) whose area is bh. Because $\triangle ABC \cong \triangle BAC'$, we get

$$\begin{aligned} bh &= \text{Area}(ACBC') \\ &= \text{Area}(\triangle ABC) + \text{Area}(BAC') \\ &= 2 \cdot \text{Area}(\triangle ABC). \end{aligned}$$

Hence $\text{Area}(\triangle ABC) = \frac{1}{2}bh$. ∎

Remark 4.3

Because any side of a triangle can be chosen as a base, the area of a $\triangle ABC$ can be determined in three ways using pairs of bases and corresponding heights. For example as Figure 4.11 indicates, $\frac{1}{2}bh_b = \frac{1}{2}ah_a = \frac{1}{2}ch_c$.

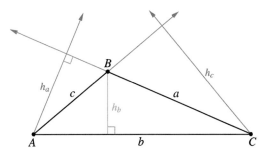

Figure 4.11

Remark 4.4

Theorem 4.3 implies that triangles with the same base and height have the same area. Figure 4.12 shows $\triangle ABC$ with base b and height h. Its area is $bh/2$. Construct line k through B parallel to the side AC. Because the distance between the lines is h, any triangle with vertex on k and base AC will have the same area $bh/2$. For example,

$$\text{Area}(\triangle AB_1C) = h_1 b/2 = hb/2 = \text{Area}(\triangle ABC).$$

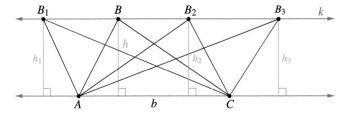

Figure 4.12

SPOTLIGHT on Teaching : Heights of Triangles and Parallelograms

Many students make mistakes when asked to identify or draw the three altitudes of an obtuse triangle or the two heights of a long and narrow parallelogram which is not a rectangle. Placing emphasis on the *definition-that the height of a triangle is the distance from a vertex to the line containing the opposite side–is* helpful for students. The distance from a point to a line needs to be defined first.

The following examples can be used to test students' understanding.

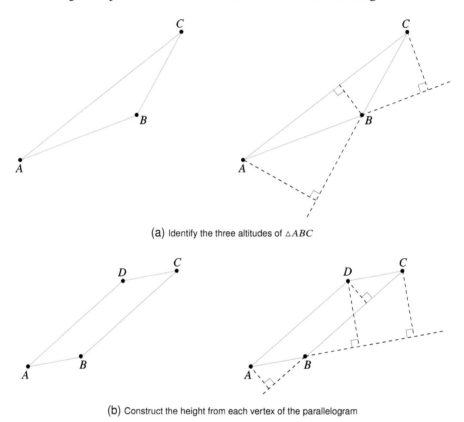

(a) Identify the three altitudes of △ABC

(b) Construct the height from each vertex of the parallelogram

Figure 4.13

Example 4.1: Redrawing a Border

Two farmers own large fields (regions I and II in Figure 4.14) divided by a common border consisting of two sections of fence as shown. They want to replace the old crooked fence with a new straight one so that the areas of the new regions are the same as the old areas. In other words, each farmer should have the same amount of land after the border is redrawn. Where should the new, straight fence be placed?

4.1 Areas of Polygons 151

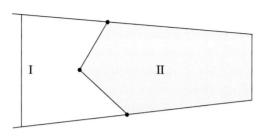

Figure 4.14

Solution. The farmer who owns region II owns $\triangle ABC$. The area of that triangle equals the area of any triangle with AC as a base and vertex on the line ED parallel to $\overline{AC}$. Thus, if point B "moves" along the segment ED, we get a variety of triangles each having the same area as the area of $\triangle ABC$. If we choose $\triangle ADC$, the area of region II does not change and the new border $\overline{CD}$ is straight. Another possible border is $\overline{AE}$.

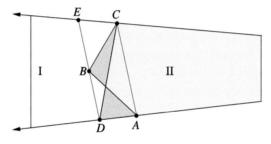

Figure 4.15

Now Solve This 4.2

Solve the "redrawing a border" problem in Example 4.1 for a border consisting of three segments like the one shown below.

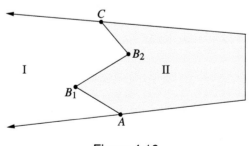

Figure 4.16

Example 4.2: Triangles Created by Medians

Prove that the medians of a triangle "divide" the triangle into six nonoverlapping smaller triangles of equal area, as shown in Figure 4.17.

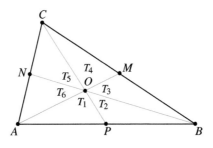

Figure 4.17

Solution. In Figure 4.17, M, N, and P are the midpoints of the sides of $\triangle ABC$. We need to show that the six triangles marked with numbers T_1 through T_6 have the same area.

We notice immediately that triangles T_1 and T_2 have the same area: Their bases AP and PB are congruent, and their corresponding heights (the distance from O to the line AB) are the same. Similarly, the pairs of triangles T_3 and T_4, as well as triangles T_5 and T_6, have equal areas.

If we could prove that one more pair such as T_1 and T_6, T_2 and T_3, or T_4 and T_5 (pairs of triangles that share $\overline{AO}$, $\overline{BO}$, and $\overline{CO}$, respectively) have equal areas then it would follow that all six triangles have equal areas. (Why?) We focus on triangles T_1 and T_6. These triangles are parts of other triangles that have the same area—namely, $\triangle ACM$ and $\triangle ABM$; these triangles have congruent bases CM and MB and share a common opposite vertex A. Hence, if we denote the area of triangle k (for $k = 1, 2, \ldots, 6$) by A_k we have

$$A_6 + A_5 + A_4 = A_1 + A_2 + A_3.$$

Because $A_4 = A_3$, it follows that

$$A_6 + A_5 = A_1 + A_2.$$

Since $A_5 = A_6$ and $A_1 = A_2$ (as shown earlier), we get

$$2A_6 = 2A_1,$$
$$A_6 = A_1.$$

In an analogous way, it follows that $A_2 = A_3$ and that $A_4 = A_5$. Consequently, the six triangles have the same area.

4.1 Areas of Polygons

> **Now Solve This 4.3**
>
> Prove that the midpoints N and P in Figure 4.17 are equidistant from the median AM. State this property of midpoints and their distance to a median in words.

Area of a Trapezoid

We will now investigate our last topic in this section—the formula for the area of a trapezoid. Such a formula can be derived in a variety of ways. Two approaches are suggested in Now Solve This 4.4.

> **Now Solve This 4.4**
>
> 1. Trace trapezoid $ABCD$ on a separate sheet of paper and retrace it next to the original trapezoid so that the lower base is up and the upper base is down, as shown in Figure 4.18(a). What kind of quadrilateral is $ABA'B'$? Find the area of that quadrilateral and then the area of the trapezoid.
>
>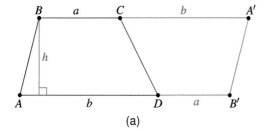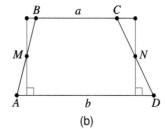
>
> (a)   (b)
>
> Figure 4.18
>
> Write a rigorous proof for finding the area of a trapezoid based on the ideas in Figure 4.18(a).
>
> 2. Derive a formula for the area of a trapezoid by the approach suggested in Figure 4.18(b).

We now state the formula for the area of a trapezoid and prove it in yet another way.

> **Theorem 4.4**
>
> *The area of a trapezoid whose bases have length a and b and whose height is h is given by $\frac{1}{2}(a+b)h$.*

Proof. We dissect the trapezoid into two triangles. This can be accomplished by drawing either of the two diagonals. In Figure 4.19, the area of $\triangle BCD$ is $ah/2$ and the area of $\triangle ABD$ is $bh/2$. (Notice that $BE = FD = h$ because the lines BC and AD are parallel.) Consequently,

$$\text{Area}(\triangle ABCD) = \text{Area}(\triangle ABD) + \text{Area}[\triangle BCD] = \frac{ah}{2} + \frac{bh}{2} = \frac{(a+b)h}{2}.$$

154   Chapter 4. Area and the Pythagorean Theorem

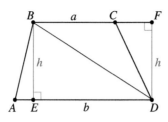

Figure 4.19

4.1.1 Problem Set

1. Find the areas of each of the following shaded figures if the distance between two adjacent dots in a row or column is one unit. In part (c), prove first that the shaded figure is a square.

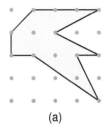
(a)

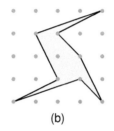
(b)

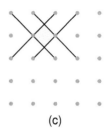
(c)

2. Prove the formula for the area of a parallelogram (Theorem 4.2) by dissecting it as suggested in the figure below.

3. Derive the formula for the area of a trapezoid by extending $\overline{BM}$ to intersect $\overleftrightarrow{AD}$ at E (where M is the midpoint of $\overline{CD}$) and showing that its area equals the area of $\triangle ABE$.

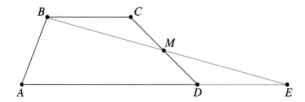

4. **a.** Find the area of a rhombus in terms of the lengths of its diagonals d_1 and d_2.
   b. Will the same formula you derived in part (a) work for kites? Why or why not?

4.1 Areas of Polygons

  c. For which other quadrilaterals will the formula you derived in part (a) work? (Find as many such quadrilaterals as you can.)

5. The formula for the area of a triangle can be derived before deriving the formula for the area of a parallelogram. To do so follow the next three suggestions.

   a. Using only the formula for the area of a rectangle, find the area of a right triangle with legs a and b.

   b. Use your result from part (a) to derive the formula for the area of a triangle by using a sum or difference of the areas of right triangles.

   c. Use the formula for the area of a triangle to derive a formula for the area of a parallelogram.

6. In a trapezoid, the parallel sides are, respectively, a and b units long. The base angles are 45° each. Find the simplest possible expression for the area of the trapezoid in terms of a and b (do not use trigonometry).

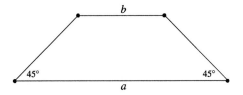

7. Suppose $ABCD$ is a parallelogram and P is a point on the diagonal AC. Prove that triangles ABP and APD have the same area.

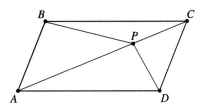

8. Let ABC be an equilateral triangle and P be any point in the interior of the triangle or on the triangle. Use the concept of area to prove that the sum of the distances from P to the three sides of the triangle is constant. That is, show that $PD + PE + PF$ is constant. How is that constant related to the height of the triangle?

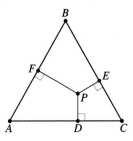

9. $ABCDEF$ is a regular hexagon. Let P be any point in the interior or on the hexagon.

a. Prove that the sum of the areas of the shaded triangles is constant (does not depend on the position of P in the interior or on the hexagon).

b. State the property in part (a) in words (without reference to the letters in the figure).

c. How does the sum of the areas in part (a) relate to the area of the hexagon? Why?

d. What kind of property follows if the point P is one of the vertices? Sketch an appropriate figure.

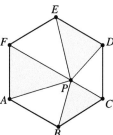

10. An arbitrary point P in the interior of a parallelogram is connected to the four vertices forming four triangles.

    a. If the areas of the triangles are A_1, A_2, A_3, and A_4 (as shown in the figure), how do the sums $A_1 + A_3$ and $A_2 + A_4$ relate to the area of the parallelogram? Prove it!

    b. Does a relationship similar to the one you found in part (a) hold if instead of a parallelogram we have a trapezoid? Justify your answer.

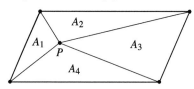

11. Construct any square, and then construct a square whose area is twice the area of the first square.

12. $\angle ACB$ is a right angle and P is a point in its interior. Through P we draw a line k intersecting the sides of the angle at A and B. If S_1 and S_2 are the respective areas of $\triangle APC$ and $\triangle BPC$, prove that $\frac{1}{S_1} + \frac{1}{S_2}$ is constant, i.e., the same for all lines k through P.

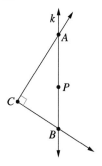

4.1 Areas of Polygons

13. In $\triangle ABC$, $\angle C$ is a right angle. $ACDE$ and $ABFG$ are squares constructed on a side and the hypotenuse, respectively. $\overline{CH}$ is perpendicular to $\overline{AB}$. The segment CH has been extended to intersect $\overline{GF}$ at I. Prove the following:

   a. $\triangle EAB \cong \triangle CAG$ and hence the triangles have the same area.

   b. The area of $\triangle EAB$ equals the area of $\triangle EAC$. Also the area of $\triangle ACG$ equals the area of $\triangle AHG$.

   c. The area of the square constructed on $\overline{AC}$ equals the area of the rectangle $AHIG$.

   d. State and justify a result similar to the one in part (c) for the square constructed on $\overline{CB}$.

   e. Using the results in parts (c) and (d), what can be said about the areas of squares constructed on the legs of a right triangle and the area of the square constructed on the hypotenuse? Justify your answer.

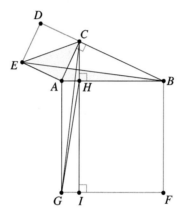

14. Squares A and B are congruent. One vertex of B is at the center of A. What is the ratio of the shaded area to the area of square A?

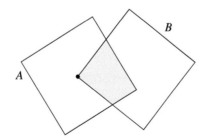

15. For each of the figures below, describe how to divide the region consisting of the interior of the squares in figure (a) and the interiors of the five congruent circles in figure (b), into two regions of equal area by a single straight line through the point P. Justify your answers. [*Note*: In figure (b), P is the center of the circle.] Show how to construct the required lines.

158   Chapter 4. Area and the Pythagorean Theorem

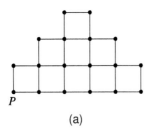
(a)

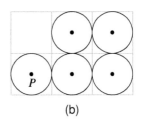
(b)

16. Given a convex quadrilateral, construct a triangle whose area is the same as the area of the quadrilateral.

17. Construct a non-isosceles trapezoid, and then construct an isosceles trapezoid with the same area. Is the solution unique?

18. Use the concept of area to demonstrate geometrically each of the following properties for positive real numbers:
   a. $a(b+c) = ab + ac$
   b. $(a+b)^2 = a^2 + 2ab + b^2$
   c. $a(b-c) = ab - ac$, where $b > c$
   d. $(a+b)(a-b) = a^2 - b^2$, for $a > b$

19. The following figures suggest a geometric approach to finding the sum of the first n consecutive numbers, $1 + 2 + 3 + \cdots + n$. Explain this approach and find the sum.

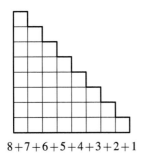
$8+7+6+5+4+3+2+1$

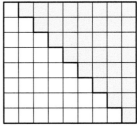
$2(8+7+6+5+4+3+2+1) = (8+1)8$

In Problems 20 and 21, assume that, as in Problem 1, the distance between two adjacent dots in a row or a column is one unit.

20. On the dot paper, find the area of the quadrilateral $ABCD$. Justify your answer.

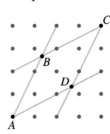

21. Consider $\triangle ABC$ and complete the following tasks:

a. Prove that it is an isosceles right triangle.
b. Find *AC* by finding the area of △*ABC* in two different ways (do not use the Pythagorean Theorem).

4.2 The Pythagorean Theorem

The Pythagorean Theorem is perhaps the most famous theorem in all of mathematics. It is simple, beautiful, and remarkably useful. William Dunham, in *The Mathematical Universe* (1994), says, "Whether regarded algebraically or geometrically, the theorem is of supreme mathematical importance." Here it is:

> **Theorem 4.5: Pythagorean Theorem**
>
> *In a right triangle with legs of length a and b and hypotenuse of length c,*
>
> $$a^2 + b^2 = c^2.$$

Figure 4.22 shows the equivalent form of the Pythagorean Theorem in terms of area:

> The area of the square constructed on the hypotenuse of a right triangle equals the sum of the areas of the squares constructed on the legs of the triangle.

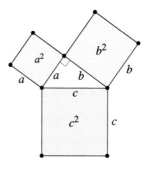

Figure 4.22

It is not clear if Pythagoras (circa 572–497 B.C.E.) had a proof of the theorem, but it is quite probable that he discovered a special case of it for isosceles triangles appearing in a floor tile pattern like the one shown in Figure 4.23. Notice that there are two congruent squares constructed on the legs of the isosceles right triangle ABC. The square on the hypotenuse $\overline{AB}$ consists of four triangles, each having an area equal to half the area of a shaded square; hence its area equals the area of the two shaded squares.

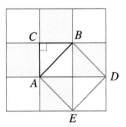

Figure 4.23

There are many known proofs of the Pythagorean Theorem. E. S. Loomis compiled 370 different proofs in his book *The Pythagorean Proposition*. The website www.cut-the-knot.org/pythagoras/ gives 98 beautifully presented proofs. We will discuss a few proofs in this section and explore others in the problem set and in Chapter 4 when similarity of triangles is introduced. Perhaps one of the simplest proofs is the following:

Proof 1 of the Pythagorean Theorem. Consider $\triangle ABC$ in Figure 4.24(a), whose legs have lengths a and b and whose hypotenuse has length c. In Figure 4.24(b) we see four triangles congruent to $\triangle ABC$ as part of a square with side $a + b$. Notice that $a^2 + b^2$—the sum of the areas of the two smaller squares in Figure 4.24(b)—is equal to the area of the large square (whose side is $a+b$) minus the area of the four congruent triangles. By contrast, in the large square with side $a + b$, the four triangles have been fitted as shown in Figure 4.24(c). Here the inner figure seems to be a square. To prove this, first notice that all of this figure's sides are c. (Why?) Second, since $\alpha + \beta = 90°$, each angle of the inner quadrilateral is $180 - (\alpha + \beta)$ or $90°$. It follows now that c^2 (the area of the inner square) is equal to the area of the large square minus the area of the four congruent triangles, which, as we showed earlier based on Figure 4.24(b), is $a^2 + b^2$. Hence $a^2 + b^2 = c^2$. ∎

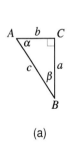
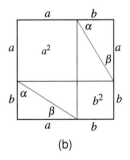
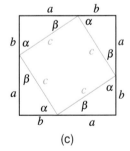

(a) (b) (c)

Figure 4.24

Our second proof of the Pythagorean Theorem is based on ideas that originated in China.

4.2 The Pythagorean Theorem

Proof 2 of the Pythagorean Theorem. We find the area of the larger square in Figure 4.24(c) in two different ways. Because the side of the square is $a+b$, its area is $(a+b)^2$. This larger square is made of four congruent triangles, each having area $\frac{1}{2}ab$ and the interior square having area c^2. Consequently,

$$(a+b)^2 = 4 \cdot \frac{1}{2}ab + c^2,$$
$$a^2 + 2ab + b^2 = 2ab + c^2,$$
$$a^2 + b^2 = c^2.$$

∎

Proof 3: Euclid's Proof of the Pythagorean Theorem. The Pythagorean Theorem appears as Proposition 47 in Book I of Euclid's *Elements*. The proof given there was the standard proof in high school textbooks for hundreds of years. (See also Problem 13 in Problem Set 4.1.1.)

In this proof $\triangle ABC$ (in Figure 4.25) is a right triangle with the right angle at C. Euclid drew a perpendicular from C to the hypotenuse AB and extended it until it intersected the side HI of the large square at L. That perpendicular split the square on the hypotenuse into two rectangles, marked I' and II'. Then Euclid proceeded to prove a remarkable property: The area of square $ACFG$ (marked I) is the same as the area of the rectangle $AKLH$ (marked I'). Similarly, the area of the square $BCED$ (marked II) is the same as the area of the rectangle $KBIL$ (marked II'). This immediately implied that $I + II = I' + II'$ = area of $ABIH$, and hence the Pythagorean Theorem.

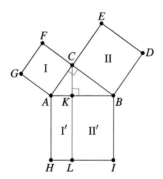

Figure 4.25

Euclid's proof that I = I' and II = II' is quite ingenious. Although we do not know how he came upon the proof, perhaps his thought process was as follows: In Figure 4.26(a) the diagonal $\overline{GC}$ of the square $ACFG$ splits the square into two congruent triangles. Similarly, the diagonal $\overline{HK}$ splits the rectangle into two congruent triangles. For that reason, it would be sufficient to prove that the areas of $\triangle ACG$ and $\triangle AKH$ are equal.

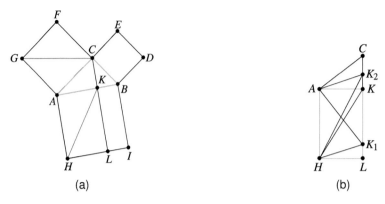

Figure 4.26

We know that given a triangle, we can create infinitely many triangles with the same base and height by moving the opposite vertex along a line through that vertex and parallel to the base. All of these triangles will have the same area (see the second remark following Theorem 4.3 and Figure 4.12). This applies to our case by focusing on $\triangle AKH$ [Figures 4.26(a) and 4.26(b)] and moving point K along the line $\overleftrightarrow{CL}$ that is parallel to the base $\overline{AH}$ of $\triangle AKH$.

Triangles AKH, AK_1H, and AK_2H in Figure 4.26(b) are three representative triangles that have the same area. Let S_1 be the set of all the triangles with base $\overline{AH}$ and opposite vertex on $\overleftrightarrow{CL}$.

Now let us focus on $\triangle ACG$. We can keep the base $\overline{AG}$ fixed and move vertex C along the line through C parallel to $\overleftrightarrow{AG}$. Notice that the points F, C, and B are collinear (since $\angle ACF$ and $\angle ACB$ are right angles); hence C can be moved along $\overleftrightarrow{FB}$. Let S_2 be the set of all triangles with base $\overline{AG}$ and opposite vertex on $\overleftrightarrow{FB}$. Perhaps we could pick two triangles, one from S_1 and the other from S_2, that are congruent. These two triangles would have the same area, and hence every triangle in S_1 would have the same area as every triangle in S_2. Consequently, $\triangle ACG$ and $\triangle AHK$ would have the same area. This approach would be especially promising if the two chosen triangles have some of the sides of $\triangle ABC$ as their sides. (Try to choose such triangles before reading on.)

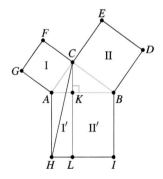

Figure 4.27

4.2 The Pythagorean Theorem

In Figure 4.27, we move point K all the way to C to get $\triangle ACH$. This triangle is a member of S_1 and its area is the same as the area of $\triangle AKH$ and hence half the area of rectangle $AKLH$. Next, we choose a triangle from S_2. Here the base is $\overline{AG}$ and the opposite vertex can move on $\overleftrightarrow{FB}$. We choose B to be the third vertex. $\triangle ABG$ will have two sides correspondingly congruent to two sides of $\triangle ABC$. The same is true of $\triangle AHC$. Moreover $\angle GAB \cong \angle CAH$ because the measure of each angle is $90° + m(\angle CAB)$. Thus $\triangle ABG \cong \triangle AHC$ by SAS. As discussed earlier, the congruence of these triangles implies I = I'. Analogously II = II' and hence I + II = I' + II' = Area(ABIH). Consequently, the Pythagorean Theorem is proved. ∎

SPOTLIGHT on Teaching

Following the CCSS recommendation of using transformations early in the curriculum we could prove that $\triangle ABG \cong \triangle AHC$ in Figure 4.27 by noticing that when $\triangle AHC$ is rotated counterclockwise by a right angle about A, its image is $\triangle ABG$. Because the image of A is A itself, the image of H is B and the image of C is G.

The preceding proof may seem long, mainly because we discussed possible motivations for taking the various steps in the proof. Next we give Euclid's proof without the motivation.

Euclid's Proof of the Pythagorean Theorem: A Short Version. Referring to Figure 4.27, we have $\overline{AH} \cong \overline{AB}$ and $\overline{AC} \cong \overline{AG}$. Also $\angle CAH \cong \angle GAB$ because the measure of each angle is $90° + m(\angle CAB)$. Consequently, by SAS

$$\triangle AHC \cong \triangle ABG. \tag{4.1}$$

Because $\overleftrightarrow{CL} \| \overleftrightarrow{AH}$, it follows from the second remark after Theorem 4.3 that the areas of $\triangle ACH$ and $\triangle AKH$ are the same. Similarly, because points B, C, and F are collinear (the angles at C are right angles) and $\overleftrightarrow{CB} \| \overleftrightarrow{AG}$, it follows that the areas of $\triangle GAB$ and $\triangle GAC$ are equal. By (4.1) above, it follows that the areas of $\triangle AKH$ and $\triangle GAC$ are equal. Because the area of $\triangle AKH$ is half the area of the rectangle $AKLH$ and the area of $\triangle GAC$ is half the area of the square $ACFG$, it follows that the area of the rectangle is the same as the area of the square. In a completely analogous way, we find that the area of $KBIL$ is the same as the area of the square $BCED$. Consequently, the sum of the areas of the squares on the legs $\overline{AC}$ and $\overline{CB}$ equals the sum of the areas of the rectangles $AKLH$ and $KBIL$. This is the area of the square on the hypotenuse $\overline{AB}$. ∎

Remark 4.5

A closer look at Figure 4.27 and Euclid's proof of the Pythagorean Theorem suggests that a relationship exists between $\overline{HC}$ and $\overline{BG}$. In addition to being congruent, these segments are perpendicular to each other. If we rotate $\triangle AHC$ about point A by $90°$ (counterclockwise), the image of $\triangle AHC$ is $\triangle ABG$ because A is mapped onto itself, H onto B, and C onto G. Because the image of $\overline{HC}$ is $\overline{BG}$, these segments are congruent and perpendicular.

Historical Note: *Pythagoras (circa 580–500 B.C.E.) and U.S. President James Garfield (1831–1881)*

Pythagoras of Samos was a Greek philosopher and mathematician. He founded a philosophical and religious school, known as the Pythagorean Brotherhood. This school formulated principles that influenced both Plato and Aristotle. Today Pythagoras is particularly remembered for the theorem that bears his name. In reality, this theorem—if not its proof—was known to the Babylonians and the Chinese about 1000 years earlier.

The followers of Pythagoras were renowned for their ethical practices, unselfishness, honesty, and love of mathematics, especially geometry. The Pythagorean Theorem very likely has more proofs than any other theorem in mathematics. The book *The Pythagorean Proposition* (Loomis, 1968) gives different proofs, among them one by James Garfield, the twentieth president of the United States. President Garfield studied mathematics at Williams College in Williamstown, Massachusetts, and after graduating taught math in public schools and at a college. He published his proof in the *Journal of Education* in 1876.

Garfield, like Pythagoras, was known for his character and honesty. Unfortunately, he served as president for only 200 days before becoming one of the four U.S. presidents who were assassinated while in office.

The Converse of the Pythagorean Theorem

Theorem 4.6

If the sides of a triangle have lengths a, b, and c, such that $c^2 = a^2 + b^2$, then the triangle is a right triangle with the right angle opposite the side of length c.

Proof. In $\triangle ABC$ in Figure 4.28(a), $c^2 = a^2 + b^2$. We need to prove that $\angle C$ is a right angle. For that purpose, we construct in Figure 4.28(b) a right triangle $A_1 B_1 C_1$ with legs a and b and a right angle at C_1. We will show that the two triangles are congruent. Then it would follow that the angles at C and C_1 are congruent and hence that $\angle C$ is a right angle.

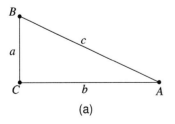
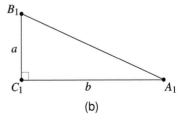

Figure 4.28

4.2 The Pythagorean Theorem

When applied to $\triangle A_1 B_1 C_1$, the Pythagorean Theorem gives $(A_1 B_1)^2 = a^2 + b^2$. Since we have $c^2 = a^2 + b^2$ (given), it follows that $c^2 = (A_1 B_1)^2$ and hence $c = A_1 B_1$. Consequently $\triangle ABC \cong \triangle A_1 B_1 C_1$ by SSS. Thus $\angle C \cong \angle C_1$ and, therefore, $\angle C$ is a right angle. ∎

Remark 4.6

The converse of the Pythagorean Theorem will be used later in this section in constructing $\sqrt{n}$, where n is a whole number and a unit segment is given.

SPOTLIGHT on Teaching

Euclid's proof of the Pythagorean Theorem has disappeared from most high school textbooks in the last several decades. Do you think it should be taught in high school? Why or why not? Should students be exposed to several proofs of the theorem or only one? Which one?

Now Solve This 4.5

Construct any segment to be one unit long, and then use the relationship $1^2 + 2^2 = 5$ to construct a segment whose length is $\sqrt{5}$.

Example 4.3

Find the area of an equilateral triangle with side a.
Solution. In Figure 4.29, ABC is an equilateral triangle. To find its area, we drop the perpendicular from a vertex to the opposite side. $\overline{BM}$ is such a perpendicular, and M is the midpoint of $\overline{AC}$. (Why?) In $\triangle ABM$, we have

$$h^2 + \left(\frac{a}{2}\right)^2 = a^2,$$

$$h^2 = \frac{3}{4}a^2,$$

$$h = \frac{a\sqrt{3}}{2}.$$

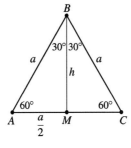

Figure 4.29

Consequently, the area of $\triangle ABC$ is

$$\frac{ah}{2} = \frac{a}{2} \cdot \frac{a\sqrt{3}}{2} = \frac{a^2\sqrt{3}}{4}.$$

Now Solve This 4.6

1. Heron (first century C.E.) derived a remarkable formula for the area of a triangle given its sides a, b, and c. He denoted half the perimeter of the triangle by p, that is, $p = (a + b + c)/2$, and proved that the area A of the triangle is $A = \sqrt{p(p-a)(p-b)(p-c)}$. What is even more remarkable is that Heron was able to derive the formula without the use of algebra, which was not known to mathematicians at his time. By answering the following questions, you will prove Heron's formula using the Pythagorean Theorem and elementary algebra.

   a. The area A of $\triangle ABC$ is $A = \frac{1}{2}ch$, where h is the length of the altitude intersecting $\overline{AB}$ at D (draw your own triangle with the greatest angle at C, $BA = c$). Let $BD = x$; then $AD = c - x$. Show that $x = \frac{a^2-b^2+c^2}{2c}$.

   b. Apply the Pythagorean Theorem in $\triangle BCD$ to express h in terms of a and x. Substitute for x the expression you obtained in part (a). You should get

   $$h^2 = a^2 - \left(\frac{a^2-b^2+c^2}{2c}\right)^2.$$

   c. Apply the difference-of-squares formula $u^2 - v^2 = (u-v)(u+v)$ to the result in part (b) to obtain

   $$4h^2c^2 = \left[2ac - (a^2 - b^2 + c^2)\right]\left[2ac + (a^2 - b^2 + c^2)\right].$$

   d. Write the expression in the pair of first brackets on the right side of the equation in part (c) as $b^2 - (c-a)^2$, and a similar expression for the expression in the second brackets. Apply the difference-of-squares formula again to obtain a product of four expressions, one of which will be $a + b - c$.

   e. One of the expressions $a + b - c$ in the product you obtained in part (d) can be written using $a + b + c = 2p$ as follows:

   $$a + b - c = (2p - c) - c = 2(p - c).$$

   Write the other three expressions in the product in a similar way using p, and thus derive Heron's formula.

2. The Hindu mathematician Brahmagupta (598–668 C.E.) found that the area of a cyclic quadrilateral (a quadrilateral that can be inscribed in a circle) with sides a,

4.2 The Pythagorean Theorem

b, c, and d is given by the formula

$$\sqrt{(p-a)(p-b)(p-c)(p-d)} \quad \text{where} \quad p = \frac{1}{2}(a+b+c+d).$$

Show that Heron's formula in part 1 follows from Brahmagupta's result.

Historical Note: *Heron (first century C.E.)*
Heron (also called Hero) of Alexandria was a first-century Greek mathematician whose treatise on geometry *Metrica* was lost until 1896, when the complete manuscript was found in Istanbul. (Constantinople, which was renamed Istanbul after World War I, is a city in present-day Turkey of about 13.5 million people, located partially in Europe and partially in Asia.) Heron also invented a jet-propelled rotary steam engine, wrote a text on mechanics for engineers, described how to dig tunnels through mountains, and formed the following principle on light reflection: *The angle of incidence is congruent to the angle of reflection.* In the delightful book *Journey Through Genius*, William Dunham discusses Heron's proof of his area formula and refers to it as a "great theorem of mathematics."

Example 4.4

Prove that it is impossible to construct an equilateral triangle on a geoboard no matter how large the geoboard is. In other words, prove that an equilateral triangle with vertices that are lattice points (have integer coordinates) does not exist.

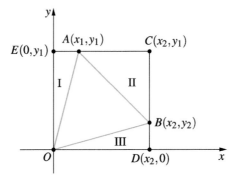

Figure 4.30

We now prove that it is impossible to have an equilateral triangle with vertices having integer coordinates. Suppose such a triangle exists. We set up a coordinate system so that one of the vertices is at the origin and the x-axis contains a row of lattice points, each one unit from its immediate neighbors (Figure 4.30). Because of our assumption that vertices are at the lattice points, the coordinates of A and B are integers. We compute the area of $\triangle OAB$ in two ways. From Example 4.3:

$$\text{Area}(\triangle OAB) = \frac{(OA)^2\sqrt{3}}{4} = \frac{(x_1^2 + y_1^2)\sqrt{3}}{4}. \tag{4.2}$$

Because x_1 and y_1, are integers and $\sqrt{3}$ is irrational, each side of Equation (4.2) is irrational. If we compute the area of $\triangle OAB$ as the difference between the area of the rectangle $OECD$ and the sum of the areas of the three triangles marked I, II, and III, we get

$$\text{Area}(\triangle OAB) = x_2 y_1 - \frac{1}{2}\Big(x_1 y_1 + (x_2 - x_1)(y_1 - y_2) + x_2 y_2\Big).$$

This is clearly a rational number, which contradicts the fact that the right side of Equation (4.2) is irrational. Consequently, it is impossible to have an equilateral triangle with all the vertices at lattice points.

The Construction of $\sqrt{n}$

If each of the legs of a right isosceles triangle is a units long, and the hypotenuse is c units long (see Figure 4.31), then the Pythagorean Theorem tells us that

$$c^2 = a^2 + a^2,$$
$$c^2 = 2a^2,$$
$$c = a\sqrt{2}.$$

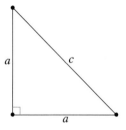

Figure 4.31

Thus, for a given segment of length a, the hypotenuse will have length $a\sqrt{2}$. If a unit segment is given, $a = 1$, then the preceding approach gives us a way to construct a segment of $\sqrt{2}$ units.

The Pythagorean Theorem can be used to construct $a\sqrt{n}$ for any positive integer n. (If n is a perfect square, the construction does not require the Pythagorean Theorem!) Let's look at a few examples. In all of these examples, a segment of length a is given.

Example 4.5

Construct a segment of length $a\sqrt{3}$.

4.2 The Pythagorean Theorem

Solution: First Approach. Let $x = a\sqrt{3}$. Then to use the Pythagorean Theorem, we write
$$x^2 = 3a^2 = a^2 + 2a^2 = a^2 + (\sqrt{2}a)^2.$$
Hence x is the hypotenuse of a right triangle with sides a and $a\sqrt{2}$. Because we have already constructed $a\sqrt{2}$, all that remains for us to do is to construct a segment of length a at a right angle to the previously constructed segment of length $a\sqrt{2}$, as shown in Figure 4.32.

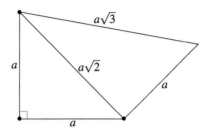

Figure 4.32

Solution: Second Approach. Notice that
$$x^2 = 3a^2 = 4a^2 - a^2 = (2a)^2 - a^2$$
and therefore
$$x^2 + a^2 = (2a)^2.$$
Thus x is a leg of a right triangle whose hypotenuse is $2a$ and one leg is a. Figure 4.33 suggests two ways to construct such a triangle and, therefore, two ways to construct $x = a\sqrt{3}$.

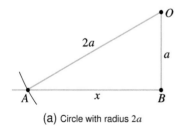

(a) Circle with radius $2a$

(b) (1) Construct segment $\overline{BO}$. (2) Construct ray ℓ perpendicular to $\overline{BO}$. (3) Construct an arc with center O and radius $2a$ to find A as the intersection of the arc with ℓ.

Figure 4.33

> **Now Solve This 4.7**
>
> Construct $a\sqrt{45}$.
>
> **Hints for a Solution.** Because $\sqrt{45} = 3\sqrt{5}$, we can write $\sqrt{45}a = 3\sqrt{5}a$. Thus we need to construct only $\sqrt{5}a$. One way to accomplish this task is to notice that $5 = 2^2 + 1^2$. Another way is to write 5 as a difference of two squares. If $1 \cdot 5 = u^2 - v^2$, then $1 \cdot 5 = (u-v)(u+v)$. Let $u - v = 1$ and $u + v = 5$; then $u = 3$ and $v = 2$. Hence $5 = 3^2 - 2^2$. Notice that any odd number can be expressed in this way as a difference of two squares.
>
> 1. Complete a construction of $a\sqrt{5}$.
> 2. Show how to construct $a\sqrt{n}$ when n is an integer and square free (i.e., every prime in the prime factorization of n appears only once).
> 3. Describe how to construct $a\sqrt{n}$ when n is even.
> 4. Explain how to construct $a\sqrt{6}$ using the factorization $\sqrt{6} = \sqrt{2} \cdot \sqrt{3}$.
> 5. Explain how to construct $a\sqrt{\dfrac{3}{2}}$.

4.2.1 Problem Set

1. The figure below suggests yet another proof of the Pythagorean Theorem. $\triangle ABC$ is a right triangle with $AB = c$, $AC = b$, and $BC = a$. Four triangles congruent to $\triangle ABC$ have been "assembled" to form a rhombus of side length c.

   a. Prove that the outer figure with side length c (the rhombus) is a square.

   b. Prove that the inner shaded figure is also a square.

   c. Compute the area of the outer square in two different ways to obtain the Pythagorean Theorem.

   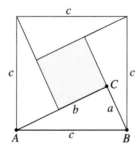

2. The following proof of the Pythagorean Theorem is attributed to James Garfield. $\triangle ABC$ is a right triangle with sides a and b and hypotenuse c. $\triangle DEB \cong \triangle CBA$ and C, B, and D are collinear. Find the area of the trapezoid $ACDE$ in two different ways

4.2 The Pythagorean Theorem

to obtain a proof of the Pythagorean Theorem.

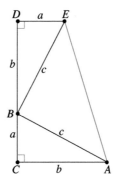

3. **a.** Prove that in a right triangle with a 30° angle, the side opposite that angle is half as long as the hypotenuse.
   b. Without using trigonometry, express the length of the legs of a 30° – 60° – 90° triangle in terms of the length of the hypotenuse c.

4. **a.** $ABCD$ is a rectangle, and O is a point in its interior. The distances from O to the vertices are x, y, z, and w, as indicated in the figure. Prove that $x^2 + z^2 = y^2 + w^2$.
   b. Pose a problem whose solution will be easy if the result in part (a) is known.

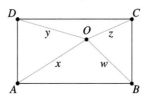

5. A boat starts at point A, moves 3 km due north, then 2 km due east, then 1 km due south, and then 4 km due east to point B. Find the distance AB.

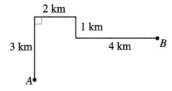

6. **a.** The distance from point A to the center of a circle with radius r is d. Express the length of the tangent segment $\overline{AP}$ in terms of r and d.

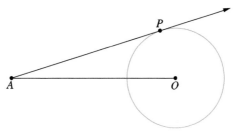

b. The distance between the centers of two circles with radii r_1 and r_2 is d, that is, $O_1O_2 = d$. Line k is a common exterior tangent to the circles at P and Q. Find PQ in terms of r_1, r_2, and d. (Assume that $r_1 > r_2$.)

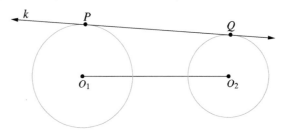

c. Line ℓ is a common interior tangent to the circles with centers O_1 and O_2 at S and T, respectively. Find ST in terms of r_1, r_2, and d, where $O_1O_2 = d$.

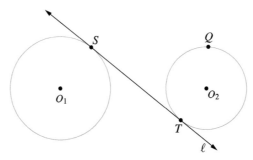

7. a. Use the segment of length a, shown below, to construct a segment of length (i) $\sqrt{19}a$ and (ii) $a\sqrt{\frac{2}{3}}$.

•—————————•
a

b. Repeat part (a) using a different construction method.

c. Given a segment of length a, explain how to construct a segment of length $\sqrt{\frac{n}{m}}a$, where n and m are positive integers. Assume we know how to construct $\sqrt{k}a$ for any positive integer k.

8. $ABCDEFGH$ is a right rectangular prism (a box) of dimensions a, b, and c as shown in the figure below. Find the length of the diagonal $\overline{DG}$ in terms of a, b, and c.

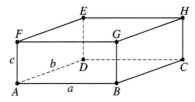

9. $\triangle ABC$ is a right triangle. Squares have been constructed on the hypotenuse and the legs of the triangle. $\triangle A'B'C'$ has been constructed so that $CC' = c$ and $\overline{CC'} \| \overline{AA'}$. Prove the following:

a. The area of the square $ABB'A'$ is equal to the sum of the areas of the parallelograms $ACC'A'$ and $CBB'C'$.

b. The height of the parallelogram $CBB'C'$ to side $\overline{CB}$ is a, and hence the area of the parallelogram is a^2. Similarly, show that the area of $ACC'A'$ is b^2.

c. Use your results from parts (a) and (b) to prove that $c^2 = a^2 + b^2$.

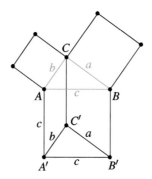

10. Leonardo da Vinci (1452–1519) proved the Pythagorean Theorem with the help of the accompanying figure, in which $DC' = AC$ and $EC' = CB$. Use the figure to prove the Pythagorean Theorem. (For an animated version search the internet.)

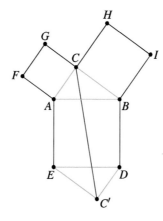

11. $ABCD$ is a rectangle with sides $AB = a$ and $AD = b$. $BEDF$ is a rhombus. Find EF in terms of a and b. (Simplify your answer.)

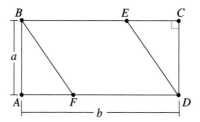

12. The Greek geometer Pappus (circa 300 C.E.) proved the following generalization of the Pythagorean Theorem: $\triangle ABC$ is any triangle (not necessarily a right triangle). Arbitrary parallelograms are constructed on the two shorter sides of the triangle (sides $\overline{AC}$ and $\overline{BC}$ in the figure). D is the intersection of the lines containing the two sides of the parallelograms. A third parallelogram is constructed on side $\overline{AB}$ such that $AE = DC$ and $\overline{AE} \| \overline{DC}$. Pappus's Theorem asserts that

$$\text{Area}(I) + \text{Area}(II) = \text{Area}(III).$$

   a. Prove Pappus's Theorem.
   b. Show that the Pythagorean Theorem follows from Pappus's Theorem and therefore that Pappus's Theorem is a generalization of the Pythagorean Theorem.

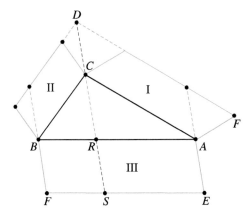

13. a. $\triangle ABC$ is a right triangle with height h to the hypotenuse. The height divides the hypotenuse into segments a_1 and b_1. Prove that $h^2 = a_1 b_1$. (This can be proved using similarity in Chapter 5, but here you should not use similar triangles.)

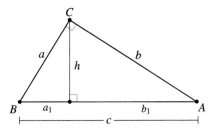

   b. Prove the converse of part (a): If it is not given that $\triangle ABC$ is a right triangle but it is known that $h^2 = a_1 b_1$, prove that $\angle C$ must be a right angle.
   c. Given a segment of length a explain how to construct a segment of length $a\sqrt{n}$ where (i) n is a whole number; (ii) n is a rational number.

14. $\triangle ABC$ is equilateral with side a. Three congruent circles in the interior of $\triangle ABC$ are tangent to each other and to the sides of the triangle.

4.2 The Pythagorean Theorem

   a. Find the radius r of the circles in terms of a.
   b. Construct an equilateral triangle and then inscribe in it three congruent circles as described above. Carry out the three-step process: investigation–construction–proof.

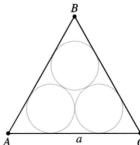

15. In the figure below, $ABCD$ is a square with side length a. $QSTP$ is also a square, and the four triangles are congruent. Congruent circles are inscribed in each triangle and in the inner square.
    a. Find the radius of the circle in terms of a.
    b. Construct your own square, the triangles, the inner square, and the inscribed circles.

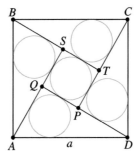

16. If a, b, and c are positive integers and $a^2 + b^2 = c^2$, then the triple (a, b, c) is called a **Pythagorean triple**. The triple $(3, 4, 5)$ is perhaps the best-known example. Use it to complete the following tasks.
    a. Prove that there are infinitely many Pythagorean triples in which a, b, c have no common factor other than 1.
    b. Find all the Pythagorean triples that consist of consecutive positive integers. (Prove that you have found them all.)
    c. Find all the Pythagorean triples that form an arithmetic sequence.

17. To find Pythagorean triples other than the ones in Problem 16, examine the following argument and answer the questions that follow.
    Notice that
    $$(u + v)^2 = (u - v)^2 + 4uv.$$

Thus, if uv is a perfect square, this equation gives us a Pythagorean triple for all positive integers u and v with $u > v$. Let $u = n^2$ and $v = m^2$, where n and m are positive integers, and $n > m$. We get

$$u - v = n^2 - m^2,$$
$$u + v = n^2 + m^2,$$
$$4uv = 4n^2 m^2 = (2nm)^2.$$

Thus, for all positive integers $n > m$ ($n^2 - m^2$, $2nm$, $n^2 + m^2$) is a Pythagorean triple.

a. Use the preceding expression to obtain the Pythagorean triples 5, 12, 13 and 7, 24, 25.

b. Does the argument in this problem prove that the preceding expression gives all the Pythagorean triples? Justify your answer.

5 Similarity

> *We could use up two Eternities in learning all that is to be learned about our own world and the thousands of nations that have arisen and flourished and vanished from it. Mathematics alone would occupy me eight million years.*
>
> Mark Twain, Notebook #22, Spring 1883–September 1884

Introduction

An undistorted photograph of an object, what we call its *image*, has the same shape as the original object. We say that the image is similar to the object. If the ratio between the height and width of a window is 2, for example, the ratio will remain 2 in the photo of the window. This suggests the following definition:

> **Definition of Similar Figures**
>
> Two figures are similar if and only if there exists a one-to-one correspondence between the figures such that the ratio between any two distances within one figure is the same as the ratio between the corresponding pair of distances in the other figure.

In Chapter 6, we will introduce the concept of size transformation and give a useful definition of similarity using transformations that is equivalent to the preceding definition.

5.1 Ratio, Proportion, and Similar Polygons

A ratio is a quotient of two real numbers. If $r \neq 0$, the number $\frac{s}{r}$ is the ratio of s to r. A proportion is an equation stating that two or more ratios are equal. We state some basic properties of ratios and proportion, most of which will already be familiar to you. (The proofs are left as exercises.)

Properties of Ratio and Proportion

Cross product property: If a, b, c, and d are real numbers, $b \neq 0$, and $d \neq 0$, then $\frac{a}{b} = \frac{c}{d}$ if and only if $ad = be$.

Reciprocal property: If a, b, c, and d are real numbers, $a \neq 0$, and $c \neq 0$, then $\frac{a}{b} = \frac{c}{d}$ if and only if and only if $\frac{b}{a} = \frac{d}{c}$.

A quick way to see that the second property is true is to notice that the cross product $ad = bc$ has not changed in the new proportion. If none of a, b, c or d are zero, then two more proportions can be derived from $\frac{a}{b} = \frac{c}{d}$, namely $\frac{a}{c} = \frac{b}{d}$ and $\frac{d}{c} = \frac{b}{a}$.

Another useful property can be obtained by adding 1 to each side of the following proportion: If $\frac{a}{b} = \frac{c}{d}$,

$$\frac{a}{b} + 1 = \frac{c}{d} + 1,$$
$$\frac{a+b}{b} = \frac{c+d}{d}.$$

We state the result obtained above and its converse more formally for easy reference.

Denominator Addition Property

If a, b, c, and d are real numbers ($b \neq 0$ and $d \neq 0$), then

$$\frac{a}{b} = \frac{c}{d} \quad \text{if and only if} \quad \frac{a+b}{b} = \frac{c+d}{d}.$$

We illustrate the next property with a story problem,

> In each of the following groups, the ratio of adults to children is the same. Suppose group A has 3 adults and 4 children, group B has 6 adults and 8 children, and group C has 15 adults and 20 children. If all the people in the three groups form one large group, without performing any computations, what do you think will be the ratio of adults to children in the new group?

We generalize this example in the following property, which we prove below.

Ratio of the Sum of Numerators to the Sum of Denominators Property

If
$$\frac{a_1}{b_1} = \frac{a_2}{b_2} = \frac{a_3}{b_3} = \cdots = \frac{a_n}{b_n} = r$$

then
$$\frac{a_1 + a_2 + a_3 + \cdots + a_n}{b_1 + b_2 + b_3 + \cdots + b_n} = r \quad (\text{if } b_1 + b_2 + b_3 + \cdots + b_n \neq 0).$$

Proof. The hypothesis implies

$$a_1 = rb_1, \quad a_2 = rb_2, \quad a_3 = rb_3, \ldots, a_n = b_n.$$

Therefore,

$$\frac{a_1+a_2+a_3+\cdots+a_n}{b_1+b_2+b_3+\cdots+b_n} = \frac{rb_1 + rb_2 + rb_3 + \cdots + rb_n}{b_1 + b_2 + b_3 + \cdots + b_n} = \frac{r(b_1 + b_2 + b_3 + \cdots + b_n)}{b_1 + b_2 + b_3 + \cdots + b_n} = r. \quad \blacksquare$$

Now Solve This 5.1

Is the converse of the "Ratio of the Sum of Numerators to the Sum of Denominators" property true? Justify your answer.

Similarity of Polygons

For polygons, our definition of similar figures given in the introduction can be shown to be equivalent to the following, more useful definition:

> **Definition of Similar Polygons**
>
> Two polygons are similar if and only if there is a one-to-one correspondence between the vertices of one polygon and the vertices of the other polygon such that the corresponding angles are congruent and the ratio of the lengths of corresponding sides is constant.

In Figure 5.1, the two quadrilaterals are similar. Corresponding vertices have been labeled by the same index. Thus A_i corresponds to B_i, for $i = 1, 2, 3, 4$. We write this as

$$A_1A_2A_3A_4 \sim B_1B_2B_3B_4.$$

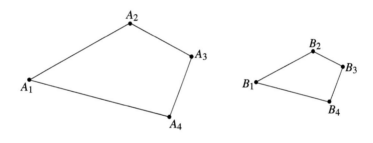

Figure 5.1

Because the quadrilaterals are similar, $\angle A_1 \cong \angle B_1$, $\angle A_2 \cong \angle B_2$, $\angle A_3 \cong \angle B_3$, and $\angle A_4 \cong \angle B_4$. Also

$$\frac{A_1A_2}{B_1B_2} = \frac{A_2A_3}{B_2B_3} = \frac{A_3A_4}{B_3B_4} = \frac{A_4A_1}{B_4B_1}.$$

If $\frac{A_1A_2}{B_1B_2} = r$, we say that the **scale factor** is r—that is, each side of the first polygon is r times as long as the side of the second polygon.

You may be aware from a high school geometry course that for triangles, it is sufficient to know that corresponding angles are congruent to conclude that the triangles are similar (the *AAA* similarity condition). In fact knowing that two corresponding angles are similar is sufficient to conclude that the triangles are similar (the *AA* similarity condition). Also, proportionality of the lengths of corresponding sides (the *SSS* similarity condition) is a sufficient condition for similarity of the triangles. These assertions will soon be stated as theorems and proved. For polygons with four or more sides, however, these conditions are not sufficient. For example, in a square and a rectangle that is not a square, all angles are congruent but the quadrilaterals are clearly not similar.

> **Remark 5.1**
>
> From this point on, instead of saying that "the lengths of the corresponding sides are proportional" we will simply say that "the corresponding sides are proportional."

> **Now Solve This 5.2**
>
> 1. Describe and sketch two quadrilaterals and two convex pentagons in which the corresponding sides are proportional but the polygons are not similar.
> 2. If two polygons P_1 and P_2 are similar and the sides of P_1 are proportional to the sides of P_2 with scale factor r, with what scale factor are the sides of P_2 proportional to the sides of P_1?

Before proving the AAA and the SSS similarity conditions for triangles, we will need the following theorem:

> **Theorem 5.1**
>
> *Parallel projection preserves ratios of lengths of segments.*

Proof. Let m and n in Figure 5.2(a) be two arbitrary lines. Let A, B, and C be any points on m [B between A and C as in Figure 5.2(a)]. Let A', B', and C' be the corresponding points on line n under parallel projection. (Lines AA', BB', and CC' are parallel.) We need to show that:

$$\frac{AB}{BC} = \frac{A'B'}{B'C'}.$$

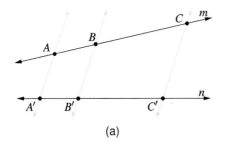
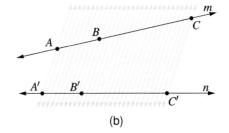

(a) (b)

Figure 5.2

Using lines parallel to line AA', divide segments AB and BC into congruent segments of the same length [Figure 5.2(b)], Suppose $\overline{AB}$ is divided into i segments each of length x and $\overline{BC}$ is divided into j segments also of length x. Next, we project each endpoint of these small segments in the direction of line AA'. Because parallel projection preserves congruence of segments, all of the small segments on line n must be the same length—say,

5.1 Ratio, Proportion, and Similar Polygons

y. Hence

$$\frac{AB}{BC} = \frac{ix}{jx} = \frac{i}{j}, \qquad \frac{A'B'}{B'C'} = \frac{iy}{jy} = \frac{i}{j}, \quad \text{and} \quad \frac{AB}{BC} = \frac{A'B'}{B'C'}.$$

Before reading on, reflect on the preceding argument. Is it a valid proof of Theorem 5.1?

The argument assumes that it is always possible to find a common measure x for any two segments, regardless of their length. This, however, is not always the case. Consider, for example, segments of length a and $a\sqrt{2}$ (a segment of length $a\sqrt{2}$ is the hypotenuse of a right triangle with legs of length a). Suppose that in Figure 5.2(a), $AB = a$ and $BC = a\sqrt{2}$. Then, if it were possible to divide the two segments into equal parts with a common measure, as in Figure 5.2(b), we would have

$$\frac{BC}{AB} = \frac{jx}{ix} = \frac{j}{i},$$

where i and j are whole numbers. But we also have

$$\frac{BC}{AB} = \frac{a\sqrt{2}}{a} = \sqrt{2}.$$

Consequently, we would get

$$\sqrt{2} = \frac{j}{i},$$

which says that $\sqrt{2}$ is a rational number, which is a contradiction. Thus the segments of length a and $a\sqrt{2}$ are not *commensurable* (they do not have a common measure). ∎

> **Remark 5.2**
>
> If each of two segments has a rational measure (that is, their measures in some units are rational numbers), then it is always possible to find a common measure for the segments. For example, if $AB = \frac{11}{3}$ and $BC = \frac{12}{7}$, then we can write each fraction with a common denominator:
>
> $$AB = \frac{77}{21} \quad \text{and} \quad BC = \frac{36}{21}.$$
>
> We can then choose $\frac{1}{21}$ of the unit to be the common measure. We would get $AB = 77 \cdot \frac{1}{21}$ and $BC = 36 \cdot \frac{1}{21}$.

> **Historical Note:**
> Pythagoreans (circa 500 B.C.E.)—that is, the disciples of Pythagoras—were both a philosophical society and a religious order. Members of this group were bound by oath not to reveal any of their discoveries or beliefs. They believed that the ratio of the lengths of any two segments is a rational number and were distressed when one of them found that this is not always the case. Hippasus, the member of the Pythagorean society who probably let out the word about the discovery, was drowned for breaking his oath of secrecy. Pythagoreans believed that the elevation of the soul to God can be attained through knowledge of mathematics.

Theorem 5.1 is of major importance. It is useful in solving problems and is the foundation for the theory of similarity of triangles. Notice that we proved Theorem 5.1 for the case in which the ratio of the length of the segments is a rational number. Because every irrational number can be approximated to any desired accuracy by rational numbers, it seems plausible that this theorem is true for any pair of real numbers. This intuitive argument can be made rigorous, and the interested reader can find a rigorous proof in Moise (pp. 167–170). In what follows we will give a rigorous proof of a special case of Theorem 5.1 using the concept of area. From this special case, a proof of Theorem 5.1 will follow.

> **Theorem 5.2: Side-Splitting Theorem**
>
> *A line parallel to a side of a triangle that intersects the other two sides in distinct points splits these sides into proportional segments.*

Proof. We have $\overleftrightarrow{PQ} \parallel \overleftrightarrow{BC}$ (Figure 5.3) and we need to prove that $\frac{AP}{PB} = \frac{AQ}{QC}$.

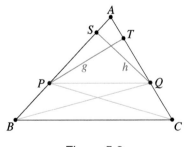

Figure 5.3

We express each ratio as a ratio of areas of triangles and show that these areas are equal.

To express $\frac{AP}{BP}$ as a ratio of areas, we are looking for two triangles with segments AP and PB as bases and a common height. Notice that $\triangle APQ$ and $\triangle PQB$ in Figure 5.3 have $\overline{AP}$ and $\overline{PB}$ as bases, a common vertex Q, and, therefore, the same height QS. For convenience,

5.1 Ratio, Proportion, and Similar Polygons

let $QS = h$. We have

$$\frac{\text{Area}(\triangle APQ)}{\text{Area}(\triangle PQB)} = \frac{AP \cdot \frac{h}{2}}{PB \cdot \frac{h}{2}} = \frac{AP}{PB}. \tag{5.1}$$

Similarly, consider two triangles with sides AQ and QC and a common height. $\triangle AQP$ and $\triangle QCP$ are such triangles, Their common vertex is at P, and hence their common height is PT ($PT = g$ in Figure 5.3). We have

$$\frac{\text{Area}(\triangle AQP)}{\text{Area}(\triangle QCP)} = \frac{AQ \cdot \frac{g}{2}}{QC \cdot \frac{g}{2}} = \frac{AQ}{QC}. \tag{5.2}$$

Focusing on Equations (5.1) and (5.2), notice that if we could prove that $\triangle PQB$ and $\triangle QCP$ have the same area, our proof would be complete, because then we could use the following string of equalities:

$$\frac{AP}{PB} = \frac{\text{Area}(\triangle APQ)}{\text{Area}(\triangle PQB)} = \frac{\text{Area}(\triangle APQ)}{\text{Area}(\triangle QCP)} = \frac{AQ}{QC}.$$

But these triangles have a common base PQ, and the vertices B and C lie on line BC parallel to line PQ. Consequently, the triangles have the same height relative to their common base PQ. From Equations (5.1) and (5.2), we get the desired assertion that

$$\frac{AP}{PB} = \frac{AQ}{QC}.$$

∎

It is now easy to prove the converse of Theorem 5.2.

> **Theorem 5.3: The Converse of the Side-Splitting Theorem**
>
> *If a line divides two sides of a triangle proportionally (the ratio of the segments on one side equals the ratio of the corresponding segments on the other side), then the line is parallel to the third side.*

Proof. We have $\frac{AP}{PB} = \frac{AQ}{QC}$ and we need to prove that $\overleftrightarrow{PQ} \parallel \overleftrightarrow{BC}$ (see Figure 5.4).

There is exactly one line through B parallel to line PQ. It intersects $\overrightarrow{AC}$ in some point C_1 (see Figure 5.4, and notice the assumption that $C_1 \neq C$). By the Side-Splitting Theorem, we have

$$\frac{AP}{PB} = \frac{AQ}{QC_1} \tag{5.3}$$

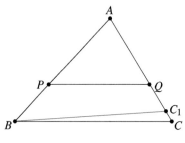

Figure 5.4

From this and the hypothesis, it follows that

$$\frac{AQ}{QC} = \frac{AQ}{QC_1}.$$

Thus $QC = QC_1$, which implies that $C = C_1$ and, therefore, $\overline{BC} \cong \overline{BC_1}$. This and the fact that line BC_1 was constructed parallel to line PQ imply that lines PQ and BC are parallel. ∎

Remark 5.3

Notice that Theorem 5.2 generalizes the midsegment theorem and Theorem 5.3 generalizes its converse.

Corollary 5.1: Corollary to the Side-Splitting Theorem

Referring to Figure 5.4 (or Figure 5.3), if $\overline{PQ} \| \overline{BC}$, then $\dfrac{AB}{AP} = \dfrac{AC}{AQ}$.

Proof. From the Side-Splitting Theorem and the reciprocal property,

$$\frac{PB}{AP} = \frac{QC}{AQ}.$$

Thus:

$$\frac{PB + AP}{AP} = \frac{QC + AQ}{AQ}, \quad \text{and} \quad \frac{AB}{AP} = \frac{AC}{AQ}.$$

∎

5.1 Ratio, Proportion, and Similar Polygons

> **Now Solve This 5.3**
>
> 1. State and prove the converse of the corollary to the Side-Splitting Theorem.
> 2. Use the approach suggested in Figure 5.5 and the Side-Splitting Theorem to prove Theorem 5.1 (i.e., that parallel projection preserves the ratio of segments).
>
>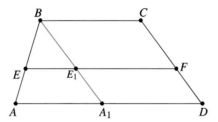
>
> Figure 5.5

Necessary and Sufficient Conditions for Triangles to Be Similar

As we already mentioned, to conclude that triangles are similar, we do not have to verify both conditions in the definition—namely, that corresponding angles are congruent and that corresponding sides are proportional. Knowing one implies the other, as we shall see in the following theorems.

> **Theorem 5.4: The AA Similarity Condition for Triangles**
>
> *If two angles of one triangle are congruent to two corresponding angles of another triangle, then the triangles are similar.*

Proof. We have $\angle A \cong \angle A_1$ and $\angle B \cong \angle B_1$ (see Figure 5.6), and we need to prove that $\triangle ABC \sim \triangle A_1 B_1 C_1$, that is,

$$\angle C \cong \angle C_1 \quad \text{and} \quad \frac{a}{a_1} = \frac{b}{b_1} = \frac{c}{c_1}.$$

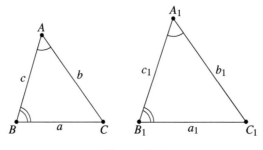

Figure 5.6

To prove that the sides are proportional, we "fit" the "smaller" triangle inside the larger triangle (as in Figure 5.7) and then use the Side-Splitting Theorem to prove part of the proportion.

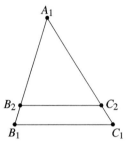

Figure 5.7

In Figure 5.7 let B_2 and C_2 be points on the sides of $\angle A_1$ such that $A_1 B_2 = AB$ and $A_1 C_2 = AC$. (Without loss of generality we assumed that $A_1 B_1 > AB$.) Then $\triangle ABC \cong \triangle A_1 B_2 C_2$ by SAS. Thus $\angle B_2 = \angle B$. Because $\angle B \cong \angle B_1$ (given), it follows by transitivity that $\angle B_1 \cong \angle B_2$. This, in turn, implies that $\overleftrightarrow{B_2 C_2} \parallel \overleftrightarrow{B_1 C_1}$ and, therefore, by the corollary to the Side-Splitting Theorem and the reciprocal property,

$$\frac{A_1 B_2}{A_1 B_1} = \frac{A_1 C_2}{A_1 C_1}.$$

Because $A_1 B_2 = AB = c$ and $A_1 C_2 = AC = b$, it follows that

$$\frac{c}{c_1} = \frac{b}{b_1}. \tag{5.4}$$

Similarly, if we fit $\triangle ABC$ inside $\triangle A_1 B_1 C_1$ such that $\angle C$ "falls" on $\angle C_1$, we get that

$$\frac{c}{c_1} = \frac{a}{a_1}. \tag{5.5}$$

From Equations (5.4) and (5.5), we get the desired proportion:

$$\frac{a}{a_1} = \frac{b}{b_1} = \frac{c}{c_1}. \tag{5.6}$$

The congruence $\angle C \cong \angle C_1$ follows from the fact that the sum of the measures of the interior angles in a triangle is 180°. ∎

> **Now Solve This 5.4**
>
> Derive the proportion stated in Equation (5.5).

Theorem 5.4 asserted that if in two triangles, corresponding angles are congruent, then the corresponding sides are proportional. The converse of this statement is stated in the following theorem:

5.1 Ratio, Proportion, and Similar Polygons

> **Theorem 5.5: The SSS Similarity Theorem**
>
> *If the corresponding sides of two triangles are proportional, then the corresponding angles are congruent and the triangles are similar.*

Proof. In Figure 5.8 we have $\frac{a}{a_1} = \frac{b}{b_1} = \frac{c}{c_1}$ and need to show that $\angle A \cong \angle A_1$, $\angle B \cong \angle B_1$, $\angle C \cong \angle C_1$.

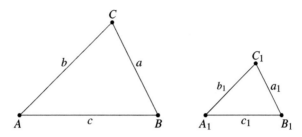

Figure 5.8

Without loss of generality we can assume that $AB > A_1B_1$; thus, let A_2 and C_2 in Figure 5.9 be points on the sides AB and BC, respectively, such that $BA_2 = B_1A_1$ and $BC_2 = B_1C_1$. (We show that $\triangle A_2BC_2 \sim \triangle ABC$ and that $\triangle A_2BC_2 \cong \triangle A_1B_1C_1$. This would imply by transitivity that $\triangle ABC \sim \triangle A_1B_1C_1$.) Because $\frac{c}{c_1} = \frac{a}{a_1}$, it follows that $\frac{BA}{BA_2} = \frac{BC}{BC_2}$ and hence using properties of proportion that $\frac{BA_2}{A_2A} = \frac{BC_2}{C_2C}$.

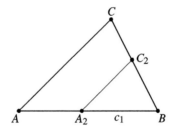

Figure 5.9

By Theorem 5.3 (the converse of the Side-Splitting Theorem), $\overline{A_2C_2} \| \overline{AC}$. Consequently, $\angle A \cong A_2$ and $\angle C \cong \angle C_2$ and, therefore, by the AA similarity condition,

$$\triangle ABC \sim \triangle A_2BC_2. \tag{5.7}$$

Remember that we want to show that $\triangle A_2BC_2 \cong \triangle A_1B_1C_1$. For that purpose, we prove that $A_2C_2 = A_1C_1 = b_1$.

From $\triangle ABC \sim \triangle A_2BC_2$ we have

$$\frac{A_2C_2}{AC} = \frac{A_2B}{AB}$$

and therefore
$$\frac{A_2C_2}{b} = \frac{c_1}{c}. \tag{5.8}$$

But the hypothesis of the theorem tells us that
$$\frac{c_1}{c} = \frac{b_1}{b}. \tag{5.9}$$

From Equations (5.8) and (5.9), we get
$$\frac{A_2C_2}{AC} = \frac{b_1}{b},$$

which implies
$$A_2C_2 = b_1.$$

Consequently, using the SSS congruency condition, we have
$$\triangle A_1B_1C_1 \cong \triangle A_2BC_2.$$

This congruence and the similarity established in Equation (5.7) imply that $\triangle ABC \sim \triangle A_1B_1C_1$. ∎

The next similarity condition for triangles is an analog of the SAS congruency condition. The proof is similar to the proof of Theorem 5.4, but a little more straightforward.

Theorem 5.6: The SAS Similarity Condition

Given two triangles, if two pairs of corresponding sides are proportional and the included angles are congruent, then the triangles are similar.

Now Solve This 5.5

Referring to Figures 5.6 and 5.7, restate and prove Theorem 5.6.

Example 5.1

In Figure 5.10, the marked angles at C and M are right angles and M is the midpoint of $\overline{AB}$. Given that $AB = c$, $BC = a$, and $AC = b$, do the following:

1. Prove that $\triangle BMN \sim \triangle BCA$.
2. Express BN and NM in terms of a, b, and c.
3. Verify that your answers in part (2) are likely correct, by checking that for the values BN and NM you get $(BN)^2 - (NM)^2 = \left(\frac{c}{2}\right)^2$.

5.1 Ratio, Proportion, and Similar Polygons

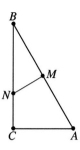

Figure 5.10

Solution

1. Because triangles ABC and BNM both have a right angle and a common angle at B, they are similar by the AA similarity condition. Because at vertices C and M the angles are right, M must correspond to C. Also, B must correspond to B and, therefore, the two other vertices N and A must correspond. Thus $\triangle BMN \sim BCA$.

2. From the similarity established in part (1), we have

$$\frac{BN}{BA} = \frac{BM}{BC} \quad \text{or} \quad \frac{BN}{c} = \frac{c/2}{a}.$$

Thus

$$BN = \frac{c^2}{2a}.$$

Similarly,

$$\frac{NM}{AC} = \frac{BM}{BC} \quad \text{or} \quad \frac{NM}{b} = \frac{c/2}{a}.$$

Thus

$$NM = \frac{bc}{2a}.$$

3. Because $\triangle ABC$ is a right triangle, $a^2 + b^2 = c^2$. Also, $\triangle BNM$ is a right triangle and hence we must have

$$(NM)^2 + (BM)^2 = (BN)^2$$

where $BM = \dfrac{c}{2}$. We verify that our answers in part 2 satisfy this equation:

$$(NM)^2 + (BM)^2 = \left(\dfrac{bc}{2a}\right)^2 + \left(\dfrac{c}{2}\right)^2$$
$$= \dfrac{b^2c^2}{4a^2} + \dfrac{a^2c^2}{4a^2}$$
$$= \dfrac{c^2(a^2+b^2)}{4a^2}$$
$$= \dfrac{c^2 \cdot c^2}{4a^2}$$
$$= \dfrac{c^4}{4a^2}$$
$$= (BN)^2.$$

Example 5.2

$ABCD$ in Figure 5.11 is a trapezoid with bases of length a and b. The diagonals of the trapezoid intersect at F. A line parallel to the bases is drawn through F. It intersects the sides of the trapezoid at points E and G, respectively. Find EF and FG in terms of a and b, and then find EG in terms of a and b.

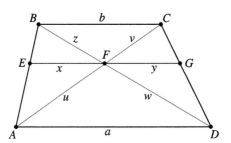

Figure 5.11

Solution. We find EF first. For convenience, we denote $EF = x$, $FG = y$, $AF = u$, $FC = v$, $BF = z$, and $FD = w$. To find x, we look for two similar triangles, one of which has a side of length x. One such pair of triangles is $\triangle AEF$ and $\triangle ABC$. From $\triangle AEF \sim \triangle ABC$ we get

$$\dfrac{EF}{BC} = \dfrac{AF}{AC} \quad \text{or} \quad \dfrac{x}{b} = \dfrac{u}{u+v}. \tag{5.10}$$

To find x in terms of a and b, it would be desirable to express u and v in terms of a and b. Notice that u, v, z and w are lengths of sides of the similar triangles $\triangle BFC$

5.1 Ratio, Proportion, and Similar Polygons

and $\triangle DFA$. From this similarity, we get

$$\frac{u}{v} = \frac{a}{b}. \tag{5.11}$$

We could now substitute $u = \frac{a}{b}v$ in Equation (5.10) or write Equation (5.10) in an equivalent form containing $\frac{u}{v}$. We will demonstrate the latter approach. We divide the numerator and the denominator of the right side of Equation (5.10) by v and then substitute $\frac{a}{b}$ for $\frac{u}{v}$:

$$\frac{x}{b} = \frac{u}{u+v} = \frac{u/v}{u/v + v/v} = \frac{a/b}{a/b+1} = \frac{a}{a+b}.$$

Hence

$$x = \frac{ab}{a+b}. \tag{5.12}$$

Analogous calculations result in $y = \frac{ab}{a+b}$ and, therefore,

$$FG = x + y = \frac{2ab}{a+b}.$$

The quantity $\frac{2ab}{a+b}$ in Example 5.2 is referred to as the **harmonic mean** of a and b. To see a connection between the harmonic mean and the arithmetic mean, let $h = \frac{2ab}{a+b}$. Then $\frac{1}{h} = \frac{1}{2}\left(\frac{1}{a} + \frac{1}{b}\right)$ and hence $\frac{1}{h}$ is the arithmetic mean of $\frac{1}{a}$ and $\frac{1}{b}$. Consequently, $\frac{1}{a}, \frac{1}{h}, \frac{1}{b}$ are in an arithmetic sequence. For example, $\frac{1}{3}, \frac{1}{4}, \frac{1}{5}$ are in a harmonic sequence because the reciprocals 3, 4, 5 are in an arithmetic sequence.

Remark 5.4

We encountered the arithmetic mean in Chapter 2 (the midsegment in a trapezoid). The harmonic mean will appear in several problems in this chapter. We will also introduce the geometric mean in this chapter. These means are used in various areas of mathematics, physics, and statistics.

Now Solve This 5.6

Example 5.2 can be solved in several different ways, one of which is suggested below.
1. Complete the following argument and derivation.
   Through F, construct a line parallel to $\overline{AB}$ intersecting the bases of the trapezoid in I and H (see Figure 5.12). If $BI = x$, then $BI = AH = x$, $IC = b - x$, and

$HD = a - x$. Since $\triangle FCI \sim \triangle FAH$ and $\triangle FBI \sim \triangle FDH$, we have

$$\frac{b-x}{x} = \frac{IF}{FH} = \frac{x}{a-x}.$$

Now solve the equation for x.

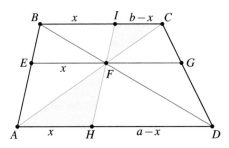

Figure 5.12

2. Solve the problem in a somewhat different way by showing that $\triangle AFD \sim \triangle CFB$ implies $\frac{AF}{CF} = \frac{a}{b}$ and $\triangle FAH \sim \triangle FCI$ implies $\frac{AF}{CF} = \frac{x}{b-x}$.
3. Which solution seems to be the simplest, the one in Example 5.2 or one of the solutions in (1) and (2) above? Why?

Example 5.3

A rhombus $EFCD$ is inscribed in $\triangle ABC$ as shown in Figure 5.13. Find the length of a side of the rhombus in terms of a, b, or c, the lengths of the sides of $\triangle ABC$.

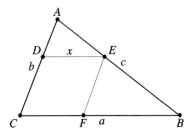

Figure 5.13

Solution. Let the length of a side of the rhombus be x. Because $\triangle DAE \sim \triangle FEB$, we have

$$\frac{DA}{EF} = \frac{DE}{FB}.$$

Because $DA = b - x$ and $FB = a - x$, the last proportion is equivalent to each of the following:

$$\frac{b-x}{x} = \frac{x}{a-x},$$
$$x^2 = (a-x)(b-x),$$
$$x^2 = ab - ax - bx + x^2, \tag{5.13}$$
$$x(a+b) = ab,$$
$$x = \frac{ab}{a+b}.$$

Reflections on Example 5.3

Notice that the expression for the length of the side of the rhombus in Example 5.3 is the same as the expression for EF in Example 5.2 (half the length of the segment obtained by the intersection of the line through the intersection of the diagonals and parallel to the bases of the trapezoid). It seems that we should be able to obtain the length of the side of the rhombus from the solution to Example 5.2. But how? Before reading on, try to answer this question on your own.

To find x in Figure 5.13 using Example 5.2, we complete the figure into a trapezoid with base $\overline{CB}$ and $\overline{AB}$ as one of the diagonals. Thus we extend $\overline{CE}$ in Figure 5.14 until it intersects the line through A, parallel to $\overline{CB}$, at G. By the result of Example 5.2,

$$x = \frac{a \cdot AG}{a + AG}. \tag{5.14}$$

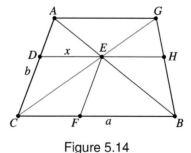

Figure 5.14

It remains now to express AG in terms of a or b. [Because we have solved the problem earlier, we know from Equation (5.13) that AG must equal b.] Because $\triangle CAG \sim \triangle CDE$, and $DE = DC$, we must have $AG = AC = b$. Thus Equation (5.14) implies that $x = \frac{ab}{a+b}$.

5.1.1 Problem Set

1. In the figure below, if ∠ABC = ∠ADE, AB = 8, BE = 14, AC = 12, CD = x, and ED = y, find x and y, if possible.

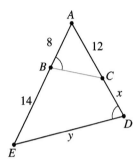

2. BD and AE are heights in △ABC. List as many pairs of similar triangles as possible. Justify your similarity statements.

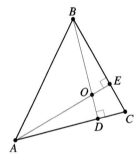

3. Let ABCD be a trapezoid with parallel bases AD and BC. The diagonals meet at O.
   a. Prove that: $\frac{AO}{CO} = \frac{OD}{BO}$.
   b. Based on the proportion in part (a) and the congruence of the vertical angles ∠BOA and ∠COD, does it follow that △BOA and △COD are similar by the SAS similarity condition (Theorem 5.6)? Explain why or why not.

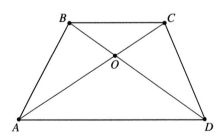

4. In the figure below, $\overline{BC} \| \overline{B_1C_1}$ and $\overline{AB} \| \overline{A_1B_1}$. Prove that
   a. △ABC ~ △$A_1B_1C_1$,

b. $\overline{AC} \| \overline{A_1C_1}$,

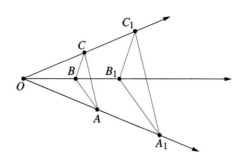

5. In the figure below, $\overline{PQ} \| \overline{BC}$.
   a. Prove that $\dfrac{PS}{SQ} = \dfrac{BT}{TC}$.
   b. If $\overline{AT}$ is a median in $\triangle ABC$, what can be concluded from the proof in part (a)? Explain.

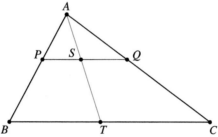

6. $ABCD$ is a trapezoid. The sides AB and CD, when extended, meet at P. The line through P and O (the intersection of the diagonals) intersects the bases of the trapezoid at M and N. Prove that M and N are the midpoints of the bases.

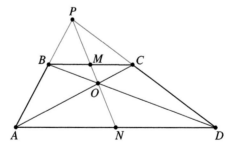

7. **a.** Use one of the conditions for similarity of triangles to prove that a midsegment of a triangle is parallel to a side of the triangle and is half the length of that side.
   b. Use similar triangles to prove that any two medians of a triangle intersect at a point that divides each median into two segments in the ratio 2 : 1.
   c. Use your result from part (b) to prove that the three medians of a triangle are concurrent.

8. $ABCD$ is a trapezoid whose diagonals intersect at O. When extended, the sides AB and CD of the trapezoid meet at E. A line parallel to the bases AD and BC has been drawn through E. That line intersects lines BD and AC at points F and G, respectively. If $AD = a$ and $BC = b$, do the following:
   a. Find FE and EG in terms of a and b.
   b. In part (a), after computing just FE, how can you argue that $EG = FE$ without performing additional computations?
   c. Choose a point P on the side AB, and draw a line parallel to the bases. That line intersects $\overline{AC}$, $\overline{BD}$, and $\overline{CD}$ at points Q, R, and S, respectively (these points are not shown in the figure). Use geometry software to conjecture a relationship between $\overline{PQ}$ and $\overline{RS}$. (Check for different trapezoids and different positions of P.)
   d. Prove your conjecture in part (c).
   e. Does a result similar to the one in parts (c) and (d) hold if P is on $\overline{FB}$? Justify your answer.

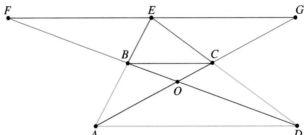

9. $ABCD$ is a rectangle. The segments DF, AC, and BE are perpendicular to line FE.
   a. Prove that $AF = AE$.
   b. If $DF = a$ and $BE = b$, find FE in terms of a and b.

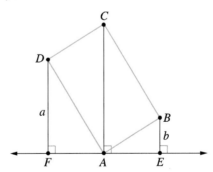

10. Suppose $\triangle ABC \sim \triangle A_1B_1C_1$ with scale factor λ; that is, $\frac{AB}{A_1B_1} = \frac{BC}{B_1C_1} = \frac{AC}{A_1C_1} = \lambda$. Prove the following:
    a. If h and h_1 are the heights to corresponding sides of the triangles, then $\frac{h}{h_1} = \lambda$.
    b. If r and r_1 are the radii of the corresponding inscribed circles, then $\frac{r}{r_1} = \lambda$.
    c. If R and R_1 are the radii of the corresponding circumscribing circles, then $\frac{R}{R_1} = \lambda$.

5.1 Ratio, Proportion, and Similar Polygons

d. If P and P_1 are the perimeters of the corresponding triangles, then $\frac{P}{P_1} = \lambda$.

11. A square $EFGH$ is inscribed in triangle ABC. Express the side of the square in terms of measurable parts of $\triangle ABC$ (such as sides or heights).

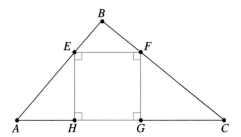

12. $MNOP$ is a square with vertices M and N on side AC of $\triangle ABC$ and vertex O on side AB. There are infinitely many such squares. (This problem will also be solved in Chapter 6 using a dilation.)

  a. Prove that the vertices P of such squares are collinear.

  b. Construct a triangle and then use your result from part (a) to construct a square inscribed in the triangle.

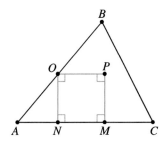

13. The right triangle in the figure below has legs of length a and b and hypotenuse of length c. Let $d_1 = a$, d_2 be the length of the perpendicular from C to the hypotenuse, and D_2 be the point where that perpendicular intersects the hypotenuse. From D_2, a perpendicular to $\overline{AC}$ is drawn; it intersects $\overline{AC}$ at D_3. Let $D_2D_3 = d_3$. We continue in this way, drawing perpendiculars from points on the hypotenuse to $\overline{AC}$, and from points on $\overline{AC}$ perpendicular to $\overline{AB}$. Denote $C = D_1$ and consider the infinite path $D_1D_2D_3D_4\ldots$ made of segments perpendicular to the hypotenuse or to $\overline{AC}$. The length of the path is the infinite sum $S = \sum_{i=1}^{\infty} d_i$. Let's explore how to find this sum using a geometrical approach.

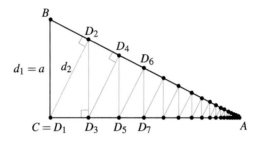

a. Prove that $\frac{d_2}{d_1} = \frac{d_3}{d_2} = \frac{d_4}{d_3}$, and express this ratio in terms of a, b, or c.

b. Explain why $\frac{d_k}{d_{k-1}}$ equals the same constant for all $k \geq 2$. Express this constant in terms of a, b, or c.

c. Prove the assertion in part (b) by mathematical induction.

d. Explain why the path associated with $\triangle CD_2A$ (that is, the path $D_2D_3D_4D_5\ldots$) is similar to the path associated with $\triangle ABC$ (that is, the path $D_1D_2D_3D_4D_5\ldots$).

e. Assume that S converges, and use your result from part (d) to show that if $\frac{d_k}{d_{k-1}} = r$ for $k \geq 2$, then

$$d_2 + d_3 + d_4 + \cdots = r(d_1 + d_2 + d_3 + \cdots),$$
$$S - d_1 = rS,$$
$$S = \frac{d_1}{1-r}.$$

14. Consider the nested equilateral triangles in the figure below, where the vertices of each triangle starting from the second are the midpoints of the sides of the previous triangle. A spiral path is created by connecting vertex A of $\triangle ABC$ with the midpoints $M_1, M_2, M_3, \ldots$ of sides in successive triangles.

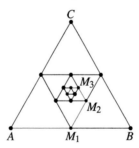

a. Prove that the lengths of the successive segments of the path constitute a geometric sequence.

b. Use the ideas from Problem 13 to find the length of the infinite path.

15. Create an infinite spiral using nested squares, rather than nested triangles as in Problem 14. Find the length of the path.

5.1 Ratio, Proportion, and Similar Polygons

16. **a.** If the ratio of the corresponding sides of similar triangles is r, what is the ratio of the corresponding areas of the triangles? Prove your answer.

    b. In $\triangle ABC$, M_1 is the midpoint of side AC. A line segment is drawn from M_1 parallel to side BC, intersecting side AB at M_2. A segment parallel to side AC is drawn from M_2, intersecting segment BM_1 at M_3. This process continues indefinitely, creating an infinite number of shaded triangles. If the area of $\triangle ABC$ is S, find the total shaded area in terms of S.

    c. If $BC = a$ and $AC = b$, find the length of the infinite path $AM_1M_2M_3M_4M_5 \ldots$ Why is the answer so simple?

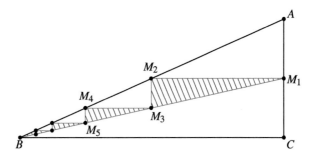

17. Construct an acute triangle ABC, and then inscribe in it a rectangle with one side on $\overline{AC}$ and the vertices of the opposite side on $\overline{AB}$ and $\overline{BC}$, respectively, such that the side of the rectangle on $\overline{AC}$ is twice as long as a side perpendicular to $\overline{AC}$. (See the figure below.) Describe your construction and prove that it is valid.

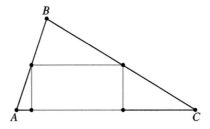

18. Given segments of length a, b, and c, the figure suggests how to construct a segment of length x such that $\frac{a}{b} = \frac{c}{x}$. Describe and justify the construction.

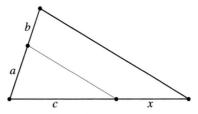

19. **a.** Define the slope of a segment, and use similar triangles to develop the equation of a line given two points on the line.

b. Prove that two lines, neither of which is vertical, are perpendicular if and only if the product of their slopes is −1.

5.2 Further Applications of the Side-Splitting Theorem and Similarity

In this section we will prove two theorems concerning the angle bisectors of interior and exterior angles of a triangle, and use these theorems to prove an unexpected characterization of a circle. We also explore similarity related to a right triangle and a circle.

> **Theorem 5.7**
>
> *The bisector of an interior angle of a triangle divides the opposite side into segments whose lengths are in the same ratio as the lengths of the other sides of the triangle.*

Proof. Referring to Figure 5.15(a), we say that $\overrightarrow{AD}$ is the angle bisector of $\angle A$. We need to show that:

$$\frac{AC}{AB} = \frac{DC}{DB}. \tag{5.15}$$

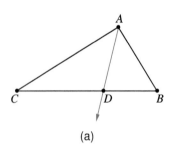
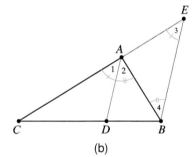

(a)   (b)

Figure 5.15

To use the Side-Splitting Theorem, we extend side CA and construct a line parallel to $\overline{AD}$ through B. That line intersects $\overleftrightarrow{CA}$ in E as shown in Figure 5.15(b). By the Side-Splitting Theorem (Theorem 5.2), we have

$$\frac{AC}{AE} = \frac{DC}{DB}. \tag{5.16}$$

This proportion will become the one we are trying to prove in Equation (5.15) if $AE = AB$. Thus we need to prove that $\triangle BAE$ is isosceles or, equivalently, that $\angle 3 \cong \angle 4$. (We are using here for the first time a different way to denote angles. Notice that

$$\angle 3 \cong \angle 1 \quad \begin{array}{l}\text{(corresponding angles formed by}\\ \text{the parallel lines } EB \text{ and } AD\text{).}\end{array} \tag{5.17}$$

5.2 Further Applications of the Side-Splitting Theorem and Similarity

We also have

$$\angle 4 \cong \angle 2 \quad \text{(alternate interior angles formed by the parallels } EB \text{ and } AD \text{ and the transversal } AB). \quad (5.18)$$

Because $\angle 1 \cong \angle 2$ (given), the relations in Equations (5.17) and (5.18) imply that $\angle 3 \cong \angle 4$. Consequently, $AE = AB$. Substituting AB for AE in Equation (5.16), we get

$$\frac{AC}{AB} = \frac{DC}{DB}.$$

∎

A Result Analogous to Theorem 5.7 for the Bisector of an Exterior Angle

A result similar to the one in Theorem 5.7 can be obtained when an exterior angle of a triangle is bisected. In Figure 5.16, our original triangle is $\triangle ABC$. $\overrightarrow{AD_1}$ is the angle bisector of the exterior angle at A that intersects line CB at D_1. To find a proportion similar to the one in Theorem 5.7, we use analogy and mimic the approach used in the proof of Theorem 5.7. In that proof, we constructed the line BE parallel to the interior angle bisector AD. For the exterior angle case in Figure 5.16, we construct through B the line parallel to the angle bisector AD_1. That line intersects $\overline{AC}$ at E_1. The proportion analogous to $\frac{AC}{AE} = \frac{DC}{DB}$ [Equation (5.16)] is

$$\frac{AC}{AE_1} = \frac{D_1C}{D_1B}, \quad (5.19)$$

which is true by the corollary to the Side-Splitting Theorem (Theorem 5.2).

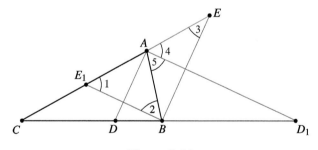

Figure 5.16

In the proof of Theorem 5.7, we showed that $AE = AB$; here we will show that $AE_1 = AB$. This result will follow if we can show that $\angle 1 \cong \angle 2$. To prove this congruence, notice that $\angle 1 \cong \angle 4$, as these corresponding angles are created by the parallel lines E_1B and AD_1, and the transversal CE. Also, $\angle 2 \cong \angle 5$, as these alternate interior angles are created by the parallel lines E_1B and AD_1 and the transversal AB. Because $\angle A_1 \cong \angle A_2$ (given), it follows that $\angle E_1 \cong \angle B_2$ and hence that $AE_1 = AB$. Substituting this for AE_1 in Equation (5.19), we get

$$\frac{AC}{AB} = \frac{D_1C}{D_1B}. \quad (5.20)$$

We have proved the following theorem.

> **Theorem 5.8**
>
> If an exterior angle of a triangle is bisected, the bisector divides the opposite sides externally into segments whose lengths are in the same ratio as the lengths of the other sides of the triangle.

Harmonic Division of a Segment

In Figures 5.15(b) and 5.16, we saw $\overline{CB}$ was divided internally and externally, respectively, by point D and extended point D_1 in the same ratio. We say that D and D_1 divide $\overline{CB}$ **harmonically**. Theorems 5.7 and 5.8 suggest that we can divide any segment harmonically in any given ratio. In Now Solve This 5.7, you are asked to accomplish the construction.

> **Now Solve This 5.7**
>
> 1. (a) Figure 5.17 suggests how to divide a given segment AB harmonically in the ratio $\frac{a}{b}$, where a and b are the lengths of given segments. Through A, construct any ray. Mark a point D such that $AD = a$ and a point C such that $DC = b$. Connect C with B. Through D, construct the line parallel to line CB. That line intersects $\overline{AB}$ at H_1. Explain why $\frac{AH_1}{BH_1} = \frac{a}{b}$.
>
>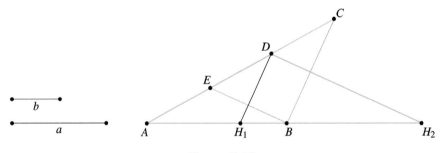
>
> Figure 5.17
>
> (b) How is point H_2 determined? Why do H_1 and H_2 divide $\overline{AB}$ harmonically?
>
> 2. (a) Segment AB can be divided harmonically in a different way by constructing a triangle with sides a, b, and $\overline{AB}$. Explain how can this be accomplished for the segment AB and the given ratio $\frac{a}{b}$.
>   (b) How can the construction be accomplished if $a < \frac{1}{2}AB$ and $b < \frac{1}{2}AB$?

The Circle of Apollonius

The Greek mathematician Apollonius (262–190 B.C.E.) found that if a point moves in such a way that its distance from one fixed point is always a constant multiple of its distance from

5.2 Further Applications of the Side-Splitting Theorem and Similarity

another fixed point, then its path is a circle or the perpendicular bisector of the segment joining the two fixed points. We will prove Apollonius's assertion for any constant using Theorems 5.7 and 5.8.

In Figure 5.18, points C and B are the two given points. Suppose P is a point such that $\frac{PC}{PB} = m$, a given constant. We have seen (from Theorems 5.7 and 5.8) that the points D and D_1, obtained by the intersection of the interior and exterior angle bisectors of $\angle A$ with line CB, are two other points on the required locus. Because $\overrightarrow{PD}$ and $\overrightarrow{PD_1}$ are angle bisectors of two supplementary angles, $\angle DPD_1$ is a right angle. Consequently, the points D, P, and D_1 are on a circle whose diameter is DD_1. We claim that this circle is the required locus. For that purpose, we prove now that every point on that circle has the property that its distance from C is m times its distance to B. Notice that the points D and D_1 which divide $\overline{CB}$ harmonically (in the ratio m) are fixed. Hence the circle in Figure 5.18 is determined by them; its center O is at the midpoint of $\overline{DD_1}$.

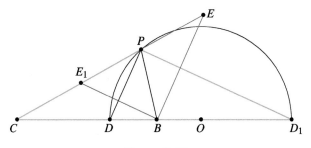

Figure 5.18

Now let P be an arbitrary point on the circle in Figure 5.18 (it is not the above point for which we assumed $\frac{PC}{PB} = m$). We need to prove that $\frac{PC}{PB} = m$. For that purpose connect P with C, D, B, and D_1. To prove that $\frac{PC}{PB} = m$, we need to show only that $\frac{PC}{PB} = \frac{DC}{DB}$. For that purpose, as in the proofs of Theorems 5.7 and 5.8, we construct $\overline{BE_1}$ parallel to $\overline{PD_1}$ and $\overline{BE}$ parallel to $\overline{PD}$. Since $\angle DPD_1$ is a right angle, it follows that $\angle E_1BE$ is also a right angle. Additionally we have

$$\frac{PC}{PE} = \frac{DC}{DB} \quad \text{(why?)}, \tag{5.21}$$

$$\frac{PC}{PE_1} = \frac{D_1C}{D_1B} \quad \text{(why?)}. \tag{5.22}$$

Since $\frac{DC}{DB} = \frac{D_1C}{D_1B} = m$, Equations (5.21) and (5.22) imply

$$PE = PE_1. \tag{5.23}$$

If we could show that $PE = PB$ (or $PE_1 = PB$), our task would be completed. For that purpose, we focus on the right triangle E_1BE. In that triangle, $\overline{BP}$ is the median to the hypotenuse [Equation (5.23)], which implies that $BP = \frac{1}{2}E_1E = PE$ (we proved this fact in Chapter 2. Thus we have proved the following theorem:

> **Theorem 5.9: The Circle of Apollonius**
>
> *The locus of all points P for which the ratio of the distances from two fixed points A and B is a constant other than 1 is the circle of diameter DD_1, where D and D_1 divide $\overline{AB}$ harmonically in the ratio of that constant.*

> **Now Solve This 5.8**
>
> Referring to Figure 5.18, let $CO = u$, $BO = v$, and $\frac{1}{2}DD_1 = R$. Prove that $uv = R^2$.

Right Triangle Similarity and Another Proof of the Pythagorean Theorem

You must have already noticed that the altitude to the hypotenuse in a right triangle dissects the triangle into two triangles, each of which is similar to the original triangle. In Figure 5.19, $\triangle ABC$ is a right triangle with legs of length a and b and hypotenuse c. As shown in the figure, the altitude $\overline{CD}$ of length h divides the hypotenuse into two segments of lengths a_1 and b_1, respectively. Notice that $\triangle ACD$ is similar to $\triangle ABC$ by the AA similarity condition since both are right triangles that share $\angle A$. Similarly, $\triangle CBD$ is similar to $\triangle ABC$. By transitivity $\triangle ACD \sim \triangle CBD$. Now, we write down the proportions resulting from the similarity conditions, paying attention to the correspondence between the set of vertices:

$$\triangle ACD \sim \triangle ABC, \qquad (5.24)$$
$$\frac{AC}{AB} = \frac{AD}{AC},$$
$$\frac{b}{c} = \frac{b_1}{b}.$$

Hence:
$$b^2 = cb_1. \qquad (5.25)$$

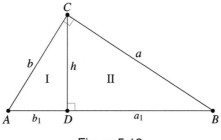

Figure 5.19

Notice that Equation (5.25) can be viewed in terms of areas: The area of a square constructed on a leg of a right triangle equals the area of a rectangle whose sides are the hypotenuse and the perpendicular projection of the leg on the hypotenuse. This is exactly the

5.2 Further Applications of the Side-Splitting Theorem and Similarity

result that was proved in Chapter 4 and used in Euclid's proof of the Pythagorean Theorem. An analogous result for side CB is

$$a^2 = ca_1. \qquad (5.26)$$

Adding Equations (5.25) and (5.26), we get

$$a^2 + b^2 = ca_1 + cb_1 = c(a_1 + b_1) = c \cdot c = c^2. \qquad \blacksquare$$

Thus we have another proof of the Pythagorean Theorem.

Next let us see what we can obtain from $\triangle ACD \sim \triangle CBD$.
We see that $\frac{CD}{BD} = \frac{AD}{CD}$ or equivalently (see Figure 5.19) that $\frac{h}{a_1} = \frac{b_1}{h}$.
Thus $h^2 = a_1 b_1$ or $h = \sqrt{a_1 b_1}$.

The number h such that $h^2 = a_1 b_1$ is called the **geometric mean** of a_1 and b_1. Just as the arithmetic mean is related to arithmetic sequences, so the geometric mean is related to geometric sequences. The nonzero numbers x, y, z are consecutive terms of a geometric sequence if and only if $\frac{y}{x} = \frac{z}{y}$; that is, if and only if $y^2 = xz$. Thus the numbers a_1, h, b_1 of Figure 5.19 are three consecutive terms in a geometric sequence.

Using the preceding terminology, we have proved the following theorem:

Theorem 5.10

In a right triangle, the height to the hypotenuse is the geometric mean of the segments into which it divides the hypotenuse.

Construction of the Geometric Mean

Given two segments of lengths a and b, we can construct their geometric mean using Theorem 5.10. We only need to construct a right triangle with hypotenuse of length $a + b$, and the altitude to the hypotenuse dividing it into segments a and b. The vertex opposite the hypotenuse will be on the circle whose diameter is the hypotenuse and also on the perpendicular to $\overline{AB}$ at D, as shown in Figure 5.20.

In Example 5.2 we have encountered the expression $\frac{2ab}{a+b}$, which was defined as harmonic mean of a and b. Another mean used in statistics is the *quadratic mean* $\sqrt{\frac{a^2+b^2}{2}}$. In the next Now Solve This, you will explore an inequality relating the four means.

Now Solve This 5.9

In Figure 5.21 $AC = a$ and $BC = b$. Use that figure to prove the following inequalities relating four means of positive real numbers a and b.

$$\frac{2ab}{a+b} \leq \sqrt{ab} \leq \frac{a+b}{2} \leq \sqrt{\frac{a^2+b^2}{2}}.$$

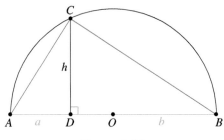

Figure 5.20

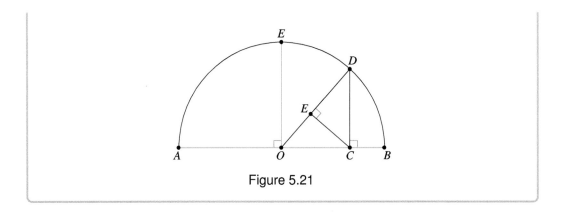

Figure 5.21

Circles and Similarity

Because all angles inscribed in the same arc (or congruent arcs) of a circle are congruent, we can create a variety of similar triangles and, therefore, proportions related to circles. In Figure 5.22(a), P is an arbitrary point in the interior of a circle with center O. We draw two arbitrary chords through P. Let A and B be the points of intersection of one of the chords with the circle, and let A_1 and B_1 be the points of intersection of the other chord with the circle.

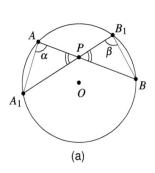

(a)

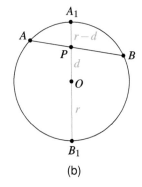

(b)

Figure 5.22

Notice that $\alpha = \beta$, as these angles are inscribed in the same arc A_1B. Also, the marked angles at P are congruent because they are vertical angles. Thus, by the AA similarity condition, we have

$$\triangle APA_1 \sim \triangle B_1PB. \tag{5.27}$$

Consequently, we get the following proportions:

$$\frac{AP}{B_1P} = \frac{PA_1}{PB} = \frac{AA_1}{B_1B}. \tag{5.28}$$

The first proportion implies that

$$AP \cdot PB = B_1P \cdot PA_1. \tag{5.29}$$

The last equation tells us that for all chords through P, the product of the distances from P to one endpoint of the chord and from P to the other endpoint of the chord is the same for all chords through P. Can we express this constant in terms of some constants clearly related to the circle?

In Figure 5.22(b), we drew a special chord: the chord through P and O (the center of the circle), which is a diameter of the circle. If the distance from P to O is d, and the radius of the circle is r, we get

$$\begin{aligned} PA_1 \cdot PB_1 &= (r-d)(r+d) \\ &= r^2 - d^2. \end{aligned} \tag{5.30}$$

We have now proved the following theorem:

Theorem 5.11

If two chords intersect in the interior of a circle, then the product of the lengths of the segments of one chord equals the product of the lengths of the segments of the other chord. Each product equals $r^2 - d^2$, where r is the radius of the circle and d is the distance from the point of intersection of the chords to the center.

Now Solve This 5.10

1. Prove Theorem 5.11.

2. You can use Theorem 5.11 to find another proof of the Pythagorean Theorem. In Figure 5.22(b), let $\overline{AB}$ be perpendicular to $\overline{A_1B_1}$. Then use Theorem 5.11 to prove that $(OP)^2 + (PB)^2 = (OB)^2$.

We now try to find a result analogous to Theorem 5.11 for a point P outside a circle. In Figure 5.23(a), two secants are drawn through P outside the circle with center O. One secant intersects the circle at A and B, and the other secant intersects it at A_1 and B_1. We can also think about this situation as having chords $\overline{AB}$ and $\overline{A_1B_1}$ that, rather than intersecting in the interior of the circle, have their extensions intersect in the exterior of the circle.

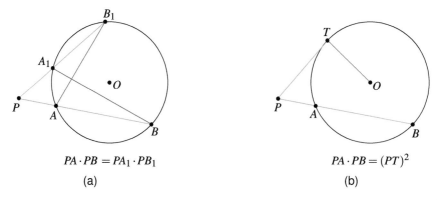

$PA \cdot PB = PA_1 \cdot PB_1$ $\quad\quad\quad\quad\quad$ $PA \cdot PB = (PT)^2$
(a) $\quad\quad\quad\quad\quad\quad\quad\quad\quad\quad\quad$ (b)

Figure 5.23

As we did in Theorem 5.11, we look at $PA \cdot PB$—that is, the product of the distance from P to one endpoint of the chord AB and the distance from P to the other endpoint. We have the following theorem:

> **Theorem 5.12**
>
> *If the lines containing two chords of a circle intersect at point P in the exterior of the circle, then the product of the distances from P to the endpoints of one chord is the same as for the second chord.*

Proof. Referring to Figure 5.23(a), we have the following information:

The lines containing chords AB and A_1B_1 of a circle intersect at P in the exterior of the circle. We need to prove that $PA \cdot PB = PA_1 \cdot PB_1$.

To obtain a proportion, we wish to identify similar triangles. For that purpose we connect some of the points in Figure 5.23(a). One option is to connect A_1 with B and A with B_1 (another option—connecting A with A_1 and B with B_1—is left for you to explore). Notice that $\angle B \cong \angle B_1$ (why?). Because $\triangle PB_1A$ and $\triangle PBA_1$ also share $\angle P$, they are similar by the AA similarity condition. We have

$$\triangle PB_1A \sim \triangle PBA_1,$$
$$\frac{PB_1}{PB} = \frac{PA}{PA_1},$$
$$PB_1 \cdot PA_1 = PB \cdot PA. \quad\blacksquare$$

Using the concept of a limit, we can obtain from Theorem 5.12 an immediate corollary concerning the length of the tangent segment from a point outside a circle. In Figure 5.23(b), the secant PB intersects the circle at points A and B. Imagine that the secant PB is rotated about P in a counterclockwise direction. The points A and B get closer to each other and

5.2 Further Applications of the Side-Splitting Theorem and Similarity

eventually coincide at T, where $\overleftrightarrow{PT}$ is tangent to the circle. In the limit, PA and PB become PT. This can also be seen in Figure 5.23(a) when A_1 and B_1 come closer to each other and in the limit the secant becomes a tangent. We get

$$PA \cdot PB = PT \cdot PT$$
$$= (PT)^2.$$

We have proved the following corollary:

Corollary 5.2

If the line *containing a chord* of a circle and a tangent intersect at point P in the exterior of the circle, then the product of the distances from P to the endpoints of the chord equals the square of the length of the tangent segment.

Now Solve This 5.11

Prove Corollary 5.2 without using Theorem 5.12 but rather by independently obtaining similar triangles.

5.2.1 Problem Set

1. $\overline{BM}$ and $\overline{NC}$ are medians in $\triangle ABC$. $\overrightarrow{BQ}$ is the angle bisector of $\angle B$, which intersects $\overline{NC}$ at P. The medians intersect at O.
   a. If $BC = 8$, $AB = 10$, and $NC = 9$, find OP.
   b. If $BC = a$, $AB = c$, and $NC = m$, express OP in terms of a, c, and m. Substitute the values of BC, AB, and NC from part (a) into the expression you obtained for OP and confirm that you get the same answer as in part (a).
   c. Without using your answer to part (b), find OP if $a = c$. Does your answer agree with the one you can obtain by substituting $a = c$ in the expression for OP you found in part (b)?

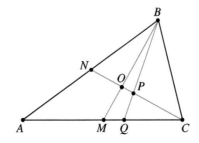

2. In the right triangle ABC, $\overrightarrow{AD}$ is the angle bisector of $\angle A$. Which is greater, BD or DC? Prove your answer.

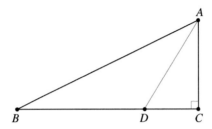

3. $\overline{BD}$ and $\overline{AC}$ are the diagonals of the convex quadrilateral $ABCD$. If $\overrightarrow{BE}$ and $\overrightarrow{BF}$ are the angle bisectors of $\angle ABD$ and $\angle DBC$, respectively, prove that $\overline{EF} \| \overline{AC}$. Is this assertion true if the quadrilateral $ABCD$ is not convex? Justify your answer.

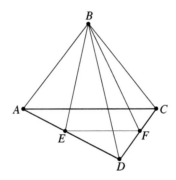

4. In $\triangle ABC$, the angle bisectors of $\angle A$ and $\angle B$ intersect at O.

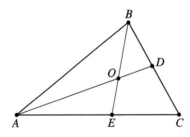

   a. If $AB = c$, $BC = a$, and $AC = b$, find $\frac{BO}{EO}$ in terms of a, b, and c.
   b. Based on your answer to part (a), and without additional computations, find $\frac{AO}{OD}$. Justify your answer.

5. A, B, and C are points on the circle with center O. The diameter GF is perpendicular to AB. Segments CF and AB intersect at D, while lines GC and AB intersect at E.

Prove that D and E divide $\overline{AB}$ harmonically.

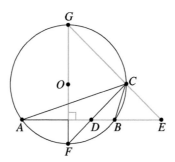

6. $ABCD$ is a trapezoid with two right angles at A and B. The diagonals intersect at P. The line through P parallel to the bases intersects the sides AB and CD at E and F, respectively. Prove that $\overrightarrow{EF}$ is the angle bisector of $\angle CED$.

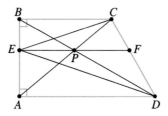

7. In $\triangle ABC$, the altitude CH and the median CM trisect $\angle C$ (divide it into three congruent angles). Find $\frac{AB}{AH}$ and $m(\angle ACB)$.

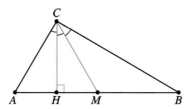

8. M, B, and T are on the circle with center O. A tangent is drawn through T that intersects the line MB at A. Answer the following:

   a. If $AM = MB = a$, find AT in terms of a.

   b. Let A be a point in the exterior of a circle. Construct a secant through A that intersects the circle in two points M and B such that M is the midpoint of $\overline{AB}$. Investigate the construction, describe the steps of the construction, and prove that the construction is valid.

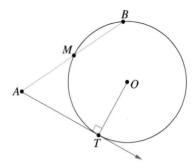

9. An isosceles triangle ABC is inscribed in a circle. $\overline{BD}$ is a chord that intersects $\overline{AC}$ at E. If $AB = BC = x$, $BE = a$, and $BD = b$, find x in terms of a and b.

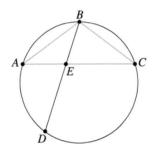

10. Right $\triangle ABC$ is inscribed in the circle with center O. $\overline{CD}$ is the altitude to the hypotenuse AB, which is also a diameter of the circle O. $\overline{BE}$ is the tangent segment to the smaller circle, whose diameter is $\overline{AD}$. Prove that $BC = BE$.

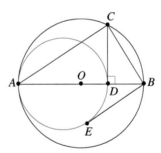

11. In the figure, $\angle BAE$ is a right angle, $\overline{AC}$ is a diameter, $AB = a$, and $AE = b$. Find BD in terms of a and b.

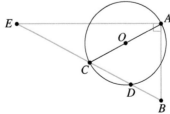

5.2 Further Applications of the Side-Splitting Theorem and Similarity

12. $\triangle ABC$ with $AB = BC$ is inscribed in a circle. Point D bisects $\overset{\frown}{BC}$. Lines BD and AC intersect at E.

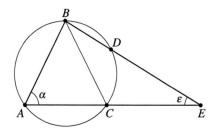

  a. Find a relationship between a and ε.
  b. If $BE = a$ and $BD = b$, express BC in terms of a and b.

13. $\triangle ABC$ is inscribed in a circle with center O. The tangent t to the circle is constructed at B. From the vertices A and C, perpendiculars to t intersect t at P and Q, respectively. If $AP = a$ and $CQ = b$, express BD in terms of a and b.

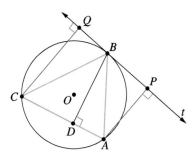

14. T is on one side of $\angle A$, while B and C are on the other side. If $(AT)^2 = AB \cdot AC$, prove that $\overleftrightarrow{AT}$ is tangent to the circle through B, C, and T.

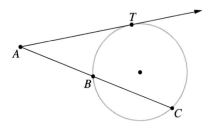

15. $\triangle ABC$ is inscribed in a circle. If the sides of the triangle have lengths a, b, and c, and if the radius of the circle is R, prove that the area of the triangle is $\frac{abc}{4R}$.

16. Two circles intersect in points A and B. If C is a point at which the tangent segments CD and CE are congruent, prove that A, B, and C are collinear.

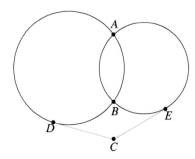

17. Two tangent segments CA and CB are drawn from C outside the circle O. From an arbitrary point P on $\overparen{AB}$, perpendiculars to the sides of $\triangle ABC$ are constructed. These perpendiculars intersect the sides of the triangle at Q, S, and T as shown in the figure. Prove that $(PT)^2 = PS \cdot PQ$.

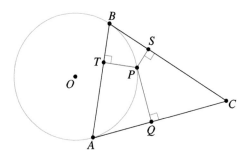

18. The figure below shows a semicircle with diameter AC and two other semicircles with diameters AB and BC (B is on $\overline{AC}$). A common tangent to the smaller semicircles intersects one at X and the other at T. Also, $\overleftrightarrow{PB}$ is a common tangent to the smaller semicircles through their common point B.

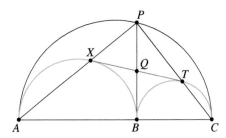

a. Prove that $XT = PB$.
b. Prove that points A, X, and P are collinear, as are points P, T, and C.
c. Parts (a) and (b) suggest a simple way to construct a common external tangent to two tangent circles. Explain how.

d. Use the ideas in part (c) to construct a common tangent to two circles intersecting in two points. Describe your construction and justify it.

19. A square with side d is inscribed in a circle. A smaller square is then inscribed in a slice of the circle (bounded by an arc and a chord), as shown in the figure below.

  a. Find the length of the side of the smaller square in terms of d.

  b. Use your answer to part (a) to construct a circle and the two squares.

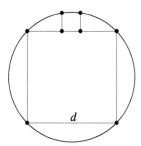

20. Construct a semicircle and two squares inscribed in the semicircle as shown. Investigate, construct, and prove that your construction is valid.

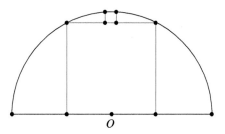

21. a. Line ℓ interests the semicircle with center O in point U anl L. From the endpoints M and D of the diameter, perpendiculars to ℓ intersect ℓ at H and K. Prove that $HU = LK$.

  b. Use part (a) to prove that $(HU)^2 + (UK)^2 = (HL)^2 + (LK)^2$.

  c. Part (b) was on the final exam of an honors high school geometry exam in Lexington, MA, in 2012. Part (a) was not on the exam. Can you prove part (b) without using (a)?

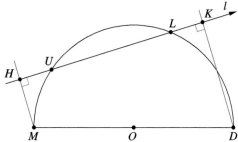

5.3 Areas of Similar Figures

The size of a rectangular television screen is commonly given as the length of the diagonal of the screen. Assuming that in two television screens the rectangular shapes are similar, how many times larger is the viewing area of a 75-inch screen than the viewing area of a 60-inch screen? To answer this question, we need to solve the following problem.

> **Problem 5.1**
>
> Given two similar triangles with similarity ratio r, what is the ratio of their areas?

Solution

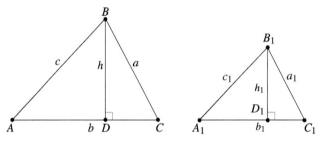

Figure 5.24

In Figure 5.24, $\triangle ABC \sim \triangle A_1 B_1 C_1$, where

$$\frac{a}{a_1} = \frac{b}{b_1} = \frac{c}{c_1} = r.$$

The ratio of the areas of the triangles can be computed in the following way. If h and h_1 are the heights corresponding to the sides AC and $A_1 C_1$, respectively, then

$$\frac{\text{Area}(\triangle ABC)}{\text{Area}(\triangle A_1 B_1 C_1)} = \frac{bh/2}{b_1 h_1/2} = \frac{b}{b_1} \cdot \frac{h}{h_1}. \quad (5.31)$$

To find how the ratio $\frac{h}{h_1}$ relates to the given ratio r, we considered triangles that have those heights as sides. One such pair is $\triangle ABD$ and $\triangle A_1 B_1 D_1$. These triangles are similar by AA (why?) and therefore

$$\frac{h}{h_1} = \frac{c}{c_1}. \quad (5.32)$$

Because $\frac{c}{c_1} = r$, we see that $\frac{h}{h_1} = r$. Substituting this last proportion in Equation (5.31), we get

$$\frac{\text{Area}(\triangle ABC)}{\text{Area}(\triangle A_1 B_1 C_1)} = \frac{b}{b_1} \cdot \frac{h}{h_1} = r \cdot r = r^2. \quad (5.33)$$

5.3 Areas of Similar Figures

We have proved the following theorem:

Theorem 5.13

The ratio of areas of similar triangles equals the square of the scale factor, i.e., the square of the ratio of corresponding sides.

Now Solve This 5.12

Is it true that the viewing area of a 75-inch television screen is about one and a half times the viewing area of a 60-inch television screen? Justify your answer.

Example 5.4

In Figure 5.25 M is any point on side AB of $\triangle ABC$. $\overline{MQ}$ and $\overline{MP}$ are parallel to the sides AC and BC, respectively. If the areas of $\triangle MBQ$ and $\triangle APM$ are S_1 and S_2, respectively, find the area of $\triangle ABC$ in terms of S_1 and S_2.

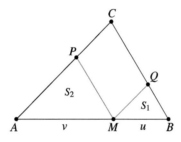

Figure 5.25

Solution. Because $\triangle APM \sim \triangle MQB$ (why?), the ratio of their areas equals the square of the ratio of corresponding sides (Theorem 5.13). Using the notation in Figure 5.25, we have

$$\frac{S_1}{S_2} = \left(\frac{u}{v}\right)^2. \tag{5.34}$$

Let S denote the area of $\triangle ABC$. Then, because $\triangle ABC \sim \triangle MBQ$, substituting for $\frac{v}{u}$ from Equation (5.34) we get

$$\frac{S}{S_1} = \left(\frac{AB}{MB}\right)^2 = \left(\frac{u+v}{u}\right)^2 = \left(1 + \frac{v}{u}\right)^2 = \left(1 + \sqrt{\frac{S_2}{S_1}}\right)^2.$$

Consequently,

$$S = S_1 \left(1 + \frac{S_2}{S_1} + 2\sqrt{\frac{S_2}{S_1}}\right) = S_1 + S_2 + 2\sqrt{S_1 S_1}.$$

Therefore, $S = S_1 + S_2 + 2\sqrt{S_1 S_2} = (\sqrt{S_1} + \sqrt{S_2})^2$ or, in the more attractive form, $\sqrt{S} = \sqrt{S_1} + \sqrt{S_2}$.

Notice that our answer implies that the area of the parallelogram $PMQC$ is $2\sqrt{S_1 S_2}$ or twice the geometric mean of S_1 and S_2.

Example 5.5

Given $\triangle ABC$, construct a line parallel to one of the sides of the triangle that divides the triangle into two regions: a triangle and a trapezoid of equal areas.
Solution. Figure 5.26 shows $\overline{DE}$, which halves the area of $\triangle ABC$. We need to find the location of D.

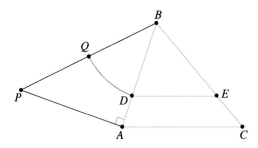

Figure 5.26

We need the area of $\triangle DEB$ to be half of the area of $\triangle ABC$. Because $\triangle DEB \sim \triangle ACB$, the ratio of the areas of the triangles is the square of the ratio of corresponding sides. Thus

$$\left(\frac{DB}{AB}\right)^2 = \frac{1}{2}, \quad \frac{DB}{AB} = \sqrt{\frac{1}{2}} = \frac{\sqrt{2}}{2}, \quad DB = \frac{AB\sqrt{2}}{2}. \tag{5.35}$$

We know that $AB\sqrt{2}$ is the length of the hypotenuse of an isosceles right triangle whose legs have length AB. Therefore in Figure 5.26, we construct a right triangle PAB such that $PA = AB$. Applying the Pythagorean Theorem, we get $PB = AB \cdot \sqrt{2}$. To obtain a segment congruent to $\overline{DB}$, Equation (5.35) suggests finding the midpoint Q of $\overline{PB}$. Then the arc with center at B and radius BQ intersects $\overline{AB}$ at the desired point D. A line through D parallel to $\overline{AC}$ intersects $\overline{BC}$ at E. Line DE is the required line.

We can now generalize Theorem 5.13 to the ratios of areas of similar polygons. Because similar polygons can be divided into corresponding similar triangles, it seems that the ratio of their areas is still the square of the ratio of corresponding sides. This is, in fact, true for both convex and nonconvex similar polygons, but we will prove the relevant theorem for convex polygons only.

5.3 Areas of Similar Figures

> **Theorem 5.14**
>
> *The ratio of areas of similar polygons equals the square of the ratio of corresponding sides.*

We prove the theorem only for similar convex pentagons. A proof for convex n-gons follows analogously.

Proof for Convex Pentagons. To use Theorem 5.13, we try to divide the similar pentagons in Figure 5.27 into similar triangles. One way to accomplish this is to connect A with D and C, and A' with D' and C'. Notice that ΔI and $\Delta I'$ are similar by the SAS similarity condition because

$$\frac{AB}{A'B'} = \frac{BC}{B'C'} \quad \text{and} \quad \angle B \cong \angle B'.$$

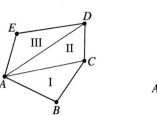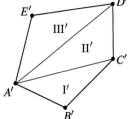

Figure 5.27

In the same way, ΔIII and $\Delta III'$ are similar. Next, ΔII and $\Delta II'$ are similar by the AA similarity condition because $\angle ACD \cong \angle A'C'D'$ and $\angle ADC \cong \angle A'D'C'$ (this follows from the just-proved fact that triangles I and I' are similar, triangles III and III' are similar, and the angles at D and C of the first pentagon correspond to the angles at D' and C' of the second pentagon). If we denote the areas of triangles I, II, and III by S_1, S_2, and S_3, respectively, and the areas of triangles I', II', and III' by $S_{1'}$, $S_{2'}$, and $S_{3'}$, respectively, then

$$\frac{S_1}{S_{1'}} = \left(\frac{BC}{B'C'}\right)^2 = \lambda^2,$$

$$\frac{S_2}{S_{2'}} = \left(\frac{DC}{D'C'}\right)^2 = \lambda^2,$$

$$\frac{S_3}{S_{3'}} = \left(\frac{DE}{D'E'}\right)^2 = \lambda^2.$$

From the ratio of the sum of numerators to the sum of denominators property introduced in Section 5.1, we have

$$\frac{S_1 + S_2 + S_3}{S_{1'} + S_{2'} + S_{3'}} = \lambda^2$$

and thus

$$\frac{S}{S'} = \lambda^2,$$

where S and S' are the areas of $AEDCB$ and $A'E'D'C'B'$, respectively.

> **Now Solve This 5.13**
>
> Construct a convex quadrilateral and then a quadrilateral similar to it whose area is three fourths of the area of the original quadrilateral. Investigate the construction, describe the construction, and prove that it is valid.

A Generalization of the Pythagorean Theorem

The Pythagorean Theorem states that the area of the square on the hypotenuse of a right triangle equals the sum of the areas of the squares on the legs. We might wonder if the assertion about the sum of the areas will hold for other figures constructed on the sides of a right triangle. In Figure 5.28, notice that we can divide each square in half by its midsegment. The resulting rectangles marked I, II, and III are figures constructed on the sides of $\triangle ABC$. Because each is half of the corresponding square, the area of the rectangle on the hypotenuse equals the sum of the areas of the rectangles on the sides of the triangle. We could now use the diagonals to divide each rectangle into two congruent triangles. The shaded triangles again satisfy the additive area property.

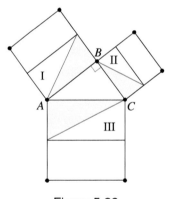

Figure 5.28

What do the squares, the rectangles, and the triangles in Figure 5.28 have in common? The three squares are similar, and so are the three rectangles and three triangles, which suggests the following theorem:

> **Theorem 5.15**
>
> *If three similar polygons are constructed on the three sides of a right triangle, then the area of the polygon on the hypotenuse equals the sum of the areas of the polygons on the legs.*

5.3 Areas of Similar Figures

Proof. We denote the areas of the polygons marked I, II, and III in Figure 5.29 by S_1, S_2, and S_3, respectively. From Theorem 5.14, we have

$$\frac{S_1}{S_3} = \left(\frac{b}{c}\right)^2,$$

$$\frac{S_2}{S_3} = \left(\frac{a}{c}\right)^2.$$

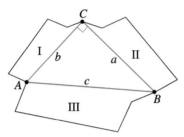

Figure 5.29

To obtain $S_1 + S_2$, we add the equations above:

$$\frac{S_1 + S_2}{S_3} = \frac{a^2 + b^2}{c^2}. \tag{5.36}$$

Because $\triangle ABC$ is a right triangle, $a^2 + b^2 = c^2$ and Equation (5.36) becomes

$$\frac{S_1 + S_2}{S_3} = \frac{c^2}{c^2} = 1,$$

$$S_1 + S_2 = S_3. \qquad \blacksquare$$

Note that the Pythagorean Theorem (where the similar polygons are squares) is a special case of Theorem 5.15.

Looking Back (at the Proof of Theorem 5.15)

Notice that up to and including Eq. (5.36), we have not used the Pythagorean Theorem. In fact, we could prove the Pythagorean Theorem from Equation (5.36) if we knew that for some similar polygons their areas are additive—that is, they satisfy the equation $S_1 + S_2 = S_3$. Such polygons are the shaded triangles in Figure 5.30. We saw earlier that the altitude to the hypotenuse ($\overline{CD}$ in Figure 5.30) divides a right triangle into two similar triangles, each of which is similar to the original triangle. These triangles satisfy the additive area property because the area of $\triangle ABC$ equals the sum of the areas of $\triangle ACD$ and $\triangle CDB$. To better see that these triangles are constructed on the sides of $\triangle ABC$, we reflect each in its hypotenuse and obtain the shaded similar triangles in Figure 5.30.

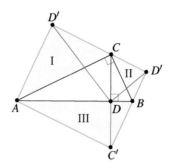

Figure 5.30

Because the shaded triangles I, II and III are congruent to $\triangle ACD$, $\triangle CDB$, and $\triangle ABC$, respectively, their areas satisfy the additive property—that is, $S_1 + S_2 = S_3$. Dividing both sides of this equation by S_3 and substituting this value in Equation (5.36), we get

$$1 = \frac{a^2 + b^2}{c^2},$$
$$c^2 = a^2 + b^2.$$

Thus we have yet another proof of the Pythagorean Theorem.

5.3.1 Problem Set

1. In the interior of $\triangle ABC$, lines parallel to the sides of the triangle are drawn through a point P. They divide the triangle into six parts, three of which are triangles. If these triangles have areas S_1, S_2, and S_3, respectively, find the area of $\triangle ABC$ in terms of S_1, S_2, and S_3.

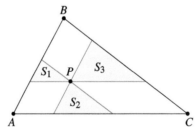

2. In $\triangle ABC$, $AB = BC$ and the angle bisector of $\angle A$ intersects $\overline{BC}$ at D. If the areas of $\triangle ABD$ and $\triangle ADC$ are S_1 and S_2, respectively, find AC in terms of S_1 and S_2. (*Hint*: Write three equations with three unknowns and solve for AC.)

3. Given a parallelogram, construct two lines parallel to one of the diagonals dividing the parallelogram into three parts of equal area.

4. Given an arbitrary triangle (construct your own scalene triangle), divide it into n parts of equal area using lines parallel to one of the sides. Solve the problem for $n = 3, 4$, and 5.

5. In a regular hexagon of side a, Kepri inscribed a second hexagon by joining the midpoints of the sides of the first hexagon. In a like manner, she inscribed a third hexagon in the second, and so on indefinitely.

   a. What is the ratio of the area of the first hexagon to the area of the sixth hexagon?

   b. Prove that the areas of the hexagons constitute a geometric sequence.

6. Given an arbitrary triangle ABC and a line n, construct a line parallel to n that will divide $\triangle ABC$ into two parts of equal area. (Investigate the construction, describe your construction, and prove that it is valid.)

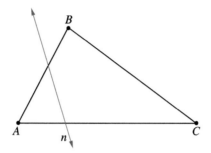

7. a. Prove that if a and b are sides of a parallelogram whose diagonals are d_1 and d_2, then $2(a^2 + b^2) = d_1^2 + d_2^2$.

   b. Explain why the result in part (a) can be regarded as a generalization of the Pythagorean Theorem.

5.4 The Golden Ratio and the Construction of a Regular Pentagon

In the world of art and architecture, not all rectangles are treated equally. One in particular is considered most pleasing to the eye. The facade of the Parthenon in Greece (built in 438 B.C.E.) has the shape of a golden rectangle. A simple way to define such a rectangle is the following.

> **Definition of a Golden Rectangle**
>
> A golden rectangle is a rectangle with the property that if it is dissected into two pieces—a square and a rectangle—then that rectangle is similar to the original rectangle.

The ratio of the longer side to the shorter side of a golden rectangle is the **golden ratio**, often denoted by the Greek letter ϕ (phi). The golden ratio is also referred to as the **divine proportion**. In the following problem, we calculate the value of ϕ.

> **Problem 5.2**
>
> Find the numerical value of ϕ.

Solution. Figure 5.31 shows a golden rectangle $ABCD$ dissected into the square $ABEF$ and the rectangle $ECDF$. The rectangles are similar if and only if the ratio of two adjacent sides in one rectangle equals the ratio of two corresponding adjacent sides in the other rectangle. Consider the ratio $\frac{a}{b}$ (see Figure 5.31), which we want to compute.

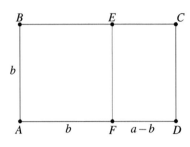

Figure 5.31

The possible ratios of adjacent sides in rectangle $ECDF$ are $\frac{a-b}{b}$ and $\frac{b}{a-b}$. Which one is equal to $\frac{a}{b}$? Notice that if $\frac{a}{b} = \frac{a-b}{b}$, then $a = a - b$, which is impossible (visually $\overline{EF}$ is the longer side of rectangle $ECDF$ and therefore is should correspond to $\overline{AD}$). Thus

$$\frac{a}{b} = \frac{b}{a-b}. \tag{5.37}$$

One way to find $\frac{a}{b}$ (which we decided earlier to call ϕ) is to divide the numerator and the denominator of the right side of Equation (5.37) by b. We get

$$\frac{b}{a-b} = \frac{\frac{b}{b}}{\frac{a-b}{b}} = \frac{1}{\frac{a}{b} - 1}.$$

Substituting this in Equation (5.37), we have

$$\frac{a}{b} = \frac{1}{\frac{a}{b} - 1}, \qquad \phi = \frac{1}{\phi - 1}, \qquad \phi^2 - \phi - 1 = 0. \tag{5.38}$$

Using the quadratic formula, we obtain $\phi = \frac{1 \pm \sqrt{5}}{2}$. Since ϕ is positive, we conclude that $\phi = \frac{1+\sqrt{5}}{2}$.

Remark 5.5

Notice that in Problem 5.2 for any given b (see Figure 5.31), we can find a segment of length a such that $\frac{a}{b}$ is the golden ratio.

5.4 The Golden Ratio and the Construction of a Regular Pentagon   225

Construction 5.4.1 Construction of a Golden Rectangle

Given the shorter side of a golden rectangle, construct the rectangle.

Solution. Given a segment of length b, we need to construct a segment of length a such that $\frac{a}{b} = \frac{1+\sqrt{5}}{2}$ or $a = \frac{b}{2} + \frac{b\sqrt{5}}{2}$. Clearly, constructing $\frac{b}{2}$ poses no difficulty. To construct a segment of length $x = \frac{b\sqrt{5}}{2}$, we try to find a right triangle whose hypotenuse or side has length x. To use the Pythagorean Theorem we wish to write $\frac{5}{4}b^2$ as a sum or difference of two squares. Because $\frac{5}{4} = 1 + \frac{1}{4}$ we have

$$x^2 = \frac{5}{4}b^2 = b^2 + \left(\frac{b}{2}\right)^2.$$

Thus x is the hypotenuse of a right triangle with legs b and $\frac{b}{2}$.

In Figure 5.32, $\overline{AB}$ has length b, $AE = \frac{b}{2}$, and $\overline{AE} \perp \overline{AB}$. Hence $BE = \frac{b\sqrt{5}}{2}$. We construct $\overline{ED} \cong \overline{BE}$ and, therefore, $AD = \frac{b}{2} + \frac{b\sqrt{5}}{2}$. All that is left is to construct the point C—the intersection of the perpendiculars at point B and D as shown in Figure 5.32. The rectangle $ABCD$ is the required rectangle.

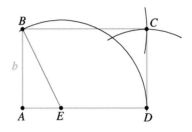

Figure 5.32

Notice that in the above construction we did not distinguish among the investigation, construction, and proof of the construction. These parts are embedded in the solution.

Now Solve This 5.14

1. Construct a segment AB of any length. Next construct the point C on the segment so that $\frac{AC}{BC} = \phi$. Write your solution in three stages: investigation, construction, proof of the construction.
2. Prove that $\frac{AB}{AC} = \phi$.
3. Using your results from parts (1) and (2), state an alternative definition of ϕ.

Properties and Construction of a Regular Pentagon

Figure 5.33 shows a regular pentagon (all sides congruent and the five interior angles congruent) and two diagonals. Each of the exterior angles of the pentagon is $\frac{360°}{5}$ or $72°$.

Hence each interior angle is 180° − 72° or 108°. Let each side be 1 unit long. If we could express the length of a diagonal using rational numbers and radicals of rational numbers, we would be able to construct $\triangle ABE$—and hence the entire pentagon.

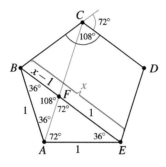

Figure 5.33

The diagonals BE and AC intersect at F. Notice that triangles ABC and BAE are congruent by SAS. To compute the length of a diagonal, we search for similar triangles and, therefore, for congruent angles. Because each of the interior angles of the pentagon is 108°, the base angles of the isosceles triangles ABC and BAE are 36° each, as shown in Figure 5.33. Also, $\angle AFE$ is an exterior angle of $\triangle BFA$; hence it measures 36° + 36° or 72°. $\angle FAE$ is 108° − 36° or 72°; consequently, $\triangle FAE$ is isosceles, which implies $FE = 1$. If $BE = x$, then $BF = x - 1$. This gives

$$\triangle ABF \sim \triangle EBA, \qquad \frac{AB}{EB} = \frac{BF}{BA}, \qquad \frac{1}{x} = \frac{x-1}{1}, \qquad \text{or equivalently,}$$
$$x^2 - x = 1,$$
$$x^2 - x - 1 = 0.$$

Solving for x, we get $x = \frac{1+\sqrt{5}}{2} = \phi$. Thus we have proved the following property:

Property of a Regular Pentagon

The ratio of the length of a diagonal of a regular pentagon to the length of its side is the golden ratio.

Remark 5.6

If the pentagon's side is a units long and a diagonal is b units long, we still have $\frac{b}{a} = \phi = \frac{1+\sqrt{5}}{2}$. Why?

Construction 5.4.2 Construction of a Regular Pentagon with a Given Side

The preceding property and remark can be used to construct a regular pentagon with a given side. To accomplish the construction, we need to construct $\triangle ABE$ as shown in Figure 5.33. We have seen that given any side of length a, the length of a diagonal is $\left(\frac{1+\sqrt{5}}{2}\right)a$. Using the construction outlined in Construction 5.4.1, we know how to construct the diagonal and, consequently, $\triangle ABE$ in Figure 5.33 by the SSS condition. To complete the construction of the pentagon, we construct vertex C (see Figure 5.33) by constructing $\triangle ABC$ congruent to $\triangle BAE$ (or by constructing $\angle B$ congruent to $\angle A$ and $\overline{BC}$ congruent to $\overline{AB}$), and similarly the rest of the vertices.

Alternative Approach for Constructing a Regular Pentagon

An alternative construction in which we find the center O of the circle that circumscribes the pentagon is shown in Figure 5.34. The main steps of the construction are as follows:

1. Given a side a of the pentagon, construct an isosceles triangle ABE with $AB = AE = a$ and $BE = a\left(\frac{1+\sqrt{5}}{2}\right)$.

2. Find the center O of the circle that circumscribes $\triangle ABE$.

3. Mark points C and D on the circle such that $BC = CD = a$.

4. Connect the points A, B, C, D, and E to obtain the required regular pentagon.

Notice that the circle with center O is not necessary for the construction. It is constructed in Figure 5.34 as a check that the construction is accurate.

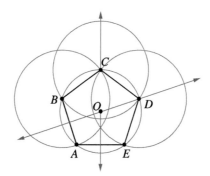

Figure 5.34

Now Solve This 5.15

1. Given a circle, construct a regular pentagon inscribed in the circle using the following approach. Describe the steps in each construction and prove that the resulting construction is valid. (Use the construction of a regular pentagon with a given side and the concept of similarity.)

2. Investigate the following construction of a regular decagon (10-sided polygon) inscribed in a given circle. Figure 5.35 shows one of the 10 congruent triangles that are created by connecting the vertices of the decagon with the center of the circle. $\angle BAO$ has been bisected, and the intersection of the bisector with $\overline{BO}$ is C. Express x, the length of the side of the regular decagon, in terms of R.

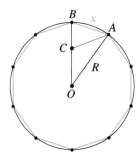

Figure 5.35

3. How can you construct a regular pentagon using the construction in part (1) above?

Construction of Other Regular Polygons

The construction of a regular n-gon is equivalent to subdividing a circle into n equal parts, which in turn is equivalent to constructing an angle of $\frac{360°}{n}$. (Why?) The constructions of an equilateral triangle, a square, and a regular hexagon are quite simple; these constructions were investigated in Chapter 2 and will be reviewed in the problem set at the end of this section. In this section, we have seen how to construct a regular pentagon and decagon. These constructions were known to Euclid. In Book IV of his *Elements*, Euclid discusses the construction of regular n-gons for $n = 3, 4, 5, 6$, and 15. By successive bisections of arcs or angles, the Greeks knew how to construct regular polygons of 2^n, $3 \cdot 2^n$, $5 \cdot 2^n$, and $15 \cdot 2^n$ sides. But despite their many attempts, no one knew how to construct other regular polygons.

In 1796, 19-year-old Gauss stated that *a regular n-gon ($n \geq 3$) can be constructed by straightedge and compass if and only if n is a prime number of the form $F(n) = 2^{2^n} + 1$, i.e., a Fermat prime or the product of distinct Fermat primes and powers of 2*. For $n = 0, 1, 2, 3, 4$, $F(n) = 3, 5, 17, 257, 65537$ are all prime. No other Fermat primes have been found.

Although Gauss proved that this condition is sufficient, he gave no explicit proof that it is necessary. Four decades later, the French mathematician Pierre Wantzel proved that the condition is necessary. These proofs are beyond the scope of this text. Gauss's proof can be found in many number theory texts (e.g., Ore, Dummit, and Foote, pp. 581–583).

> **Historical Note:**
> **Carl Friedrich Gauss** (1777–1855), a German mathematician, astronomer, and physicist, made major contributions to geometry and other branches of mathematics as well as to statistics, electricity, magnetism, and gravitation. He completed his doctoral thesis when he was only 22. At age 19, he proved the result mentioned previously about constructability of regular polygons. While a student, Gauss also showed how to construct a regular 17-gon with only a compass and straightedge; he was so proud of this discovery that he asked that a regular 17-gon be engraved on his tombstone.
>
> Many mathematical and physical terms are named after Gauss, including Gaussian curvature, Gaussian distribution (commonly called normal distribution), Gaussian domain, Gaussian elimination, Gaussian field, Gaussian function, Gaussian integer, Gaussian plane, Gauss–Jordan elimination, Gauss's lemma, Gauss–Seidel iteration, Gauss's test, and Gauss's fundamental theorem of electrostatics. When Napoleon invaded Germany, he decided to spare Göttingen—the city where Gauss lived (and was a professor at the university there)—commenting that "the foremost mathematician of all times lives there."
>
> In 1837, **Pierre Wantzel** (1814–1848), a French mathematician, became the first person to prove the impossibility of trisecting any given angle by means of only a straightedge and compass. He also proved the impossibility of several other constructions. At age 15, Wantzel gave a proof of a method for finding square roots that was widely used but previously unproved. In 1832, he placed first in the entrance examinations to both Ecole Polytechnique and Ecole Normale (two of the most prestigious institutions of higher learning in France)—something that no one else had ever achieved. His friend Saint-Yenant wrote the following about Wantzel:
>
>> "One could reproach him for having been too rebellious against those counseling prudence. He usually worked during the evening, not going to bed until late in the night, then reading, and got few hours of agitated sleep, alternatively abusing coffee and opium, taking his meals, until his marriage, at odd and irregular hours."

5.4.1 Problem Set

All constructions should include an investigation, a description of the construction, and a proof that the construction is valid, i.e., that it satisfies the required conditions.

1. Prove that the ratio of the length of the longer side to the length of the shorter side in a triangle with angles 36°, 72°, and 12° is the golden ratio. (*Hint*: Bisect one of the 72° angles or consider Figure 5.35.)

2. **a.** Given a segment AB, construct the point P on the segment that divides it into two segments such that the shorter is to the longer as the longer is to the whole. (You need P so that $\frac{PB}{AP} = \frac{AP}{AB}$.)

b. Use your answer in part (a) to give an alternative definition of the golden ratio (see Now Solve This 5.14).

3. If ϕ is the golden ratio, show that $\frac{1}{\phi} = \phi - 1$.

4. An equilateral triangle ABC is inscribed in a circle. Points M and N are midpoints of sides AC and AB, respectively, and line MN meets the circle at P and Q. Prove that $\frac{MN}{PQ}$ is the golden ratio.

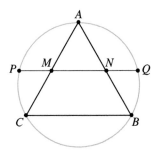

5. $ABCDE$ is a regular pentagon. Sides CB and EA have been extended to meet at P. If each side of the pentagon is a units long, find BP in terms of a.

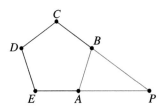

6. The following is an alternative construction of a regular pentagon inscribed in a circle. Use the step-by-step instructions along with the figure to prove that the construction is valid.

   a. Choose any point A on the circle and connect it to the center O.

   b. Through O, construct the perpendicular line to $\overline{OA}$. Let P be one of the points where the perpendicular line intersects the circle.

   c. Find the midpoint M of $\overline{OA}$. Connect M with P.

   d. Construct the angle bisector of $\angle OMP$, and let N be the point where the angle bisector meets $\overline{OP}$.

5.4 The Golden Ratio and the Construction of a Regular Pentagon 231

e. At N, construct the perpendicular to $\overline{OP}$ intersecting the circle at B as shown. The segment BP is a side of the pentagon.

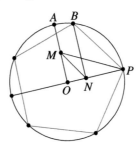

7. Because we know how to construct an equilateral triangle and a regular pentagon, it is possible to construct a regular 15-gon. Explain how to accomplish the construction.

8. Take a strip of paper, make a knot, and tighten and flatten the knot as shown in the figure. Prove that the shape of the flattened knot is a regular pentagon.

9. Draw all the diagonals of a regular pentagon.
    a. Prove that the shaded figure is also a regular pentagon.
    b. Find the ratio of the area of the larger pentagon to the area of the smaller pentagon. Justify your answer.

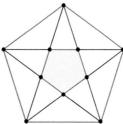

10. If you had a course in number theory or abstract algebra, you know that if m and n are positive integers whose greatest common divisor is 1, then it is possible to find integers x and y such that $mx + ny = 1$. Suppose that m and n are such integers and that it is possible to construct a regular m-gon and a regular n-gon. Prove that it is possible to construct a regular mn-gon. (Use the fact that a regular k-gon can be constructed if and only if an angle of $\frac{360°}{k}$ can be constructed.)

11. This problem investigates the connection between the golden ratio and the **Fibonacci sequence**. The Fibonacci sequence 1, 1, 2, 3, 5, 8, 13, 21, . . . whose nth term is denoted by F_n, is defined by $F_1 = F_2 = 1$ and for $n \geq 3$, $F_n = F_{n-1} + F_{n-2}$.

a. If ϕ is the golden ratio, show that

$$\phi^2 = 1 + 1 \cdot \phi,$$
$$\phi^3 = 1 \cdot \phi + 1 \cdot \phi^2,$$
$$\phi^4 = 1 \cdot \phi + 2 \cdot \phi^2,$$
$$\phi^5 = 2 \cdot \phi + 3 \cdot \phi^2,$$
$$\vdots$$

Conjecture an expression for ϕ^n in terms of ϕ, ϕ^2, and Fibonacci numbers.

b. Prove your conjecture in part (a).

c. Compute the sequence of ratios $\frac{F_n}{F_{n-1}}$ for $n = 2, 3, 4, \ldots, 10$. What number does the sequence seem to approach?

d. Assume that the sequence in part (c) has a limit and calculate that limit. (*Hint*: Use the relationship $F_n = F_{n-1} + F_{n-2}$.)

12. Given a rectangle, cut off three triangles from three corners of the rectangle such that the three cut-off triangles have the same area. Include an investigation, construction and a proof.

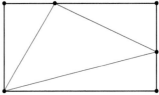

13. In $\triangle ABC$, $BC = AC$. Also, D is a point on side AC such that $BD = AB$. Find the ratio $\frac{AB}{AD}$. Justify your answer.

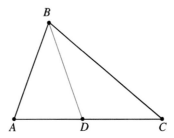

5.5 Circumference and Area of a Circle

In one of the propositions in the *Elements*, Euclid states, "The areas of any two circles are to each other as the squares of their diameters." In other words, if one circle has area A_1 and diameter d_1, and another circle has area A_2 and diameter d_2, then

$$\frac{A_1}{A_2} = \frac{d_1^2}{d_2^2}. \tag{5.39}$$

This implies that $\frac{A_1}{d_1^2} = \frac{A_2}{d_2^2}$; that is, the ratio of the area of any circle to the square of its diameter is a constant. (See Now Solve This 5.16.) If we denote this constant by k, then for any circle with diameter D and area A, we have

$$\frac{A}{d^2} = k. \tag{5.40}$$

It has also been known that the circumferences of two circles are related to each other as their corresponding diameters. That is, if C_1 and C_2 are the circumferences of any two circles, then

$$\frac{C_1}{C_2} = \frac{d_1}{d_2}. \tag{5.41}$$

Consequently, $\frac{C_1}{D_1} = \frac{C_2}{D_2}$ and the ratio of the circumference of a circle to its diameter is constant. If we denote this constant by k', we find that for any circle with circumference C and diameter D:

$$\frac{C}{d} = k'. \tag{5.42}$$

> **Now Solve This 5.16**
>
> Use the fact that a circle can be approximated by an inscribed (or circumscribed) regular polygon of many sides, your knowledge about the ratios of areas of similar polygons, and the ratio of their perimeters to give a plausible justification for Equations (5.39) and (5.41).

Around 225 B.C.E., Archimedes in his treatise *Measurement of a Circle* proved that the constants $\frac{k}{4}$ and k' are equal. Each is, of course, is the famous constant π. Here we will give a plausible argument for obtaining the formula for the area of a circle from the formula $C = 2\pi r$ for the circumference of a circle given its radius. We will also explore Archimedes's ingenious method for approximating the value of π and discuss other methods for evaluating π to any desired degree of accuracy.

$C = 2\pi r$ Implies $A = \pi r^2$

To obtain the formula for the area of a circle from its circumference, we consider a regular n-gon inscribed in the circle. We first express the area of the polygon in terms of its perimeter. For large n, the area of the n-gon approaches the area of the circle and the perimeter of the n-gon approaches the circumference of the circle. Figure 5.38 shows a regular octagon of side length a inscribed in a circle with center O and radius r. The area of the octagon is 8 times the area of $\triangle OAB$ or $8 \cdot \frac{ah}{2}$, where h is the height of the triangle from O. Similarly, the area of a regular n-gon inscribed in the circle is $n \cdot \frac{a_n \cdot h_n}{2}$, where a_n is the length of a side of the n-gon and h_n is the distance from O to a side. (Notice that the distance from O to each side of a regular n-gon is the same.) Let A_n be the area of the n-gon and p_n be its perimeter. Then

$$A_n = n\frac{a_n h_n}{2} = \frac{(na_n)h_n}{2} = \frac{p_n h_n}{2}.$$

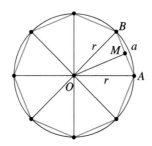

Figure 5.36

For large n, A_n approaches the area A of the circle, p_n approaches the circumference C of the circle, and h_n approaches the radius r of the circle. Hence, for the area of circle, we get $A = \frac{C \cdot r}{2}$. Substituting $2\pi r$ for C, we get $A = \frac{2\pi r \cdot r}{2} = \pi r^2$.

> **Now Solve This 5.17**
>
> Notice that we have defined neither the circumference nor the area of a circle. Try to give the definition of each using the following concepts:
>
> 1. Limit
> 2. Least upper bound
> 3. Greatest lower bound

Archimedes's Approximation of π

Archimedes showed that $\frac{223}{71} < \pi < \frac{22}{7}$. Using decimal notation (which was not known in Archimedes's time), this becomes

$$3.140845\ldots < \pi < 3.142857\ldots$$

5.5 Circumference and Area of a Circle

Thus π was computed to two decimal places accurately as 3.14.

Archimedes knew that $C = 2\pi r$ and hence $\pi = \frac{C}{2r}$, where C is the circumference of a circle and r is its radius. He noticed that the perimeters of regular polygons inscribed in a circle as well as the perimeters of those circumscribed about a circle approximate the circumference of a circle; he also observed that this approximation gets better as the number of sides increases. Assuming that the circumference of a circle is greater than the perimeter of any inscribed polygon and smaller than the perimeter of any circumscribed polygon, we get

$$\frac{\text{Perimeter of inscribed polygon}}{2r} < \pi < \frac{\text{Perimeter of circumscribing polygon}}{2r}. \tag{5.43}$$

Starting with an inscribed regular hexagon whose sides must equal the radius of the circle (why?), Archimedes calculated the perimeters of 12-, 24-, 48-, and 96-gons. He then developed formulas for the perimeters of circumscribing polygons in terms of the perimeters of the inscribed ones.

Before Archimedes, geometers gave a fairly crude approximation of π by calculating the area or perimeter of particular inscribed polygons. Archimedes' approach represented a completely new iterative approach in which any desired level of accuracy can be achieved by repeating the process and getting a new approximation using the previous one.

We will now investigate Archimedes's approach by starting in Figure 5.37 with a circle of radius 1. This choice is convenient but also legitimate because the ratio of the circumference to the diameter is constant and, therefore, does not depend on the size of the diameter.

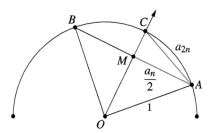

Figure 5.37

Let $AB = a_n$ be the length of a side of a regular n-gon inscribed in a circle of radius 1. The perpendicular bisector of $\overline{AB}$ contains the center O (why?), and bisects $\overline{AB}$ at M, and the arc AB at C. Because $BC = CA$, $\overline{CA}$ is the side of a $2n$-gon inscribed in the circle. We express a_{2n} in terms of a_n by applying the Pythagorean Theorem first in $\triangle AMC$ and then in $\triangle OMA$:

$$CA^2 = CM^2 + MA^2, \tag{5.44}$$

$$a_{2n}^2 = CM^2 + \left(\frac{a_n}{2}\right)^2,$$

$$CM = CO - OM = 1 - \sqrt{1 - MA^2} = 1 - \sqrt{1 - \left(\frac{a_n}{2}\right)^2}.$$

Substituting the last expression for CM in Equation (5.44), we get

$$a_{2n}^2 = \left(1 - \sqrt{1 - \left(\frac{a_n}{2}\right)^2}\right)^2 + \left(\frac{a_n}{2}\right)^2$$

$$= \left(1 - 2\sqrt{1 - \left(\frac{a_n}{2}\right)^2} + 1 - \left(\frac{a_n}{2}\right)^2\right) + \left(\frac{a_n}{2}\right)^2$$

$$= 2 - \sqrt{4 - a_n^2}.$$

Thus

$$a_{2n} = \sqrt{2 - \sqrt{4 - a_n^2}}. \tag{5.45}$$

Using this recursive formula repeatedly and the fact that the length of the side of a regular hexagon inscribed in a circle equals the radius of the circle (we took $r = 1$), we get

$$a_6 = 1,$$
$$a_{12} = \sqrt{2 - \sqrt{3}},$$
$$a_{24} = \sqrt{2 - \sqrt{2 + \sqrt{3}}},$$
$$a_{48} = \sqrt{2 - \sqrt{2 + \sqrt{2 + \sqrt{3}}}},$$
$$a_{96} = \sqrt{2 - \sqrt{2 + \sqrt{2 + \sqrt{2 + \sqrt{3}}}}},$$
$$\vdots$$
$$a_{6 \cdot 2^n} = \sqrt{2 - \underbrace{\sqrt{2 + \sqrt{2 + \cdots + \sqrt{3}}}}_{n + 1 \text{ radicals}}}. \tag{5.46}$$

Based on the discussion leading up to Equation (5.43), π can be approximated by dividing the perimeters of the inscribed polygons by the diameter of the circle, which is 2 in our case. Assuming that the perimeters of the polygons converge to the circumference of the circle, we get

$$\pi = \lim_{k \to \infty} \frac{p_k}{2},$$

5.5 Circumference and Area of a Circle

where p_k is the perimeter of a regular k-gon. If $k = 6 \cdot 2^n$, we can use equation (5.46) and get

$$\pi = \lim_{k \to \infty} \frac{p_k}{2} = \lim_{n \to \infty} 6 \cdot 2^n \underbrace{\sqrt{2 - \sqrt{2 + \sqrt{2 + \cdots \sqrt{3}}}}}_{n+1 \text{ radicals}}. \tag{5.47}$$

> **Now Solve This 5.18**
>
> By starting with an inscribed square rather than a hexagon, derive a formula similar to Equation (5.47).

Not having computers, calculators, or decimal notation, Archimedes stopped calculating after he reached the perimeter of a 96-gon. He also used circumscribing polygons to obtain the right side (upper bound) of the inequality $\frac{223}{71} < \pi < \frac{22}{7}$ (He estimated the involved square roots.) Using only inscribed polygons, we would not be able to tell with certainty which digits of the approximation are the correct digits of π. To overcome this problem, we also use circumscribing polygons as in Table 5.1.

Figure 5.38 shows a side of an inscribed regular n-gon and half a side of a circumscribing regular n-gon. The circumscribing polygon is constructed by drawing a tangent to the circle at each vertex of the inscribed polygon. The intersections of the tangents at the vertices of the regular inscribed n-gon determine the vertices of the circumscribing polygon.

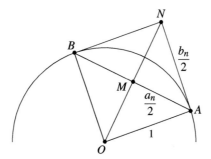

Figure 5.38

From $\triangle AMN \sim \triangle OMA$ (why?), we get

$$\frac{AM}{OM} = \frac{AN}{OA}.$$

Therefore

$$\frac{\frac{a_n}{2}}{\sqrt{1 - \left(\frac{a_n}{2}\right)^2}} = \frac{\frac{b_n}{2}}{1}, \quad b_n = \frac{2a_n}{\sqrt{4 - a_n^2}}. \tag{5.48}$$

Table 5.1. Approximation of π

Sides	$\frac{1}{2}$ of Perimeter of Inscribed Polygon	$\frac{1}{2}$ of Perimeter of Circumscribing Polygon	Approximation of π
6	3.0000000000000000000	3.4641016151377545871	3.
12	3.1058285412302491482	3.2153903091734724777	3.
24	3.1326286132812381972	3.1596599420975004833	3.1
48	3.1393502030468672071	3.1460862151314349711	3.1
96	3.1410319508905096381	3.1427145996453682982	3.14
192	3.1414524722854620755	3.1418730499798238717	3.141
384	3.1415576079118576455	3.1416627470568485262	3.141
768	3.1415838921483184087	3.1416101766046895388	3.1415
1536	3.1415904632280500957	3.1415970343215261520	3.14159
3072	3.1415921059992715505	3.1415937487713520280	3.14159
6144	3.1415925166921574476	3.1415929273850970335	3.141592
12288	3.1415926193653839552	3.1415927220386138183	3.141592
24576	3.1415926450336908967	3.1415926707019980479	3.1415926
49152	3.1415926514507676517	3.1415926578678444198	3.14159265
98304	3.1415926530550368417	3.1415926546593060325	3.14159265
196608	3.1415926534561041393	3.1415926538571714369	3.141592653
393216	3.1415926535563709637	3.1415926536566377881	3.141592653
786432	3.1415926535814376698	3.1415926536065043759	3.141592653
1572864	3.1415926535877043463	3.1415926535939710228	3.1415926535
3145728	3.1415926535892710154	3.1415926535908376846	3.1415926535
6291456	3.1415926535896626827	3.1415926535900543500	3.1415926535
12582912	3.1415926535897605995	3.1415926535898585163	3.141592653589
25165824	3.1415926535897850787	3.1415926535898095579	3.141592653589
50331648	3.1415926535897911985	3.1415926535897973183	3.14159265358979
100663296	3.1415926535897927285	3.1415926535897942584	3.14159265358979
201326592	3.1415926535897931110	3.1415926535897934935	3.141592653589793
402653184	3.1415926535897932066	3.1415926535897933022	3.141592653589793
805306368	3.1415926535897932305	3.1415926535897932544	3.1415926535897932
1610612736	3.1415926535897932365	3.1415926535897932424	3.1415926535897932
3221225472	3.1415926535897932380	3.1415926535897932395	3.1415926535897932
6442450944	3.1415926535897932383	3.1415926535897932387	3.141592653589793238

Using Equations (5.45) and (5.48) and starting with $a_6 = 1$, we can compute the perimeters of the inscribed regular polygons in the unit circle and the perimeters of the corresponding circumscribing polygons with $6, 6 \cdot 2, 6 \cdot 2^2, 6 \cdot 2^3, \ldots, 6 \cdot 2^n$ number of sides.

5.5 Circumference and Area of a Circle

When we divide the perimeter by the diameter of the circle (i.e., by 2), we get a lower and upper approximations of π; see Equation 5.43 and Table 5.1.

> **Now Solve This 5.19**
>
> Use Equations (5.45) and (5.48) to show that
> 1. $3 < \pi < 2\sqrt{3}$,
> 2. $3.11 \approx 6\sqrt{2 - \sqrt{3}} < \pi < 12(2 - \sqrt{3}) \approx 3.22$,
> 3. $3.1326 \approx 12\sqrt{2 - \sqrt{2 + \sqrt{3}}} < \pi < \dfrac{24\sqrt{2 - \sqrt{2+\sqrt{3}}}}{\sqrt{4 - (2 - \sqrt{2+\sqrt{3}})}} \approx 3.1596$,
> 4. Conclude that 3.1 is an approximation of π that is correct to one decimal point.

> **Historical Note:** *Archimedes*
>
> Archimedes, born in Syracuse, on the island of Sicily (circa 287–212 B.C.E.), was a Greek mathematician, physicist, and inventor. He is regarded by many as the greatest mathematician of antiquity, indeed, as one of the greatest mathematicians of all time. Archimedes invented mechanical devices that enabled the besieged city of Syracuse to delay the Roman conquest of the city for several years. Archimedes also discovered the law of the lever. Based on his "law," Archimedes presumably said, "Give me a place to stand on and a lever long enough and I will move the Earth." Another famous story tells of an occasion when he discovered the principle of buoyancy while in his bathtub: In his excitement, Archimedes got out of the tub and ran through the streets naked shouting, "Eureka! Eureka!" ("I have found it!"). In his work on areas and volumes, Archimedes used methods similar to the ones used in integral calculus.
>
> Many historical organizations and publications are named after Archimedes. For example, the Cambridge Society (in the United Kingdom) have named themselves the Archimedeans and publishes the journal *Eureka*. Information related to Archimedes can be viewed at the Archimedes' home page at https://math.nyu.edu/~crornes/Archimedes/contents.html and the program description on the Public Broadcasting services Nova website.

More Modern Ways of Evaluating π

The Chinese mathematician Tsu Chung Chi (430–501 C.E.) gave the rational approximation for π as $\frac{355}{113}$, which is correct to six decimal places. Thus he held the world record for the longest approximation of π for 800 years. He also proved that $3.1415926 < \pi < 3.1415927$. Details on how Tsu Chung Chi proved his results are unknown because his manuscript, which he co-wrote with his son, has been lost.

The next improvement was accomplished in 1424 by the Moslem mathematician Jamshid al-Kashi (born about 1380 in Kashan, Iran; died in 1429 in Sumarkand, now Uzbekistan), who used a method similar to Archimedes's to compute the value of π to 14 correct digits. Ludolf van Ceulen (1540-1610 C.E.) used Archimedes's polygon method to calculate π to 35 decimal places. In Europe, π was called the *Ludolphine number* for a long time.

In the 1600s the discovery of calculus by Newton and Leibniz enabled the finding of new formulas for π. The Taylor formula applied to arctan x gives

$$\arctan x = x - \frac{x^3}{3} + \frac{x^5}{5} - \frac{x^7}{7} + \cdots = \sum_{n=0}^{\infty} \frac{(-1)^n x^{2n+1}}{2n+1}. \tag{5.49}$$

Substituting $x = 1$, we get

$$\frac{\pi}{4} = 1 - \frac{1}{3} + \frac{1}{5} - \frac{1}{7} + \frac{1}{9} - \frac{1}{11} + \cdots = \sum_{n=0}^{\infty} \frac{(-1)^n}{2n+1}.$$

In spite of its elegant form, this series for $\frac{\pi}{4}$ is not useful because it converges very slowly. To obtain π accurate to only two digits, about 1000 terms are required. Nevertheless, the series in Equation (5.49) converges more rapidly for values smaller than 1. Using the identity

$$\frac{\pi}{4} = \arctan \frac{1}{2} + \arctan \frac{1}{3}. \tag{5.50}$$

and the series in Equation (5.49) for $x = \frac{1}{2}$ and $x = \frac{1}{3}$, we can obtain a faster and better approximation of π. In 1706, the British mathematician John Machin (1680–1751) discovered the identity

$$\frac{\pi}{4} = 4 \arctan \frac{1}{5} - \arctan \frac{1}{239}$$

and with Equation (5.49) used it to calculate the first 100 digits of π.

Euler (1707–1783), one of the most prolific mathematicians of all time, derived many expressions for π, including the following:

$$\frac{\pi^2}{6} = \sum_{k=1}^{\infty} \frac{1}{k^2} \quad \text{and} \quad \frac{2^2 \pi^4}{5!3} = \sum_{k=1}^{\infty} \frac{1}{k^4}.$$

In 1910, the prodigious Indian mathematician Ramanujan (1887-1920) discovered the following remarkable formula for π:

$$\frac{1}{\pi} = \frac{2\sqrt{2}}{9801} \sum_{k=0}^{\infty} \frac{(4k)!(1103 + 26390k)}{(k!)^4 396^{4k}}.$$

Each term in this series produces eight additional correct digits, In 1985, the formula was used to compute π to 17 million correct digits. In 1976, Salamin and Brent independently discovered a new algorithm that produces approximations of π that converge more rapidly

5.5 Circumference and Area of a Circle

than in any previous approach. The algorithm may be stated as follows (see Bailey et al., 1997). We let $a_0 = 1$, $b_0 = \frac{\sqrt{2}}{2}$, and $s_0 = \frac{1}{2}$:

$$a_k = \frac{a_{k-1} + b_{k-1}}{2},$$
$$b_k = \sqrt{a_{k-1} b_{k-1}},$$
$$c_k = a_k^2 - b_k^2,$$
$$s_k = s_{k-1} - 2^k c_k,$$
$$p + k = \frac{2 a_k^2}{s_k}.$$

Each iteration of this algorithm approximately doubles the number of correct digits. Only 25 iterations are necessary to compute π to more than 45 million correct digits. In 1994, the Chudnovsky brothers computed π to more then 4 billion digits using a formula for $\frac{1}{\pi}$ similar to Ramanujan's formula given above. For a very informative article about π, its history and latest calculations, see https://en.wikipedia.org/wiki/pi.

5.5.1 Problem Set

1. In the following figure, find the length of the indicated path from A to B in terms of r, the radius of the largest semicircle.
   a. The first path along semicircle $\widehat{AB}$.
   b. The path along the two semicircles $\widehat{AO}$ and $\widehat{OB}$.
   c. The third path along the four congruent semicircles.
   d. The nth path along 2^{n-1} congruent semicircles.

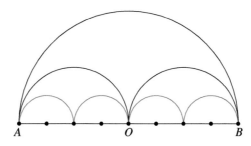

2. Assuming that the Earth is a perfect sphere, imagine that a rope is tightly stretched around the Earth's equator, then that the rope is cut and 100 feet of new rope are added to it. The extended rope is put back around the equator so that an evenly spread "gap"

created between the new circle and the equator is the same all around.

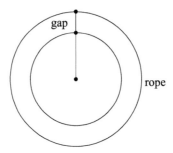

    a. Without making any calculations, guess whether a cat could squeeze under the rope.

    b. Find the size of the "gap"—that is, the distance between the two circles in the figure. Will the answer be different if a rope is stretched around the equator of a different size planet?

3.   **a.** Four congruent circles are cut out from a square sheet of tin as shown. What percentage of the tin is wasted?

    b. What percentage of the tin is wasted if nine congruent circles are cut out of the sheet?

    c. What percentage of the tin is wasted if n^2 congruent circles are cut out of the sheet? Prove your answer.

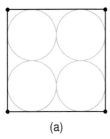

(a)

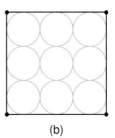

(b)

4.   For each of the following, find the ratio of the area of the circumscribing regular polygon to the area of the inscribed circle.

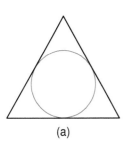

(a)

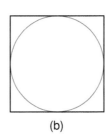

(b)

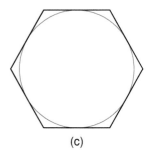

(c)

5.5 Circumference and Area of a Circle

5. Prove that the sum of the areas of the semicircles constructed on the legs of a right triangle equals the area of the semicircle constructed on the hypotenuse of the triangle.

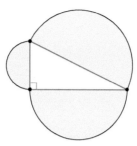

6. $\triangle ABC$ is inscribed in a semicircle with diameter AB. With the legs of the triangle as diameters, two semicircles are constructed. Prove that the shaded area between the semicircles (called **lunes**) equals the area of the triangle.

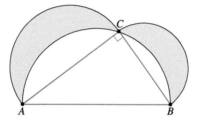

7. **a.** Find formulas involving radicals for the side of a regular 48-gon inscribed in a unit circle and the side of a regular 48-gon circumscribing the unit circle.

b. Use your results from part (a) to write inequalities for π involving perimeters of inscribed and circumscribed 48-gons.

8. Each of the three congruent circles shown below passes through the centers of the other two circles. Find the shaded area in terms of the radius r of the circles.

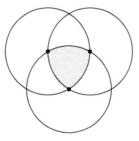

9. In the regular hexagon with side a, six petals are drawn. Each petal is made of two arcs of congruent circles whose centers are the vertices of the hexagon. All of the arcs intersect at the center of the hexagon (the center of the circumscribing circle).

Find the shaded area in terms of a.

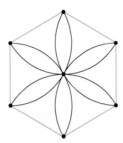

10. The four arcs in the interior of a square of side a are centered at the vertices of the square. Find the shaded area in terms of a. The answer may surprise you. Why?

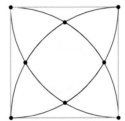

11. Draw a circle and then construct two concentric circles that will divide the original circle into three nonoverlapping regions of equal area. (Include an investigation, construction and a proof.)

12. In a circle, an arc that subtends a chord of length a is twice the length of an arc that subtends a chord of length b. Find the radius of the circle in terms of a and b.

13. Buffon's Needle Problem is a remarkable way to estimate the value of π using probability. Investigate this problem using references from the Internet and write a paper describing the problem, its solution, and its history.

6 Isometries and Size Transformations

> *To gain access to a greater number of designs, I used transformational geometry techniques including reflections, glide reflections, translations, and rotations.*
>
> M. C. Escher, 1898–1972, artist and mathematician at heart

Introduction

In *The Elements* (about 300 B.C.E.), Euclid attempted to prove the congruence of two triangles by moving one triangle so that it coincides with the other. However, such "motions" were not defined and had only an intuitive meaning for a given figure. In 1827, the German astronomer and mathematician A. F. Mobius (1790–1868) gave a meaning to the composition of motions by extending the idea of motion to the whole plane. In this chapter, we define a "motion" as an isometry in the plane and investigate properties of three types of isometries: translations, reflections, and rotations. We also discover a fourth isometry—glide reflection—and use isometries to solve problems that were investigated in the eighteenth and nineteenth centuries and that are much more difficult to solve using traditional Euclidean geometry.

6.1 Reflections, Translations, and Rotations

Symmetry is perhaps one of the most recognizable properties of figures. One type of symmetry is shown in Figure 6.1. We say that each figure is symmetrical about the respective line, or that it has a line symmetry. (A formal definition will be given soon.)

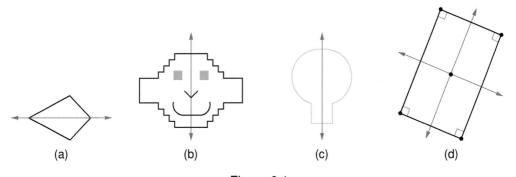

Figure 6.1

If a mirror were placed perpendicular to the plane of a drawing, then the image of each drawing would be the drawing itself. A convenient device that behaves like a mirror is a **Mira**. A Mira is a transparent plastic device that enables one to see the reflection of a given figure behind the Mira. (For more information on the Mira, search the Internet.)

Reflections

To investigate symmetries, and in particular line symmetry, we need a definition of a reflection in a line. In Figure 6.2(a), point P is reflected in line ℓ, and the image P' is such that $\overline{PP'}$ is perpendicular to ℓ and P' is the same distance from ℓ as is P; in other words, ℓ is the perpendicular bisector of $\overline{PP'}$. If P is on ℓ as in Figure 6.2(b), its image P' is P itself.

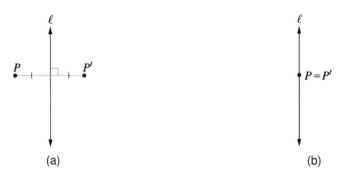

Figure 6.2

The preceding discussion leads to the following definition of a **reflection in a line**.

Definition of a Reflection in a Line

A reflection in a line ℓ is a correspondence that pairs each point P in the plane and not on ℓ with point P' such that ℓ is the perpendicular bisector of $\overline{PP'}$. If P is on ℓ, then P is paired with itself; that is, $P' = P$.

Notice that reflection in a line is a function that assigns points in the plane to other points in the plane. The domain of this function is the set of points in the plane, and the values of the function are also points in the plane. In geometry and other areas of mathematics, a function is frequently referred to as a **mapping**. Thus reflection is a mapping from the plane to the plane.

If we denote the reflection in line ℓ by M_ℓ (M for mirror and ℓ for the line of reflection), then if P is assigned to P' we may write $M_\ell(P) = P'$. The point P' is the value of the function at P and is commonly referred to as the **image** of P. The point P is called the **preimage** of P'. Given any point B in the plane, we can always find its preimage—that is, a point A whose image under M_ℓ is B or $M_\ell(A) = B$. Because every point in the plane has a preimage, the **range** of M_ℓ (the set of all images) is the whole plane and the mapping is **onto**. Consequently, M_ℓ maps the plane onto the entire plane. Also, notice that different points in the plane have different images under M_ℓ; that is, whenever $A \neq B$, we have $M_\ell(A) \neq M_\ell(B)$ (or equivalently $A' \neq B'$). Thus M_ℓ is a **one-to-one** mapping. A reflection in a line is an example of a **transformation of the plane**.

6.1 Reflections, Translations, and Rotations

> **Definition of a Transformation of the Plane**
>
> A transformation of the plane is a one-to-one and onto mapping of the plane to the plane.

From now on, we will refer to transformations of the plane simply as transformations.

If P is a point in the plane and T is some transformation, we say that $T(P)$ is the image of P under the transformation T. We frequently denote the image by P'. Thus $T(P) = P'$ and the point P is the preimage of P'. We also say that T **transforms** P to P'. As for any function, $T_1 = T_2$ if and only if $T_1(P) = T_2(P)$ for all points P in the plane. Because transformations are one-to-one and onto functions from the plane to the plane, the composition of two transformations is a transformation, and each transformation has an inverse.

If I denotes the identity function for the plane, then under I each point in the plane is transformed onto itself. We have the following definition:

> **Definition of the Identity Function for the Plane**
>
> I is the identity function for the plane if $I(P) = P$, for all P in the plane.

Notice that T and I are names of functions, whereas $T(P)$ and $I(P)$ are values of the corresponding functions. If T_1 and T_2 are two transformations, then the **composition of** T_1 **with** T_2 is a transformation written as $T_2 \circ T_1$. Notice that first T_1 acts on a point, and then T_2 acts on its image.

> **Definition of $T_2 \circ T_1$: The Composition of T_1 Followed by T_2**
>
> $(T_2 \circ T_1)(P) = T_2(T_1(P))$, for all P in the plane.

If T is a transformation, then the **inverse of** T (denoted by T^{-1}) is also a transformation. If $T(P) = Q$, then T^{-1} sends Q back to P. More formally, we have the following definition:

> **Definition of the Inverse T^{-1} of T**
>
> $T^{-1}(Q) = P$ if and only if $Q = T(P)$.

We can picture the actions of T and T^{-1} in Figure 6.3, where the arrow at Q indicates that P is sent to Q by T and the arrow below it indicates that Q is sent back to P by T^{-1}. Notice that

$$(T^{-1} \circ T)(P) = T^{-1}(T(P)) = T^{-1}(Q) = P, \quad \text{for every } P \text{ in the plane.}$$

Hence $T^{-1} \circ T$ is the identity I. Similarly, $T \circ T^{-1}$ is the identity. Notice that because the range of T is the same as the domain (i.e., the plane), $T^{-1} \circ T = T \circ T^{-1} = I$.

What properties does a reflection in a line have? It seems that the length of the image of a segment is always the same as the length of the original segment. In other words,

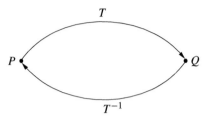

Figure 6.3

the distance between the images of any two points is the same as the distance between the original points. More succinctly, we say that reflection **preserves distance**. It also seems that the image of the segment $\overline{AB}$ is the segment $\overline{A'B'}$ and the image of the line $\overleftrightarrow{AB}$ is the line $\overleftrightarrow{A'B'}$. (The image of a geometric figure is defined as the set of all images of the points of the original figure.) In the problem set at the end of this section, we will investigate proofs of the fact that if A, B, and C are collinear (on the same line), their respective images A', B', and C' are also collinear.

Because a reflection preserves distance, the image of a triangle under a reflection is a congruent triangle. As we shall soon see, this is the case for any transformation that preserves distance. We will study such transformations and give them a special name: **isometries**.

Definition of Isometry

An isometry is a transformation that preserves distance. In other words, an isometry is a transformation T of the plane such that for any two points A and B, the distance between A and B equals the distance between their images $T(A)$ and $T(B)$.

Notice that $T(A)$, the image of A, can also be written as A'. Using this notation, a transformation is an isometry if and only if for any two points A and B we have $A'B' = AB$. If the distance between two points P and Q is denoted by $d(P, Q)$, then a transformation T of the plane is an isometry if for any two points A and B in the plane, $d(T(A), T(B)) = d(A, B)$. The word "isometry" is derived from the Greek words *isos* (meaning "equal") and *metron* (meaning "measure"), and is also commonly referred to as **rigid motion**.

Example 6.1

Given nonperpendicular intersecting lines m and n as in Figure 6.4, a mapping F from the plane to the plane is defined as follows: If P is on n, then $P' = P$. If P is not on n, then P' is the point for which $\overline{PP'}$ is parallel to m and bisected by n.

1. Is F a transformation of the plane?
2. Is F an isometry?

6.1 Reflections, Translations, and Rotations

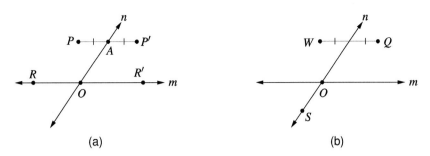

Figure 6.4

Solution

1. F is a transformation if it is onto and one-to-one. To show that F is onto, we need to demonstrate that every point on the plane has a preimage. In Figure 6.4(b), the preimage of Q is W (and vice versa), and the preimage of S on n is S itself. Hence F is onto. It is easy to see that different points have different images; thus F is one-to-one. Consequently, F is a transformation of the plane.

2. A careful drawing of two points P and R and their images P' and R' [see Figure 6.4(a)] suggests that $PR \neq P'R'$ and hence that F is not an isometry. To prove that this is the case, suppose the contrary: $P'R' = PR$. This implies that $RPP'R'$ is an isosceles trapezoid. Thus $\overline{AO}$ is a segment connecting the midpoints of opposite bases of an isosceles trapezoid. As such, $\overline{AO}$ must be perpendicular to the bases of the trapezoid. (This fact is left for you to prove.) Because A and O are on n, it follows that $n \perp m$, which contradicts the hypothesis that n and m are not perpendicular.

   Another way to prove that F is not an isometry is to choose two points, one of which is the point O, whose image is itself. Then if $PO = P'O$, O would be equidistant from the endpoints of $\overline{PP'}$ and hence on the perpendicular bisector of $\overline{PP'}$. Because A is the midpoint of $\overline{PP'}$, A is also on the perpendicular bisector of $\overline{PP'}$ and hence $\overleftrightarrow{OA}$ is the perpendicular bisector of $\overline{PP'}$, which contradicts the hypothesis.

Now Solve This 6.1

1. Show that a composition of two isometries is an isometry.
2. Show that the inverse of an isometry is an isometry.
3. What kind of single isometry is $M_\ell \circ M_\ell$?
4. What kind of isometry is M_ℓ^{-1}?

Translations

Informally, a translation is a transformation of the plane that moves every point in the plane a specified distance in a specified direction along a straight line. A segment $\overline{AB}$ determines the translation in the following way: If P is any point in the plane not on $\overleftrightarrow{AB}$, then its image P' is found by moving P along the line ℓ parallel to $\overleftrightarrow{AB}$ in the direction from A to B, so that $PP' = AB$ [see Figure 6.5(a)]. But how do we define "in the direction from A to B? Notice that $ABP'P$ is a parallelogram.

Figure 6.5

If P is on $\overleftrightarrow{AB}$, then P' is found on $\overleftrightarrow{AB}$ so that $PP' = AB$ [see Figure 6.5(b)]. Since there are two such points P' we define the location of P' using the concept of a **degenerate parallelogram**. If A, B, C, and D are collinear, then $ABCD$ is a degenerate parallelogram if $AB = CD$ and $BC = AD$. By themselves, the conditions $AB = CD$ and $BC = AD$ imply that $ABCD$ is either a parallelogram or a degenerate parallelogram. We need the collinearity of A, B, C, and D to establish that the parallelogram is degenerate, however. We are now ready for the following definition:

> **Definition of Translation**
>
> A translation from A to B is a transformation of the plane that assigns to every point P in the plane the point P', which is determined as follows: If P is not on $\overleftrightarrow{AB}$, then P' is the point in the plane for which $ABP'P$ is a parallelogram; if P is on $\overleftrightarrow{AB}$, then P' is the point such that $ABP'P$ is a degenerate parallelogram.

We will denote the translation from A to B by τ_{AB}. Notice that τ_{AB} is determined not only by the **directed segment** AB (the segment AB and the direction from A to B), but also by infinitely many directed segments having the same length and direction, such as PP' in Figure 6.5(a) or $\overline{CD}$ and $\overline{EF}$ in Figure 6.6.

A directed line segment is also referred to as a **vector**. Thus the concept of a vector is related to a translation. We say that τ_{AB} is a translation determined by the vector AB, which is also denoted by $\overrightarrow{AB}$.

In Figure 6.6, the segments have the same length and are all parallel; therefore $\tau_{AB} = \tau_{CD} = \tau_{EF}$. Similarly, we say that the three vectors are equal; that is, $\overrightarrow{AB} = \overrightarrow{CD} = \overrightarrow{EF}$.

6.1 Reflections, Translations, and Rotations

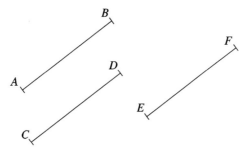

Figure 6.6

A vector like $\overrightarrow{AB}$ is pictured by a segment $\overline{AB}$ with an arrow at B, as shown in Figure 6.7. Point A is referred to as the **tail** of the vector $\overrightarrow{AB}$; point B is called the **head** of the vector.

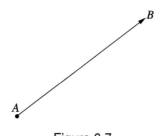

Figure 6.7

> **Now Solve This 6.2**
>
> 1. Show that a translation is an isometry.
> 2. What is the image of a point P under τ_{AA}?
> 3. Based on your answer to part 2, what kind of mapping is τ_{AA}?
> 4. Find τ_A^{-1}.

Rotations

Informally, we can think about a rotation by thinking about a phonograph record on a turntable, the hands of a clock or a rotating restaurant. The center of the record is the center of rotation, and the entire plane rotates around it like an infinitely large turntable. Thus, to describe a rotation, we need to specify the center of the rotation and the amount and direction of the rotation. For this purpose we use the concept of a **directed angle**. A positive angle indicates a counterclockwise rotation, whereas a negative angle indicates a clockwise rotation. (Notice that we have not given rigorous definitions of clockwise and counterclockwise directions.)

> **Definition of Rotation**
>
> If O is any point and α is a real number, then the rotation about O (or with center O) through (or by) α, which is denoted by $R_{O,\alpha}$, is a function from the plane to the plane that maps O onto itself and any other point P onto point P' such that $OP = OP'$ and $m(\angle POP') = \alpha$.

> **Remark 6.1**
>
> We use degree and radian measures interchangeably.

Figure 6.8 illustrates the images of two points P and Q under a rotation by an angle with a positive measure α. The angle of rotation can be measured in either radians or degrees.

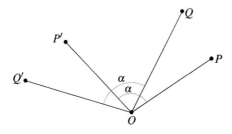

Figure 6.8

A rotation through an angle of 180° is of particular interest and is called a **half-turn**. A half-turn about O is denoted by H_O. (Notice that $R_{O,\pi} = H_O$.) Figure 6.9 shows the images of two points P and Q under a half-turn about O.

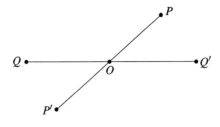

Figure 6.9

In Now Solve This 6.3, you will encounter a surprising property of half-turns. You are invited to perform the indicated experiment and make a conjecture. (A compass and straightedge are sufficient for performing the experiment, but you can also use geometry software.) Why the conjecture is true will be explored in Chapter 7.

6.1 Reflections, Translations, and Rotations

> **Now Solve This 6.3**
>
> 1. A treasure was buried on a deserted island that contained a rock R and three easily identifiable trees A, B, and C. The map for locating the treasure gave the following directions: Start at R and go to the point R_1 which is the image of R under H_A. Continue to R_2, which is the image of R_1, under H_B. Then go to R_3, which is the image of R_2 under H_C. Continue by finding the images under H_A, H_B, and H_C until you reach R_6, where the treasure is buried. (Points R_1 through R_5 are shown in Figure 6.10.)
>
>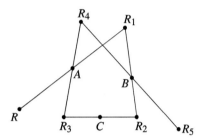
>
> Figure 6.10
>
> A treasure-hunting expedition found the island and the rock R but not the trees, as they had long since rotted. The expedition dug all over the island, but the island was large, and the treasure seekers finally gave up.
>
> To get an idea where the treasure was buried, choose arbitrary positions for A, B, and C and follow the directions. Repeat the experiment for several different locations for points A, B, and C. Based on your results, make a conjecture about the location of the treasure.
>
> 2. A half-turn can be defined without reference to a rotation. Write such a definition.
>
> 3. Prove that $H_A^{-1} = H_A$.

As for reflections and translations, it is a straightforward task to show that a rotation is an isometry. The proof of this fact is left as an exercise for you. It is also not difficult to show that reflections, translations, and rotations preserve other properties, such as collinearity.

At this juncture, you may be wondering whether reflections, translations, and rotations are the only isometries or whether other isometries exist. This question will be explored and answered later in this section and independently in Chapter 7.

Invariant Properties of a Transformation

Now we will state several properties of isometries. These properties can be proved using the definition of an isometry. Their proofs will be explored in the problem set at the end of this section.

Properties of isometries

1. If A and B are any two points and A' and B' are their respective images, then $A'B' = AB$. (This is part of the definition of an isometry.)
2. The image of a line is a line, and the image of a segment is a segment.
3. Isometry preserves collinearity. That is, if A, B, and C are collinear, so are their respective images A', B', and C'.
4. Betweenness is preserved: If A-B-C (B is between A and C), then A'-B'-C'.
5. The image of a triangle is a triangle congruent to the original triangle.
6. The absolute value of angle measure is preserved: The absolute value of the measure of the image of an angle equals the absolute value of the measure of the angle.

Properties such as those listed above, which stay unchanged under a transformation, are called **invariant properties** of the transformation.

A point is invariant under a transformation; that is, it is a **fixed point** under the transformation if its image is the point itself, that is, $P' = P$. Thus, if T is a transformation, P is a fixed point under T if $T(P) = P$.

What are the fixed points (if any) of the isometries we have encountered so far? It follows from the definition of a reflection that all points on the line of reflection are fixed points under the reflection. This definition also implies that these points are the only fixed points. If a translation is not the identity (a translation τ_{AB} for which $A \neq B$), then the definition of a translation implies that it has no fixed points. Lastly, each rotation by α which is not a multiple of $360°$ (why?) has a single fixed point—its center.

Sometimes an entire figure remains unchanged under a transformation. In such a case, we say that the figure is invariant under the transformation or that the transformation is a **symmetry** of the figure. For example, under any rotation about the center of the circle that is not the identity, the image of a point on the circle is a different point on the circle, but the set of all images of points on the circle is the same circle. Thus the circle is invariant under any rotation about its center, and any such rotation is a symmetry of the circle. Consequently, a circle has infinitely many rotational symmetries.

The **image** of a figure (a set of points in the plane) under a transformation is the set of all images of the points of the figure. If S is a set of points in the plane, T is a transformation, and $T(S)$ is the image of S under the transformation, then using set notation we have $T(S) = \{T(P) \mid P \in S\}$. We are ready now for a more formal definition of symmetry of a figure.

> **Definition of Symmetry of a Figure**
>
> Given a figure S, transformation T is a symmetry of the figure if $T(S) = S$.

6.1 Reflections, Translations, and Rotations

> **Remark 6.2**
>
> If T is a symmetry of S, we also say that S is invariant under T. Also notice that the composition of any two symmetries of a figure is a symmetry of the figure.

> **Example 6.2**
>
> List all the symmetries of a rectangle.
>
> **Solution**
>
> Let O be the center of the rectangle $ABCD$ (the intersection point of the diagonals), and let k and ℓ be lines through O that are parallel to the sides of the rectangle (see Figure 6.11). Then the following isometries leave the rectangle invariant and hence are symmetries of the rectangle:
>
> Reflections in k and ℓ—that is, M_k and M_ℓ
> The rotation about O by $180°$—that is, the half-turn H_O
> The identity transformation
>
>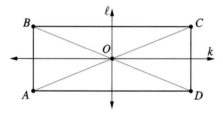
>
> Figure 6.11
>
> It seems that these isometries are the only symmetries of the rectangle, but we have not yet proved this fact.

> **Now Solve This 6.4**
>
> List all of the symmetries of the following figures:
> 1. An equilateral triangle
> 2. A square
> 3. A regular pentagon
> 4. A regular hexagon
> 5. A circle
> 6. A segment
> 7. A straight line

Another concept useful in investigating transformations is **orientation**. If A, B, and C are three points on a circle, we say that $\triangle ABC$ has a clockwise (or counterclockwise) orientation if, when moving on the circle from A to B and then to C (in that order), the direction traversed is clockwise (or counterclockwise.) In Figure 6.12, $\triangle ABC$ has a clockwise orientation and $\triangle BAC$ has a counterclockwise orientation.

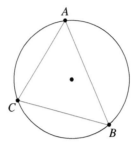

Figure 6.12

> **Remark 6.3**
>
> This description of orientation is not a rigorous definition—clockwise and counterclockwise are not rigorously defined. Instead, we used our experience of knowing the direction of a clock rather than previously defined terms or the terms we agreed to leave undefined.

Figure orientation is **invariant under a transformation** if for all triples of noncollinear points A, B, and C of the figure, the orientation of $\triangle ABC$ is the same as that of $\triangle A'B'C'$, where A', B', and C' are the images of points A, B, and C, respectively, under the transformations. If figure orientation is invariant under a transformation, the transformation is called an **orientation-preserving transformation**; otherwise, it is called an **orientation-reversing transformation**. Translations and rotations are orientation-preserving isometries, whereas reflection in a line is an orientation-reversing isometry, as can be seen by looking into a mirror.

> **Now Solve This 6.5**
>
> 1. Use the concept of orientation and the fact that a rectangle has only four symmetries (listed in Example 6.2) to show that $M_k \circ M_\ell = M_\ell \circ M_k = H_O$, and similarly to find $M_\ell \circ H_O$ and $H_O \circ M_\ell$. Also show that $M_\ell^2 = M_\ell \cdot M_\ell = I$, $M_k^2 = I$, and $H_O^2 = I$, where I is the identity function.
> 2. In a Cartesian coordinate plane, let M_k be the reflection in the x-axis, M_ℓ be the reflection in the y-axis, and H_O be the half-turn about the origin. Show that
>   (a) $M_k(x, y) = (x, -y)$.
>   (b) $M_\ell(x, y) = (-x, y)$.
>   (c) $H_O(x, y) = (-x, -y)$.

3. Assume that the lines k and ℓ in Example 6.2 are the x-axis and y-axis, respectively, and notice that for all points (a, b) in the plane:

$$(M_\ell \circ M_k)(a, b) = M_\ell(M_k(a, b)) = M_\ell(a, -b)$$
$$= (-a, -b)$$
$$= H_O(a, b),$$

and hence that $M_\ell \circ M_k = H_O$.
In a similar way show that
(a) $M_k \circ M_\ell = H_O$.
(b) $M_\ell \circ H_O = M_k$.
(c) $H_O M_\ell = M_k$.
(d) $H_O^2 = M_k^2 = M_\ell^2 = I$.

Glide Reflections

Are the isometries we have encountered so far—reflections, translations and rotations – the only isometries? In Chapter 7, we will investigate in detail the composition of these basic isometries and, in the process of doing so, discover a new isometry that we initially introduce next. For now, let's consider the composition of a reflection and a translation. In particular, let the translation be determined by a vector parallel to the line of reflection. In Figure 6.13, quadrilateral $ABCD$ is reflected in line ℓ, and then its image $A'B'C'D'$ is translated by vector $\overrightarrow{PQ}$ (on ℓ or parallel to ℓ) onto quadrilateral $A''B''C''D''$. Thus the final image $A''B''C''D''$ is obtained from $ABCD$ by the composition $\tau_{PQ} \circ M_\ell$, where M_ℓ is the reflection in ℓ and τ_{PQ} is the translation by $\overrightarrow{PQ}$.

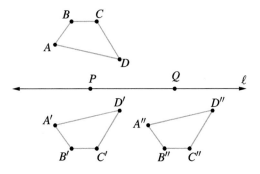

Figure 6.13

Notice that $\tau_{PQ} \circ M_\ell$ (think about it as a new single function), being a composition of two isometries, is itself an isometry. Let us see if $\tau_{PQ} \circ M_\ell$ could be one of the three familiar isometries. Because a reflection reverses orientation and a translation preserves it, the composition reverses orientation. Therefore, $\tau_{PQ} \circ M_\ell$ could not be a rotation or a translation, as both preserve orientation. Could it be a single reflection in some line? We know that reflections have infinitely many fixed points (all the points on the line of reflection), but $\tau_{PQ} \circ M_\ell$ does not have any fixed points (you will be asked to justify this fact in Now Solve This 6.6). Consequently, $\tau_{PQ} \circ M_\ell$ is a new isometry called a **glide reflection**.

> **Definition of Glide Reflection**
>
> A glide reflection is an isometry that is a composition of a reflection in a line, followed by a translation by a nonzero vector parallel to the line.

We have shown that a glide reflection in which the translation is by a nonzero vector is a fourth isometry that is not a reflection, a translation, or a rotation. In Chapter 7, we will prove that there are no other isometries in the plane. We will then show that a composition of a reflection in a line followed by a translation by a vector not parallel to the line is equal to a glide reflection (in a new line and by a new vector).

> **Now Solve This 6.6**
>
> 1. Consider the glide reflection $\tau_{PQ} \circ M_\ell$, where $P \neq Q$. Choose the line ℓ to be the x-axis, and show that the image of any point (x, y) under the glide reflection is $(x + a, -y)$ for some fixed, nonzero real number a. This is frequently written as $(x, y) \to (x + a, -y)$.
> 2. Use your result from part 1 to prove that a glide reflection in which the translation is not by a zero vector has no fixed points.
> 3. Prove that if $\tau_{PQ} \circ M_\ell$ is a glide reflection, then $\tau_{PQ} \circ M_\ell = M_\ell \circ \tau_{PQ}$.
> 4. Prove that a glide reflection composed with itself is a translation.
>   (a) One of the band ornaments in Figure 6.14 was created by a glide reflection. Identify the line of reflection and the translation vector.
>
>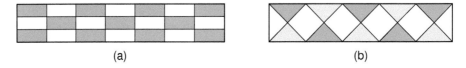
>
> Figure 6.14
>
>   (b) Assume that each figure is extended indefinitely to the right and to the left and identify the symmetries of each of the figures.
> 5. Create your own band ornament using a glide reflection. Indicate the line of reflection and the translation vector.

6.1 Reflections, Translations, and Rotations

6.1.1 Problem Set

In this problem set, "construct" unless noted otherwise, means using only a compass and straightedge.

1. Draw a line and a point not on the line. Use each of the following to construct the image of the point under a reflection in the line.
   a. Paper folding
   b. Paper tracing
   c. Mira (if available)
   d. Compass and straightedge

2. a. Construct the image of $\triangle ABC$ under a reflection in line ℓ. Describe your procedure and state the assumptions you are making in your solution.
   b. Find the image of $\triangle ABC$ in part (a) without actually constructing the image of point C. Explain why your procedure works.

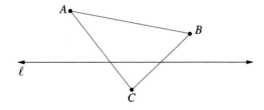

3. Draw a circle and a line that does not contain a diameter of the circle. Construct the image of the circle under the reflection in the line. Describe the construction in words and state any assumptions you are using in your construction.

4. When the image of a figure under a reflection in a line is the figure itself, the line is referred to as the **line of symmetry** for the figure. For example, a rectangle has two lines of symmetry, each of which is the perpendicular bisector of the opposite sides.
   a. For each of the following figures, describe and draw all the lines of symmetry of the figure. If a figure has no line of symmetry, say so.
      (i) Segment
      (ii) Line
      (iii) Square
      (iv) Rhombus
      (v) Equilateral triangle
      (vi) Trapezoid inscribed in a circle
      (vii) Circle
      (viii) Kite
   b. Find three figures not mentioned in part (a) and describe all their lines of symmetry.

c. Describe all the lines of symmetry of a regular n-gon.

5. When a ray of light bounces off a surface, or when a billiard ball bounces off the rail of a billiard table, the angle between the surface and the incoming ray equals the angle between the surface and the reflected ray. That is, in the figure below, $\alpha = \beta$. Assuming this fact, show the following (note that parts of this problem were done in Chapter 2):

  a. Prove that the path that the light will travel from a source A to a mirror surface x and then to B can be obtained by first finding B', the reflection of B in x and then the intersection M of $\overleftrightarrow{AB'}$ with x.

  b. Prove that the path A-M-B is the shortest path connecting A to B and passing through a point on x. That is, show that if M_1 is any point on x such that $M_1 \neq M$, then $AM_1 + M_1B > AM + MB$. [*Hint*: Use your answer from part (a) and the triangle inequality.]

  c. Suppose a coordinate system is introduced such that the line x is the x-axis, point A has coordinates (x_1, y_1), and point B has coordinates (x_2, y_2). Prove that x_M, the x-coordinate of M, equals $x_M = \frac{x_1 y_2 + y_1 x_2}{y_1 + y_2}$.

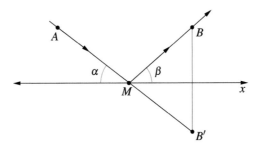

6. Consider a ray of light bouncing off the two perpendicular surfaces shown in the following figure.

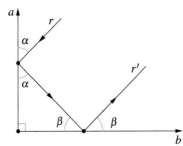

  a. Draw two perpendicular lines a and b, choose an initial ray r, and construct (using any tools) the corresponding outgoing ray r'. Repeat your construction for different angles α that the initial ray r makes with line a, and in each case construct the outgoing ray r'. State a conjecture concerning the relationship between r and r'.

  b. Prove the conjecture you made in part (a).

6.1 Reflections, Translations, and Rotations

7. In Example 6.1, we defined a transformation F using lines m and ℓ as follows: Any point on ℓ is mapped to itself; if a point P is not on ℓ, its image P' is such that $\overline{PP'}$ is parallel to m and bisected by ℓ.

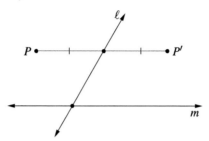

a. We have shown that F is not an isometry; this means that it is not always true that $PQ = P'Q'$. However, the transformation may preserve the distance between certain points. Do such points exist? If so, describe them and justify your answer.

b. Prove that F does not preserve equality of distance. That is, prove that it is not always the case that $PQ = RS$ implies $P'Q' = R'S'$. (*Hint*: Demonstrate a counterexample in which one of the segments is located in a convenient location.)

c. Does the transformation preserve the midpoints of segments? That is, if M is the midpoint of $\overline{AB}$, is it always the case that M' is the midpoint of $\overline{A'B'}$? Either prove this statement or give a counterexample.

d. Does the transformation preserve parallelism? That is, is it true that whenever two lines are parallel, their images are also parallel? Either prove this statement or give a counterexample.

e. Prove that the transformation preserves the area of a triangle.

f. Does your proof in part (e) imply that the transformation preserves the area of any figure? Justify your answer.

g. Suppose a coordinate system is introduced such that line m is the x-axis and the origin is at the point of intersection of ℓ with the x-axis. Assume that α is an angle between the x-axis and ℓ, and express x' and y' (the coordinates of P') in terms of x and y (the coordinates of P).

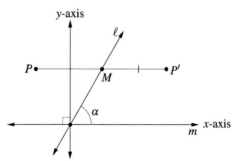

h. (*For those who studied linear algebra*) The area of a triangle with coordinates $A(x_1, y_1)$, $B(x_2, y_2)$, and $C(x_3, y_3)$ can be expressed as the absolute value of the determinant:

$$\frac{1}{2} \begin{vmatrix} x_1 & y_1 & 1 \\ x_2 & y_2 & 1 \\ x_3 & y_3 & 1 \end{vmatrix}$$

Use your answer from part (g) and properties of determinants to prove part (e)—that is, that the area of a triangle is an invariant property of the transformation.

i. Let α be the angle between m and ℓ, and let β be the angle between m and $\overleftrightarrow{AB}$ as shown in the figure below. Express $d' = A'B'$ in terms of $d = AB$, α, and β. (Use trigonometric functions.) Check that your answer is plausible by checking three special cases: (i) $\alpha = 90°$, (ii) $\alpha = \beta$, and (iii) $\beta = 0$. For each of those special cases, confirm that your formula results in what can be found from the geometry of the corresponding case.

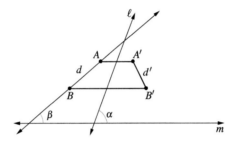

8. Prove that a function from the plane to the plane that preserves distance is a transformation of the plane and hence an isometry.

9. Prove the six properties of isometries listed on page 255.

10. a. Use tracing paper to find the image of point P under the translation from M to N.

b. Use properties of a parallelogram to construct $\tau_{MN}(P)$. Explain your procedure and prove that it works.

c. Find $\tau_{NM}(\tau_{MN}(P))$. What does your result tell you about $\tau_{NM} \circ \tau_{MN}$?

d. Based on your answer to part (c), what is τ_{MN}^{-1}?

11. Classify each of the following statements as true or false. If a statement is true, prove it; if it is false, give a counterexample.

a. $\tau_{AB} \circ \tau_{BC}$ is a translation.

b. $\tau_{AB} \circ \tau_{CD}$ is a translation.

c. $\tau_{AB} \circ \tau_{CD} = \tau_{CD} \circ \tau_{AB}$.

12. Use each of the following methods to construct the image of a point P under a rotation about O by the indicated angle and direction (counterclockwise):
   a. Paper tracing
   b. Compass and straightedge

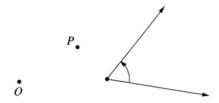

13. a. Draw a line ℓ, a point O not on the line, and a directed angle α. Construct the image of ℓ under $R_{O,\alpha}$.

   b. Justify the following method for constructing the image of ℓ under $R_{O,\alpha}$. Construct $\overline{OP} \perp \ell$ such that P is on ℓ. Find P', the image of P under $R_{O,\alpha}$. Construct the line ℓ' through P' and perpendicular to $\overline{OP'}$. The line ℓ' is the image of ℓ under $R_{O,\alpha}$.

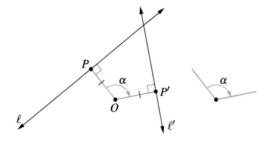

   c. If ℓ' is the image of ℓ under the rotation $R_{O,\pi/2}$, how are ℓ' and ℓ related? Prove your answer.

   d. Two intersecting lines form two pairs of vertical angles. The angle from ℓ_1 to ℓ_2 in a clockwise direction is shown in the figure below as θ_1. The angle from ℓ_1 to ℓ_2 in a counterclockwise direction is shown as θ_2. Notice that θ_1 is negative and θ_2 is positive. Use these notions to prove the following: If ℓ' is the image of ℓ under $R_{O,\alpha}$ and $|\alpha| < \pi$, then one of the angles from ℓ to ℓ' equals α.

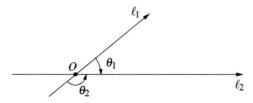

   e. If ℓ' is the image of ℓ under a half-turn, what is the relationship between ℓ' and ℓ?
       (i) Give a plausible justification of your answer based on the result in part (d).

(ii) Give an independent proof of your answer.

14. Recall the Treasure Island Problem from Chapter 1 and Problem Set 1.3. Here it is again for your convenience:

Prove that the location of the treasure is independent of the location of the gallows by considering two positions for the gallows: one at one of the trees Γ^* and the other Γ in an arbitrary position. (*Hint*: Prove that the segments $S_1 S_2$ and $S_1^* S_2^*$ bisect each other.)

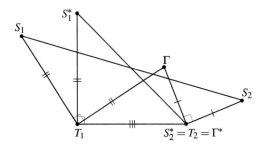

Now you can use rotations and your result from Problem 13(c) to simplify your proof. Show that $S_1 S_1^* S_2 S_2^*$ is a parallelogram by proving that $S_1 S_1^* = S_2 S_2^*$ and then proving that line $\Gamma\Gamma^*$ is perpendicular to each of the lines $S_1 S_1^*$ and $S_2 S_2^*$. For that purpose first notice that $\overline{S_2 S_2^*}$ is the image of $S_1 S_1^*$ under rotation by $-90°$ about T_1, followed by a rotation by $-90°$ about T_2.

15. In Now Solve This 6.5 we pointed out that when the point (x, y) is reflected in the x-axis in a Cartesian coordinate system, its image is $(x, -y)$. If M_x denotes the reflection in the x-axis, we say that M_x maps (x, y) to $(x, -y)$ and write $M_x: (x, y) \to (x, -y)$ or $M_x(x, y) = (x, -y)$. Find the image of (x, y) under each of the following isometries:

   a. Reflection in the y-axis
   b. Reflection in the line $y = x$
   c. A half-turn about the origin
   d. $R_{O, \pi/2}$ where O is the origin
   e. τ_{AO} where O is the origin and $A(x_0, 0)$
   f. τ_{OB} where O is the origin and $B(0, y_0)$
   g. τ_{OC} where O is the origin and $C(x_0, 0)$

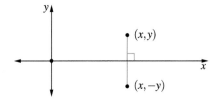

6.1 Reflections, Translations, and Rotations

16. Consider the mapping F defined for all points (x, y) in the plane as follows: If $y \geq 0$, then $F: (x, y) \to (x, y + 1)$; if $y < 0$, then $F: (x, y) \to (x, y - 1)$.
    a. Find the domain and the range of F.
    b. Is F one-to-one?
    c. Is F a transformation?
    d. Is F an isometry?

17. Which of the following mappings F from the plane to the plane are transformations? Which are isometries?
    a. $F: (x, y) \to (x, y + 1)$
    b. $F: (x, y) \to (x - 1, y + 1)$
    c. $F: (x, y) \to (|x|, |y|)$
    d. $F: (x, y) \to (x^2, y^2)$
    e. $F: (x, y) \to \left(\frac{1}{x}, \frac{1}{y}\right)$

18. a. Prove that $F: (x, y) \to \left(\frac{x-y}{\sqrt{2}}, \frac{x+y}{\sqrt{2}}\right)$ is an isometry.
    b. Let F be a mapping of the plane given by $F: (x, y) \to (ax + by, cx + dy)$. Find as simple necessary and sufficient conditions as possible on a, b, c, and d such that
       (i) F is a transformation.
       (ii) F is an isometry.

19. Explain why the equation of the image of the circle $x^2 + y^2 = R^2$ under the translation $(x, y) \to (x + h, y + k)$ can be found as follows. Let $x_1 = x + h$, $y_1 = y + k$. Substitute $x = x_1 - h$ and $y = y_1 - k$ into the equation of the circle to obtain $(x_1 - h)^2 + (y_1 - k)^2 = R^2$. Next, drop the subscripts to obtain the equation of the image circle as $(x - h)^2 + (y - k)^2 = R^2$.

20. Use your answers to Problems 15 and 19 to answer each of the following.
    a. Find the image of $y = ax^2$ under the following transformations
       (i) The translation $(x, y) \to (x + k, y + k)$
       (ii) Reflection in the x-axis
       (iii) Reflection in the y-axis
       (iv) Reflection in the line $y = x$
    b. Find the equation of the image of $|y| = |x|$ under each of the isometries given in part (a).

21. Recall that a curve is said to have line symmetry if its reflection in the line is the curve itself. It has rotational symmetry if its image under the rotation is the curve itself. Similarly, it has translational symmetry if the image of the curve under the translation

is the curve itself. Find all the symmetries of the following curves and justify your answers.
- **a.** $y = kx^2$
- **b.** $y = kx^3$
- **c.** $y = |x|$
- **d.** $|y| = |x|$
- **e.** $|x| + |y| = 1$
- **f.** $x^3 + y^3 = a^3$
- **g.** $y = \sin x$
- **h.** $y = \cos x$
- **i.** $y = \tan x$

22. Using only elementary algebra, find each of the following. Justify your answers,
    - **a.** Find the equation of the line of symmetry of the function $f(x) = ax^2 + bx + c$.
    - **b.** In Problem 21, you found that the curve $y = kx^3$ has a half-turn symmetry about the origin. Similarly, show that the curve $y = ax^3 + bx^2 + cx + d$ has a half-turn symmetry about a certain point on the curve. Find that point.

23. Let f be a function whose domain is $\mathbb{R}$ (the set of real numbers) and whose values are in $\mathbb{R}$. What kind of symmetry does each of the following functions have? Justify your answers.
    - **a.** $g(x) = \dfrac{f(x) + f(-x)}{2}$
    - **b.** $h(x) = \dfrac{f(x) - f(-x)}{2}$

24. **a.** Prove that a rotation R by 90° counterclockwise about the origin is given by $R(x, y) = (-y, x)$. Check that $R^2 = R \circ R = H_0$.
    - **b.** Find the equations of the images of the curves in Problem 21 under the rotation given in part (a) of this problem.
    - **c.** Use your result from part (a) to prove that the product of the slopes of two perpendicular lines, neither of which is vertical, is -1.

25. **a.** Consider all lines $y = ax + b$. Which of these lines have the line $y = x$ as their line of symmetry?
    - **b.** Find all curves $y = \dfrac{ax+b}{cx+d}$ whose line of symmetry is the line $y = x$.

6.2 Congruence and Euclidean Constructions

Earlier, we defined two polygons as being congruent if there exists a one-to-one correspondence between their vertices such that the corresponding sides are congruent and the corresponding angles are congruent. This definition does not generalize to figures that are not polygons, such as the two smooth curves in Figure 6.15.

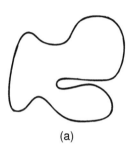
(a)

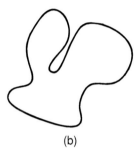
(b)

Figure 6.15

Euclid's idea of congruence was based on the assumption that figures can be moved without changing their size or shape. Thus, if we could move one of the figures so that, when one is superimposed on the other, the figures coincide, then the figures would be congruent. As an example, trace Figure 6.15(a) on a tracing paper and try to move it by sliding and rotating this figure so that when superimposed on Figure 6.15(b), it will coincide with the latter figure. You will find that this is impossible to accomplish. However, if you flip the tracing paper, you should be able to move it so that it will coincide with Figure 6.15(a). Because a formal concept of a function, and hence a transformation, was developed only in the 19th century, Euclid did not have the means to rigorously define his intuitively appealing approach to congruence. By contrast, our study of transformations—and in particular isometries—enables us to define congruence between any two figures.

Flipping the paper is equivalent to reflecting a figure in a line. Thus it seems that in Figure 6.15 one figure can be made to coincide with the other by reflecting one of the figures in a line if necessary, and then translating or rotating it appropriately. Thus we could define figure S to be congruent to figure S' if there exists a sequence of isometries that maps S onto S'. Because any composition of isometries is an isometry, we could simply require the existence of an isometry that maps S onto S'. Also notice that if an isometry maps S onto S', then its inverse maps S' onto S. Thus, if S is congruent to S', then S' is congruent to S. We summarize this discussion in the following definition.

> **Definition of Congruence**
>
> Two figures are congruent if there exists an isometry that maps one onto the other.

> **Now Solve This 6.7**
>
> In Figure 6.16, the ellipse E can be mapped onto the ellipse E' by a composition of two isometries. First use tracing paper to superimpose E onto E', and then write a composition of two familiar isometries that maps E onto E' (use the centers O and O' of the ellipses in your description of the isometries).
>
>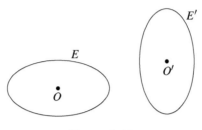
>
> Figure 6.16

Applications of Isometries to Construction Problems

Transformations can be used to solve geometrically a variety of maximum and minimum problems without the help of calculus. In fact, the transformational approach is often simpler and shorter. Earlier you may have worked on the problem of connecting two points on the same side of a line via a shortest path that goes through a point on the line. Because the approach in the solution of this problem is used in subsequent problems, we reintroduce the problem in the following example.

> **Example 6.3**
>
> Given points A and B on the same side of line ℓ (see Figure 6.17), find the point X on ℓ such that $AX + XB$ is minimum.
>
>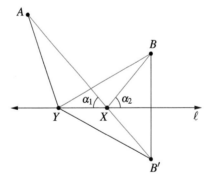
>
> Figure 6.17
>
> **In Search of a Strategy.** A simpler, though trivial, question would be to look at the case where A and B are on opposite sides of ℓ, as in Figure 6.18. In this case, the line segment would intersect ℓ at X and the path $A - X - B$ would be the required

path because any other path $A - Y - B$ would be longer (by the triangle inequality applied to $\triangle AYB$).

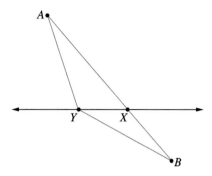

Figure 6.18

This simpler case suggests how to solve the original problem in Figure 6.17. We reflect B in ℓ and obtain its image B'. Because a reflection preserves distance, $XB = XB'$. Consequently, the problem is equivalent to finding a point X on ℓ such that the path $A - X - B'$ is minimal.

Solution. Connect A with B' (the image of B under reflection in ℓ). The intersection of $\overline{AB'}$ with ℓ is the required point X. Figure 6.17 shows why any other path is longer: $AY + YB = AY + YB' > AB'$ by triangle inequality. Because $AB' = AX + XB'$, it follows that $AY + YB' > AX + BX$.

In Figure 6.17, $\alpha_1 = \alpha_2$. This follows immediately from the fact that a reflection preserves the angle measure and that vertical angles are congruent. In fact, the proof of the converse statement—that is, if $\alpha_1 = \alpha_2$, then the path $A - X - B$ (where X is the point on ℓ for which $\alpha_1 = \alpha_2$) is the shortest—can be found by creating a figure similar to Figure 6.17. (This will be explored in Now Solve This 6.8.) This converse statement and its proof were discovered by Heron of Alexandria in the first century B.C.E. The Greeks knew that a ray of light r from a point A meeting a plane mirror m in a point X is reflected in the direction of a point B such that $\overline{AX}$ and $\overline{XB}$ form equal angles with the mirror (Figure 6.19). Heron proved that the path $A - X - B$ is the shortest possible path between A and B by way of the mirror.

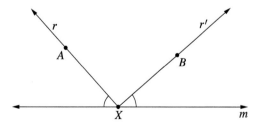

Figure 6.19

The shortest-path problem in Example 6.3 can be generalized to include several lines, as shown in Example 6.4.

> **Now Solve This 6.8**
>
> Prove that if X in Figure 6.17 is a point on line ℓ such that $\alpha_1 = \alpha_2$, then the path $A - X - B$ is the shortest among all paths from A to a point on ℓ and then to B.

> **Example 6.4**
>
> Given an acute angle A and an arbitrary point P in the interior of the angle, construct a triangle with one vertex at P and the other two vertices on each of the sides of the angle so that the perimeter of the triangle is minimum. See Figure 6.20(a).
>
>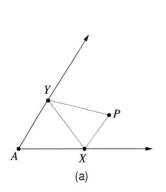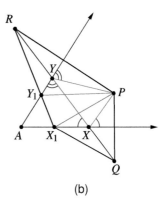
>
> (a)   (b)
>
> Figure 6.20
>
> **Solution.** Following the ideas in Example 6.3, we reflect P in each of the sides of the angle and obtain the respective images R and Q as shown in Figure 6.20(b). The segment $\overline{RQ}$ intersects the sides of the angle in points Y and X. We claim that $\triangle PXY$ is the triangle with minimal perimeter. To prove that this is the case, consider any other triangle PX_1Y_1 in Figure 6.20(b). The perimeter of that triangle is $RY_1 + Y_1X_1 + X_1Q$, while the perimeter of $\triangle PXY$ is $RY + YX + XQ$, which equals RQ. Because the segment connecting R and Q is the shortest path between R and Q, it follows that $RY_1 + Y_1X_1 + X_1Q > RQ$ and hence that the perimeter of $\triangle PY_1X_1$ is greater than the perimeter of $\triangle PYX$.

> **Remark 6.4**
>
> Notice that our construction implies that the equally marked angles in Figure 6.20(b) are congruent. Consequently, if a ray of light emanates from a source P inside an angle whose sides act like mirrors, the light will be reflected from both sides of the angle and will return to P if the light is directed toward point Y (found as described in Example 6.4).

6.2 Congruence and Euclidean Constructions

Example 6.5: The Shortest Highway Problem

A highway connecting two cities A and B as in Figure 6.21(a) needs to be built so that part of the highway is on a bridge perpendicular to the parallel banks b_1 and b_2 of a river. Where should the bridge be built so that the path $AXYB$ is as short as possible?

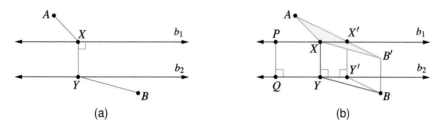

Figure 6.21

Solution. First we look at a special case. If the distance between lines b_1 and b_2 is 0, then the shortest path is the segment connecting A and B.

Returning to the original problem, we notice that regardless of where the bridge is located, its length is always the same: It is equal to the width of the river. Consequently, to make $AX + XY + YB$ [see Figure 6.21(b)] as short as possible, we need simply make $AX + BY$ as short as possible. Applying the "wishful thinking strategy," we wish that Y and X were the same point. Then the problem would be reduced to its special case. This can be achieved by moving $\overline{YB}$. Notice that Y can be moved to X by the translation τ_{YX}, which equals τ_{QP} where PQ is the width of the river. The image of $\overline{YB}$ in Figure 6.21(b) under τ_{QP} is $\overline{XB'}$. Consequently, $YB = XB'$ and therefore $AX + YB = AX + XB'$.

To solve the problem, we need to find the point X for which $AX + XB'$ is minimal. Thus we have reduced the problem to its special case. To find the shortest path, we connect A with B'. The point X' where $\overline{AB'}$ intersects b_1 is the location of X for which the highway is the shortest. The point Y' where the perpendicular to b_2 through X' intersects b_2 is the other end of the bridge. The proof that the highway along $\overline{AX'} - \overline{X'Y'} - \overline{Y'B}$ is the shortest possible path is embedded in the preceding discussion (consider the shaded $\triangle AXB'$).

Other Construction Problems

All of our examples so far have involved solving minimum-type problems. The next example is somewhat different.

Example 6.6

Given three parallel lines a, b, and c and a point P on one of the lines, construct an equilateral triangle PQS that has one vertex on each of the lines.

Investigation. Tyto, who loves constructions, proceeded as follows. Suppose P is on b and $\triangle PQS$ is equilateral, as required and as shown in Figure 6.22. We need to find the locations of Q and S. What do we know about Q? Because $\triangle PQS$ is equilateral, if we rotate Q about P by $60°$ clockwise, we get the point S—that is, the image of Q under $R_{P,\pi/3}$ is S.

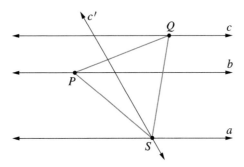

Figure 6.22

Thus, if we knew Q, we could find S. Using a "trial and error" approach, we could try many different points on c, rotate each about P by $60°$ clockwise, and see which one has its image on a. Of course, it would be better if we could rotate all the points on c about P by $60°$ clockwise. If c' is the image of c under this rotation, we can check which point on c' is also on a by finding the intersection of a with c'. The point of intersection is S. This rotation is accomplished in Figure 6.22. Point Q can now be found in several ways, including by finding the image of S under the rotation $R_{P,\pi/3}$.

Construction. As shown in Figure 6.22, we construct c', which is the image of c under $R_{P,-\pi/3}$. We then find the intersection of c' with a. Next we find Q, the image of S under $R_{P,\pi/3}$. Triangle PQS is the required triangle.

Proof. Because c' is the image of c under $R_{P,-\pi/3}$, the point S described in the construction is the image of a unique point on line c. That point is the preimage of S under $R_{P,-\pi/3}^{-1}$. Because $R_{P,-\pi/3}^{-1} = R_{P,\pi/3}$, the preimage is the point Q described in the construction. This implies that $\angle SPQ$ is $60°$ and $PS = PQ$. Consequently, $\triangle SPQ$ is an isosceles triangle with a vertex angle of $60°$. Hence all the angles of $\triangle SPQ$ must be $60°$ and the triangle is equilateral as required. ■

Now Solve This 6.9 includes some observations and questions related to Example 6.6 and its solution.

6.2 Congruence and Euclidean Constructions

Now Solve This 6.9

1. In the solution, we assumed that the given point is on line b, which is between a and c. Will a similar construction work if the given point is on any of the parallel lines?
2. A moment of thought will probably convince you that there is actually another triangle that satisfies the conditions of the problem and has point P as a vertex. Can you find it?
3. It is possible to find an equilateral triangle with a vertex on each of any three given curves such as lines or circles. Under what conditions will a solution exist?
4. Suppose four lines are given. Under what conditions is it possible to construct a square that has one vertex on each of the lines?

Example 6.7

In Figure 6.23(a) $\triangle ABC$ is an equilateral triangle inscribed in a circle and P is any point on the minor arc AC. Prove that $PA + PC = PB$. (This problem was first introduced in Chapter 3. We solve it here using a transformational approach.)

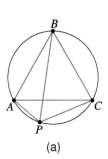 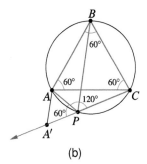

(a)   (b)

Figure 6.23

Proof. To obtain a single segment whose length is $PA + PC$, we extend the segment PC as shown in Figure 6.23(b). Because $ABCP$ is a quadrilateral inscribed in a circle, the opposite angles of the quadrilateral are supplementary. Hence $m(\angle APC) = 120°$ and the supplementary angle $\angle APA'$ is $60°$. Consequently, if we rotate point A by $60°$ about P, its image A' is on line PC and $PA = PA'$. Thus $PA + PC = PA' + PC = A'C$. It remains to prove that $A'C = PB$. To achieve this, we will show that the image of segment $A'C$ under an appropriate isometry is segment PB. Notice that under the rotation $R_{A,60°}$ the image of C is B. Thus we need simply show that the image of A' under that rotation is P. Because $\triangle AA'P$ is equilateral, this is the case and the theorem is proved. (Notice that the image of $\triangle ACA'$ is $\triangle ABP$.) ∎

6.2.1 Problem Set

In the following construction problems you should construct your own figures.

1. In the following figure, $ABCD$ is a rectangular swimming pool. Elysian is at a well-marked point P in the pool, and she needs to swim to the bank $\overline{BC}$, then to $\overline{CD}$, then to $\overline{DA}$, and finally back to the original point P. Elysian wants to minimize her total swimming distance around the path from P and back to P. The figure suggests how to accomplish this goal and how to justify that the path $PQST$ is the shortest. Examine the figure and answer the questions that follow.

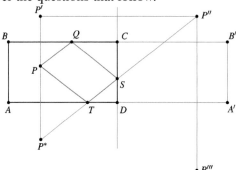

   a. Describe how to construct the required path $PQST$.

   b. Prove that the path $PQST$ is shorter than any path from P to a point on $\overline{BC}$, then to a point on $\overline{CD}$, then to a point on $\overline{AD}$, and back to P.

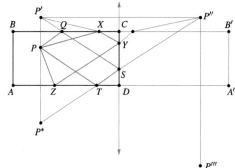

   c. If P''' is the image of P'' under reflection in line AD, prove that the intersection of lines PP''' and AD is the point T.

   d. Prove that $PQST$ is a parallelogram.

   e. Suppose the sides of the rectangle $ABCD$ have lengths a and b, where $BC = a$, $BA = b$, and $a > b$. Make a graph and a coordinate system so that A is at the origin, AD is on the x-axis, and AB is on the y-axis. Prove that $PQST$ is a rectangle if and only if P in the interior of the rectangle is at a distance $a - b$ from the side AB or the side CD.

   f. Using the coordinate system you developed in part (e), let the coordinate of P be (x_0, y_0) and let P be $a - b$ units away from side AB. If $PQST$ is a rectangle, find the coordinates of Q, S, and T in terms of a, b, x_0, or y_0.

g. Show that there is only one point P in the interior of $ABCD$ such that the path $PQST$ is a square. Where is that point?

h. Suppose the swimming pool is in the form of a square. For which points P will $PQSR$ be a rectangle and for which points will it be a square? Justify your answer.

i. Prove that if P is at the distance $a - b$ from side AB of the rectangle $ABCD$, and a line through P that is perpendicular to side AD is then drawn, the boundary of the shaded region below is a square. Use this result to prove part (h).

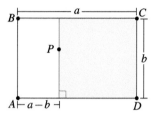

2. In the following figure, B is a ball and H is a hole on a miniature golf course. If you want the ball to bounce off all three walls, describe how to find the points P, Q, and S in the diagram. Assume that the angle of incidence equals the angle of reflection. Prove that the path $B - P - Q - S - H$ is the shortest path connecting B with H through points on all three walls.

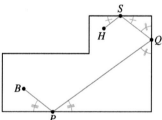

3. A kaleidoscope is in the shape of a prism and has a base in the form of an equilateral triangle whose side is a units. A beam of light is sent from a point P on the base of the prism at a 60° angle and is reflected from the mirrored sides, which are perpendicular to the base.

   a. Construct the path that the beam of light follows.

   b. Find, in terms of a, the length of the path that the beam of light follows from P to the moment it reaches P again.

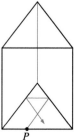

4. Given a point P, a line ℓ, and a circle as shown, construct points X and Y (X on ℓ and Y on the circle) such that P is the midpoint of segment XY. Motivate your steps in the construction (that is, explain how you knew to take the major steps) and prove that your construction is correct.

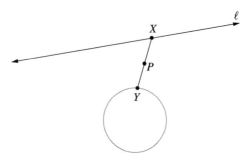

5. $ABCD$ is an a-unit by b-unit rectangular pool table. A ball positioned on side AB is hit toward side BC at a $45°$ angle. It bounces off the sides as shown in the figure and returns to its original position at P.

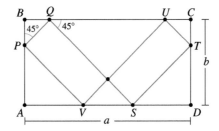

  a. Find all the values of a and b for which this phenomenon happens for every point on side AB. [*Hint*: Let $P(0, h)$, where $0 < h < b$, and calculate the coordinates of points Q, U, T, S, and V in terms of h, a, and b.]

  b. Suppose that P is at (k, h) in the interior of the rectangle. Use your result from part (a) to find a necessary and sufficient condition (relating k, h, a, and b) such that a ball sent from point P at a $45°$ angle toward side BC will follow a similar path to the one shown in the figure and return to P.

6. A highway connecting two cities A and B needs to be constructed such that a part of the highway is on a bridge perpendicular to the parallel banks a_1 and a_2 of a river and another part of the highway is on a second bridge perpendicular to the parallel banks b_1 and b_2 of a second river. Where should the bridges be built so that the highway is as short as possible? Describe the construction, construct the bridges and the highway, and prove that the highway you constructed is the shortest option.

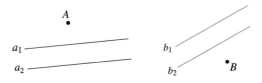

6.3 More on Extremal Problems

7. A new road connecting streets *A* and *B* needs to be constructed. The road, marked *XY* in the figure below, needs to be parallel to *PQ* and the same length as *PQ*. Construct the road. Motivate your construction and prove that it is correct.

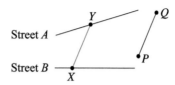

8. Given two circles and point *P*, construct a segment *AB* such that $AP = PB$, *A* is on circle O_1, and *B* is on circle O_2. How many solutions are there? Is there always a solution?

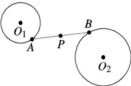

9. Construct an equilateral triangle such that there is exactly one vertex on each of the concentric circles shown.

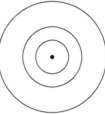

10. Construct two chords of equal length through points *A* and *B* that are perpendicular to each other. (Hint: If a line is rotated 90° about a point, the line and its image are perpendicular.)

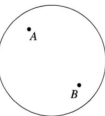

6.3 More on Extremal Problems

In the 19th century, several mathematicians investigated the solution of maximum and minimum problems by synthetic methods—that is, without using coordinates. The Swiss mathematician Jacob Steiner (1796–1863) was one of the greatest synthetic geometers of all time. Using purely geometric methods, he solved maxima and minima problems that had previously required the use of calculus, including some problems that even required the use of calculus of variations (a higher branch of mathematics that is concerned with finding maxima or minima of definite integrals). Perhaps one of the most famous problems that Steiner investigated was the **isoperimetric theorem**: Of all plane figures with a given perimeter, the circle encloses the greatest area. The famous German mathematician H. A. Schwarz (1843–1921) gave a complete proof of the isoperimetric problem in three dimensions; he also solved synthetically the following problem.

Inscribed Triangle with Minimum Perimeter

Given an acute-angled triangle, inscribe in it a triangle whose perimeter is as small as possible. In Figure 6.24, $\triangle DEF$ is inscribed in the given acute-angled triangle ABC. There are infinitely many triangles with vertices on each of the sides of $\triangle ABC$; our problem is to find the one with minimum perimeter.

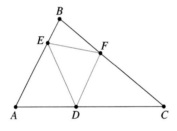

Figure 6.24

This problem was originally proposed by the Italian mathematician J. F. de'Toschi di Fagnano (1715–1797), who solved it in 1775 using calculus. Herman Schwarz (1843–1921) gave an ingenious synthetic solution that uses reflections and compositions of reflections. (See cut-the-knot.)

Another approach to the problem was given by the Hungarian mathematician Leopold Fejér (1880–1959), who was a student of Schwarz. His solution was praised by Schwarz as being especially simple and elegant; the following approach is based on that solution.

Investigation: In Figure 6.25, $\triangle ABC$ is an acute-angled triangle.

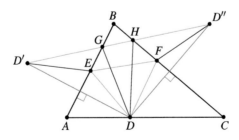

Figure 6.25

6.3 More on Extremal Problems

Let $\triangle DEF$ be an arbitrary triangle inscribed in $\triangle ABC$ (with D on side AC, E on side AB, and F on side BC). To find which inscribed triangle will have the smallest perimeter, we will represent the perimeter of $\triangle DEF$ by the length of a path connecting two points. For that purpose, we reflect one of the vertices of $\triangle DEF$ in the two sides of $\triangle ABC$ that do not contain the vertex. We reflect D in side AB to obtain the image D' and in side BC to obtain the image D'', as shown in Figure 6.25. Because $ED = ED'$ and $FD = FD''$, the perimeter of $\triangle EDF$ equals $D'E + EF + FD''$, the length of the path $D'EFD''$. Certainly, the shortest path between D' and D'' is the segment $D'D''$. That segment intersects the sides of $\triangle ABC$ at points G and H. The perimeter of $\triangle DGH$ is $DG + GH + HD = D'G + GH + HD'' = D'D$. Because the segment $D'D''$ is shorter than the length of the path $D'EFD''$, the perimeter of $\triangle DGH$ is less than the perimeter of $\triangle DEF$.

The trouble is that we don't know the position of D, and hence we don't know the positions of D' and D'' or the positions of G and H. But let us see what we have proved so far.

If we choose a point D (on side AC) and keep it fixed, then we have shown that among all inscribed triangles with one fixed vertex (at D), $\triangle DGH$ is the triangle with the minimal perimeter. We will refer to such a triangle as a **minimal triangle**. Now we need simply compare the various minimal triangles (for various positions of D between A and C) and pick the one with the smallest perimeter. This is equivalent to finding the shortest of the segments $D'D''$. Because the segment $D'D''$ is a side in $\triangle D'BD''$ (see Figure 6.26), we look more closely at that triangle.

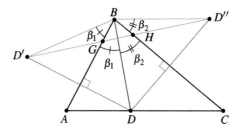

Figure 6.26

Notice that $BD' = BD$ (D' is the image of D under the reflection in side AB). Similarly, $BD'' = BD$. Thus $\triangle D'BD''$ is isosceles. Also, because a reflection preserves the angle measure, we have two pairs of equal-size angles with vertices at B—namely, two angles whose measure is denoted by β_1 in Figure 6.26 and two angles with measure β_2.

Thus $m(\angle D'BD'') = 2\beta_1 + 2\beta_2 = 2(\beta_1 + \beta_2) = 2 \cdot m(\angle ABC)$. For any position of D on side AC, $m(\angle D'BD'')$ is always the same: it is twice the measure of $\angle B$ in $\triangle ABC$.

Let us get back now to our main goal: finding the location of D on side AC for which $D'D''$ is the smallest. Because the angle at B in $\triangle D'BD''$ is the same for all D, the segment $D'D''$ will be the shortest when the congruent sides BD' and BD'' are shortest. (This fact should be intuitively obvious and not hard to prove.) Because $BD' = BD'' = BD$, we need simply find the point D on side AC for which BD is smallest. This happens if and only if D is the foot U of the altitude from B to side AC, as shown in Figure 6.27.

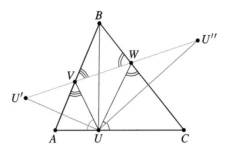

Figure 6.27

Thus the vertices of the required minimal triangle are point U and the points V and W where the line $U'U''$ intersects the sides AB and BC, respectively. (U' and U'' are the images of U under the reflections in sides AB and BC, respectively.) We have shown that the problem of finding an inscribed triangle with minimal perimeter has only one solution—that is, the minimal triangle UVW is unique. Had we started our solution by considering various triangles with vertices on side BC, we would have found that the minimal triangle must have as its vertex the foot of the altitude from A to side BC. Because the minimal triangle is unique, its vertices must be the feet of the three altitudes to the three sides of the original triangle. The triangle whose vertices are the feet of the three altitudes of a given triangle is called the **pedal triangle**.

The preceding investigation also proved that the minimal triangle and hence the pedal triangle share the property that at each vertex, the similarly marked angles in Figure 6.27 are congruent. This follows immediately from the congruency of vertical angles and the fact that reflection preserves angle measure. We will soon state this fact as a corollary. In the meantime, notice that the congruency of the previously mentioned angles implies that if three mirrors are placed on the sides of a given triangle and positioned perpendicular to the plane of the triangle, then whenever a beam of light emanates from any vertex of the pedal triangle toward another vertex of the pedal triangle, it will return to the original vertex from which it emanated. Using the principle that a beam of light travels via a path of shortest distance, it should be clear why the minimal triangle should have the property that at each vertex, the angles that its sides make with a side of the original triangle are equal.

We now state the theorem that follows from this investigation. The proof is embedded in the investigation. In the proof that follows, we omit the explorations and motivations we discussed earlier.

Theorem 6.1

Of all triangles inscribed in a given acute-angled triangle, the pedal triangle (the triangle whose vertices are the feet of the three altitudes of the given triangle) has the least perimeter.

Proof. Let EFD be a triangle inscribed in $\triangle ABC$ as shown in Figure 6.28. Reflect D in line AB to get D'. Also reflect D in line BC to get D''. From the properties of a reflection,

6.3 More on Extremal Problems

it follows that the perimeter of $\triangle DEF$ equals the length of the path $D'EFD''$. For a fixed D, that path will be the smallest if D' is connected to D'' by a straight segment. Then the vertices of the minimal triangle (for a fixed D) are the points G and H where the line $D'D''$ intersects the sides AB and BC of the given triangle, respectively (see Figure 6.29). Then the distance $D'D''$ equals the perimeter of $\triangle GHD$. Because reflection preserves distance, $D'B = BD$ and $BD'' = BD$. Consequently, $\triangle D'BD''$ is isosceles. Reflection also preserves angle measure, so $m(\angle D'BD'') = 2x + 2y = 2(x+y) = 2 \cdot m(\angle ABC)$. Thus, even though $\triangle D'BD''$ varies with the location of D, its vertex angle always remains the same. Hence $D'D''$—the length of the base of the isosceles triangle $D'BD'''$—will be the least when the legs $D'B$ and BD'' are shortest. Because each of these legs has the same length as BD, the length $D'D''$ (and therefore the perimeter of $\triangle GHD$) will be minimal when BD is shortest. That happens when $\overline{BD}$ is perpendicular to $\overline{AC}$ and hence when D is the foot of the altitude from B to $\overline{AC}$. Our argument, therefore, shows that the minimal triangle is unique.

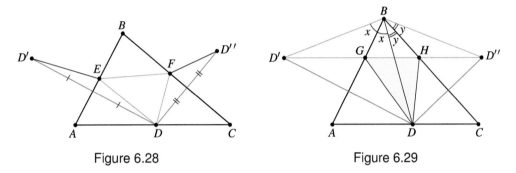

Figure 6.28   Figure 6.29

If we follow the same approach starting with another vertex of $\triangle ABC$, we find that the other vertices of the minimal triangle must be the feet of the altitudes from those vertices.

Corollary 6.1

At each vertex of a pedal triangle, both sides of the pedal triangle triangle make equal angles with a side of the original triangle.

Proof. This corollary follows from the construction in Figure 6.27 and the fact that reflection preserves angle measure.

The following minimum problem involving distances in a triangle was posed by Pierre de Fermat (1601–1665) to Evangelista Torricelli (1608–1647), a student of Galileo who is best known for his invention of the barometer.

Shortest-Network Problem. Given a triangle in which each angle measures less than 120°, construct a point in the interior of the triangle (or on the triangle) such that the sum of the distances from that point to the vertices is minimum.

Solution. Our solution is based on a solution given by J. E. Hoffman in 1929.

Consider $\triangle ABC$ in Figure 6.30, in which each angle measures less than 120°, and a point P in the interior of the triangle. The three segments connecting P with the vertices

A, B, and C will be referred to as a network. We need to find the point P for which the corresponding network is the shortest.

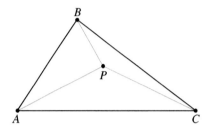

Figure 6.30

As in previous minimum problems, we try to find a pair of points and a path connecting them that is the same length as $PA + PB + PC$. In fact, we hope to find such a path that corresponds to every network. If the pair of points that are endpoints of the path turn out to be fixed (that is, the same for all choices of P), then the segment connecting the points will be the shortest path. Reflecting any point P will not be useful because P' will not be fixed. The ingenuity of Hoffman's approach was to use a rotation by 60° counterclockwise about any of the vertices. (It is not clear what motivated Hoffman to use such a rotation, other than perhaps the creation of equilateral triangles and hence congruent segments.) Figure 6.31(a) shows $\triangle ABC$ along with the corresponding images P' and A' of points P and A, respectively, under rotation with center B by 60° counterclockwise.

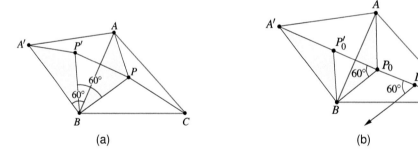

Figure 6.31

By connecting P with P' and A with A', we create two equilateral triangles BPP' and BAA'. Consequently, $AP + BP + CP = A'P' + P'P + CP$. Notice that regardless of the position of P in the interior of the triangle, A' is always at the same location. Thus, to find the shortest network, we need to find the shortest path connecting A' with C—that is, we need to draw the segment $A'C$. The position of the minimal point P_0 (the point that corresponds to the minimal network) is determined by the fact that $\angle A'P_0B$ is 60°. (We can pick any point D on the segment $A'C$, construct a 60° angle with a vertex at D and one side at the ray DA', and then construct a line through B parallel to the other side of the 60° angle.)

6.3 More on Extremal Problems

Because $m(\angle A'P_0B) = 60°$, the measure of the supplementary angle BP_0C is $120°$. Also notice that the minimal point P_0 is unique because the length of the network corresponding to any other point is greater than $AP_0 + BP_0 + CP_0$. Consequently, the same approach with rotation about B replaced by rotation about another vertex will result in the same minimal point P_0 and in angles BP_0C and CP_0A also being $120°$. A $60°$ rotation about A of vertex C in Figure 6.32 will result in the equilateral triangle $CC'A$, with the minimal point P_0 on the segment BC'. A rotation about C would result in the equilateral triangle CBB' and therefore in P_0 being on segment AB'. The minimal point P_0 is named **Fermat's point** after Pierre de Fermat (1601–1665).

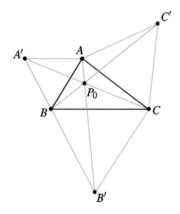

Figure 6.32

The fact that this solution is not valid for a triangle with an angle measure greater than $120°$ will be explored in the problem set at the end of this section. Meanwhile, we summarize our findings in the following theorem.

Theorem 6.2: Fermat's Point

1. *Given a triangle with no angle having measure greater than $120°$, the point in the interior of the triangle (Fermat's point) from which the sum of the distances to the three vertices of the triangle is minimum is the point at which each of the sides of the triangle subtends a $120°$ angle.*

2. *If an equilateral triangle is constructed (externally) on each of the sides of the triangle, then the three segments connecting a vertex of the given triangle with the vertex of the equilateral triangle constructed on the opposite side (and which is not a vertex of the original triangle) are concurrent at the Fermat's point of the original triangle.*

A Surprising Connection

The shortest-network problem for a triangle in which each angle measures less than $120°$ can be related to the theorem that states that the sum of the distances from every point in the interior of an equilateral triangle to the sides of the triangle is a constant, equal to the height of the triangle.

In Figure 6.33, $\triangle ABC$ is a triangle in which each angle measures less than $120°$ and P is the point in the interior of the triangle for which the segments PA, PB, and PC make $120°$ angles with each other. We will use the previously mentioned theorem about equilateral triangles to prove that the point P minimizes the sum $PA + PB + PC$.

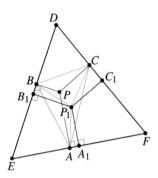

Figure 6.33

To connect this problem to the sum of the distances from a point in the interior of an equilateral triangle to its sides, we construct lines through A, B, and C perpendicular to the segments AP, BP, and CP, respectively. These lines intersect at D, E, and F and form $\triangle DEF$. We first show that this triangle is equilateral. To find the measure of the angle at E, consider the quadrilateral $EBPA$. The angles of the quadrilateral at A and B are $90°$ each, and the angle of the quadrilateral at P is $120°$. Hence $m(\angle E) = 60°$. Similarly, the angles at D and F are $60°$ each. As a consequence, $PA + PB + PC$ is the sum of the distances from P to the sides of the equilateral triangle EDF and hence a constant. We need to show now that if P_1 is another point in the interior of $\triangle ABC$, then $PA + PB + PC < P_1A + P_1B + P_1C$. For that purpose, notice that our property of equilateral triangles implies that $PA + PB + PC = P_1A_1 + P_1B_1 + P_1C_1$, where the quantities on the right side of the equation are the distances from P_1 to the three sides of $\triangle EDF$, respectively. However, $P_1A_1 < P_1A$, $P_1B_1 < P_1B$, and $P_1C_1 < P_1C$ (a leg in a right triangle is shorter than the hypotenuse). Thus $PA + PB + PC < P_1A + P_1B + P_1C$.

Shortest-Network Problems for Polygons

A natural generalization of the minimal network problem for three points is to find the point P in the plane for which the sum of the distances from P to n given points is a minimum. For four points that are vertices of a convex quadrilateral, the minimal point is the point of intersection of the diagonals. (The proof uses only the triangle inequality and will be explored in Problem 5 in the problem set at the end of this section.)

Jacob Steiner investigated a more useful generalization of the minimal network problem. He abandoned the search for a single point P and instead looked for a kind of "street network" of shortest total length. The problem can be expressed as follows.

6.3 More on Extremal Problems

Steiner's Shortest-Network Problem

Given n points $A_1, A_2, \ldots, A_n$, find a connected set of line segments of shortest total length such that any two of the given points can be joined by a polygonal path consisting of some of the line segments.

The solution to the problem is called **Steiner networks**. Analogous to his solution to the three-points problem, for four points Steiner constructed equilateral triangles on two opposite sides of the quadrilateral $ABCD$, as shown in Figure 6.34.

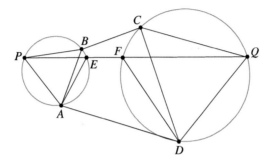

Figure 6.34

He then constructed the circles that circumscribe the triangles and connected the vertices P and Q (the vertices that are also not the vertices of the quadrilateral) by a straight line. The line intersects the circles at points E and F. Steiner showed that $AE + BE = PE$ and, similarly, that $CF + FD = FQ$. Consequently, the length of the network connecting the vertices of the quadrilateral with E and F is $AE + BE + EF + CF + FD = PQ$.

Next Steiner created an analogous construction on the other two opposite sides of the quadrilateral, $\overline{BC}$ and $\overline{AD}$, and obtained a corresponding network (not shown in Figure 6.34). He proved that the shorter of the two networks is the solution to the problem; that is, it is a Steiner network for four points. (Some quadrilaterals have two different Steiner networks of the same length.) A partial proof will be investigated in the problem set. The interested reader can find the full details in Steiner's original works (Steiner, 1882, 1951).

Steiner networks for regular pentagons, regular hexagons, and even some polygons with a greater number of sides have been found, but exact solutions for arbitrarily arranged n points are not known. Nevertheless, in this and other extremal problems, it is possible to see the solution of a mathematical problem as an interpretation of a physical phenomenon. Steiner networks can be realized using the following physical principle: Because of surface tension, a film of liquid will be in a stable condition if its surface area is minimal. For example, if two parallel plastic plates joined by four pins perpendicular to the plates are immersed in a soap solution and then withdrawn, the resulting film forms a system of vertical planes between the plastic plates joining the four pins. The projection of the system of planes is a Steiner network for the four points. Shortest networks have applications in engineering and computer science.

Historical Note: *Jacob Steiner*

The Swiss mathematician Jacob Steiner (1796–1863) is considered one of the greatest geometers of all times. Steiner was born at Utzensdorf on March 18, 1796. He grew up without schooling and did not learn to write until age 14. Against the wishes of his parents, at age 18 Steiner enrolled in the innovative Pestalozzi boarding school at Yverdon, Switzerland, where his exceptional geometric talent was discovered. [The school was established by the Swiss educator Johann Heinrich Pestalozzi (1746–1827), who was revered in Switzerland for his dedication to orphans and underprivileged children. His school, founded in 1805, became a model for teaching and teacher training throughout Europe.] Later, Steiner studied at universities in Heidelberg and Berlin, where he successfully supported himself by tutoring other students. He became well known when his articles were published in a newly founded mathematical journal, *Crell's Journal*. In 1834, Steiner was appointed extraordinary professor at the University of Berlin. He possessed unusual talent in the synthetic treatment of geometry and made major contributions to many areas of geometry. In particular, he greatly contributed to projective geometry, where many of the basic concepts and results are attributed to him. Steiner disliked analytic and algebraic methods, arguing that calculations replace thinking. He made so many new discoveries that he did not always have time to record their proofs. For this reason, many of his theorems were later proved by others.

6.3.1 Problem Set

1. If all the angles of $\triangle ABC$ are acute, prove each of the following:
   a. The altitudes of $\triangle ABC$ are the angle bisectors of its pedal triangle.
   b. The perimeter of the pedal triangle is less than twice any height of the original triangle.
   c. The perimeter p of the pedal triangle is given by $p = 4R \sin A \sin B \sin C$, where R is the radius of the circle that circumscribes $\triangle ABC$.
   d. The ratio between the perimeters of $\triangle ABC$ and its pedal triangle equals the ratio between the circumferences of the circle that circumscribes $\triangle ABC$ and the circle that is inscribed in that triangle. (This statement is equivalent to $P/p = R/r$, where P is the perimeter of the original triangle, p is the perimeter of its pedal triangle, and R and r are the radii of the circumscribing circle and the inscribed circle of $\triangle ABC$, respectively.)

2. In this problem you are asked to investigate what happens to Fagnano's problem (the inscribed triangle with minimum perimeter) when the original triangle is not an acute-angled triangle.
   a. What happens when one of the angles of the triangle is a right angle? Why?

6.3 More on Extremal Problems

b. Prove that if one of the angles of $\triangle ABC$ (the angle at C in the figure) is obtuse, the perimeter of the pedal triangle is greater than twice the smallest height of $\triangle ABC$ and hence is not a solution to Fagnano's problem.

c. Conjecture what the solution to Fagnano's problem is when one of the angles of the triangle is obtuse.

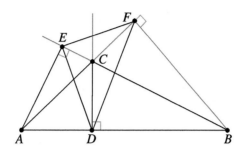

3. Let $\triangle ABC$ be an acute-angled triangle, and let $A_1B_1C_1$ be its pedal triangle. Let A_2 be a point between A_1 and B such that $A_1A_2 < A_1C$. A beam of light is sent from A_2 parallel to the side A_1C_1 of the pedal triangle, and is reflected from the sides of $\triangle ABC$ (which act like mirrors).

a. Use a compass and a straightedge or geometry computer software to confirm that the path repeats itself after six reflections.

b. Choose different positions for A_2 and, in each case, measure the length of the closed path A_2-C_2-B_2-A_3-C_3-B_3-A_2. What do you notice? Write your answer in the form of a conjecture.

c. Conjecture a relationship between the perimeter of $\triangle A_1B_1C_1$ and the length of the path in part (b).

d. Prove that $A_1A_2 = A_1A_3$, $B_1B_2 = B_1B_3$, $C_1C_2 = C_1C_3$, and so on.

e. Use your result from part (d) to prove your conjecture in part (b).

f. Use your result from part (d) to justify the assertion in (a).

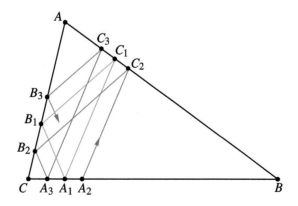

4. In $\triangle ABC$, each angle measures less than $120°$. On each of the sides of the triangle, equilateral triangles have been constructed externally, as shown in the figure. Also, circles circumscribing the triangles have been constructed.

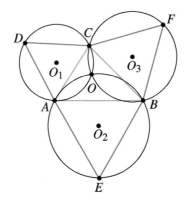

  a. Prove that the three circles are concurrent (intersect in a single point). What is the significance of the point O at which the circles intersect? Justify your answer.

  b. Use rotations to prove that the segments joining each vertex of $\triangle ABC$ with the remote vertex of the corresponding equilateral triangle constructed on the opposite side of $\triangle ABC$ are congruent.

  c. Draw a triangle in which one of the angles is greater than $120°$. What happens with the construction in part (a) for this triangle?

  d. Is it possible to construct Fermat's point for the triangle in part (c)? Why or why not?

  e. Prove that the centers O_1, O_2, and O_3 of the three circles determine an equilateral triangle. (We will later prove this statement using a newly introduced size transformation; for now, you can prove it using similar triangles.)

5. Given a convex quadrilateral, prove that the point determined by the intersection of the diagonals is the minimum distance point for the quadrilateral—that is, the point from which the sum of the distances to the vertices is minimal.

6. After completing Problem 5, determine whether the following construction of the minimum distance point for a convex quadrilateral is valid. Justify your answer.

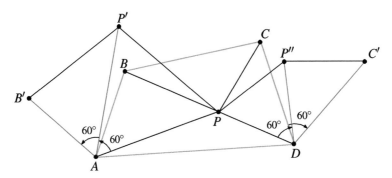

6.3 More on Extremal Problems

Given a convex quadrilateral $ABCD$, let P be a point in its interior, let B' be the image of B under a rotation with center A by $60°$ counterclockwise, and let P' be the image of P under the same rotation. Then $B'P' = BP$ and $AP = PP'$. Similarly, consider the rotation with center D by $60°$ clockwise. Let C' be the image of C and P'' be the image of P under this rotation. Then $DP = PP''$ and $CP = C'P''$. Thus $BP + AP + PD + PC = B'P' + P'P + PP'' + P''C'$. Consequently, for every point P in the interior of $ABCD$, there exists a corresponding route connecting the fixed points B' and C' whose length is the sum of the distances from P to the four vertices of the quadrilateral. The shortest such route will be the one connecting B' to C' by a straight line. Thus P will be on the segment $B'C'$.

Next we make a similar construction for the other pair of opposite sides of the quadrilateral by connecting C'' with D' (not shown in the figure), where C'' is the image of C under a rotation with center B, $60°$ counterclockwise, and D' is the image of D under a rotation with center A, $60°$ clock wise. Because P must also be on the segment $C''D'$, it is determined by the intersection of the segments $B'C'$ and $C''D'$.

Does this construction result in the same point as the construction in Problem 5? Why or why not?

7. Let $ABCD$ be a square with side a. Find, in terms of a, the length of the network connecting the vertices of the square shown in the figure. In other words, show that this network is shorter than the path determined by the intersection of the diagonals.

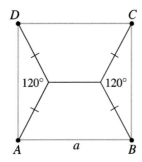

6.4 Similarity Transformation with Applications to Constructions

A similarity of the plane is a transformation S_k of the plane that multiplies all distances by the same positive constant k called the scale factor.[1] That is, for all points A and B in the plane, if $S_k(A) = A'$ and $S_k(B) = B'$ then $A'B' = k \cdot AB$.

Notice that every isometry is a similarity with scale factor 1, but not every similarity is an isometry. We next define a special similarity; a **dilation**, also called **central similitude** or **homothety**.

Definition of Dilation

A dilation with center O and scale factor k is a transformation of the plane that, for $k > 0$, assigns to each point P in the plane a point P' on the ray $\overrightarrow{OP}$ such that $OP' = k \cdot OP$. If $k < 0$, then P' is on the ray opposite $\overrightarrow{OP}$ (so that O is between P and P') and $OP' = |k|OP$. Point O is called the **center of the dilation**.

Now Solve This 6.10

Justify each of the following statements:

1. A dilation is a similarity transformation.
2. Under a dilation:
   a. Segment AB not belonging to a line through the center O is transformed onto the segment $A'B'$ parallel to segment AB.
   b. A line through O is transformed onto itself, and a line not through O is transformed onto a parallel line.
   c. Every triangle is transformed onto a similar triangle, and consequently angles are transformed to congruent angles.
   d. A circle is transformed onto a circle. The image of the center of the circle is the center of the transformed circle, and the ratio of the radius of the transformed circle to the radius of the given circle is equal to the scale factor of the dilation.
   e. In a coordinate plane, if the center of the dilation is at the origin, then the image of point (x, y) is (kx, ky), that is, $(x, y) \rightarrow (kx, ky)$.

Using similarity transformations, we can define any two figures to be **similar** if there exists a similarity transformation that maps one figure onto the other. Two figures are called **centrally similar** if there exists a dilation that maps one figure onto the other. In Chapter 7, we will prove that every similarity transformation is a composition of a dilation followed by an isometry. You may want to try to prove this statement as well as another property of dilations in Now Solve This 6.11.

---
[1]The terminology related to similarity transformations is not standardized. For example, some texts use the term "similarity ratio" instead of "scale factor."

6.4 Similarity Transformation with Applications to Constructions

> **Now Solve This 6.11**
>
> 1. **a.** Let S_k be a similarity transformation with scale factor k, and let $C_{1/k}$ be a dilation about some point O. Prove that $S_k \circ C_{1/k}$ is an isometry and hence that any similarity transformation is a composition of a dilation followed by an isometry. (Notice that point O can be chosen anywhere in the plane.)
>   **b.** Is it true that a similarity transformation is a composition of an isometry followed by a dilation? Justify your answer.
> 2. Prove that the centers of dilation of three triangles that are pairwise centrally similar are collinear. You may want to proceed as follows: In Figure 6.35, O_{12}, O_{13}, and O_{23} are the respective centers of the dilations that take $\triangle P_1Q_1S_1$ onto $\triangle P_2Q_2S_2$, $\triangle P_1Q_1S_1$ onto $\triangle P_3Q_3S_3$, and $\triangle P_2Q_2S_2$ onto $\triangle P_3Q_3S_3$. Let ℓ be the line $O_{23}O_{13}$, and show that ℓ goes through O_{12}. For that purpose, show that the image of ℓ under the dilation with center O_{13} is ℓ itself and that the image of ℓ under the dilation with center O_{23} is ℓ itself.
>
>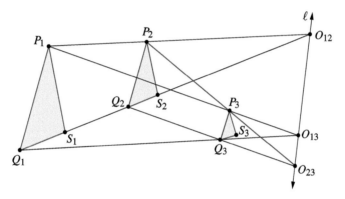
>
> Figure 6.35

Similarity of Circles

Any two circles with different radii can be shown to be centrally similar. We first consider nonconcentric circles. Let P_1 and P_2 be the centers of two circles with radii r_1 and r_2, respectively, as shown in Figure 6.36. We want to find the dilation that will transform the circle with center P_2 onto the circle with center P_1. That is, we want to find the center of the dilation and the scale factor. For that purpose, we can construct any diameter through P_2 intersecting the circle with center P_2 at A_2 and B_2 as shown in Figure 6.36. The required dilation should map the center P_2 onto P_1 and the radius A_2P_2 to a parallel radius A_1P_1. If the image of A_2 is to be A_1 (under the required dilation), then the center of the dilation is O_1, which is the point of intersection of the lines A_1A_2 and P_1P_2. However, if the image of B_2 is to be A_1, then the center of the dilation is the point O_2. For each of the dilations transforming circle P_2 onto circle P_1, the scale factor $k = \frac{A_1P_1}{A_2P_2} = \frac{r_1}{r_2}$.

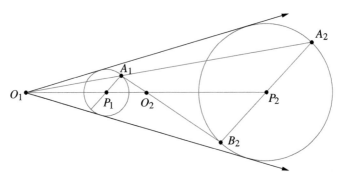

Figure 6.36

6.4.1 Construction of a Common Tangent

Figure 6.36 and our earlier discussion suggest a new way to construct a common tangent to two circles with different radii (when one circle is not in the interior of the other). We first find O_1, the center of dilation; we then construct the tangent to one of the circles through O_1 (an easy task). That tangent must also be the tangent to the other circle because one point of tangency must be the image or preimage of the other point of tangency under the dilation. The internal tangent can be constructed in a similar way.

Compare this construction of a common tangent to the one presented in Chapter 3. Which seems to be easier? Why? Next we prove a celebrated theorem attributable to Leonhard Euler.

> **Theorem 6.3: Euler Line**
>
> *The centroid G (the intersection of the medians), the orthocenter H (the intersection of the altitudes), and the circumcenter O (the center of the circumscribing circle) of a triangle are collinear. (This line is called the Euler line.) Moreover $OG = \frac{1}{3}OH$.*

Proof. In Figure 6.37, A', B', and C' are the midpoints of the sides of $\triangle ABC$ opposite the vertices A, B, and C, respectively. The point O is the orthocenter of $\triangle A'B'C'$. Under dilation with center G and dilation factor $-\frac{1}{2}$, the image of $\triangle ABC$ is $\triangle A'B'C'$. Because under the dilation the orthocenters must correspond, the image of H is O. By definition of a dilation O, G, and H are collinear. Notice also that the medians of $\triangle ABC$ bisect the sides of $\triangle A'B'C'$. Therefore the centroids of the triangles are the same point, G. (This should also be clear because the dilation with center G takes G to itself.) If we name the dilation f, we have

$$f(G) = G, \quad f(H) = O.$$

6.4 Similarity Transformation with Applications to Constructions

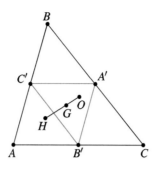

Figure 6.37

Since the absolute value of the dilation factor is $\frac{1}{2}$, we have $OG = \frac{1}{2}HG$.

> **Historical Note:** *Leonhard Euler*
> Leonhard Euler (1707–1783) was the most prolific mathematician of all time. Euler was born in Basel, Switzerland, and finished his studies at the University of Basel at age 15.
> Euler published more than 800 books and articles. His collected works are published in 75 substantial volumes. Euler's name is attached to concepts and theorems in almost all branches of mathematics. Among them are *Euler's characteristic, Euler circle, Euler circuit, Euler's phi function, Euler's formula,* and *Euler multiplier*. Euler introduced the notation $f(x)$ (in 1734), e for the base of the natural logarithm (in 1727), i for $\sqrt{-1}$ (in 1777), Σ for summation (in 1755), R and r for the radii of circumscribing and inscribed circles, and $\sin x$ and $\cos x$ for values of the sine and cosine functions, among many other notations.
> Euler lost the sight in one of his eyes at age 31; soon after, he became totally blind. He produced about half of his work while being blind. Euler had 13 children and many grandchildren. He claimed that he did some of his best work while holding a baby, with other children playing around his feet.

Constructions Using Dilation

The definition and properties of dilation can simplify many construction problems, as we shall see next.

Construction 6.4.2

Construct a circle through a point in the interior of a given angle that is not on the angle bisector of the angle, and is tangent to the sides of the angle. In the following solution we combine the three stages of investigation, construction, and proof into one. In Figure 6.38, $\angle A$ and the point P_1 in the interior of the angle are given. We need to construct a circle through P_1 tangent to the sides of the angle. We imagine that such a circle is already constructed. The center of the desired circle must be on the angle bisector of ∠A. We can easily construct a different circle with center O at some point on the angle bisector of ∠A, tangent to the sides of ∠A (the radius of this circle is OH, the distance from O to a side of the angle, as shown in Figure 6.38). The desired circle through P_1 and center O_1 can be regarded as the image of the circle O under a dilation with center of similitude at A. To find the point on the circle O that is mapped to P_1, we connect A with P_1. The point P that is the preimage of P_1 under the dilation must be on $\overrightarrow{AP_1}$ as well as on the circle O. Two such points, P and P^*, exist as shown in Figure 6.38.

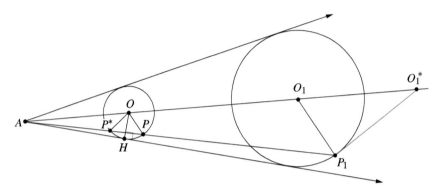

Figure 6.38

We know that under dilation $\overline{O_1P_1}$, the image of $\overline{OP}$ and $\overline{OP}$ itself are parallel radii. Thus we can find the center of the desired circle by drawing a line through P_1, parallel to line OP. That line intersects the angle bisector at O_1, which is the center of the required circle (whose radius is O_1P_1).

Another circle also satisfies the requirements of this problem. If we draw a line parallel to line OP^* through P_1, the point where that line intersects the angle bisector at O_1^* is the center of the second circle that satisfies the requirements of the problem (that circle is not shown in Figure 6.38).

The next construction was introduced in Chapter 5. Here we solve it using a similarity transformation.

Construction 6.4.3

Construct a square inscribed in a given triangle such that one of the sides of the square is on a side of the triangle.

We imagine that the required square $E'F'G'H'$ in Figure 6.39(a) is already constructed and that it is the image of another square $EFGH$ under a dilation with center A. We can easily construct the square $EFGH$ with F on line AB and G as well as H on line AC. Because E' is on $\overline{BC}$ as well as on $\overrightarrow{AE}$, we can find it by identifying the point of intersection of $\overrightarrow{AE}$ with side BC. Dropping the perpendicular from E' to side AC, we obtain the vertex H'. The vertex F' is the point of intersection of the line through E' parallel to line AC with side AB. The perpendicular from F' to line AC determines the fourth vertex, G'. Because

6.4 Similarity Transformation with Applications to Constructions

under dilation the sides of $EFGH$ are mapped to segments parallel to the corresponding sides of $EFGH$ and each segment is enlarged by the same factor, $E'F'G'H'$ is, indeed, a square.

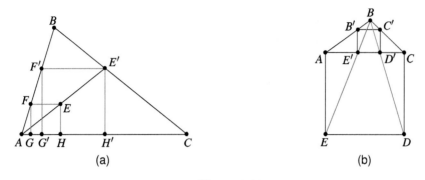

Figure 6.39

Figure 6.39(b) shows a dilation with center at B that achieves the same construction.

6.4.2 Problem Set

In your answers to the following problems, include an investigation, a construction, and a proof.

1. In a semicircle with diameter AB, construct a square with two vertices on the semicircle and two vertices on the diameter.

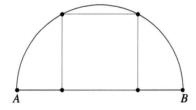

2. a. In a given circle, inscribe a square $ABCD$ and a smaller square $EFGH$ with two vertices on the arc BC and the other two vertices on side BC.
   b. If the side of the larger square is s, find the side of the smaller square in terms of s.

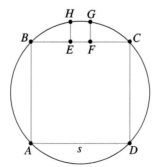

3. Given three lines a, b, and c intersecting at point O, construct a circle tangent to line a, having its center on b, and cutting off on line c a chord congruent to a given segment.

4. Inscribe a rectangle in which one side is twice as long as the other side in each of the following:

   a. A triangle

   b. A semicircle

5. Two circles with centers O_1 and O_2, and a point P are given. Construct a third circle such that the line through A and B the points of tangency with the two given circles, passes through point P. (Notice that points A, B and X are not known; they appear in the figure to clarify what is needed.)

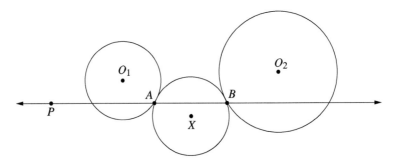

6. Through a point in the interior of a given circle, construct a chord that is bisected by the point.

7. In a given circle, inscribe a triangle similar to a given triangle.

8. Are all parabolas similar to each other? Justify your answer.

9. Consider the ellipses $\frac{x^2}{a^2} + \frac{y^2}{b^2} = 1$ and $\frac{x^2}{a_1^2} + \frac{y^2}{b_1^2} = 1$. Find a necessary and sufficient condition for the ellipses to be similar. (The condition should relate a, b, a_1, and b_1.)

10. Are all hyperbolas $xy = c$, where c is any real number, similar to each other? Justify your answer.

7 Composition of Transformations

The pursuit of truth and beauty is a sphere of activity in which we are permitted to remain children all our lives.

Albert Einstein (1879–1955) *American physicist and Nobel laureate*

7.1 Introduction

In Chapter 6, we saw that a composition of two transformations is a transformation and that composition of two isometries is an isometry. We have encountered four types of isometries: reflections, translations, rotations, and glide reflections. Are these the only isometries, or do others exist?

Because a composition of two isometries is an isometry, it is natural to ask whether a composition of any of the isometries we have encountered so far results in one of those isometries or, perhaps, in a new isometry. We will study compositions of reflections, translations, rotations and glide reflections, which were first introduced in Chapter 6. We will prove that there are no other isometries. At the same time, we will discover that reflections are the building blocks for the plane isometries in the sense that every isometry is a composition of at most three reflections. The information about the results of various compositions of isometries will enable us to solve some geometric problems in simple and efficient ways. The Treasure Island Problem introduced in Chapter 1 and revisited in subsequent chapters is one such problem. Knowing the result of composition of two rotations about two different points will not only give us a surprisingly simple solution and proof, but will also show similar "treasure island" problems that can be created and solved.

In geometry, crucial roles are played not only by individual transformations but also by sets of transformations satisfying prescribed conditions, called **transformation groups**. The concept of a group is a unifying one for all of mathematics, and in geometry in particular, it will enable us to investigate symmetry of figures.

In 1827, the German astronomer and mathematician A. F. Möbius gave a meaning to the composition of isometries by extending the idea of transformation to the whole plane. This extension enabled him to apply the concept of a *group* to geometry, and to define notions such as congruence and similarity as properties that were invariant (unchanged) under certain groups of transformations of the plane. These concepts did not gain wide acceptance among mathematicians until 45 years later, when another German mathematician, Felix Klein, extended Möbius's ideas in his inaugural lecture, in honor of his appointment to a chair at Erlanger University. In that lecture, which became known as the *Erlanger Programme*, Klein introduced a unifying principle for classifying Euclidean and various non-Euclidean geometries: He defined **geometry** as the study of those properties of a given set that remain invariant under some group of transformations. Klein's approach has had an enormous impact on the development of modern mathematics.

> **Historical Note:**
> **August Ferdinand Möbius** (1790–1868) was a student of Gauss and became a professor of astronomy at Leipzig University in 1815. In a paper presented to the French Academy and discovered only after his death, Möbius describes the properties of one-sided surfaces, including the now-famous Möbius strip. Around 1850, Möbius pioneered the study of transformations in the geometry of complex numbers. A transformation named after him has important applications in both pure and applied mathematics.
>
> At age 23, **Felix Klein** (1849–1925) became a professor at the University of Erlanger, where he delivered his famous Erlanger Programme. In 1886, Klein accepted a position at Göttingen University, which, under his leadership, became a world center for mathematical research. Klein was a gifted teacher and personally supervised 48 doctoral dissertations. In 1895, the English mathematician Grace Chisholm, who, as a woman, was not permitted to attend graduate school in England, was awarded a Ph.D. in mathematics under Klein. Her doctorate was the first in any field to be awarded to a woman in Germany. In 1897, the American mathematician Mary Winston also earned her doctorate under Klein. She was the first American woman to obtain a Ph.D. in mathematics from a European university.

7.2 In Search of New Isometries

Chapter 6 introduced three basic types of isometries: reflections, translations, and rotations (we are assuming here that the glide reflection has not yet been discovered). To find other possible isometries, we will look at various compositions of the isometries we have encountered so far. For that purpose, it is useful to know which qualities determine an isometry. For example is there a unique isometry that maps a given point A to point A'? The reflection M_ℓ, where ℓ is the perpendicular bisector of the segment AA', is one such isometry because $M_\ell(A) = A'$. But the translation $\tau_{AA'}$ (the translation from A to A') also maps A to A'. Moreover, any rotation with center O on the perpendicular bisector of $\overline{AA'}$ (see Figure 7.1), and by $\angle AOA'$ in an appropriate direction (counterclockwise in Figure 7.1), will map A onto A'.

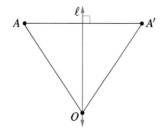

Figure 7.1

7.2 In Search of New Isometries

We have just seen that given two points there are infinitely many isometries that map one point onto the other. What if two points A, B and two other points A', B' are given so that $A'B' = AB$? Is there a unique isometry that map A to A' and B to B'? (Why do we need $A'B' = AB$?) We will show that at least two different isometries map A to A' and B to B'. In Figure 7.2, A is mapped to A' by the reflection M_k in the perpendicular bisector k of $\overline{AA'}$. Under M_k, the point B is mapped to B^*. Next B^* is mapped to B' by the reflection M_ℓ in the perpendicular bisector ℓ of $\overline{B^*B'}$. Thus the composition $M_\ell \circ M_k$ maps B to B'. [Notice that $(M_\ell \circ M_k)(B) = M_\ell(M_k(B)) = M_\ell(B^*) = B'$.] We will show that $M_\ell \circ M_k$ maps A to A'.

First A is mapped by M_k to A'. Careful drawing as in Figure 7.2 suggests that A' is on ℓ.

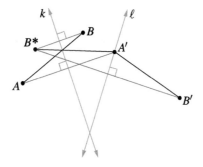

Figure 7.2

If that were the case, then we would have $(M_\ell \circ M_k)(A) = M_\ell(M_k(A)) = M_\ell(A') = A'$. Because ℓ is the perpendicular bisector of $\overline{B^*B'}$, to prove that A' is on ℓ we need simply to show that A' is equidistant from the endpoints of $\overline{B^*B'}$—that is, that $B^*A' = A'B'$. To show that this is the case, notice that $\overline{B^*A'}$ is the image of $\overline{BA}$ under M_k and hence $B^*A' = BA$. It was given that $BA = B'A'$, so $B^*A' = B'A'$ and, therefore, A' is on ℓ. Thus we have shown that $M_\ell \circ M_k$ maps A to A' and B to B'.

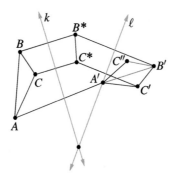

Figure 7.3

To show that another isometry maps A to A' and B to B', we use the fact that an isometry maps a triangle onto a congruent triangle. We pick any point C not collinear with A and

B, as in Figure 7.3. and obtain its image C' under the isometry $M_\ell \circ M_k$ that mapped the points A, B to A', B', respectively. For convenience, we denote $M_\ell \circ M_k$ by T. We will express the new isometry in terms of T. Notice that reflecting $\triangle A'B'C'$ in line $A'B'$ results in $\triangle A'B'C''$ congruent to $\triangle A'B'C'$. For convenience we denote line $A'B'$ by n. Because A' and B' are fixed points under M_n, the transformation $M_n \circ T$ still maps A to A' and B to B'. However, $M_n \circ T \neq T$ because T maps C to C' and $M_n \circ T$ maps C to C''. (Recall that two functions with the same domain are equal if and only if they give the same output for every input.) Thus we have found two different isometries that map A to A' and B to B'.

Are these two isometries the only isometries that map A, B onto A', B' respectively? You will soon be able to answer this question.

> **Now Solve This 7.1**
>
> We have seen that if $A'B' = AB$, then there are at least two different isometries, each a composition of reflections, that map A to A' and B to B'. It is easy to find two isometries, each a composition of translation followed by a rotation, that accomplish the same task. They are:
>
> 1. Translation from A to A' followed by a rotation about A'.
> 2. Let M be the midpoint of $\overline{AB}$ and M' the midpoint of $\overline{A'B'}$; then the translation from M to M' followed by a rotation about M' will also map A to A', and B to B'.
>
> Are the two isometries different? (Remember that two functions may be equal in spite of looking different. You may like to use the theorem (which will be proved soon) that if two isometries agree on three noncollinear points, then they are equal.

Necessary and Sufficient Conditions for Three Points and Their Images to Determine a Unique Isometry

You may wonder if three noncollinear points and their images determine a unique isometry, or if you need even more points for an isometry to be unique. Given three noncollinear points A, B, C and the points A', B', C', is there a unique isometry that maps A to A', B to B', and C to C'? Since we are looking for an isometry, distances must be preserved:

$$A'B' = AB,$$
$$B'C' = BC, \qquad (7.1)$$
$$A'C' = AC.$$

Thus the given points must satisfy condition (7.1). If the points A, B, C are collinear, then under every isometry the images of these points will also be collinear (why?). Thus A', B', C' must be collinear and, if B is between A and C, B' must be between A' and C' as in Figure 7.4 (why?).

7.2 In Search of New Isometries

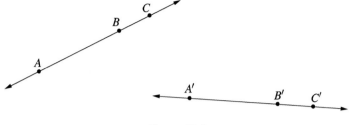

Figure 7.4

Let us focus for a moment only on A, B and A', B'. We have seen that there exist two different isometries that map A, B onto A', B', respectively. We can show that these isometries—and, in fact, any isometry that maps A, B onto A', B', respectively—will automatically map C onto C' as long as C and C' satisfy Condition (7.1). (You will explore the proof of this fact in Problem 3 of Problem Set 6.1.) Thus the two different isometries that map A, B onto A', B' will also map C to C'. Consequently, there are two different isometries that map A, B, C onto A', B', C', respectively [satisfying Condition (7.1) when the points are collinear]. Thus noncollinearity of the three points is necessary for obtaining a unique isometry. The next theorem shows that this condition is sufficient.

Theorem 7.1: The First Fundamental Theorem of Isometries

If A, B, C are three noncollinear points, and A', B', C' are also noncollinear such that $A'B' = AB$, $B'C' = BC$, and $A'C' = AC$, then there exists a unique isometry that maps A to A', B to B', and C to C'. This isometry can be expressed as a composition of at most three reflections.

Proof.

Part I: Existence. First we need to prove that there exists an isometry that maps the noncollinear points A, B, C onto the points A', B', C'. If $A \neq A'$, $B \neq B'$, and $C \neq C'$, the isometry $M_\ell \circ M_k$, described in Figure 7.2, will map A to A' and B to B'. Assume that the image of C under $M_\ell \circ M_k$ is not C'. Before reading on, try to find the line m such that $M_m \circ (M_\ell \circ M_k)$ will map $\triangle ABC$ onto $\triangle A'B'C'$.

In Figure 7.5, points A, B are mapped to points A', B' by $M_\ell \circ M_k$ (this was done in Figure 7.2). Now, however, we also have the points C and C', and we need to find an isometry that will not only map A, B onto A', B' but also map C onto C'. Let's see what $M_\ell \circ M_k$ does to C.

Notice that $(M_\ell \circ M_k)(C) = M_\ell(C^*) = C^{**}$. If $C^{**} = C'$, we are done, and $M_\ell \circ M_k$ is the required isometry. Suppose that $C^{**} \neq C'$. Figure 7.5 suggests that the line $A'B'$ (denoted by m) is such that the reflection of C^{**} in it is C'. If this were true, it would follow that $M_\ell \circ M_k$ which maps $\triangle ABC$ onto $\triangle A'B'C^{**}$ followed by M_m will map $\triangle ABC$ onto $\triangle A'B'C'$ (since M_m will not change A' and B' and will map C^{**} to C'). To show that $M_m(C^{**}) = C'$, we need to prove that A' and B' are equidistant from C' and C^{**}. We first

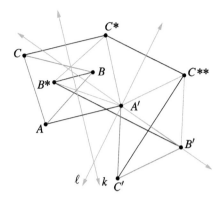

Figure 7.5

show that $A'C' = A'C^*$. From the hypothesis,

$$A'C' = AC. \tag{7.2}$$

When we consider the reflection in ℓ, we see that the image of $\overline{A'C^*}$ is $\overline{A'C^{**}}$ and hence that

$$A'C^{**} = A'C^*. \tag{7.3}$$

Finally, reflection in k implies

$$A'C^* = AC. \tag{7.4}$$

The last two equations imply

$$A'C^{**} = AC.$$

This, along with Equation (7.2), implies

$$A'C' = A'C^{**}.$$

The last equation tells us that A' is equidistant from the endpoints of the segment $\overline{C'C^{**}}$. In the same way, we can show that B' is also equidistant from the endpoints of $\overline{C'C^{**}}$. Thus the line m through A' and B' is the perpendicular bisector of $\overline{C'C^{**}}$. Consequently, $M_m \circ M_\ell \circ M_k$ maps A, B, C, onto A', B', C', respectively. ■

In the preceding argument, we assumed that $A \neq A'$, $B \neq B'$, and $C \neq C'$. You will investigate the cases when one or more of the points are the same as their corresponding images in the problem set at the end of this section.

Part II: Uniqueness. We need to prove that the isometry we have found is unique—that is, that no other isometry will accomplish the required mapping. Suppose F and G are two isometries such that $F(A) = G(A) = A'$, $F(B) = G(B) = B'$, and $F(C) = G(C) = C'$, where A, B, C are noncollinear. We need to prove that $F = G$, i.e., $F(P) = G(P)$ for all points P in the plane.

7.2 In Search of New Isometries

Suppose $F \neq G$. Then there must exist a point P in the plane such that

$$F(P) \neq G(P).$$

Let $F(P) = P'$ and $G(P) = P''$ (see Figure 7.6). Consider the distance AP. Because $F(A) = A'$, $F(P) = P'$, and F is an isometry, we have

$$AP = A'P'. \tag{7.5}$$

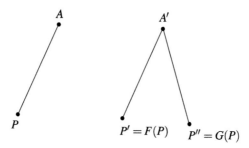

Figure 7.6

We apply now the same reasoning to the isometry G. Because $G(A) = A'$ and $G(P) = P''$ we get

$$A'P'' = AP. \tag{7.6}$$

Thus $A'P' = A'P''$, which in turn implies that A' is on the perpendicular bisector of $\overline{P'P''}$. If we apply the same reasoning to B rather than A, we find that B' must also be on the perpendicular bisector of $\overline{P'P''}$. Likewise, C' must be on the perpendicular bisector of $\overline{P'P''}$. Consequently, A', B', C' are on the same line (the perpendicular bisector of $\overline{P'P''}$) and, therefore, are collinear. But if A', B', C' are collinear, then A, B, C must also be collinear, because an isometry preserves collinearity. This contradicts the fact that A, B, C are not collinear and hence $F = G$. Consequently, the isometry we found in Part I of the proof is unique. ∎

We state the result just proved separately from Theorem 7.1 as follows:

Corollary 7.1

Two isometries that have the same values (images) at three noncollinear points are equal.

Corollary 7.1 and Theorem 7.1 imply:

Theorem 7.2: The Second Fundamental Theorem of Isometries

Every isometry is equal to a composition of at most three reflections.

Proof. Try to write your own proof before reading on. Let F be an isometry. Pick any three noncollinear points A, B, C. Let A', B', C' be their corresponding images under F. Because F is an isometry and A, B, C are noncollinear, their corresponding images A', B', C' are noncollinear and $A'B' = AB$, $A'C' = AC$, and $B'C' = BC$. By the first part of Theorem 7.1, there exists a succession of at most three reflections that maps the points A, B, C onto A', B', C'. By the uniqueness part of Theorem 7.1 (or Corollary 7.1), F equals the composition of these reflections. ∎

Theorem 7.2 tells us that all the possible isometries including rotations and translations are either reflections, the composition of two reflections, or the composition of three reflections. Thus, to find all possible isometries, we need simply investigate the compositions of two or three reflections.

We next investigate the composition of two reflections in two intersecting lines.

Composition of Two Reflections in Two Intersecting Lines

In Figure 7.7, the lines k and ℓ intersect at O. We reflect figure A in line k and obtain its image A'. Then, we reflect A' in line ℓ to obtain figure A''. Since reflection in a line reverses orientation and we reflected figure A in two lines, the composition of two reflections preserves orientation. The only isometries we have encountered that preserve orientation are translations and rotations. It looks as if we could obtain A'' directly from A by rotating A about the point O.

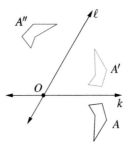

Figure 7.7

Thus it seems that a composition of two reflections in two intersecting lines is a rotation about the point O, where the lines intersect. To prove that this is the case, let P be any point in the plane and P' be its image under reflection in k. Then let P'' be the image of P' under reflection in ℓ. We need to show that for every point P in the plane, $OP = OP' = OP''$, and that the measure of $\angle POP''$ is always the same. What do we know about P, P', and P''?

In Figure 7.8, line k is the perpendicular bisector of $\overline{PP'}$. Because O is on k, $OP' = OP$. Similarly, $OP' = OP''$ and hence $OP = OP''$. Also, the pairs of angles marked in the same way in Figure 7.8 are congruent. (The measure of the angles in the first pair is denoted by α and in the second by β.) Furthermore,

$$m(\angle POP'') = 2\alpha + 2\beta = 2(\alpha + \beta) = 2\theta$$

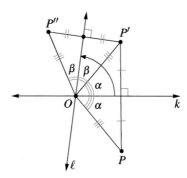

Figure 7.8

where θ is the measure of the angle between lines k and ℓ, as shown in Figure 7.8. Notice that θ is a fixed angle. Hence it seems that P'' can be obtained from P by rotation with center O by an angle whose measure is twice the angle between the lines of reflection in the direction from line k to line ℓ.

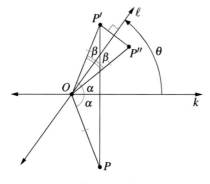

Figure 7.9

Upon closer examination of the preceding argument, you might see that we have not proved the key assertion for all points P in the plane. For example, if P is situated as shown in Figure 7.9 (so that P' lies on the other side of ℓ), the argument showing that $m(\angle POP'') = 2\theta$ is somewhat different.

From Figure 7.9, we have

$$m(\angle POP'') = 2\alpha - 2\beta = 2(\alpha - \beta) = 2\theta.$$

To continue proving in this manner that $m(\angle POP'') = 2\theta$ for all points in the plane, we would need to consider all possible positions of P in relation to the lines k and ℓ. There are only a small number of cases to consider, so we could complete our argument on a case-by-case basis, although it would certainly be tedious. There is also the danger that we might miss a case. Fortunately for us, Theorem 7.1 (or Corollary 7.1) comes to rescue, as we can see in the following proof.

A Shorter Proof. In addition to point P in Figure 7.8, we can choose two other points, Q and R, such that the images of these points behave like the images of P and such that P, Q, and R are noncollinear. (We can choose Q and R to be close to line k in the half-plane determined by line k and P.) $M_\ell \circ M_k$, the reflection in line k followed by the reflection in line ℓ, is an isometry (as a composition of two isometries) and, therefore, maps the noncollinear points P, Q, R to the noncollinear points P'', Q'', R''. We have seen from our previous explanation that the rotation $R_{O, 2\theta}$ (where θ is the angle between the lines k and ℓ and the rotation is in the direction from line k to line ℓ) also maps P, Q, R onto P'', Q'', R''. But Theorem 7.1 (or Corollary 7.1) assures us that the isometry that maps P, Q, R onto P'', Q'', R'' is unique. Consequently,

$$M_\ell \circ M_k = R_{O, 2\theta}.$$

We have proved the following theorem.

Theorem 7.3

If k and ℓ are two lines that intersect at O, then the composition $M_\ell \circ M_k$ (the reflection in line k followed by the reflection in line ℓ) is a rotation about O, with the angle of rotation being twice the directed angle formed by the intersecting lines in the direction from k to ℓ.

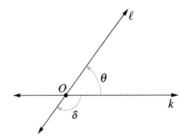

Figure 7.10

Remark 7.1

Theorem 7.3 tells us that $M_\ell \circ M_k = R_{O, 2\theta}$ where θ is the directed angle formed by lines k and ℓ in the direction from k to ℓ. But the angle whose measure is δ (see Figure 7.10) is another angle between the lines k and ℓ that is in the direction from k to ℓ.

Thus we could also conclude from Theorem 7.3 that $M_\ell \circ M_k = R_{O, 2\delta}$, and we had better have $R_{O, 2\theta} = R_{O, 2\delta}$. To see that this is the case, notice that

$$|\theta| + |\delta| = \pi.$$

If $\theta > 0$ and $\delta < 0$ as in Figure 7.10, then $|\theta| = \theta$ and $|\delta| = -\delta$. Hence

$$\theta - \delta = \pi.$$

7.2 In Search of New Isometries

This implies

$$2\theta - 2\delta = 2\pi,$$
$$2\theta = 2\pi + 2\delta.$$

Consequently,

$$R_{O,2\theta} = R_{O,2\pi+2\delta}.$$

Because rotating a point about O by $2\delta + 2\pi$ is the same as rotating the point by 2δ, $R_{O,2\pi+2\delta} = R_{O,2\delta}$ and hence $R_{O,2\theta} = R_{O,2\delta}$.

Let's take a closer look at Theorem 7.3. It says that a reflection in line k followed by a reflection in line ℓ is a rotation about the point O (where the lines intersect) by twice the angle between the lines in the direction from k to ℓ. Of course, there are infinitely many different pairs of lines through O such that the angle between them is θ in the direction from k to ℓ. Figure 7.11 shows a pair of such lines that are different from the pair k and ℓ. Notice how $M_\ell \circ M_k = M_n \circ M_m$. In fact, one of the lines through O, say line m, could be chosen at will. The other line here, line n, is the line that makes angle θ with line m in the same direction as the direction from k to ℓ. This discussion along with Theorem 7.3 implies the following corollary.

Corollary 7.2

Any rotation $R_{O,\theta}$ (the rotation about a point O through a directed angle θ) can be expressed in infinitely many ways as the composition of two reflections. The lines of reflection can be any pair of lines through O such that the directed angle between the lines is $\theta/2$. (See Figure 7.11.)

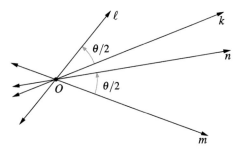

Figure 7.11

Example 7.1

Are there two intersecting lines for which the composition of reflections in these lines is commutative? If so, find all such lines.

Solution. Let k and ℓ be two lines intersecting at O. We want to find all the lines for which

$$M_\ell \circ M_k = M_k \circ M_\ell. \tag{7.7}$$

That is, every point in the plane has the same image under each composition. On the one hand, by Theorem 7.3, $M_\ell \circ M_k = R_{O,2\theta}$, where the rotation is about the directed angle in the direction from line k to line ℓ. On the other hand, $M_k \circ M_\ell$ is a rotation about O by the same angle (in absolute value) but in the direction from line ℓ to line k (that is, in the opposite direction). Thus $M_k \circ M_\ell = R_{O,-2\theta}$. Consequently, Equation (7.7) is true if and only if

$$R_{O,2\theta} = R_{O,-2\theta}. \tag{7.8}$$

For which θ does this equation hold? We know that $R_{O,\pi} = R_{O,-\pi}$. Consequently, if $2\theta = \pi$ (that is, $\theta = \pi/2$), then Equation (7.8) holds and therefore Equation (7.7) is true. Because θ is the angle between the lines, we see that if the lines are perpendicular, then the composition of reflections in the lines is commutative.

To show that if the lines are not perpendicular, then the composition is not commutative, we can find all θ for which Equation (7.8) is true. Because $R_{O,-2\theta}$ is the inverse of $R_{O,2\theta}$, each of the following is equivalent to Equation (7.8):

$$\begin{aligned} R_{O,2\theta} \circ R_{O,-2\theta} &= R_{O,2\theta} \circ R_{O,2\theta}, \\ R_{O,4\theta} &= I. \end{aligned} \tag{7.9}$$

Equation (7.9) is true if and only if $4\theta = 2\pi k$, where k is an integer. If $k = 0$, then $\theta = 0$, which is impossible because the lines are distinct. If $k = 1$, then $4\theta = 2\pi$ and $\theta = \pi/2$, and hence the lines are perpendicular. Because the smaller of the two angles between two intersecting lines is less than or equal to $\pi/2$, there is no need to consider any other values of k.

> **Now Solve This 7.2**
>
> Show that for any two lines k and ℓ (intersecting or parallel), the following three equations are equivalent:
>
> $$M_\ell \circ M_k = M_k \circ M_\ell \quad \text{(which is in general not true)}$$
> $$(M_\ell \circ M_k)^2 = I,$$
> $$(M_\ell \circ M_k)^{-1} = M_\ell \circ M_k.$$
>
> Recall that we write $T \circ T$ as T^2.

Composition of Two Reflections in Two Parallel Lines

In Figure 7.12, lines k and ℓ are parallel. To determine the isometry type of $M_\ell \circ M_k$, is we pick any three convenient noncollinear points P, Q, and S so that their reflections in k fall between the two lines. (This is always possible.) P'', Q'' and S'' are then as shown in Figure 7.12. It seems that $\triangle P''Q''S''$ can be obtained from $\triangle PQS$ by a translation. To see that this is indeed the case, look at P, P', and P''.

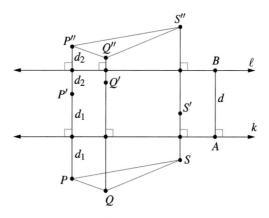

Figure 7.12

Let d be the distance between the lines. Referring to the notation in Figure 7.12, we have

$$PP'' = 2d_1 + 2d_2 = 2(d_1 + d_2) = 2d.$$

Because the line PP'' is perpendicular to the given lines, we can obtain P'' from P by a translation in the direction from line k to line ℓ perpendicular to these lines. Notice that the same is true for Q'' and Q, as well as for S'' and S. If A is on k, B is on ℓ, and $\overline{AB}$ is perpendicular to the lines, we have

$$M_\ell \circ M_k(P) = \tau_{2AB}(P),$$
$$M_\ell \circ M_k(Q) = \tau_{2AB}(Q),$$
$$M_\ell \circ M_k(S) = \tau_{2AB}(S).$$

In other words, $M_\ell \circ M_k$ and τ_{2AB} have the same values (images) at three noncollinear points. But, by Corollary 7.1, if two isometries have the same values at three noncollinear points, then they have the same values at all points, i.e., are equal. We can write this fact as $M_\ell \circ M_k = \tau_{2AB}$. We state this result in the following theorem.

> **Theorem 7.4**
>
> *The composition of two reflections in two parallel lines is a translation by twice the distance between the lines in the direction from the first to the second.*

Like Corollary 7.2 to Theorem 7.3, which stated that a rotation can be written in infinitely many ways as a composition of two reflections in two intersecting lines, we have an analogous corollary to Theorem 7.4.

> **Corollary 7.3**
>
> A translation τ_{PQ} can be expressed in infinitely many ways as a composition of two reflections in two lines, each of which is perpendicular to the line PQ such that the distance between the lines is $\frac{PQ}{2}$.

Notice that we can choose one of the lines of reflection in Corollary 7.3 arbitrarily. For example, given τ_{PQ}, if we choose line k to be perpendicular to line PQ at P, as in Figure 7.13, then line ℓ must be the line parallel to k through M, the midpoint of $\overline{PQ}$. Then $\tau_{PQ} = M_\ell \circ M_k$. If we choose line ℓ to be our first line of reflection, then the second line of reflection must be line m through Q, which is parallel to lines k and ℓ. We have $\tau_{PQ} = M_\ell \circ M_k = M_m \circ M_\ell$.

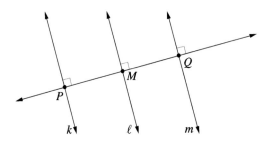

Figure 7.13

> **Remark 7.2**
>
> We have seen that if two lines k and ℓ are parallel, then $M_\ell \circ M_k$ is the translation by twice the distance between the lines in the direction from k to ℓ. Similarly, $M_k \circ M_\ell$ is the translation by twice the distance between the lines in the direction from ℓ to k. Thus, if $M_\ell \circ M_k = \tau_{PQ}$, then $M_k \circ M_\ell = \tau_{QP} = \tau_{PQ}^{-1}$. Consequently, $M_\ell \circ M_k$ and $M_k \circ M_\ell$ are inverses of each other.

By Theorem 7.2, we know that we can express any isometry as a composition of at most three reflections. So far, we have investigated compositions of two reflections in two distinct lines. To account for all possible isometries, we next consider composition of three reflections in three distinct lines. Three lines can be concurrent, all parallel, two parallel and the third intersecting each of the other two or in a configuration, where any two lines intersect in exactly one point not on the third line.

Composition of Reflections in Three Concurrent Lines

Suppose lines k, ℓ, and m intersect at O as shown in Figure 7.14. Our goal is to express $M_m \circ M_\ell \circ M_k$ as a single familiar isometry. Because each reflection reverses orientation,

7.2 In Search of New Isometries

three reflections will also reverse orientation. Among reflections, translations, and rotations, only a reflection reverses orientation. (Although glide reflection was introduced in Chapter 6, we are not assuming a familiarity with this isometry.) Thus, if $M_m \circ M_\ell \circ M_k$ is one of these isometries, it must be a reflection. To prove that $M_m \circ M_\ell \circ M_k$ is a single reflection, we use Theorem 7.3 and Corollary 7.2, which tell us that we can express a composition of any two reflections in infinitely many ways as a composition of two reflections. For example, $M_m \circ M_\ell = M_y \circ M_x$, where either line x or line y could be chosen at will. Therefore we have

$$M_m \circ M_\ell \circ M_k = (M_m \circ M_\ell) \circ M_k = (M_y \circ M_x) \circ M_k = M_y \circ (M_x \circ M_k).$$

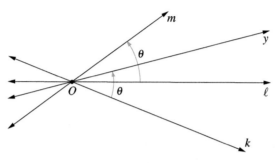

Figure 7.14

The result will be especially simple if we choose x to be line k. Then

$$M_x \circ M_k = M_k \circ M_k = I$$

and hence

$$M_m \circ M_\ell \circ M_k = M_y \circ (M_x \circ M_k) = M_y \circ I = M_y.$$

Thus a composition of three reflections in three concurrent lines is a reflection. We can identify that reflection by finding y such that $M_m \circ M_\ell = M_y \circ M_k$. In Figure 7.14, y is the line for which the directed angle from k to y is the same as the directed angle from ℓ to m (angle θ in Figure 7.14). In a completely analogous way, we can show that a composition of three reflections in three parallel lines is a reflection. The details are left as an exercise.

Composition of Reflections in Three Lines That Are Neither Concurrent Nor All Parallel

Perhaps the simplest case is when two lines are parallel and the third line is perpendicular to them, as shown in Figure 7.15. Consider

$$M_m \circ M_\ell \circ M_k, \quad \text{where } k \| \ell \text{ and } m \perp k. \tag{7.10}$$

Is this composition of three reflections a familiar isometry? Notice that $M_m \circ M_\ell \circ M_k$ reverses orientation and hence cannot be a rotation or a translation. Is it a reflection?

A reflection has infinitely many fixed points (the points on the line of reflection). The composition of the three reflections in (7.10), however, does not seem to have any fixed points. (The justification of this fact will be explored in Now Solve This 7.3.) Thus the transformation in Equation (7.10) is not a rotation, a translation, or a reflection in a single line. Consequently, it is a new isometry. Because $k \| \ell$, $M_\ell \circ M_k$ is a translation. Referring to Figure 7.15, $M_\ell \circ M_k = \tau_{2AB}$. Thus $M_m \circ M_\ell \circ M_k = M_m \circ \tau_{2AB}$. Because m is perpendicular to k and ℓ, m must be parallel to $\overline{AB}$. Consequently, $M_m \circ \tau_{2AB}$ is a translation followed by a reflection in a line parallel to the direction of the translation. Such a composition of a translation followed by a reflection is a **glide reflection**.

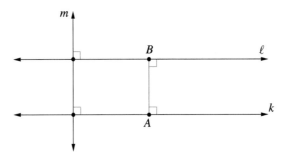

Figure 7.15

Now Solve This 7.3

1. Prove that a glide reflection has no fixed points by showing that any point in one half-plane determined by line m in Figure 7.15 is transformed by the glide reflection to a point in the other half-plane. Also show that every point on m is transformed to a different point on m.

2. In Figure 7.16, P_2 is the image of P_1 under reflection in m, $PP_1 = 2AB$, and $PP_1P_2P_3$ is a rectangle. Use the figure below to show that $M_m \circ \tau_{2AB} = \tau_{2AB} \circ M_m$.

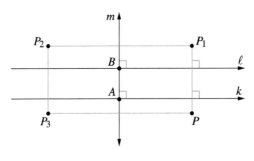

Figure 7.16

Did part 2 of Now Solve This 7.3 convince you that in a glide reflection, instead of performing the translation (glide) first and then the reflection, we obtain the same result if we reverse the order, and perform the reflection first and then the translation? We can prove this fact using the result of Example 7.1—namely, if k and ℓ are perpendicular, then $M_\ell \circ M_k = M_k \circ M_\ell$. This is done in Example 7.2 using the algebra of the composition of functions.

Example 7.2

If $k \| \ell$ and m is perpendicular to k and ℓ, use Example 7.1 to prove that

$$M_m \circ (M_\ell \circ M_k) = (M_\ell \circ M_k) \circ M_m.$$

Solution. Our strategy is to gradually move M_m all the way to the right.

$$\begin{aligned}
M_m \circ (M_\ell \circ M_k) &= (M_m \circ M_\ell) \circ M_k && \text{(associativity)} \\
&= (M_\ell \circ M_m) \circ M_k && \text{(because } m \perp \ell\text{)} \\
&= (M_\ell \circ M_k) \circ M_m && \text{(because } k \perp m\text{)} \\
&= (M_\ell \circ M_k) \circ M_m. && \text{(associativity)}
\end{aligned}$$

Thus $M_m \circ (M_\ell \circ M_k) = (M_\ell \circ M_k) \circ M_m$.

We summarize the result of Now Solve This 7.3 and Example 7.2 in the following theorem:

Theorem 7.5

If τ_{PQ} is a translation defined by the vector PQ, and m is parallel to line PQ, then $M_m \circ \tau_{PQ} = \tau_{PQ} \circ M_m$.

Now Solve This 7.4

The reflections M_k, M_ℓ, and M_m of Example 7.2 can be composed in $3! = 3 \cdot 2 \cdot 1 = 6$ different ways. In Example 7.2, we saw that $M_m \circ M_\ell \circ M_k = M_\ell \circ M_k \circ M_m$. There are four other composition arrangements. Write them down and show that each is a glide reflection equal either to $M_m \circ M_\ell \circ M_k$ or to its inverse.

Composition of Reflections in Three Nonconcurrent Lines, at Least Two of Which Intersect

Let the lines k, ℓ, and m intersect at the points A, B, and C as shown in Figure 7.17. (The case when m and ℓ do not intersect does not require a separate investigation because we will not use point C in the following argument.) To determine the isometry type of $M_m \circ M_\ell \circ M_k$, our strategy will be to consider $M_m \circ (M_\ell \circ M_k)$ and use Theorem 7.3 and Corollary 7.2 as before. We replace lines k and ℓ with lines k_1 and ℓ_1, respectively, in Figure 7.17 so that the angle α from k_1 to ℓ_1 is the same as the angle between k and ℓ in the direction from k to ℓ. Because we can choose one of the new lines through A at will, we choose ℓ_1 to be perpendicular to m. (Such a choice will get us closer to the form where the composition of three reflections is a glide reflection, but reflection in two perpendicular lines is equal to the reflection in any other two perpendicular lines through the same intersection point.) Next we find k_1 (see Figure 7.17) through A so that the angle from k_1 to ℓ_1 is also α. We have

$$\begin{aligned} M_m \circ (M_\ell \circ M_k) &= M_m \circ (M_{\ell_1} \circ M_{k_1}) \\ &= (M_m \circ M_{\ell_1}) \circ M_{k_1}. \end{aligned} \tag{7.11}$$

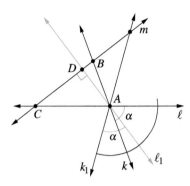

Figure 7.17

Because $\ell_1 \perp m$, we can replace ℓ_1 and m with any two perpendicular lines through D, the point where ℓ_1 and m intersect. As shown in Figure 7.18, ℓ_1 is replaced with ℓ_2 so that $\ell_2 \| k_1$ (because we want to obtain the glide for a glide reflection). We replace m with m_1 so that $m_1 \perp \ell_2$.

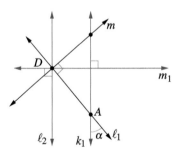

Figure 7.18

We get $M_m \circ M_{\ell_1} = M_{m_1} \circ M_{\ell_2}$. This, along with Equation (7.11), implies

$$\begin{aligned} M_m \circ (M_\ell \circ M_k) &= M_m \circ (M_{\ell_1} \circ M_{k_1}) \\ &= (M_{m_1} \circ M_{\ell_2}) \circ M_{k_1} \\ &= M_{m_1} \circ (M_{\ell_2} \circ M_{k_1}). \end{aligned}$$

Because $\ell_2 \| k_1$ and m_1 is perpendicular to these lines, the resulting isometry and thus the original $M_m \circ M_\ell \circ M_k$ are glide reflections. We have thus proved the following theorem.

7.2 In Search of New Isometries

> **Theorem 7.6**
>
> *A composition of reflections in three nonconcurrent lines, at least two of which intersect, is a glide reflection.*

7.2.1 Problem Set

1. Consider two perpendicular congruent segments AB and $A'B'$ such that $A = A'$. Describe two different isometries, each being a reflection or composition of reflections and each mapping A, B onto A', B' respectively.

   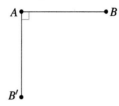

2. Construct any two congruent segments AB and $A'B'$, and then find two different isometries, each as a composition of reflections and each mapping A, B onto A', B', respectively. Construct the lines of reflections, and explain why the isometries are different.

3. In this problem, points A, B, C are on line k and points A', B', C' are on line ℓ. In addition, $A'B' = AB$, $A'C' = AC$, and $B'C' = BC$.

   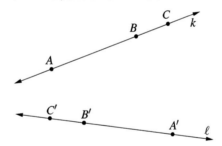

   a. Find two different isometries that map A, B onto A', B'. Check experimentally (using a compass and a straightedge or geometry software) that each isometry maps C onto C'.

   b. Prove that every isometry that maps A, B onto A', B' also maps C to C'. [Hint: Let F be an isometry that maps A to A' and B to B', and let $F(C) = C^*$. You need to show that $C^* = C'$. Because every isometry preserves betweenness and B is between A and C, B' must be between C^* and A'. Consequently, C' and C^* must be on the ray $B'C'$. Show that this implies that $C^* = C'$.]

4. **a.** Construct two congruent triangles ABC and $A'B'C'$ so that the orientation $A' - B' - C'$ is the same as the orientation $A - B - C$. Find an isometry in the form of a composition of reflections that maps A, B, C onto A', B', C', respectively. How many reflections are needed?

b. Repeat part (a) but with the opposite orientations. (If $A - B - C$ is clockwise, $A' - B' - C'$ is counterclockwise.)

5. a. Suppose triangles ABC and $A'B'C'$ are congruent and $A = A'$. What is the smallest number of reflections whose composition will map A, B, C onto A', B', C', respectively? Justify your answer.

b. Suppose triangles ABC and $A'B'C'$ are congruent with $A = A'$ and $B = B'$. Will one reflection map triangle ABC onto triangle $A'B'C'$? Justify your answer.

6. Suppose F is an isometry, and A, B, and C are three noncollinear points for which $F(A) = A$, $F(B) = B$, and $F(C) = C$. What kind of isometry must F be? Prove your answer in two different ways. One way should use Corollary 7.1 and the other should start as follows: Suppose there exists a point P such that $F(P) = P' \neq P$. Then each of the points A, B, and C must be equidistant from P and P' because..."

7. Let point P be in the interior of angle θ formed by the lines k and ℓ as shown. Construct P' and P'' defined by $(M_\ell \circ M_k)(P) = M_\ell(P') = P''$. Show that in this case, $m(\angle POP'') = 2\theta$ and, consequently, that if P is in the interior of θ, P'' can be obtained from P by the rotation $R_{O,2\theta}$.

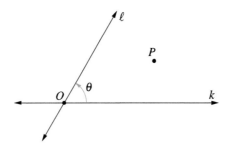

8. Lines k and ℓ intersect at O.

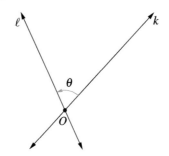

a. Construct though O line x such that

$$M_\ell \circ M_k = M_x \circ M_\ell.$$

Is line x unique? Justify your answer.

b. Given the lines in part (a), construct and describe line y through O such that

$$(M_\ell \circ M_k) \circ (M_\ell \circ M_k) = M_y \circ M_k.$$

7.2 In Search of New Isometries

9. a. In the accompanying figure, $x \perp y$ and k makes a 45° angle with line x. Prove that $M_y \circ M_k \circ M_x$ can be expressed as a reflection in a single line. Find that line.

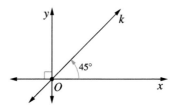

b. The accompanying figure suggests a generalization of the result in part (a). State this generalization and justify it.

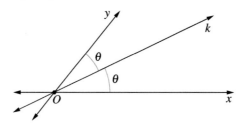

10. What kind of isometry is the composition of three reflections in three parallel lines? Prove your answer.

11. What kind of isometry is a composition of a glide reflection with itself? Prove your answer in two different ways:
  a. Geometrically; looking at the effect of a glide reflection applied twice on a point P
  b. Algebraically; by using the fact that $M_k \circ \tau_{AB} = \tau_{AB} \circ M_k$, where $\overline{AB}$ is parallel to the line of reflection k.

12. For lines k, ℓ, and m (where k is parallel to ℓ, but m is not), in the accompanying figure, prove that $M_m \circ M_\ell \circ M_k$ is a glide reflection. Determine the translation vector and the line of reflection in the glide reflection.

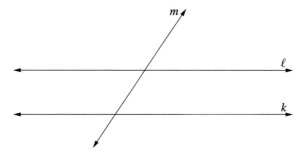

13. Suppose that lines k, x, and y form an isosceles right triangle as shown in the figure, where A is at $(1, 0)$ and B at $(0, 1)$. Write $M_k \circ M_y \circ M_x$ in a standard form $M_k \circ \tau_{OP}$, where O is at the origin and line OP is parallel to k. Find the equation of k and the coordinates of P.

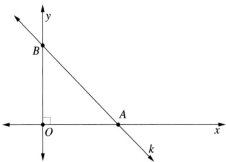

14. The lines a, b, and c form an equilateral triangle.

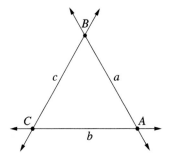

   a. Prove that
   $$M_c \circ M_b \circ M_a = M_n \circ \tau_{\frac{3}{2}CA},$$
   where n is the line through the midpoints of the two sides of the triangle which meet at B, and $\tau_{\frac{1}{2}CA}$ is the translation defined by the vector $\frac{3}{2}\overrightarrow{CA}$ (the translation in the direction from C to A by the distance $\frac{3}{2}CA$).

   b. Keeping in mind the notation in part (a), find an analogous way to write $M_a \circ M_b \circ M_c$ as a glide reflection in the standard form. Check that your answer, when composed with the right side of the equation in part (a), gives the identity. [Notice $M_a \circ M_b \circ M_c$ is the inverse of $M_c \circ M_b \circ M_a$, the left side of the equation in part (a).]

15. Investigate the types of isometries resulting from composition of reflections in

   a. n concurrent lines

   b. n parallel lines

16. In the xy-coordinate system, a reflection M_x in the x-axis can be described as $M_x(x, y) = (x, -y)$. Similarly describe a reflection in the line $y = b$ using the following two approaches.

   a. Let M_b be the reflection in the line $y = b$ and τ be the translation τ_{OP}, where O is the origin and $P(0, b)$. Explain why $M_b = \tau \circ M_x \circ \tau^{-1}$ and use this fact to find $M_b(x, y)$.

7.3 Composition of Rotations, the Treasure Island Problem...

b. Use the midpoint formula $x_N = \frac{x_A+x_B}{2}$, $y_N = \frac{y_A+y_B}{2}$ for the coordinates of the midpoint N of a segment AB, where $A = (x_A, y_A)$, $B = (x_B, y_B)$.

17. Follow the steps below to provide a proof that the composition of a translation followed by a reflection in a line is a glide reflection.

    a. Prove that $M_\ell \circ \tau_{AB}$, where $\overline{AB}$ is perpendicular to ℓ, is a reflection. Identify the line of reflection.

    b. Let τ_{CD} be a translation from C to D and ℓ be a line not parallel to $\overline{CD}$. Write τ_{CD} as a composition of two translations, one parallel to ℓ and the other perpendicular to ℓ.

    c. Use parts (a) and (b) to prove that $M_\ell \circ \tau_{CD}$ is a glide reflection.

18. Let k and ℓ be two lines intersecting at O and θ be the least positive angle by which ℓ needs to be rotated about O so that its image is line k. If R_θ is the rotation about O by θ, prove that $M_\ell = R_{-\theta} \circ M_k \circ R_\theta$.

7.3 Composition of Rotations, the Treasure Island Problem, and Other Treasures

The result of composition of two rotations about the same point is quite obvious: If we rotate a point about O by angle α and then rotate its image about O by angle β, the result is the same as if we had applied a single rotation about O by angle $\alpha + \beta$. In fact, the order of the rotations is immaterial. We write this relationship as follows:

$$R_{O,\beta} \circ R_{O,\alpha} = R_{O,\alpha} \circ R_{O,\beta} = R_{O,\alpha+\beta}.$$

What if the rotations are about different points? Is the result always a rotation and, if so, about what point? We state our question more formally in the following problem.

> **Problem 7.1**
>
> What kind of isometry is a composition of two rotations about two different points?

Figure 7.19 demonstrates what happens to a point P under the composition of two rotations: with center A and positive angle α, and the other with center B and positive angle β.

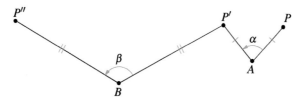

Figure 7.19

Because $R_{A,\alpha}(P) = P'$ and $R_{B,\beta}(P') = P''$, we have

$$(R_{B,\beta} \circ R_{A,\alpha})(P) = R_{B,\beta}(R_{A,\alpha}(P)) = R_{B,\beta}(P') = P''.$$

We want to find a single isometry that will have the same effect on all points of the plane, as $R_{B,\beta} \circ R_{A,\alpha}$. Notice that composition of rotations preserves orientation, whereas the only isometries that preserve orientation are rotations and translations. Thus $R_{B,\beta} \circ R_{A,\alpha}$ is either a rotation or a translation. To find out what kind of isometry $R_{B,\beta} \circ R_{A,\alpha}$ is, we write each rotation as a composition of reflections. We know from Corollary 7.2 that a rotation can be written as a composition of two reflections in many ways with the following two conditions: The lines of reflection must go through the center of rotation, and the angle formed by the lines must be equal to half the angle of the rotation. Let $R_{A,\alpha} = M_\ell \circ M_k$ and $R_{B,\beta} = M_n \circ M_m$, where the lines k, ℓ, m, and n are still to be determined. We can choose one of the lines in each pair in any way we want. Before we decide, though, let's see what we want to accomplish. We have

$$R_{B,\beta} \circ R_{A,\alpha} = (M_\ell \circ M_k) \circ (M_n \circ M_m). \tag{7.12}$$

To make the right side of Equation (7.12) as simple as possible, it would be advantageous to make $M_k = M_n$; that is, we choose $k = n$ (then $M_k \circ M_k = I$). But line n goes through A and line k goes through B (see Figure 7.19), so $k = n$ if and only if the line goes through A and B. This line of reflection, along with lines m and ℓ, is shown in Figure 7.20.

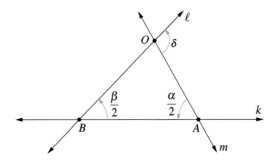

Figure 7.20

Because α and β were chosen to be positive, we use Equation (7.12) for $n = k$ and the associative property of composition of isometries to obtain

$$R_{B,\beta} \circ R_{A,\alpha} = (M_\ell \circ M_k) \circ (M_k \circ M_m)$$
$$= M_\ell \circ (M_k \circ M_k) \circ M_m$$
$$= M_\ell \circ I \circ M_m$$
$$= M_\ell \circ M_m.$$

Consequently,

$$R_{B,\beta} \circ R_{A,\alpha} = M_\ell \circ M_m. \tag{7.13}$$

7.3 Composition of Rotations, the Treasure Island Problem...

If lines m and ℓ intersect (in point O in Figure 7.20) and the angle from m to ℓ is δ, then $M_\ell \circ M_m$ is a rotation with center O by angle 2δ. How is 2δ related to the given angles α and β? Notice that δ is an exterior angle in $\triangle OAB$, so $\delta = \alpha/2 + \beta/2$ and, therefore, $2\delta = \alpha + \beta$. Thus, if the lines m and ℓ intersect in O, then $R_{B,\beta} \circ R_{A,\alpha}$ is a rotation with center O (as found in Figure 7.20) by the angle $\alpha + \beta$.

Lines ℓ and m in Figure 7.20 intersect if and only if a triangle is formed. Because the sum of the interior angles in a triangle is $180°$, the lines in Figure 7.20 will not form a triangle if $\alpha/2 + \beta/2 = 180°$ (that is, if $\alpha + \beta = 360°$). Thus, if $\alpha + \beta \neq 360°$, the composition of the two rotations is a rotation. If $\alpha + \beta = 360°$, lines m and ℓ are parallel and Equation (7.13) implies that the composition of the rotations is a translation.

> **Now Solve This 7.5**
>
> Express each of the following as a single rotation or translation.
> 1. $R_{B,\frac{\pi}{2}} \circ R_{A,-\frac{2\pi}{3}}$
> 2. $R_{B,\frac{\pi}{2}} \circ R_{A,\frac{\pi}{2}}$

Based on the preceding discussion, we have the following theorem:

> **Theorem 7.7**
>
> If $R_{A,\alpha}$ and $R_{B,\beta}$ are two rotations with centers at A and B and angles of rotation α and β, respectively (α and β may be positive or negative), then $R_{B,\beta} \circ R_{A,\alpha}$ is a rotation by the angle $\alpha + \beta$ if and only if $\alpha + \beta$ is not a multiple of $360°$; it is a translation otherwise.

> **Remark 7.3**
>
> If $\alpha > 0$ and $\beta > 0$ and $\alpha + \beta \neq 360°$, the center of rotation O can be found as in Figure 7.20.

> **Problem 7.2: The Treasure Island Problem**
>
> We introduced the Treasure Island Problem in Chapter 1, where we suggested that you investigate it with geometry software, and later we proposed various proofs. We restate the problem here for easy reference.
> Tava found a parchment describing the location of a hidden pirate treasure buried on a deserted island. The island contained a coconut tree, a banana tree, and a gallows where traitors were hanged. A reproduction of the map appears in Figure 7.21. It was accompanied by the following directions:
> Walk from the gallows to the coconut tree, counting the number of steps. At the coconut tree, turn $90°$ and go to the right. Walk the same distance, and put a spike in the ground. Return to the gallows and walk to the banana tree, counting your steps.

At the banana tree, turn 90° and go to the left. Walk the same number of steps, and put another spike in the ground. The treasure is halfway between the spikes.

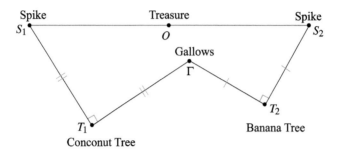

Figure 7.21

Tava found the island and the two trees but no trace of the gallows or the spikes, which had probably rotted. In desperation, Tava and friends began to dig at random, but soon they realized that the exact location of the treasure can be found using a transformational, approach. Devise a plan for finding the treasure and prove your answer.

Solution. This problem should remind you of composition of rotations. No matter where Γ is, we can obtain S_2 from S_1 by rotating S_1 by $-90°$ about T_1 and then rotating its image Γ by $-90°$ about T_2. We can write this as follows:

$$\left(R_{T_2,-\frac{\pi}{2}} \circ R_{T_1,-\frac{\pi}{2}}\right)(S_1) = S_2.$$

As was done in the proof of Theorem 7.7 (see Figure 7.20), we have

$$R_{T_2,-\frac{\pi}{2}} \circ R_{T_1,-\frac{\pi}{2}} = (M_m \circ M_k) \circ (M_k \circ M_\ell)$$
$$= M_m \circ M_\ell$$
$$= R_{O,\pi}.$$

Thus, no matter where Γ is, we can obtain the corresponding spike S_2 from S_1 by a half-turn about a fixed point O. Thus O is the midpoint of the segment S_1S_2 resulting from any location of Γ and, therefore, is the location of the treasure.

The above solution suggest at least three ways to find the treasure (point O):

- Start with any location for Γ, find the corresponding spikes, and then construct the midpoint O of the segment S_1S_2.
- Construct triangle T_1T_2O as shown in Figure 7.22.
- O is the vertex of an isosceles right triangle and hence on the perpendicular bisector of $\overline{T_1T_2}$ at a distance $\frac{1}{2}T_1T_2$ from line T_1T_2 (why?).

7.3 Composition of Rotations, the Treasure Island Problem... 323

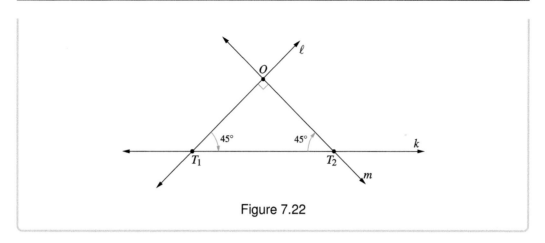

Figure 7.22

Now Solve This 7.6

To generalize the Treasure Island Problem, we can turn at T_1 by an angle α (rather than $\frac{\pi}{2}$) and at T_2 by angle β. For which α and β will the solution be the same as in the original problem? Why?

Composition of Half-Turns

Rotations by 180° about a given point deserve separate attention because they have special properties. A rotation by 180° about a point A is called a half-turn about A and is denoted by H_A, as mentioned earlier. By Theorem 7.7, a composition of two half-turns is a translation. To gain a better understanding, we will independently verify that this is the case.

Problem 7.3: Composition of two half-turns

Show that $H_B \circ H_A$ is a translation and identify that translation.
Solution. In Figure 7.23, A and B are two arbitrary points. By Theorem 7.3, H_A can be written as a composition of two reflections in any pair of perpendicular lines through A, and H_B can also be written as a composition of two reflections through B. As in the composition of two rotations (Figure 7.20), we use the lines shown in Figure 7.23 as the lines of reflection.

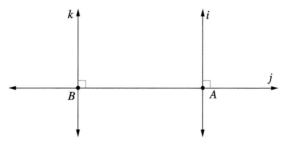

Figure 7.23

Thus

$$H_B \circ H_A = M_k \circ M_j \circ M_j \circ M_i$$
$$= M_k \circ M_i$$
$$= \tau_{2AB}.$$

Problem 7.4: Composition of three half turns

What kind of isometry is a composition of three half-turns $H_C \circ H_B \circ H_A$, where A, B, C are non-collinear?

Solution. We have seen that $H_B \circ H_A = \tau_{2AB}$, a translation in the direction from A to B by twice the distance AB. We also know that $\tau_{2AB} = M_q \circ M_p$, where lines p and q are any two lines perpendicular to line AB, that are distance AB apart. Lines p and q must also be such that the direction from p to q is the same as from A to B. We can express H_C as a composition of two reflections in any two perpendicular lines through C. If r and s are such lines through C, we have

$$H_C \circ H_B \circ H_A = (M_s \circ M_r) \circ (M_q \circ M_p).$$

To simplify this composition of four reflections, we choose the lines r and q such that $r = q$. Then $M_r \circ M_q = I$. But where are those lines? Because r goes through C, and q must be perpendicular to line AB, the line $r = q$ is uniquely determined, and the location of q determines p. The arrangement of the lines is shown in Figure 7.24.

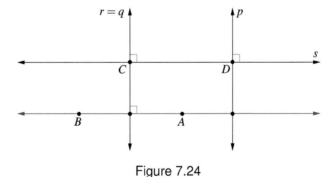

Figure 7.24

Because s must be perpendicular to r through C, it is also uniquely determined. Let the point where s and p intersect be D. Then

$$H_C \circ H_B \circ H_A = (M_s \circ M_q) \circ (M_q \circ M_p)$$
$$= M_s \circ M_p$$
$$= H_D.$$

7.3 Composition of Rotations, the Treasure Island Problem...

Thus a composition of three half-turns is a half-turn. If A, B, and C are noncollinear, the center of the half-turn $H_C \circ H_B \circ H_A = H_D$ is determined as the fourth vertex of the parallelogram with vertices A, B, C, and D, where D is such that the direction from D to C is the same as from A to B. What is the composition $H_C \circ H_B \circ H_A$ when A, B, C are collinear?

Now Solve This 7.7

Zeya found a map describing the location of a buried treasure on a deserted island. There were originally three landmarks on the island located at points A, B, and C. The pirates put a spike P in the ground and then a spike Q in the ground so that A is the midpoint of $\overline{PQ}$. Then they put a spike R in the ground so that B is the midpoint of $\overline{RQ}$. Then they put a fourth spike S in the ground so that C is the midpoint of $\overline{RS}$. Finally, they connected S to P and buried the treasure at the midpoint M of the segment $\overline{PS}$. (See Figure 7.25.)

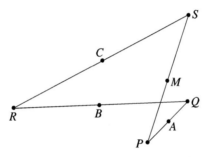

Figure 7.25

Zeya arrived at the island and found no trace of the four spikes, but the landmarks A, B, C were there. How did Zeya find the treasure?

Now Solve This 7.8

1. Prove that for every point C, there exist points A and B such that:
$$(H_C \circ H_B \circ H_A)^2 = I.$$

2. In Figure 7.26, P_1 is the image of P under H_A, P_2 is the image of P_1 under H_B, and P_3 is the image of P_2 under H_C. We then find P_4, the image of P_3 under H_A, then P_5 the image of P_4 under H_B (not shown in the figure); and finally P_6, the image of P_5 under H_C. How is P_6 related to P? Prove your answer. (This problem was introduced earlier in Now Solve This 6.3, where the approach was experimental.)

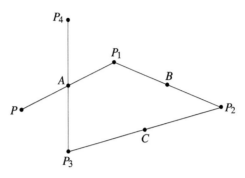

Figure 7.26

3. Write a "buried treasure" problem based on what you found in part 1.

Problem 7.5: Squares of the Sides of a Convex Quadrilateral

1. Use geometry software or straightedge and compass to experimentally verify (for several different quadrilaterals) a surprising property of quadrilaterals. Let $ABCD$ be a quadrilateral as shown in Figure 7.27. Outside the quadrilateral, construct squares on a pair of opposite sides of the quadrilateral. Let Q and S be the centers of the squares (where the diagonals intersect). Similarly, construct such squares on the pair of the other opposite sides, and let T and P be the centers of these squares. Check that no matter which quadrilateral you start with, the segments QS and TP are congruent and perpendicular.

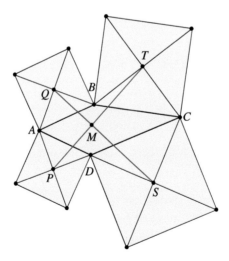

Figure 7.27

2. Prove that $\overline{QS}$ and $\overline{TP}$ are congruent and perpendicular.

7.3 Composition of Rotations, the Treasure Island Problem...

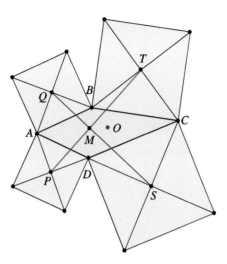

Figure 7.28

Solution to Part 2. Part of the assertion would follow if we could show that $\overline{TP}$ is the image of $\overline{QS}$ under an isometry. Because we want to show that the segments are perpendicular, the desirable isometry should be rotation by 90°. Recall that the image of a line under a rotation by an angle θ is a line that makes angle θ (or $180° - \theta$) with the original line. Under such rotation, the image of Q could be T and the image of S could be P. Figure 7.28 shows the point O we wish we had, so that $OQ = OT$, $OS = OP$, and both $\angle QOT$ and $\angle SOP$ measure 90°. If we could find such a point O, the image of $\overline{QS}$ under a rotation with center O by 90° would be $\overline{TP}$, and our proof would be complete.

If you focus on points Q, T, A, B, C, and O, the corresponding part of Figure 7.28 may remind you of a composition of two rotations about points Q and T; in particular, the rotations in the Treasure Island Problem (Q and T correspond to the trees, B to the gallows, A and C to the spikes, and the midpoint of $\overline{AC}$ to the treasure). By Theorem 7.7 (or the Treasure Island Problem), the composition $R_{T,\frac{\pi}{2}} \circ R_{Q,\frac{\pi}{2}}$ is a rotation by 180° about a point that is the vertex of an isosceles triangle with the base QT and base angles of 45°. This is the point O we wished we had in Figure 7.28. As in the Treasure Island Problem, O is the midpoint of segment AC.

Another way to see that O is the midpoint of $\overline{AC}$ is to find the image of A under $R_{T,\frac{\pi}{2}} \circ R_{Q,\frac{\pi}{2}}$. Because the first rotation takes A to B and the second rotation takes B to C, the image of A under $R_{T,\frac{\pi}{2}} \circ R_{Q,\frac{\pi}{2}}$ is C; but because $R_{T,\frac{\pi}{2}} \circ R_{Q,\frac{\pi}{2}} = R_{O,\pi}$, we must have $R_{O,\pi}(A) = C$. Hence O is the midpoint of $\overline{AC}$.

In a completely analogous way, we can show that $R_{P,\frac{\pi}{2}} \circ R_{S,\frac{\pi}{2}}$ is a rotation by π about a point O_1, which is the vertex of an isosceles right triangle with base $\overline{PS}$. We would like to show that $O_1 = O$. This is the case because $R_{S,\frac{\pi}{2}}$ takes C also D, while

$R_{P,\frac{\pi}{2}}$ takes D to A. Because $R_{P,\frac{\pi}{2}} \circ R_{S,\frac{\pi}{2}} = R_{O_1,\pi}$, we have $R_{O,\pi}(C) = A$. Thus O_1 is also the midpoint of $\overline{AC}$. Hence $O_1 = O$, and the proof is complete.

Now Solve This 7.9

1. State Problem 7.4 as a theorem and prove it. (Write a concise proof without investigation and motivation.)

2. There is a theorem similar to Problem 7.5 for triangles. Sketch a figure analogous to Figure 7.27. What happens when B gets closer and closer to C? (The squares with base $\overline{BC}$ should get smaller and smaller.) State the theorem based on your findings and write its proof.

Napoleon's Theorem

The French emperor and conqueror Napoleon I (1769–1821) is credited with the following theorem.

Theorem 7.8

In Figure 7.29, ABC is an arbitrary triangle. Equilateral triangles have been constructed on the sides of the triangle as shown. If P, Q, and S are the centers of the equilateral triangles (that is, the centers of the circumscribing and inscribed circles of the equilateral triangles), then $\triangle PQS$ is equilateral.

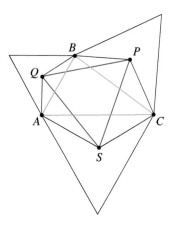

Figure 7.29

Examine the proof in Problem 7.4 and then try to prove Napoleon's Theorem before reading on.

Proof. To prove that $\triangle PQS$ in Figure 7.29 is equilateral, it will suffice to show that one of the vertices of this triangle is the image of another vertex under a rotation by 60° in

7.3 Composition of Rotations, the Treasure Island Problem...

an appropriate direction about the third vertex. We will show that Q is the image of P under a rotation about S by $60°$ (counterclockwise). Because P, Q, and S are the centers of equilateral triangles, the triangles AQB, BPC, and CSA are isosceles triangles with angles of $120°$ at the vertices Q, P, and S. Analogous to what we did in Problem 7.4, we will consider $R_{P,120°} \circ R_{Q,120°}$. By Theorem 7.3,

$$R_{P,120°} \circ R_{Q,120°} = R_{O,240°} \tag{7.14}$$

where O is a point determined by lines i, j, and k as shown in Figure 7.30.

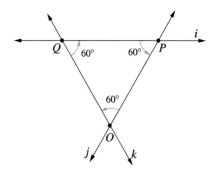

Figure 7.30

For a better understanding, it is helpful to revisit the proof of Theorem 7.3. We write each rotation as a composition of reflections:

$$R_{P,120°} \circ R_{Q,120°} = (M_j \circ M_i) \circ (M_i \circ M_k) = M_j \circ M_k = R_{O,240°}.$$

We now show that point O is actually point S, the third vertex of $\triangle PQS$ in Figure 7.29. One way to do so is to show that

$$R_{P,120°} \circ R_{Q,120°} = R_{S,240°}. \tag{7.15}$$

This would imply that $R_{O,240°} = R_{S,240°}$, and hence that $O = S$. We compose each side of Equation (7.15) with $R_{S,-240°}$ and obtain an equivalent equation

$$R_{S,-240°} \circ R_{P,120°} \circ R_{Q,120°} = I. \tag{7.16}$$

Because $R_{S,-240°} = R_{S,120°}$, we can write

$$R_{S,120°} \circ R_{P,120°} \circ R_{Q,120°} = I. \tag{7.17}$$

Because in (7.17) all the angles add up to $360°$, this composition of rotations is, by Theorem 7.7, a translation. It is the identity I since the image of A is the point itself. (A translation that fixes a point must be the identity.) Indeed referring to Figure 7.29, we find that $R_{Q,120}$ takes A to B, $R_{P,120°}$ takes B to C, and $R_{S,120}$ takes C to A. Thus the image of A under the composition of the three rotations in Equation (7.17) is A and hence Equation (7.17) is true. Equation (7.17) is equivalent to Equation (7.15), which along with Equation (7.14) implies that $R_{O,240°} = R_{S,240°}$, which in turn implies that $O = S$, as in Figure 7.30.

7.3.1 Problem Set

1. Using an approach similar to the one we used in proving Theorem 7.7, prove that $R_{B,\pi/2} \circ R_{A,\pi}$ is a rotation. Specify the center of the rotation and the angle of the rotation.

2. In the figure below, lines h and k are parallel, and line AB is perpendicular to both lines (A is on h and B is on k). $H_A(P) = P'$ and $H_B(P') = P''$. Write a "buried treasure" problem based on the figure, solve it, and write your answer in the form of a proof.

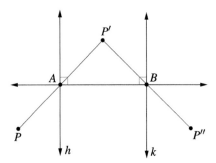

3. In Section 7.3, we showed that a composition of three half-turns is a half-turn. Our proof was based on expressing each half-turn as a composition of two reflections. It is also possible to give a synthetic proof of this fact—that is, a proof that uses only classical Euclidean geometry. For that purpose, consider three noncollinear points A, B, and C, and let P be an arbitrary point. The accompanying figure shows $H_A(P) = P'$, $H_B(P') = P''$, and $H_C(P'') = P'''$. Thus $(H_C \circ H_B \circ H_A)(P) = P'''$. Let O be the midpoint of the segment PP'''. We want to show that for all points P in the plane, the midpoint of $\overline{PP'''}$ is the same point O. Use the properties of midsegments to prove this fact.

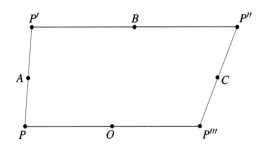

4. In the figure below, squares with centers O_1 and O_2 were constructed on sides AB and BC of an arbitrary $\triangle ABC$. On side AC, an arbitrary point D was chosen, and squares with centers O_3 and O_4 were constructed on $\overline{AD}$ and $\overline{DC}$ as sides.

7.3 Composition of Rotations, the Treasure Island Problem... 331

Prove that the segments O_1O_3 and O_2O_4 are congruent and perpendicular.

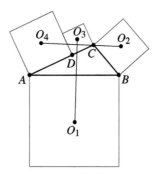

5. In the figure below, squares were constructed on the sides of a concave quadrilateral. If the centers of the squares are O_1, O_2, O_3, and O_4, is it true that segments O_1O_3 and O_2O_4 are congruent and perpendicular? Justify your answer.

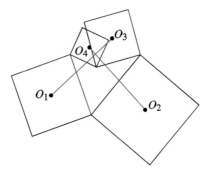

6. In the figure below, $ABCD$ is a parallelogram. Squares are constructed on its sides with centers O_1, O_2, O_3, and O_4. Prove that $O_1O_2O_3O_4$ is a square.

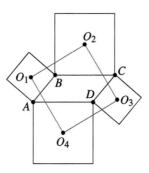

7. **a.** Use a compass and straightedge or geometry software to conjecture whether Napoleon's Theorem remains true when equilateral triangles are constructed on the sides of an arbitrary triangle ABC, all inward rather than outward (in such a way that each equilateral triangle is on the same half-plane determined by the line containing a side of $\triangle ABC$ and the third vertex of that triangle).

b. Justify your conjecture in part (a).

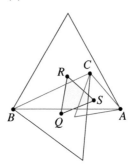

8. *ABCD* is an arbitrary quadrilateral. Equilateral triangles have been constructed on its sides so that one pair of triangles on opposite sides of the quadrilateral is constructed exterior to the quadrilateral (△*GDC* and △*BEA* in the figure), while the other pair of equilateral triangles is constructed so that their interiors intersect the interior of *ABCD* (△*CBF* and △*DHA*).
   a. Construct an arbitrary quadrilateral and verify that the vertices *E*, *F*, *G*, and *H* of the equilateral triangles form a parallelogram.
   b. Prove that *EFGH* is a parallelogram.

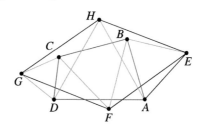

9. On two sides of an arbitrary triangle *ABC*, two equilateral triangles have been constructed outwardly as shown in the figure below. *M*, *P*, and *Q* are the midpoints of sides *AB*, *DC*, and *CE*, respectively.
   a. Construct a triangle and verify that △*PQM* is equilateral.
   b. Prove the assertion in part (a).

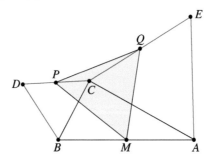

7.3 Composition of Rotations, the Treasure Island Problem...

10. On the sides AC and BC of an arbitrary triangle ABC, squares $BCDE$ and $ACFG$ have been constructed outwardly as shown. P and Q are centers of the squares, and N and M are the midpoints of $\overline{EG}$ and $\overline{DF}$, respectively.
    a. Experiment with different triangles ABC to conjecture the most that can be said about triangles ABN and PQM.
    b. Prove your conjecture in part (a).

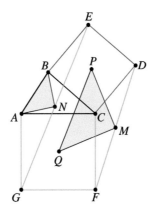

Problems 11–16 involve the concept of a **transformation group**. A set S of transformations from the plane onto the plane is a group under composition of transformations if the following properties hold:

a. **Closure:** If T_1 and T_2 are two transformations in S, then $T_2 \circ T_1$ is in S.
b. **Associative property:** For every three transformations $T_1, T_2,$ and T_3,

$$(T_3 \circ T_2) \circ T_1 = T_3 \circ (T_2 \circ T_1).$$

c. **Existence of identity transformation I** ($I(P) = P$, for all P in the plane) satisfying $I \circ T = T \circ I = T$ for all T in S.
d. **Existence of Inverse Transformation:** If T is in S, so is T^{-1}.

If S_1 is a subset of S and is itself a group, we say that S_1 is a **subgroup** of S. Notice that the set E of all the isometries of the plane is a subgroup of the group of all transformations of the plane.

11. Let $T_1, T_2,$ and T_3 be any three transformations of the plane. Prove the *associative property* (stated above) by showing that for all points P in the plane

$$((T_3 \circ T_2) \circ T_1)(P) = (T_3 \circ (T_2 \circ T_1))(P).$$

12. Which of the following are groups and which are not? Justify your answers.
    a. The set of all translations.
    b. The set of all direct isometries (translations and rotations).

c. The set of all indirect isometries (reflections and glide reflections).

   d. The set of all half-turns and all translations.

13. A subgroup S of a group G is **normal** if for all $s \in S$ and $g \in G$, $g^{-1}sg$ is an element of S. Use this definition to prove or disprove the following.

    a. The group N of all translations is a normal subgroup of E, the set of all isometries in the plane; that is, if τ is any translation, then for all f in E $f^{-1} \circ \tau \circ f$ is a translation.

    b. The group T of all translations is a normal subgroup of the group of all direct isometries (rotations and translations); that is, if τ is any translation, then for all f that is a direct isometry, $f^{-1} \circ \tau \circ f$ is a translation.

14. Recall that a *symmetry* of a figure is an isometry that maps the figure onto itself, and answer the following.

    a. Explain why the set of all symmetries of a figure is a group.

    b. The symmetry group of a rectangle that is not a square is the set $\{I, H, V, R\}$ where H is the reflection in a line through the midpoints of two opposite sides, V is the reflection in the line through the midpoints of the other pair of opposite sides, and R is a rotation about the center of the rectangle (the intersection of the diagonals) by 180°. Prove that $H^2 = V^2 = R^2 = I$, $H \circ V = V \circ H = R$, which can be displayed in the following **Cayley table**:

$\circ$	I	R	H	V
I	I	R	H	V
R	R	I	V	H
H	H	V	I	R
V	V	H	R	I

    c. List the symmetries of an equilateral triangle and construct a Cayley table for them.

15. a. Describe the symmetry group of a circle.

    b. Use the concept of a group to define what it means to say that one figure is more symmetric than another.

    c. Why is a circle more symmetrical than any regular n-gon?

16. a. Prove that the set of all dilations with a common center of similitude constitutes a group.

    b. Show that the set of all possible dilations in a plane (with all points of the plane as centers of similitude) does not constitute a group.

8 More Recent Discoveries

> *Everybody knows that mathematics is about miracles, only mathematicians have a name for them: theorems.*
>
> Roger Howe, invited MAA address, Baltimore, Maryland, January 9, 1998

Introduction

In this chapter we prove a few of the more recent results in geometry, some of which were presented in Chapter 1. In Section 8.2, we discuss complex numbers and their use in proving geometrical theorems, including the Treasure Island Problem. We also show how the most important trigonometric formulas can be derived from properties of complex numbers.

8.1 The Nine-Point Circle and Other Results

The Nine-Point Circle was introduced without proof in Chapter 1 to arouse readers' curiosity. We will now prove the theorem. But first we will briefly survey the theorem's historical background.

During the 19th century, interest in classical Euclidean geometry experienced a resurgence. Probably the most spectacular discovery was the **Nine-Point Circle**, which was investigated simultaneously by French mathematicians Charles Jules Brianchion (1785–1864) and Jean Victor Poncelet (1788–1867). They published their findings jointly in 1821; however, the theorem is commonly attributed to the German mathematician and school teacher Karl Wilhem Feurbach (1800–1834), who independently discovered the theorem and published it along with some related results in 1822.

> **Theorem 8.1: The Nine-Point Circle**
>
> Given a triangle ABC, the following nine points lie on a circle called the Nine-Point Circle: (1) The three midpoints of the sides; (2) The midpoints of the segments joining the vertices of the triangle with the orthocenter and the points of intersection of each altitude with the corresponding side of the triangle.

Proof of the Nine-Point Circle Theorem. In the proof, we will repeatedly use the Midsegment Theorem from Chapter 2. It states that the segment joining the midpoints of two sides of a triangle is parallel to the third side and half as long as the third side.

We first show that the midpoints M_1, M_2, and M_3 of the sides of $\triangle ABC$ and F_1, the foot of the altitude from A to the side BC, lie on a circle. For that purpose, consider Figure 8.2. We have $\overline{M_2M_3} \| \overline{BC}$, $\overline{M_1M_2} \| \overline{AC}$, and $M_1M_2 = \frac{1}{2}AC$. Because $\overline{F_1M_3}$ is a

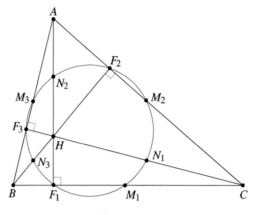

Figure 8.1

median to the hypotenuse in the right $\triangle AF_1C$, $F_1M_3 = \frac{1}{2}AC$. Thus $M_1M_2 = F_1M_3$ and, therefore, $M_1M_2M_3F_1$ is an isosceles trapezoid and hence cyclic (why?). Consequently, M_1, M_2, M_3, and F_1 lie on a single circle. In the same way, we can show that M_1, M_2, M_3, and F_2 as well as M_1, M_2, M_3, and F_3 lie on the same circle. Thus we have six points on the same circle.

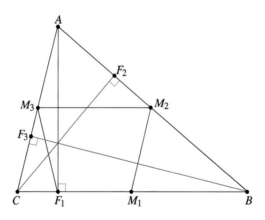

Figure 8.2

Next we focus on the seventh point N_2, the midpoint of $\overline{AH}$. Notice that $\overline{M_3N_2}\|\overline{CF_2}$ and $\overline{M_1M_3}\|\overline{AB}$. Because $\overline{CF_2}$ is perpendicular to $\overline{AB}$, $\overline{M_3N_2}$ is perpendicular to $\overline{M_1M_3}$ as shown in Figure 8.3. Thus F_1 and M_3 are on the circle whose diameter is $\overline{M_1N_2}$ (why?). Because this circle passes through M_1, M_3, and F_1, it is the same circle that went through the six points M_1, M_2, M_3, F_1, F_2, and F_3. In an analogous way, we can show that the points N_1 and N_3 also lie on that circle. ∎

8.1 The Nine-Point Circle and Other Results

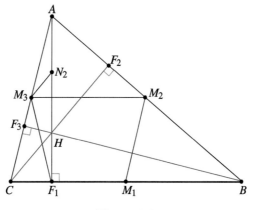

Figure 8.3

Now Solve This 8.1

The following questions will lead you through a different proof of the Nine-Point Circle Theorem, although the result in question 2 has its own merit as an interesting property.

1. Let O be the center of the circle that circumscribes $\triangle ABC$ as in Figure 8.4. Let H be the point of intersection of the altitudes. Prove that $AH = 2HF_1$, by showing that $ALCH$ is a parallelogram, $LC = AH$, and $LC = 2(OM_1)$.
2. Referring to Figure 8.4 again, prove that $HF_1 = F_1P$.

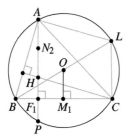   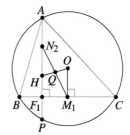

Figure 8.4   Figure 8.5

3. In Figure 8.5 N_2 is the midpoint of $\overline{AH}$. Prove that (a) $\overline{M_1N_2}$ and $\overline{OH}$ bisect each other, (b) $N_2Q = QM_1 = F_1Q$, and (c) $QF_1 = \frac{1}{2}OP$ (hint: $\overline{QF_1}$ is a midsegment in $\triangle OPH$).
4. Prove that the circle centered at Q with radius $\frac{1}{2}OP$ passes through F_1, M_1, and N_2.
5. Argue that the circle in part 4 passes through the other six points of the Nine-Point Circle.

Morley's Theorem

In 1899, Frank Morley, an English-born American mathematician, discovered an amazing property of triangles:

> **Theorem 8.2**
>
> *The points of intersection of the adjacent angle trisectors of the angles of any triangle ($\triangle ABC$ in Figure 8.6) are the vertices of an equilateral triangle ($\triangle PQS$).*
>
>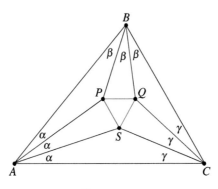
>
> Figure 8.6

The following argument is based on a proof by H. D. Grossman (1943) and the article "Morley's Triangle" (M. E. Barnes, 2002). Our motivation, however, does not appear in either of these sources.

We begin our investigation by assuming that $\triangle PQS$ in Figure 8.7 is equilateral; we then search for properties that follow from this assumption. We extend the trisectors to intersect at H, K, and L and prove that $\triangle HPS$, $\triangle KSQ$, and $\triangle LQP$ are isosceles. To prove that $\triangle HPS$ is isosceles, notice that Q is the point in the interior of $\triangle HBC$ where the angle bisectors of $\angle HBC$ and $\angle BCH$ intersect. Because angle bisectors are concurrent, $\overrightarrow{HQ}$ must be the angle bisector of $\angle CHB$ and, therefore, Q is equidistant from the sides BH and HC; that is, $QT = QV$ (T and V are the feet of the perpendiculars dropped from Q to the sides). Also, $HV = HT$ because $\triangle HTQ \cong \triangle HVQ$. Since $HP = HT - PT$ and $HS = HV - SV$, we only need to show that $PT = SV$. This fact follows from the congruence of the shaded triangles PTQ and SVQ (hypotenuse–leg congruence condition). In a similar manner, we can show that $\triangle KSQ$ and $\triangle LQP$ are isosceles.

We will now find some of the angles in Figure 8.7 in terms of α, β, and γ. We find $\angle ASP$ as follows. First notice that

$$m(\angle ASP) = m(\angle ASH) + m(\angle HSP). \tag{8.1}$$

Because $\angle ASH$ is an exterior angle in $\triangle ASC$,

$$m(\angle ASH) = \alpha + \gamma. \tag{8.2}$$

8.1 The Nine-Point Circle and Other Results

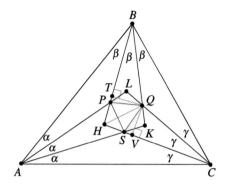

Figure 8.7

Because $\triangle HPS$ is isosceles $m(\angle HSP) = \frac{180° - m(\angle SHP)}{2} = 90° - \frac{1}{2}m(\angle SHP)$. But from $\triangle HBC$, we know that $m(\angle SHP) = 180° - (2\beta + 2\gamma)$. Since $3\alpha + 3\beta + 3\gamma = 180°$, we have $\alpha + \beta + \gamma = 60°$ and, therefore, $2\beta + 2\gamma = 120° - 2\alpha$. Consequently,

$$m(\angle SHP) = 180° - (120° - 2\alpha) = 60° + 2\alpha. \tag{8.3}$$

Thus $m(\angle HSP) = 90° - \frac{1}{2}m(SHP) = 90° - \frac{1}{2}(60° + 2\alpha) = 60° - \alpha$. Given that $\alpha + \beta + \gamma = 60°$, $60° - \alpha = \beta + \gamma$ and

$$m(\angle HSP) = \beta + \gamma. \tag{8.4}$$

Substituting $60° - \alpha$ for $m(\angle HSP)$ in Equation (8.1) and $\alpha + \gamma$ [from Equation (8.2)] for $m(\angle ASH)$, we get

$$\angle ASP = (\alpha + \gamma) + (60° - \alpha) = 60° + \gamma. \tag{8.5}$$

Similarly, we get that

$$\angle CSQ = 60° + \alpha. \tag{8.6}$$

With the preceding results in mind, we now provide the actual proof. Let $\triangle ABC$ be an arbitrary triangle with angles 3α, 3β, and 3γ. Then $3\alpha + 3\beta + 3\gamma = 180°$ and therefore

$$\alpha + \beta + \gamma = 60°. \tag{8.7}$$

We extend the trisector CS and construct the line through S that makes an angle equal to $\beta + \gamma$ with $\overrightarrow{CS}$. That line intersects the trisector of $\angle A$ closest to side AB at P. Looking at Equation (8.4) for motivation, we construct

$$m(\angle SPH) = \beta + \gamma. \tag{8.8}$$

Thus we have constructed the isosceles triangle SHP. Similarly, we construct $\triangle KSQ$ so that

$$m(\angle KSQ) = m(\angle SQK) = \alpha + \beta. \tag{8.9}$$

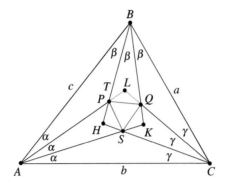

Figure 8.8

Our next goal is to show that $\triangle PQS$ is equilateral. First we find that

$$M(\angle PSQ) = 360° - (\angle PSH + \angle HSA + \angle KSC + \angle KSQ + \angle ASC)$$
$$= 360° - \left[(\beta + \gamma) + (\alpha + \gamma) + (\alpha + \gamma) + (\alpha + \beta) + (180° - (\alpha + \gamma))\right]$$
$$= 180° - 2(\alpha + \beta + \gamma)$$
$$= 180° - 120° = 60°.$$

We proceed with the rest of the proof in Now Solve This 8.2.

> **Now Solve This 8.2**
>
> 1. Show that $m(\angle APS) = m(\angle CQS)$ by showing that each equals $60° + \beta$.
> 2. Notice that S is the intersection of the angle bisectors of $\triangle ALC$. Show that $PS = SQ$ and hence that $\triangle PQS$ is equilateral.
> 3. Prove that $m(\angle APH) = \alpha + \beta$. Show that the angle formed by line c (see Figure 8.8) and line KQ is 2β, and therefore that line HP bisects this angle. Similarly, show that line KQ bisects the angle formed by line a and line HP.
> 4. To complete the proof, show that lines c, a, HP, and KQ are concurrent.

8.1.1 Problem Set

1. Prove that the center of the Nine-Point Circle is located at the midpoint of the segment whose endpoints are the orthocenter (the intersection of the altitudes) and the circumcenter (the center of the circumscribing circle) of the triangle.
2. Prove that the radius of the Nine-Point Circle is one half of the radius of the circumcircle.
3. Prove that the centroid C of a triangle (the intersection of the medians) trisects the segment connecting the orthocenter and the circumcenter. (This fact was proved in 1765 by Leonhard Euler and the segment is called *Euler Line*).
4. Prove **Feurbach's Theorem** (proved by Karl Wilhelm Feurbach in 1822): The nine-point circle is tangent to the incircle (inscribed circle) and the three excircles (an

8.1 The Nine-Point Circle and Other Results

excircle is a circle tangent to one side of a triangle and the extensions of the other two sides).

You may want to convince yourself of the theorem using geometry software.

5. Read and supply the details of a proof of Morley's Theorem given in 1909 by M. T. Naraniengar and found in Honsberger (pp. 92-95). A good exposition is also given on line at http://www.cut-the-knot.org/triangle/Morley/Naraniengar.shtml.

The following problems have been discovered recently. The proofs are challenging.

6. (M. J. Zerger, MME *Journal*, II(7), Fall 2002, p. 392, problem 1051.) The two squares in the figures below are congruent. In the figure on the left, the octagon is formed by joining the midpoints of the sides of the square to vertices as shown. In the figure on the right, the trisection points of the sides are used instead.

   a. Show that the octagons are similar and equilateral, but not equiangular.

   b. Find the ratios of the areas.

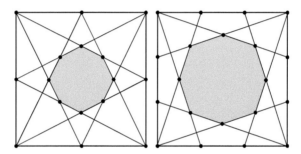

7. (Leon Bankoff, IIME *Journal*, II(6), Spring 2002, p. 328, problem 1041.) The figure below shows a quarter circle with smaller circles inside.

   a. Prove that the three larger circles have radii of equal length.

   b. Prove that the other six smaller shaded circles have radii of equal length.

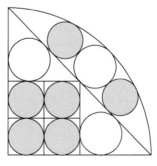

8. Prove a theorem due to Marion Walter (1992): If the trisection points of the sides of any triangle are connected with the opposite vertices, the resulting hexagon has area

one tenth the area of the original triangle.

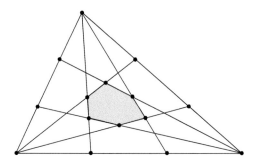

9. **A minimal inscribed quadrilateral** (G. Y. Sosnow, and R. Honsberger). $ABCD$ is a cyclic quadrilateral whose diagonals meet at X. The points P, Q, R, and S are the feet of the perpendiculars from X to the sides of $ABCD$. Prove that, among all quadrilaterals having a point on each side of $ABCD$, $PQRS$ has minimum perimeter.

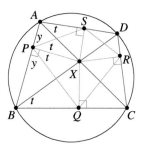

8.2 Complex Numbers and Geometry

If we extend the integers to rational numbers, then the equation $ax + b = 0$, where a and b are integers and $a \neq 0$, always has a solution. We can also extend the real numbers to include the solution of $x^2 + 1 = 0$. For that purpose, the symbol i was introduced. **A complex number** is then defined as $a + bi$, where a and b are real numbers and $i^2 = -1$. Complex number equality, addition, and multiplication are the same as for polynomials, where i^2 is replaced by -1. Complex numbers have many applications in mathematics, electrical engineering, and physics. In this section we focus on some of the applications of complex numbers to geometry.

Real numbers correspond to points on a line. In the xy-coordinate system, the real number a corresponds to the point $(a, 0)$ on the x-axis. If we view a as a complex number $a + 0 \cdot i$, we have the following correspondence:

$$a + 0 \cdot i \longleftrightarrow (a, 0).$$

8.2 Complex Numbers and Geometry

Thus it seems natural that the complex number $a + bi$ should correspond to the point (a, b) in the plane:

$$a + bi \longleftrightarrow (a, b). \tag{8.10}$$

Consequently, we have a one-to-one correspondence between the complex numbers and points in the plane. For convenience, it is customary to simply refer to the point that corresponds to a complex number as the complex number, and vice versa. For example, since $i = 0 + 1 \cdot i$, corresponds to $(0, 1)$, we often say that *i is the point* $(0, 1)$.

We have seen that $(a, 0)$ on the x-axis corresponds to $a + 0 \cdot i$, that is, to a real number. Similarly, the point $(0, b)$ on the y-axis corresponds to the complex number $0 + ib = ib$. For historical reasons (see the historical note later in this section), i is called the **imaginary unit** and if b is real, the complex number ib is called an **imaginary number**, and the y-axis is called the **imaginary axis**. The xy-plane whose points correspond to complex numbers is referred to as the **complex plane**.

Complex numbers are commonly denoted by z, w, or the Greek letter ζ. If $z = a + ib$, where a and b are real, we say that z is in **standard form**. We call a the **real part** of z and b to be the **imaginary part** of z and write:

$$\begin{aligned} \operatorname{Re}(z) &= \operatorname{Re}(a + bi) = a, \\ \operatorname{Im}(z) &= \operatorname{Im}(a + bi) = b. \end{aligned} \tag{8.11}$$

We also define the **conjugate** $\bar{z}$ of $z = a + ib$ as

$$\bar{z} = a - ib. \tag{8.12}$$

If z corresponds to point $P(a, b)$, then $\bar{z}$ corresponds to the point $P'(a, -b)$, where P' is the reflection of P in the x-axis, shown in Figure 8.9(a). We say that $\bar{z}$ is the reflection of z in the x-axis and draw a diagram like the one in Figure 8.9(b). Of course, if z is above the x-axis, then $\bar{z}$ is below the x-axis.

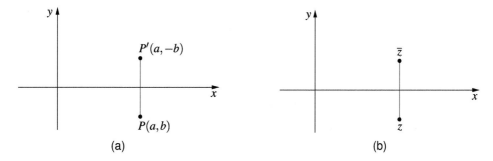

Figure 8.9

Recall that for real number a, $|a|$ is the distance from the point $(a, 0)$ to the origin. Similarly the **absolute value** of z is defined as the distance from the point that corresponds to z to the origin. That is, if $z = a + ib$, then

$$|z| = \sqrt{a^2 + b^2}. \qquad (8.13)$$

Notice that

$$|z|^2 = z \cdot \bar{z}. \qquad (8.14)$$

> **Now Solve This 8.3**
>
> 1. Why is the horizontal x-axis often called the real axis and the vertical axis is called the imaginary axis?
> 2. Is the definition of the absolute value of a complex number valid if the complex number is real? Are Equations (8.13) and (8.14) valid if $b = 0$?

> **Historical Note:** *Complex Numbers*
>
> Like negative numbers, complex numbers were not accepted right away. The Italian mathematician Girolamo Cardano (1501–1576) considered several forms for the quadratic equations to avoid using negative numbers. Cardano, known for formula for solving any cubic equation, encountered square roots of negative numbers when applying his formula to cubic equations that he knew had real roots. His contemporary Rafael Bombelli (1526–1572), in his treatise *Algebra*, used complex numbers reluctantly, in his treatise.
>
> René Descartes called a negative solution of an equation "false" and a solution of $x^2 = -1$ "imaginary." Leonhard Euler used complex numbers extensively, and introduced the symbol i for a solution of $x^2 = -1$. Nevertheless, he called complex numbers "impossible numbers."
>
> The full geometric and vector representation of complex numbers was first given by the Norwegian mathematician and surveyor Caspar Wessel (1745–1818). Gauss obtained results about integers in *Number Theory* using complex numbers, "which gave a tremendous boost to the acceptance of complex numbers in the mathematical community" (Klein, p. 593).
>
> William Rowan Hamilton (1805–1865), the famous Irish mathematician, viewed complex number as ordered pairs of real numbers and hence put complex numbers on a firm "real" foundation. Many mathematicians say that complex numbers come "alive" only after studying abstract algebra.

In addition to ordered pairs, it is also useful to see complex numbers as vectors. In Section 6.1, we viewed vectors as directed segments. There we introduced a translation τ_{AB} that was determined both by the directed segment AB and by infinitely many directed segments having the same length and direction.

In the plane, we define a vector as an ordered pair (a, b), where a and b are real numbers. If O is the origin and $P = (a, b)$, then the vector $\overrightarrow{OP}$ is called the **position vector**; its **head** P

gives the position of the point P. As shown in Figure 8.10, there are infinitely many vectors equivalent to $\overrightarrow{OP}$ but $\overrightarrow{OP}$ is the unique vector showing the position of the point P.

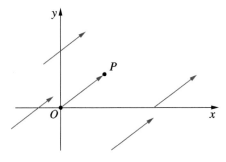

Figure 8.10

The point $P(a, b)$ corresponds to the position vector $\overrightarrow{OP}$. Because each complex number $a + ib$ corresponds to the unique point (a, b), each complex number corresponds to a unique position vector.

We define **vector addition** of as follows:

$$(a, b) + (c, d) = (a + c, b + d). \tag{8.15}$$

Notice that this operation corresponds to the addition of complex numbers:

$$(a + ib) + (c + id) = (a + c) + i(b + d). \tag{8.16}$$

(In abstract algebra, we say that the set of vectors under addition is isomorphic to the set of complex numbers under addition.)

Vector addition is shown visually in Figure 8.11(a). The sum is the vector $\overrightarrow{OP}$, which is the diagonal of the parallelogram $OAPB$ starting at O. The vertex P of the parallelogram has coordinates $(a + c, b + d)$. Instead of drawing the entire parallelogram, we could draw, at the head A of $\overrightarrow{OA}$, the vector $\overrightarrow{AP}$ equivalent to $\overrightarrow{OB}$ and obtain $\overrightarrow{OP}$. For the corresponding complex numbers in Figure 8.11(b), the sum $z + w$, corresponds to the diagonal of the parallelogram starting at O. $z + w$ can also be obtained by drawing the vector z (that is, the vector that corresponds to the complex number z) with the initial point at the origin and then drawing the equivalent vector w whose initial point is at the terminal point (head) of vector z.

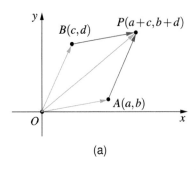 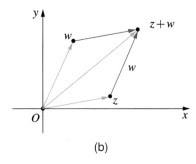

Figure 8.11

Vector multiplication by a real number λ is defined by

$$\lambda(x, y) = (\lambda x, \lambda y).$$

The **additive inverse** of (a, b) is defined as

$$-(a, b) = (-a, -b)$$

Subtraction of vectors is defined as

$$(a, b) - (c, d) = (a, b) + \bigl[-(c, d)\bigr]$$
$$= (a - c, b - d).$$

We understand subtraction of complex numbers in the same way: $z - w = z + (-w)$. Graphically subtraction can be performed using addition or simply by connecting the points corresponding to z and w and obtaining the vector whose initial point is at w and whose terminal point is at z. (However, in this way, $z - w$ does not represent the corresponding position vector.) The position vector is the vector equivalent to $z - w$ whose initial point is at O. The validity of the vector $z - w$ can be confirmed by the addition of complex numbers (or the addition of vectors): $w + (z - w) = z$.

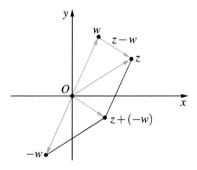

Figure 8.12

8.2 Complex Numbers and Geometry

Example 8.1

If z_1 and z_2 are two points in the complex plane, find the complex number z that corresponds to:

1. The midpoint of the segment connecting z_1 and z_2.
2. The point between z_1 and z_2 that divides the segment connecting z_1 and z_2 into ratio $m : n$.
3. The centroid (the intersection of the medians) of a triangle with vertices at $z_1, z_2,$ and z_3.

Solution

1. One approach is to use the fact that the diagonals of a parallelogram bisect each other. As shown in Figure 8.13, $z_1 + z_2$ corresponds to the diagonal with initial point at the origin. Therefore $\frac{z_1+z_2}{2}$ corresponds to its midpoint M.

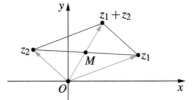

Figure 8.13

A second approach is to notice that the midpoint M of the segment connecting z_1 and z_2 corresponds to the vector $\overrightarrow{OM}$:

$$\overrightarrow{OM} = \overrightarrow{Oz_1} + \overrightarrow{z_1M}$$
$$= z_1 + \frac{1}{2}(z_2 - z_1)$$
$$= \frac{z_1 + z_2}{2}.$$

2. In Figure 8.14, because point P satisfies $\frac{PA}{PB} = \frac{m}{n}$, we have $\frac{|z-z_1|}{|z_2-z_1|} = \frac{m}{m+n}$. Consequently, $z - z_1 = \frac{m}{m+n}(z_2 - z_1)$, which implies

$$z = \frac{nz_1 + mz_2}{m+n}. \qquad (8.17)$$

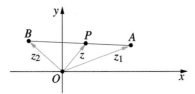

Figure 8.14

3. We can obtain the answer from Equation (8.17) or proceed independently as follows. Since the centroid G lies on all the three medians, in particular on the median $z_1 M$ (Figure 8.15), two thirds of the way from a vertex, we have

$$\overrightarrow{z_1 G} = \frac{2}{3}\left(\frac{z_2 + z_3}{2} - z_1\right).$$

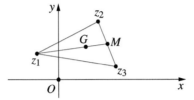

Figure 8.15

Notice that $\overrightarrow{z_1 G}$ is not a position vector. We can obtain the position vector $\overrightarrow{OG}$ by adding z_1 (since $z_1 + \overrightarrow{z_1 G} = \overrightarrow{OG}$):

$$\overrightarrow{OG} = z_1 + \frac{2}{3}\left(\frac{z_2 + z_3}{2} - z_1\right) = \frac{z_1 + z_2 + z_3}{3}.$$

Geometric Interpretation of Complex Multiplication

Before addressing multiplication of complex numbers is general, let's investigate first the special case of multiplying a complex number z by -1, i, and $-i$. Notice that

$$(-1)z = -z = -x - iy \longleftrightarrow (-x, -y).$$

Hence the image of (x, y) under multiplication by -1 is $(-x, -y)$. Next let's see what happens when we multiply by i.

$$iz = i(x + iy) = ix - y = -y + ix,$$

which corresponds to $(-y, x)$. Hence under multiplication by i the image of (x, y) is $(-y, x)$. Similarly,

$$(-i)z = -(iz) = y + (-i)x.$$

Hence under multiplication by $-i$ the image of (x, y) is $(y, -x)$.

Figure 8.16 shows that

$z \longrightarrow (-1)z$ amounts to rotating z about the origin by $180°$,	(8.18)
$z \longrightarrow iz$ amounts to rotating z counterclockwise by $90°$ about the origin, and	(8.19)
$z \longrightarrow -iz$ amounts to rotating z clockwise by $90°$ about the origin.	(8.20)

8.2 Complex Numbers and Geometry

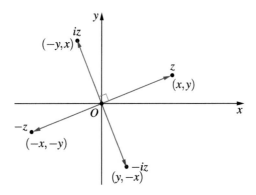

Figure 8.16

Example 8.2

In Figure 8.17, $ABCD$ is a square whose diagonals intersect at O. If P is the midpoint of $\overline{OB}$ and Q is the midpoint of $\overline{CD}$, prove that $\overrightarrow{AP}$ and $\overrightarrow{PQ}$ are congruent and perpendicular to each other.

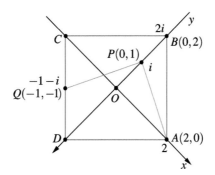

Figure 8.17

Solution. Because the diagonals of a square are congruent perpendicular bisectors of each other, we can set up a coordinate system such that O is at the origin, diagonal $\overline{AC}$ is the x-axis, and diagonal $\overline{DB}$ is the y-axis. Because any point on the x-axis can correspond to 1, we choose the unit to be $\frac{1}{2}OA$. Then we have $A(2,0), B(0,2), P(0,1)$, and $Q(-1,-1)$. The complex number corresponding to $A(2,0)$ is 2, to $P(0,1)$ is i, and to $Q(-1,-1)$ is $-1-i$.

Because $\overrightarrow{AP} = \overrightarrow{OP} - \overrightarrow{OA}$, the complex number corresponding to $\overrightarrow{AP}$ is $-2+i$. We write

$$\overrightarrow{AP} = -2 + i. \qquad (8.21)$$

Similarly, because $\overrightarrow{QP} = \overrightarrow{OP} - \overrightarrow{OQ}$, we get $\overrightarrow{QP} = i - (-1-i) = 2i+1$. Thus

$$\overrightarrow{QP} = 1 + 2i. \qquad (8.22)$$

To show that segments QP and AP are congruent and perpendicular, it suffices to show that one of the vectors is i or is $-i$ times the other [see Equations (8.19) and (8.20)]:
$$i \cdot \overrightarrow{QP} = i(1+2i) = -2+i = \overrightarrow{AP}.$$

Example 8.3

In Figure 8.18, $\triangle ABC$ is an arbitrary triangle with squares $ACDE$ and $AFGB$ constructed (externally) on two of its sides. Prove that the line containing the median AM is perpendicular to $\overline{EF}$.

Solution. We choose a coordinate system with the origin at A. Let the complex numbers corresponding to B and C be z_1 and z_2, respectively. We find the complex numbers corresponding to the vectors $\overrightarrow{AM}$ and $\overrightarrow{EF}$ and show that one is λi times the other for some real number λ. Indeed, $\overrightarrow{AM} = \frac{z_1+z_2}{2}$. Now we can obtain $\overrightarrow{AF}$ from $\overrightarrow{AB}$ by a rotation of 90° clockwise about A. Thus $\overrightarrow{AF} = -iz_1$. Similarly, $\overrightarrow{AE} = iz_2$. Therefore $\overrightarrow{EF} = \overrightarrow{AF} - \overrightarrow{AE} = -iz_1 - iz_2 = -i(z_1+z_2)$, and we see that $\overrightarrow{EF} = -2i\overrightarrow{AM}$. This implies that segments AM and EF are perpendicular and, in addition, that $AM = \frac{1}{2}EF$.

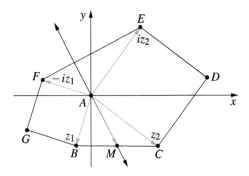

Figure 8.18

Example 8.4

Let's revisit the Treasure Island Problem from Chapter 1 (and other chapters).

Solution. Let the x-axis go through the trees, and let the origin O be the midpoint of $\overline{T_1 T_2}$ (see Figure 8.19). We can choose $OT_2 = 1$, so that the trees correspond to the complex numbers 1 and -1. Let $z = x + iy$ be the complex number corresponding to the gallows Γ. We will try to find, in terms of z, the complex number that corresponds to the treasure T. Because T is at the midpoint of $\overline{S_1 S_2}$,

$$\overrightarrow{OT} = \frac{\overrightarrow{OS_1} + \overrightarrow{OS_2}}{2}. \tag{8.23}$$

8.2 Complex Numbers and Geometry

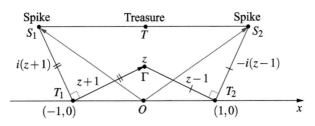

Figure 8.19

Because $\overrightarrow{T_1 \Gamma} = z - (-1) = z + 1$ can be rotated counterclockwise about T_1 to obtain $\overrightarrow{T_1 S_1}$, we have

$$\overrightarrow{T_1 S_1} = i(z+1). \tag{8.24}$$

To obtain the position vector $\overrightarrow{OS_1}$, we proceed as follows:

$$\overrightarrow{OS_1} = \overrightarrow{OT_1} + \overrightarrow{T_1 S_1} = -1 + i(z+1). \tag{8.25}$$

Similarly, $\overrightarrow{T_2 S_2} = -i\overrightarrow{T_2 \Gamma} = -i(z-1)$, and so

$$\overrightarrow{OS_2} = \overrightarrow{OT_2} + \overrightarrow{T_2 S_2} = 1 - i(z-1). \tag{8.26}$$

From Equations (8.23), (8.25), and (8.26), we get

$$\overrightarrow{OT} = \frac{\overrightarrow{OS_1} + \overrightarrow{OS_2}}{2} = \frac{-1 + i(z+1) + 1 - i(z-1)}{2} = i = 0 + 1 \cdot i \longleftrightarrow (0, 1).$$

This means that the treasure is at the point corresponding to i, that is, the point $(0, 1)$. Consequently, the position of the treasure is independent of the position of the gallows and can be found by choosing any point for the gallows and then following the given directions. Alternatively, we can find the treasure by constructing the point $(0, 1)$—that is, by constructing the perpendicular bisector of $\overrightarrow{T_1 T_2}$ and finding T (above $\overrightarrow{T_1 T_2}$) such that $OT = \frac{1}{2} T_1 T_2$.

Now Solve This 8.4

1. Prove that $|z_1 z_2| = |z_1| \cdot |z_2|$.
2. On the sides of an arbitrary quadrilateral, squares have been constructed as shown in Figure 8.20. The centers of the squares are C_1, C_2, C_3, and C_4. Use complex numbers to prove that the segments $C_1 C_3$ and $C_2 C_4$ are congruent and perpendicular.

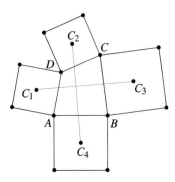

Figure 8.20

3. In this problem, we investigate the effect of multiplying any complex number $z = x + iy$ by a constant complex number $z_0 = a + ib$; that is, we consider the function $f: z \to z_0 z$ or, equivalently, $f(z) = z_0 z$. Notice that

$$z_0 z = (a + ib) \cdot (x + iy) = ax - by + i(bx + ay).$$

Therefore the image of (x, y) is $(ax - by, bx + ay)$. If we denote the image of (x, y) by (x_1, y_1), we have

$$\begin{cases} x_1 = ax - by, \\ y_1 = bx + ay. \end{cases} \quad (8.27)$$

Using matrix notation, we can write this image as follows:

$$\begin{pmatrix} x_1 \\ y_1 \end{pmatrix} = \begin{pmatrix} a & -b \\ b & a \end{pmatrix} \begin{pmatrix} x \\ y \end{pmatrix}.$$

Suppose that $z_1 = z_0 z$ is transformed to $w_0 z_1$, where $w_0 = c + id$ (a constant). In other words, we have a new function $g: z \to w_0 z$ [that is, $g(z) = w_0 z$] and we want to find $g \circ f$, the composition of f and g.

a. Show that $(g \circ f)(z) = (w_0 z_0) z$, where $w_0 z_0 = (a + ib)(c + id) = (ac - bd) + i(bc + ad)$.

b. Write the matrix that corresponds to the transformation in part (a).

c. Explain why the matrix you found in part (b) is equal to the product

$$\begin{pmatrix} a & -b \\ b & a \end{pmatrix} \begin{pmatrix} c & -d \\ d & c \end{pmatrix}.$$

8.2 Complex Numbers and Geometry

Polar (or Trigonometric) Representation of Complex Numbers

Figure 8.21 shows the polar coordinates (r, θ) of a point $P(x, y)$. Notice that $r = OP$ and that θ is the angle between $\overrightarrow{OP}$ and the positive x-axis. By definition we know that $\cos \theta = \frac{x}{r}$ and $\sin \theta = \frac{y}{r}$. Thus

$$x = r \cos \theta,$$
$$y = r \sin \theta. \tag{8.28}$$

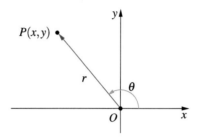

Figure 8.21

Consequently, we may write any complex number $z = x + iy$ as follows:

$$z = x + iy = r(\cos \theta + i \sin \theta). \tag{8.29}$$

Equation (8.29) is called the **polar** or **trigonometric representation** of z. Notice that $r = \sqrt{x^2 + y^2} = |z|$. It is common practice to refer to θ as the **argument** of z and to write

$$\theta = \arg(z) \quad \text{or simply arg } z. \tag{8.30}$$

The polar representation of z is very convenient when multiplying complex numbers, because $\arg(z_1 z_2) = \arg z_1 + \arg z_2$. To see why, consider Figure 8.22, where we illustrate what happens when z is multiplied by z_0. Let $\alpha = \arg z$ and $\beta = \arg z_0$. We will show that $\arg z_0 z = \alpha + \beta$. (Why is OA chosen as 1?)

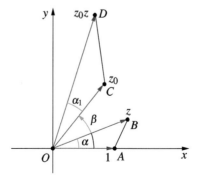

Figure 8.22

Notice that

$$\arg(z_0 z) = m(\angle AOD) = \alpha_1 + \beta \quad \text{(see Fig. 8.22)}.$$

To show that $\alpha = \alpha_1$ we only need to prove that $\triangle COD \sim \triangle AOB$. The triangles will be similar if

$$\frac{CO}{AO} = \frac{OD}{OB} = \frac{DC}{BA}. \tag{8.31}$$

Referring to Figure 8.22, we see that Equation (8.31) is equivalent to

$$\frac{|z_0|}{1} = \frac{|zz_0|}{|z|} = \frac{|z_0 z - z_0|}{|z - 1|}. \tag{8.32}$$

Using the fact that $|z_1 z_2| = |z_1| \cdot |z_2|$ (see Now Solve This 8.4), we immediately see that the ratio in (8.32) is equivalent to $|z_0|$, so the relationship in Equation (8.32) holds.

We have proved the following theorem:

> **Theorem 8.3**
>
> If $z_1 = r_1(\cos \theta_1 + i \sin \theta_1)$ and $z_2 = r_2(\cos \theta_2 + i \sin \theta_2)$, then
>
> $$z_1 z_2 = r_1 r_2 \Big(\cos(\theta_1 + \theta_2) + i \sin(\theta_1 + \theta_2) \Big).$$

Theorem 8.3 implies the following (why?):

> **Theorem 8.4: De Moivre's Theorem**
>
> For all positive integers n,
>
> $$(\cos \theta + i \sin \theta)^n = \cos n\theta + i \sin n\theta.$$

> **Remark 8.1**
>
> Theorem 8.4 holds for negative integers as well.

Using the Taylor series expansion for $\cos x$, $\sin x$, and e^x, we can be show that

$$\cos x + i \sin x = e^{ix}. \tag{8.33}$$

Substituting $x = \pi$, we have Euler's famous formula, which some say is one of the most beautiful equations in mathematics:

$$-1 = e^{i\pi} \quad \text{or} \quad e^{i\pi} + 1 = 0. \tag{8.34}$$

8.2 Complex Numbers and Geometry

> **Now Solve This 8.5**
>
> 1. Use Theorem 8.3 to prove the following fundamental trigonometric formulas:
>   a. $\cos(\alpha + \beta) = \cos\alpha\cos\beta - \sin\alpha\sin\beta$,
>   b. $\sin(\alpha + \beta) = \sin\alpha\cos\beta + \sin\beta\cos\alpha$.
> 2. Using part 3 of Now Solve This 8.4, show the following:
>   a. The matrix that corresponds to multiplying a complex number by $\cos\theta + i\sin\theta$ is
> $$\begin{pmatrix} \cos\theta & -\sin\theta \\ \sin\theta & \cos\theta \end{pmatrix}.$$
>   b.
> $$\begin{pmatrix} \cos\alpha & -\sin\alpha \\ \sin\alpha & \cos\alpha \end{pmatrix} \begin{pmatrix} \cos\beta & -\sin\beta \\ \sin\beta & \cos\beta \end{pmatrix} = \begin{pmatrix} \cos(\alpha+\beta) & -\sin(\alpha+\beta) \\ \sin(\alpha+\beta) & \cos(\alpha+\beta) \end{pmatrix}.$$
>   c. Finally, use the product of the matrices in part (b) to derive the formulas (a) and (b) in part 1.
> 3. Convince yourself of the validity of de Moivre's Theorem using the following semi-legitimate approach. Let $z = \cos x + i\sin x$. Then
> $$\frac{dz}{dx} = -\sin x + i\cos x = i(\cos x + i\sin x) = iz.$$
>   Hence,
> $$\frac{dz}{z} = i\,dx.$$
>   Complete the "proof" by integrating both parts of the equation above (don't forget the constant).
> 4. Why did we call the approach in question 3, part (b), "semi-legitimate"?
> 5. Prove De Moivre's Theorem for negative integers.
> 6. Show that $\arg\left(\dfrac{z_1}{z_2}\right) = \arg z_1 - \arg z_2$.

> **Example 8.5**
>
> Without using a calculator, prove that if $AE = ED = DC = BC$ in Figure 8.23, then $\alpha + \beta + \gamma = 90°$.

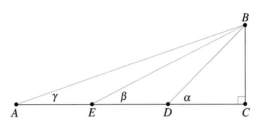

Figure 8.23

Solution. We create a coordinate system with C as the origin, $\overleftrightarrow{CB}$ as the y-axis, and $\overleftrightarrow{CA}$ as the x-axis. We choose $DC = 1$. If $B(0,1)$, $D(-1,0)$, $E(-2,0)$, and $A(-3,0)$, then the complex numbers representing these points are $i, -1, -2$, and -3, respectively. Notice that $\alpha, \beta,$ and γ are the arguments of the complex numbers represented by the vectors $\overrightarrow{DB}, \overrightarrow{EB},$ and $\overrightarrow{AB}$, respectively. These complex numbers are $i - (-1), i - (-2),$ and $i - (-3)$, that is, $1+i, 2+i,$ and $3+i$, respectively. The sum $\alpha + \beta + \gamma$ is the argument of the product $(1+i)(2+i)(3+i)$. The argument of this product will be $90°$ (as anticipated) if the product is ki for some real number k (then ki will correspond to a point on the y-axis). Indeed,

$$(1+i)(2+i)(3+i) = (1+i)(5+5i) = 10i.$$

> **Now Solve This 8.6**
>
> Use an approach similar to the one in Example 8.5 to show that if α and β are angles as shown in Figure 8.24, then $\alpha + 2\beta = 45°$.
>
>
>
> Figure 8.24

8.2.1 Problem Set

All problems should be completed using complex numbers.

1. Let $z_1, z_2, z_3,$ and z_4 be the complex numbers corresponding to the vertices of a convex quadrilateral. Find each of the following in terms of the following:

   a. The complex number corresponding to the intersection of the two segments connecting the midpoints of the opposite sides of the quadrilateral.

   b. The complex number corresponding to the midpoint of the segment joining the midpoints of the diagonals.

   c. State the theorem based on your answers to parts (a) and (b).

8.2 Complex Numbers and Geometry

2. Let z_1, z_2, and z_3 be any three complex numbers. Show geometrically that

$$|z_1 + z_2| \leq |z_1| + |z_2|.$$

3. Let $a_1, a_2, \ldots, a_n$ and $b_1, b_2, \ldots, b_n$ be real numbers. Justify the following inequality geometrically using complex numbers:

$$(a_1^2+b_1^2)^{\frac{1}{2}}+(a_2^2+b_2^2)^{\frac{1}{2}}+\cdots+(a_n^2+b_n^2)^{\frac{1}{2}} \geq \left[(a_1+a_2+\cdots+a_n)^2+(b_1+b_2+\cdots+b_n)^2\right]^{\frac{1}{2}}.$$

4. Prove that the centers of the four squares constructed on the sides of a parallelogram are vertices of a square.

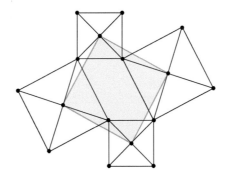

5. Prove Napoleon's Theorem: If Q, P, and S are the centroids (the points where medians intersect) of equilateral triangles constructed on the sides of an arbitrary $\triangle ABC$ (as shown in the figure), then $\triangle QPS$ is equilateral.

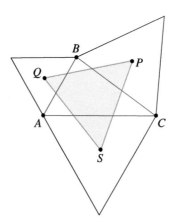

6. The figure below shows a right triangle ABC and squares constructed on its sides. Prove that
   a. The quadrilaterals $JLDC$ and $BEFG$ are parallelograms.
   b. $\triangle ALF$ is isosceles.

c. Is $\triangle ALF$ also a right triangle? Justify your answer.

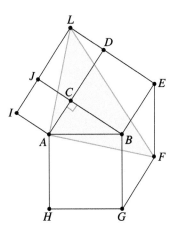

7. Suppose $ABCD$ is a square, and $A_1B_1C_1D_1$ is another square in the interior of $\triangle ABCD$ as shown in the figure. The points P, Q, R, and S are the midpoints of $\overline{AA_1}$, $\overline{BB_1}$, $\overline{CC_1}$, and $\overline{DD_1}$, respectively.

   a. Prove that $PQRS$ is a square (regardless of the location of the smaller square in the interior of the larger one).

   b. Will $PQRS$ still be a square even if $A_1B_1C_1D_1$ is not entirely in the interior of $ABCD$? Justify your answer.

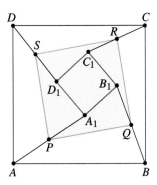

8. Prove that a triangle inscribed in a unit circle with vertices at z_1, z_2, and z_3 is equilateral if and only if $|z_1| = |z_2| = |z_3| = 1$ and $z_1 + z_2 + z_3 = 0$.

Problems 9–19 are about isometries and similarity transformations expressed via complex numbers. All functions are from complex numbers to complex numbers.

9. Prove that $f(z) = az + b$ represents an isometry if and only if $|a| = 1$.
10. Justify the following:

8.2 Complex Numbers and Geometry

a. Every translation can be represented by the function $T(z) = z + z_0$, where z_0 is a complex number.

b. Every rotation about the origin by O can be represented by $R_\theta(z) = az$, where $a = e^{i\theta}$.

c. A rotation about z_0 by θ can be represented by $R_{z_0,\theta}(z) = a(z - z_0) + z_0$, where $a = e^{i\theta}$. (The diagram should be helpful.)

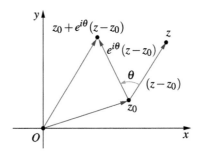

d. Obtain the result in part (c) by showing first that the rotation about z_0 by θ equals $T_{z_0} \circ R_{0,\theta} \circ T_{-z_0}$, where T_{z_0} is the translation by the vector z_0, T_{-z_0} is the translation by the vector $-z_0$ and $R_{O,\theta}$ is the rotation about the origin by θ.

e. A reflection in line ℓ through the origin making an angle θ with the positive x-axis can be achieved by the composition $R_\theta \circ M_x \circ R_{-\theta}$. Thus the reflection in line ℓ is given by

$$z \to ze^{-i\theta} \to \overline{ze^{-i\theta}} \to (e^{i\theta}\overline{z})e^{i\theta} = e^{2i\theta}\overline{z} \quad \text{or} \quad M_\ell(z) = e^{2i\theta}\overline{z}.$$

f. A reflection in line k intersecting the x-axis at $(h, 0)$ and parallel to line ℓ of part (e) is given by

$$M_k(z) = (\overline{z} - h)e^{2i\theta} + h.$$

11. Use part (e) of Problem 10 to find the matrix that corresponds to the reflection in a line through the origin that makes angle θ with the positive x-axis.

12. Show that the functions in Problem 10, parts (e) and (f), are reflections by proving that the composition of each with itself is the identity.

13. Let $f(z) = \overline{z} + ik$, where k is a real number. Prove that $f(z)$ is a reflection, and find the line of reflection.

14. Let $g(z) = a\overline{z} + b$, where $|a| = 1$. Find necessary and sufficient conditions on a and b for $g(z)$ to be a reflection in a line.

15. Using the fact that a rotation that is not the identity must have one fixed point and a translation that is not an identity has no fixed points, find necessary and sufficient conditions on a and b for $f(z) = az + b$ with $|a| = 1$, to be

a. a rotation. Find the center of the rotation.

b. a translation.

16. Complete the following part of a proof that an isometry with at least two fixed points is either the identity or a reflection in a line:

    Let f be an isometry for which $f(A) = A$ and $f(B) = B$, with $A \neq B$. If f is not the identity, then there exists a point P such that $f(P) = P'$ and $P' \neq P$ be distinct points.

    a. Let P and P' be distinct points. Prove that $AP' = AP$ and $BP' = BP$. Conclude that A and B must be on the perpendicular of $\overline{PP'}$, which is $\overleftrightarrow{AB}$.

    b. Conclude that any isometry with at least two fixed points is either the identity or a reflection in a line.

17. Prove that every plane isometry is represented by either $f(z) = az + b$ or $a\bar{z} + b$, where $|a| = 1$, following the steps below:

    a. Let $g(z)$ be an isometry, and let $f(z) = az + b$. Notice that $f(0) = b$. Choose b such that $b = g(0)$. Notice that $f(1) = a + b$. Choose a such that $a = g(1) - b = g(1) - g(0)$. Show that for these choices of a and b, $g(0) = f(0)$ and $g(1) = f(1)$.

    b. Consider the function $f^{-1} \circ g$. Show that it has two fixed points at 0 and 1 and, by part (a) that it is either the identity or reflection in the line through 0 and 1 (i.e., the x-axis). Thus $f^{-1} \circ g = I$ or $f^{-1} \circ g = M_x$. Show that in the first case $g(z) = f(z)$ and in the second case $g(z) = f(\bar{z})$.

18. In Problem 15 you may have proved that $f(z) = az + b$ is a rotation if and only if $|a| = 1$. Use this fact to prove that the composition of two rotations about two different points is either a rotation or a translation. [*Hint:* Let $f_1(z) = e^{i\theta_1}z + b_1$ and $f_2(z) = e^{i\theta_2}z + b_2$.] Finally, generalize the Treasure Island Problem by considering turning at the trees by angles α and β (put a condition on $\alpha + \beta$).

19. Prove that every similarity transformation can be achieved by $f(z) = az + b$ or $f(z) = a\bar{z} + b$, where $a \neq 0$.

Problems 20–23 involve the concept of a group (see Problems 11–16 in 6.3.1 Problem Set). The variables are complex numbers and so are the constants a and b.

20. **a.** Let $f(z) = az + b$, where $a \neq 0$. Find $f^{-1}(z)$ in terms of a and b.
    b. In part (a), if $|a| = 1$, show that $f^{-1}(z) = \bar{a}z - \bar{a}b$.

21. **a.** Prove that the set of functions $f(z) = az + b$, where $a \neq 1$, forms a group with respect to composition of functions.
    b. Prove that the set of functions $f(z) = az + b$, where $|a| = 1$, is a subgroup of the set in part (a).
    c. Prove that the set of functions $T(z) = z + b$ is a subgroup of the set in part (b).
    d. Prove that the set of functions $H(z) = -z + b$ is a subgroup of the group in part (b).
    e. Prove that the set of functions $\mathcal{F}$ such that $\mathcal{F} = \{f \mid f(z) = az + b \text{ or } f(z) = a\bar{z} + b, \text{ where } a \neq 0\}$ is a group with respect to composition of functions and that each of the sets in parts (a) through (d) is a subgroup of $\mathcal{F}$.

8.2 Complex Numbers and Geometry

22. Consider the set of functions

$$G = \{f \mid f(z) = -z + b \text{ or } f(z) = z + b\}.$$

Prove that G is a group.

23. Let $H = \{f \mid f(z) = z + b \text{ or } f(z) = \bar{z} + b\}$. Is H a group? Justify your answer.

9 Inversion

9.1 Introduction

Among the isometries we studied earlier are those that are their own inverse. For example, if M_l is a reflection in line l, then $M_l^{-1} = M_l$, or equivalently $M_l \circ M_l = M_l^2 = I$, where I is the identity transformation on the plane. Similarly, if H_O is a half-turn (rotation by 180°) about point O, then $H_O^{-1} = H_O$, or equivalently $H_O \circ H_O = H_O^2 = I$. Reflections and half-turns are examples of what is known as *involutions*. Formally, an *involution* is a function f which is its own inverse, that is, $f(f(x)) = x$ for all x in the domain of the function, or equivalently $f^2 = I$, where I is the identity function on the domain of f.

In the complex plane, perhaps the simplest reflection is the reflection in the x-axis given by $M_x(z) = \bar{z}$, where $\bar{z}$ is the complex conjugate of z (if $z = a + ib$, then $\bar{z} = a - ib$). Also, the simplest half-turn is the half-turn whose center is the origin, which is given by $H_O(z) = -z$. Other simple complex functions that are involutions are $f(z) = 1/z$ and the composition (in any order) or f with M_x, that is, $g(z) = 1/\bar{z}$. The latter is geometrically appealing as the point corresponding to z and the image point corresponding to $1/\bar{z}$ are on the same ray with endpoint at the origin. Furthermore, if you consider the polar form of z which is $re^{i\theta}$, then $1/\bar{z} = \frac{1}{r}e^{i\theta}$, which shows that the modulus of z times the modulus of $1/\bar{z}$ is equal to $r \cdot \frac{1}{r} = 1$. In Problem 15 at the end of this section, the reader will be asked to prove that z and $1/\bar{z}$ are the inverse images of each other under inversion in the unit circle centered at the origin.

Given a center O and a distance r, any two points P and P', which are collinear with O and lie on the same side of O, such that the product of their distances from O is equal to r^2, are inverse points with respect to the center O and the distance r [70]. In the plane, this represents what is known as *inversion in a circle* or *circle inversion*. In space, this represents what is known as *inversion in a sphere* or *sphere inversion*.

It is believed that inversion was likely first introduced by Apollonius of Perga (225–190 B.C.E.). However, the systematic investigation of inversion began with Jakob Steiner (1796–1863) in the 1820s [38], who succeeded in using it to make many geometric discoveries by the age of 28. During the following decades, many physicists and mathematicians independently rediscovered inversion, focusing their attention on properties that were most relevant and applicable to their particular fields. For example, William Thomson used inversion in a sphere to calculate the effect of a point charge on a nearby conductor made of two intersecting planes [61].

In 1855, August F. Möbius gave the first comprehensive treatment of inversion. Subsequently, Mario Pieri developed the subject axiomatically and systematically in *New Principles of the Geometry of Inversions, Memoirs I and II* in the early 1910s [60]. In those memoirs, Pieri presented the geometry of inversions as its own geometry (*inversive geometry* [18]) independent of Euclidean geometry.

Some of the material in this chapter was adapted from [51]. The main focus of this chapter is circle inversion; after we formally define it, we'll simply refer to it as inversion. Informally, circle inversion is a transformation of the plane that takes points inside the circle (except for the center of the circle) in a one-to-one fashion to points outside the circle and vice versa, while points on the circle remain where they are. In other words, the bounded region that lies inside the circle is mapped onto the unbounded region that lies outside the circle and vice versa. Special attention is required when determining the image of the center of the circle of inversion as demonstrated below.

> **Definition: Circle Inversion**
>
> Let C be a circle with radius r and center O. Let T be the map that takes a point P to a point P' on the ray OP such that $OP \cdot OP' = r^2$. Then, T is an inversion in the circle C.

Notice that the definition of inversion implies that points inside the circle that are close to the center O get mapped to points that are very far from the circle. In addition, it can be easily verified that points on C are fixed by inversion in C. However, there is one point, namely O, in the plane that does not have an image under inversion. As P gets closer to O, P' gets farther away from O, so in some sense, we can think of O as being mapped to a point infinitely far from O, often referred to as "the point at infinity." We will explore this in more detail in Section 9.5 with a geometric interpretation on the sphere. For now, we will simply say that the center of the circle of inversion, O, is mapped to the "point at infinity" on the extended plane, and O will be referred to as the *center* of the inversion. Note that if P maps to P', then P' also maps to P, and we say that P and P' are *inverses* with respect to the inversion in C.

Given a circle of inversion C centered at O of radius r and an arbitrary point P, how can we determine P', the image of P under inversion in circle C? That is, how do we construct P' on the ray $\overrightarrow{OP}$ such that $OP \cdot OP' = r^2$? Well, there are several ways to do this including the two constructions described below. Construction I requires a compass and straightedge and its proof only requires basic knowledge of similar triangles. In this construction, the method for determining P' depends on whether the point P lies inside or outside the circle of inversion C, hence the need for the two cases. Construction II on the other hand, see [26], is not dependent on where P lies and uses reflection about a line to determine P'.

Construction I: Before describing the details of this construction, we will discuss a basic result from Euclidean geometry needed for demonstrating the validity of this construction. This result states that when the altitude is drawn to the hypotenuse of a right triangle, then the resulting two right triangles are similar to the original triangle. For instance, in Figure 9.1, $\triangle ACD \sim \triangle ABC$, which follows from the fact that $\triangle ACD$ and $\triangle ABC$ are right triangles and share the acute $\angle A$, hence $\angle ACD \cong \angle ABC$. Therefore, $\frac{AD}{AC} = \frac{AC}{AB}$, and $AD \cdot AB = (AC)^2$. By the definition of inversion, this implies that B is the image of D under inversion in the circle centered at A and going through C.

9.1 Introduction

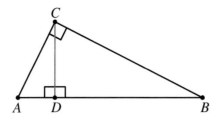

Figure 9.1

Similarly, $\triangle CBD \sim \triangle ABC$, which leads to $BD \cdot BA = (BC)^2$. This implies that A is the image of D under inversion in the circle centered at B and going through C.

Case 1. *The point P is inside the circle of inversion.*

Consider a circle C centered at O of radius r and a point P lying inside C, as in Figure 9.2. To determine P' we first construct line l through P and perpendicular to $\overline{OP}$, and let E be one of the points of intersection of l with the circle C. We then construct line k tangent to the circle C at point E by constructing the perpendicular to the radius $\overline{OE}$. Let P' be the point of intersection of line k and the ray $\overrightarrow{OP}$. Note that $\overline{EP}$ is the altitude to the hypotenuse of the right triangle $\triangle OP'E$. Therefore, using the basic result discussed above gives $OP \cdot OP' = (OE)^2 = r^2$. Thus, P' is the image of P under inversion in the circle C.

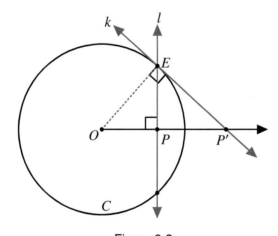

Figure 9.2

Case 2. *The point P is outside the circle of inversion.*

Consider a circle C centered at O of radius r and an arbitrary point P lying outside C, as in Figure 9.3. To construct P' we first construct the circle C_1 centered at M, the midpoint of $\overline{OP}$, and going through O. We then let E and F be the points of intersection of the circles C_1 and C, and P' be the intersection of $\overline{OP}$ and $\overline{EF}$. Note that $\overline{EP'}$ is the altitude to the hypotenuse of the right triangle $\triangle OPE$. Therefore, using the basic result discussed above gives $OP \cdot OP' = (OE)^2 = r^2$. Thus, P' is the image of P under inversion in the circle C.

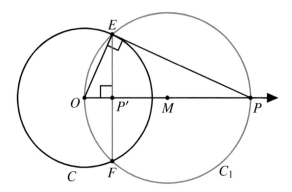

Figure 9.3

Construction II: This construction uses reflection about a line to determine the image of a point under inversion [26]. To be more precise, consider the circle of inversion C centered at O of radius r, and let P be a point inside this circle, as in Figure 9.4. Let the ray $\overrightarrow{OP}$ intersect the circle C in the point E, and choose a point F on the circle C distinct from E. Finally, let $\overleftrightarrow{FP'}$, the reflection of $\overleftrightarrow{FP}$ about $\overleftrightarrow{FE}$, intersect $\overrightarrow{OP}$ in the point P'. The reader is encouraged to justify how to construct $\overleftrightarrow{FP'}$ using Euclidean tools and to show that it has to intersect $\overrightarrow{OP}$. This point P' is the image of P under inversion in the circle C. Note that there is nothing to prevent us from interchanging the roles of P by P' in the preceding discussion, which implies that this construction is also valid if P lies outside the circle C. A brief proof of the validity of this construction is presented below.

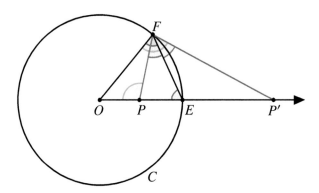

Figure 9.4

Proof. Our goal is to show that the point P' described above and shown in Figure 9.4 satisfies $OP \cdot OP' = r^2$. To do so, note that $\triangle OEF$ is an isosceles triangle and hence $\angle OEF \cong \angle EFO$. Now since $\overleftrightarrow{FP'}$ is the reflection of $\overleftrightarrow{FP}$ about $\overleftrightarrow{FE}$, we have $\angle EFP \cong \angle P'FE$. The fact that $\angle OPF$ is exterior to $\triangle PEF$ implies that $m(\angle OPF) = m(\angle PEF) + m(\angle EFP) =$

9.1 Introduction

$m(\angle EFO)+m(\angle P'FE) = m(\angle P'FO)$. Therefore, $\triangle OPF$ and $\triangle OFP'$ are similar triangles and hence $\frac{OP}{OF} = \frac{OF}{OP'}$. Therefore, $OP \cdot OP' = (OF)^2 = r^2$, which completes the proof. ∎

> **Remark 9.1**
>
> It's worth noting that the circle going through the points P, P', and F in Figure 9.4 (circle not shown) inverts onto itself. This circle turns out to be orthogonal to C (see Theorem 9.5 in Section 9.2).

We will now introduce some basic notation to aid in our study of inversion. If C is a circle, we denote inversion with respect to C by I_C, so if the point P' is the inverse of P, then $I_C(P) = P'$. Notice that $I_C(P') = P$ as well, and hence $I_C(I_C(P)) = I_C(P') = P$. Therefore, $I_C^2 = I_C \circ I_C$ is the identity function on the plane.

Recall that reflection M_l in a line l also has the same property, that is, M_l^2 is the identity transformation. We now show that reflection in a line can in fact be realized as a limit of circle inversions. Let l be a line, $P' = M_l(P)$ be the reflection of P in l, and Q be the point of intersection of l and $\overleftrightarrow{PP'}$. Now consider circles $C_1, C_2, C_3 \ldots$ tangent to l at Q with centers $O_1, O_2, O_3, \ldots$ lying on $\overleftrightarrow{PP'}$ such that P is between O_i and l for each i, as in Figure 9.5. For $i = 1, 2, 3, \ldots$, construct $P_i = I_{C_i}(P)$, the inverse of P under inversion in circle C_i. Notice that as i goes to infinity, P_i appears to approach P', the reflection of P in l.

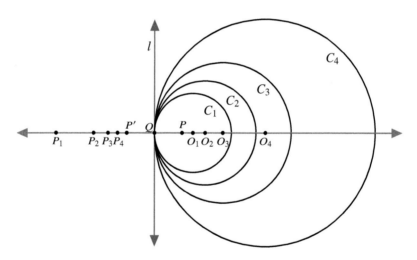

Figure 9.5

Before formally proving that P_i approaches P' as i approaches infinity, let's consider the simple case where l coincides with the y-axis and $\overleftrightarrow{PP'}$ coincides with the x-axis. If we let $P = (a, 0)$, then $P' = (-a, 0)$. Next, let $O_i = (a + i \cdot c, 0)$ and $P_i = (-s_i, 0)$, where c and s_i are positive constants. Now by the definition of inversion $O_iP \cdot O_iP_i = r_i^2$, which in terms of Cartesian coordinates means $(i \cdot c)(a + i \cdot c + s_i) = (a + i \cdot c)^2$. Solving for s_i in this equation gives $s_i = \frac{a^2}{i \cdot c} + a$. But $i \cdot c$ approaches infinity as i approaches infinity, hence s_i

approaches a as i approaches infinity. This shows that the sequence of points P_i approaches $P' = (-a, 0)$ as i goes to infinity.

We now proceed to formally prove that P_i does in fact approach P' as i approaches infinity. In the following proof, l and $\overleftrightarrow{PP'}$ are perpendicular lines but are not necessarily the x-axis and y-axis, and hence Q is not necessarily the origin.

Proof. First, note that by the definition of inversion, $O_i P \cdot O_i P_i = r_i^2$ for all i, where r_i is the radius of C_i. Thus, $O_i P_i = \frac{r_i^2}{O_i P}$. By collinearity, $O_i P = O_i Q - QP = r_i - QP$, where Q is the intersection of l and $\overleftrightarrow{PP'}$. Hence, $O_i P_i = \frac{r_i^2}{r_i - QP}$. We now want to see what happens to QP_i as i goes to infinity. By collinearity and using the expression we just arrived at for $O_i P_i$ we have

$$QP_i = O_i P_i - O_i Q = O_i P_i - r_i = \frac{r_i^2}{r_i - QP} - r_i = \frac{r_i QP}{r_i - QP} = \frac{QP}{1 - \frac{QP}{r_i}}.$$

As i get larger and larger, so does r_i, and hence $\frac{QP}{r_i}$ goes to zero. Therefore, QP_i approaches QP, and hence P_i approaches P'. ∎

> **Remark 9.2**
>
> In order to gain experience and intuition with this interesting and powerful plane transformation, the reader is encouraged to use freely available dynamic geometry software such as GeoGebra to investigate how various figures are transformed under inversion.

9.1.1 Problem Set

1. Show that under inversion in a circle centered at (a, b) of radius r, a point (x, y) gets mapped onto $(x', y') = \left(\frac{(x-a)r^2}{(x-a)^2 + (y-b)^2} + a, \frac{(y-b)r^2}{(x-a)^2 + (y-b)^2} + b \right)$. Use this to show that the image of the line $x = \frac{1}{2}$ under inversion in the circle $x^2 + y^2 = 1$ is the circle $(x-1)^2 + y^2 = 1$.

2. Is there any circle C in the plane whose center is inverted onto the center of its image C' under inversion in a circle K? Explain your answer.

3. Prove that inversion is not an isometry (i.e., the distance between any two points is not preserved under inversion).

4. Let C be a circle with center O. Let P' and Q' be the images of points P and Q, respectively, under inversion in C. Show that $\triangle OPQ$ is similar to $\triangle OQ'P'$.

5. Let A and B be two distinct points on a circle C. Prove that if A and B are fixed under inversion in a circle K, then this inversion takes C onto C.

6. Let A and B be two distinct points on a circle C. Assume that the tangents to C at A and B meet at point E, and let F be the midpoint of the chord $\overline{AB}$. Prove that E and F are the images of each other under inversion in C.

9.1 Introduction

7. Prove that if the points A' and B' are the respective images of A and B under inversion in a circle centered at O of radius r, then $A'B' = \frac{r^2}{OA \cdot OB} \cdot AB$.

8. Use dynamic geometry software to show that the image of a parabola under inversion in a circle centered at the focus of the parabola is a cardioid [92].

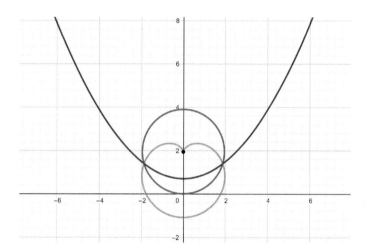

9. Use dynamic geometry software to show that the image of a hyperbola under inversion in a circle centered at the center of a hyperbola is a lemniscate [92].

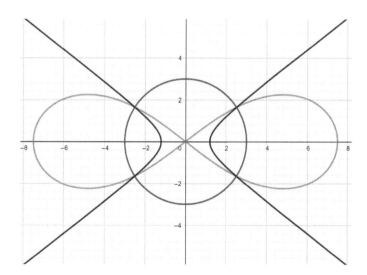

10. Use dynamic geometry software to show that the image of a hyperbola under inversion in a circle centered at one of the foci is a limaçon [92].

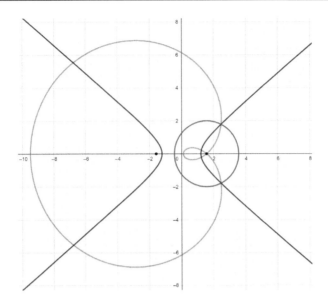

11. Use dynamic geometry software to verify that the blue curve ψ' (in the figure below) is the image of the orange curve ψ, under inversion in the circle centered at $(0,0)$ with radius 3. The parametric equations of the orange curve are $x(t) = 2\cos(t) + \cos(5t)$ and $y(t) = 2\sin(t) + \sin(5t)$ for $t \in [0, 2\pi]$.

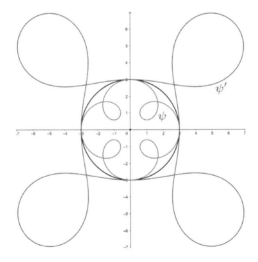

12. Inversion of polar curves:
   a. Show that, under inversion in a circle centered at $(0,0)$ with radius k, the image of a curve whose polar equation is $r = f(\theta)$ is the locus of points satisfying the polar equation $r = \frac{k^2}{f(\theta)}$.
   b. Use dynamic geometry software to verify that the image of the curve $r = 3\cos(4\theta)$ (orange rose with 8 petals in the figure below) is the set of blue curves.

c. Do you notice anything special about the image of two adjacent petals?

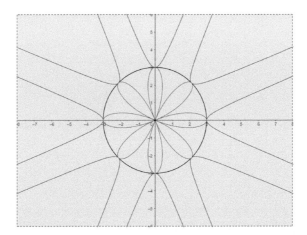

13. Consider a circle C centered at O of radius r and an arbitrary point P lying outside C as shown in the figure below. To construct P', the image of P under inversion in C, we first construct the circle C_1 centered at P and going through O, and let E be one of the points of intersection of the circles C_1 and C. We then construct the circle C_2 centered at E and going through O. The point P' where C_2 intersects $\overrightarrow{OP}$ is the image of P under inversion in the circle C. Prove that this construction is valid.

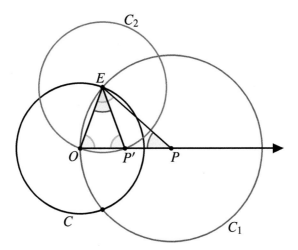

14. An alternative construction of the image of P under inversion in a circle C can be based on the fact that the measure of the altitude drawn from the vertex of the right angle of a right triangle to its hypotenuse is the geometric mean of the measures of the two segments into which the hypotenuse is divided. If P is inside the circle C, find its image P' under inversion in circle C by first finding its image P^* under a half-turn about O, and then finding the point P' such that $OP^* \cdot OP' = r^2$ (see the figure below). Complete and describe this approach and verify that this construction is also valid if P lies outside the circle C.

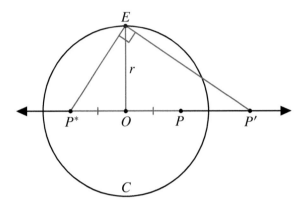

15. **Inversion in the Complex Plane.** Another way of viewing inversion is by defining it as a function on the complex plane.

    a. Show that the mapping $F(z) = \frac{1}{z}$ satisfies the equation
    $$|z| \cdot |z'| = 1,$$
    where $z' = F(z)$ and $|z|$ is the modulus of z, that is, $|z| = \sqrt{a^2 + b^2}$ if $z = a + ib$.

    b. Show that there exists a point $z \in \mathbb{C}$ such that $z' = F(z)$ is not on the ray from the origin to z.

    c. Notice that part (b) shows that $F(z)$ is not an inversion map since z and z' do not lie on the same ray emanating from the origin. We can fix this by defining
    $$z' = \bar{F}(z) = \frac{1}{\bar{z}}$$
    as our inversion mapping, where $\bar{z} = a - ib$ if $z = a + ib$. Show that z and z' lie on the same ray emanating from the origin and that $\bar{F}$ still satisfies the equation $|z| \cdot |z'| = 1$. Thus, $\bar{F}$ is inversion in a circle of radius 1 centered at the origin.

    d. Show that inversion in a circle of radius r centered at the origin of the complex plane can be written as $D_r \circ \bar{F} \circ D_r^{-1}(z)$, where $D_r(z) = rz$ is dilation by a factor of r.

    e. Use translations and the previous parts of this problem to write the equation for inversion in a circle of radius r centered at $z_0 = a_0 + b_0 i$.

9.2 Properties of Inversions

Having defined inversion and described how to construct the image of an arbitrary point under inversion in a given circle, we now turn our attention to studying how basic geometric

9.2 Properties of Inversions

figures such as lines and circles are transformed under inversion. This will prove very helpful in solving several challenging geometry problems such as the Shoemaker's Knife problem, Apollonius problem, Nine-Point Circle problem, etc., which will be investigated in the following section. Such problems are typically difficult to solve when inversion is not available as a problem solving tool. As we shall see, inversion turns these problems into much simpler problems, which shows the power and versatility of this tool.

To begin with, let us consider the image of lines under inversion. Here we will deal with two cases: Lines that go through the center of inversion and lines that do not. We will show that the former invert onto themselves and the latter invert onto circles going through the center of inversion.

> **Theorem 9.1**
>
> *The image of a line through the center of the inversion is the line itself.*

Proof. Let C be a circle centered at O of radius r, and l be a line through O. We will show that $l' = I_C(l)$, the image of l under inversion in C, is a subset of l and that l is a subset of l'. Let P be a point on l and P' be the image of P under inversion in circle C. Then by the definition of inversion, P' is on the ray $\overrightarrow{OP}$ such that $OP \cdot OP' = r^2$, and hence P' is also on l. This implies that l' is a subset of l. Moreover, if P is a point on l, then $P' = I_C(P)$ is also on l and so is $P = I_C(P')$. This implies that P is on l' and hence l is a subset of l'. Therefore, a line l through the center of inversion is inverted onto itself. ∎

Having seen that a line through the center of inversion inverts onto itself, we next investigate the effect of inversion on other lines and on circles. This will allow us to determine the inverse images of more complicated objects such as those in Figure 9.6, where objects and their images share the same color.

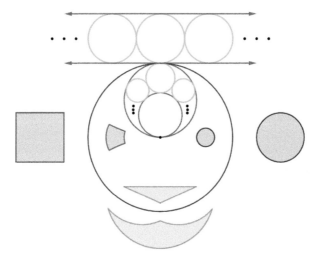

Figure 9.6

Let l be a line not through O, as in Figure 9.8. Using dynamic geometry software such as GeoGebra, we can see and hence conjecture that the image of l is a circle through O. To find the exact location of the circle without relying on the software, and to motivate a rigorous proof, recall that under inversion the image of any line through O is the line itself. The fact that under inversion the absolute value of the measure of the angle between two curves is preserved (see Theorem 9.4 in the next section) allows us to conclude that the measure of the angle between l and a line through O will equal the absolute value of the angle between the image of l (the circle through O) and the line through O that we chose. It will be especially convenient to choose the line m through O perpendicular to l. Assuming that the image of l is a circle through O, we conclude that the angle between that circle and line m is a right angle. This can be true if and only if the diameter of the circle (the image of l) lies on m (why?). If the image of an arbitrary point B on l is B' and the image of A (the intersection of m and l) is A', the diameter of that circle must be $\overline{OA'}$. To prove that the image of l is indeed the circle with diameter $\overline{OA'}$ (without the assumptions we made), we just have to show that $\angle OB'A'$ is a right angle. This reasoning uses *Thales' Theorem*, which states that *an angle inscribed in a semicircle is a right angle*. This theorem is the subject of Now Try This 9.1.

> **Now Try This 9.1**
>
> Figure 9.7 shows a circle with an angle whose sides intersect the circle in two points A and C such that $\overline{AC}$ is a diameter of the circle. We refer to the $\angle ABC$ as the angle subtended by the diameter $\overline{AC}$ and whose vertex is B. Show that the point B lies on the circle if and only if $\angle ABC$ is a right angle.
>
>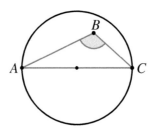
>
> Figure 9.7

We are now ready to give a formal proof of the following theorem.

> **Theorem 9.2**
>
> Let K be a circle with center O and radius r.
> 1. The image under inversion in K of an "extended line" l (a line together with the point at infinity) that does not pass through O is a circle that passes through O.
> 2. The image of a circle that passes through O is an "extended line" not passing through O.

9.2 Properties of Inversions

Proof. Consider a circle of inversion K centered at O of radius r and a line l not passing through O. Let the perpendicular to l from O intersect l at point A, as in Figure 9.8. Let B be an arbitrary point on l other than A or the point at infinity. Furthermore, let $A' = I_K(A)$ and $B' = I_K(B)$ be the respective images of A and B under inversion in the circle K. Now by the definition of inversion, $OA \cdot OA' = OB \cdot OB' = r^2$, which implies that $\frac{OA}{OB'} = \frac{OB}{OA'}$. Hence $\triangle AOB$ and $\triangle B'OA'$ share $\angle O$ and their corresponding sides that include $\angle O$ are proportional. By a previously proven result, these triangles are similar and therefore $\angle OB'A'$ is a right angle. Hence B' lies on a circle C whose diameter is $\overline{OA'}$ (why?). Since this is true for all points B on l, each finite point on the line l is inverted onto a point on the circle C. By convention, the point at infinity is inverted onto O, the center of inversion. This proves that the image of l under inversion in K is a subset of C.

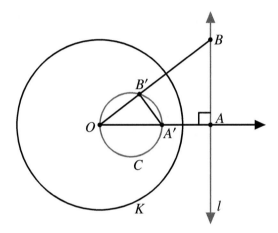

Figure 9.8

Having shown that $I_K(l)$, the image of line l under inversion in the circle K, is a subset of the circle C whose diameter is $\overline{OA'}$, we next show that C is a subset of $I_K(l)$, that is, we show that every point on the circle C is an image of some point on l. To do so, consider an arbitrary point B' on C other than A' and O, and let B be the intersection of $\overline{OB'}$ and l. Choosing B' as an arbitrary point on C allows us to continue to use Figure 9.8 for this proof. The reader is cautioned not to assume that B' here is the image of B under inversion in the circle K; as a matter of fact this is what we need to show next. Since B' is on the circle C, $\angle OB'A'$ is a right angle as it subtends the diameter $\overline{OA'}$. Consequently, the corresponding angles of $\triangle OA'B'$ and $\triangle OBA$ are congruent, and hence the triangles are similar. Therefore $\frac{OB'}{OA} = \frac{OA'}{OB}$, which implies that $OB \cdot OB' = OA \cdot OA' = r^2$. This shows that B and B' are inverse images of each other under inversion in the circle K, and hence for any point B' on C (other than O) there is a unique point B on l such that $I_K(B) = B'$ [recall that $I_K(A') = A$]. This shows that the circle C is a subset of $I_K(l)$, the image of the line l under inversion in the circle K (by convention I_K maps O onto the point at infinity and vice versa). This completes the proof that $I_K(l) = C$.

We now proceed to prove part (2) of the theorem, that is, the image of a circle that passes through the center of inversion O is an "extended line" l not passing through O. By definition, if P' is the image of P under inversion in a circle K, then P is the image of P' under the same inversion. Hence I_K^2 is the identity function and I_K is its own inverse. Let C be a circle going through O and A' be the endpoint of its diameter through O. Let $A = I_K(A')$ and l be the line perpendicular to $\overrightarrow{OA'}$ and going through A. By part (1) of the theorem, $I_K(l) = C$ and $I_K^2(l) = l = I_K(C)$. Hence the "extended line" l is the image of the circle C under inversion in the circle K. This completes the proof of part (2). ∎

> **Now Prove This 9.1**
>
> Give a direct proof of part (2) of Theorem 9.2 that does not use the fact that $I_K^{-1} = I_K$.

Having proved that the image of a circle passing through the center of inversion is a line not passing through that center, we next investigate the images of circles not passing through the center of inversion.

> **Now Try This 9.2**
>
> 1. Explain why the image of a circle not passing through the center of inversion cannot be a line.
>
> 2. Use dynamic geometry software to conjecture the image of a circle not passing through the center of inversion.

It's clear from the definition of inversion that the circle of inversion C centered at O inverts onto itself. In fact, the inverse image of any point P on C is the point P itself. Furthermore, if C_0 is a circle with radius r_0 that is concentric with C, that is, C_0 has O as its center, then points on C_0 are equidistant from O, so their images under inversion will also be equidistant from O. Therefore, the image of C_0 is a circle centered at O of radius $\frac{r^2}{r_0}$. This is true since if P is a point on C_0, then $OP \cdot OP' = r^2$. But $OP = r_0$, hence $r_0 \cdot OP' = r^2$, which implies that $OP' = \frac{r^2}{r_0}$.

In the special cases described above we have seen that circles not passing through O invert onto circles. Surprisingly enough, this fact is true for all circles that do not pass through O, as we shall prove in the next theorem.

> **Theorem 9.3**
>
> *The image under inversion of a circle not passing through the center of inversion is a circle not passing through the center of inversion.*

Proof. We will divide the proof into two parts:
Part I: Consider a circle of inversion K centered at O of radius r and a circle C centered at M and not going through O. Let the ray emanating from O and passing through M intersect

C in the points E and G, and let F be an arbitrary point on C (other than E and G), as in Figure 9.9. Furthermore, let E', F', and G' be the respective images of E, F, and G under inversion in the circle K.

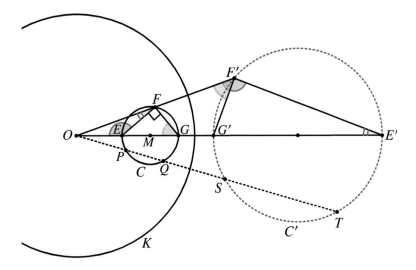

Figure 9.9

As in the proof of Theorem 9.2 we look at the given circle with center M as the locus of all points (angle vertices) for which the subtended angle by the diameter $\overline{EG}$ is a right angle and show that when an arbitrary point F on circle C is inverted onto F' and $\overline{EG}$ is inverted onto $\overline{E'G'}$, then $\angle G'F'E'$ is a right angle. To achieve this, we look for similar triangles that would allow us to conclude the measure of $\angle G'F'E'$. Since $OF \cdot OF' = OE \cdot OE' = r^2$, we have $\frac{OF}{OE'} = \frac{OE}{OF'}$. This together with the fact that $\angle O$ is a shared angle implies that the triangles $\triangle OFE$ and $\triangle OE'F'$ are similar. Similarly, we can show that $\triangle OFG$ is similar to $\triangle OG'F'$. Hence $\angle OEF \cong \angle OF'E'$ and $\angle OGF \cong \angle OF'G'$. Note that $\angle GFE$ is a right angle since it subtends the diameter $\overline{EG}$ of the circle C. Note also that $\angle OEF$ is an exterior angle to $\triangle EFG$, so we have $m(\angle OEF) = 90° + m(\angle EGF) = m(\angle E'F'G') + m(\angle OF'G')$. Now since $\angle OF'G' \cong \angle OGF \cong \angle EGF$, we can conclude that $\angle E'F'G'$ is a right angle. Furthermore, since an angle inscribed in a semicircle is a right angle, we know that for all points F on C, $F' = I_K(F)$ lies on a circle C' having $\overline{G'E'}$ as a diameter. Hence $I_K(C)$, the image of circle C under inversion in circle K, is a subset of the circle C'.

Part II: To show that every point on C' is the image of a point on C under inversion in the circle K, let S be an arbitrary point on C', and let the ray $\overrightarrow{OS}$ intersect C at the points P and Q and intersect C' at the points S and T. Now suppose the image of Q under inversion in the circle K is Q' (not shown in Figure 9.9) and show that $Q' = S$. Reasoning exactly as in Part I of the proof, we can conclude that Q' is on a circle whose diameter is $\overline{G'E'}$ which is the circle C'. Since Q' is on the ray $\overrightarrow{OQ}$, it must be the point S. The reader is encouraged to justify that T is not the image of Q; rather it is the image of P.

Analogous arguments can be applied to prove the theorem for circles whose locations relative to K are different from C's location relative to K in Figure 9.9. For instance, Figure 9.10 and Figure 9.11 illustrate how such arguments can be applied when the center of inversion O lies within the circle C and when C intersects K, the circle of inversion. This completes the proof of Theorem 9.3. ∎

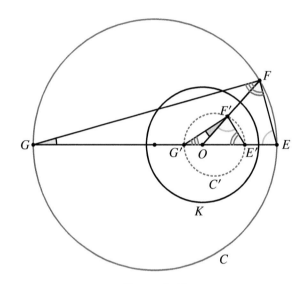

Figure 9.10

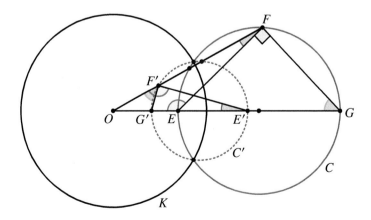

Figure 9.11

9.2 Properties of Inversions

> **Now Try This 9.3**
>
> Let K be a circle centered at O of radius r. Prove that the image of circle C (see Figure 9.12) under inversion in circle K can be obtained using a dilation by showing the following:
>
> 1. Let A be an arbitrary point on C such that $\overrightarrow{OA}$ intersects C in a second point B. Let A' be the image of A under inversion in K. Show that $OA' = \frac{r^2}{OA}$ and $OA \cdot OB = s^2$, where s is the length of the tangent segment from O to C (not shown), and hence $OA' = \frac{r^2}{s^2} \cdot OB$.
>
> 2. Use the equation in (1) to explain why the image of C under inversion is the image of C under dilation with center O and ratio $\frac{r^2}{s^2}$.
>
> 3. Does (2) imply that inversion is actually a dilation? Why or why not?
>
>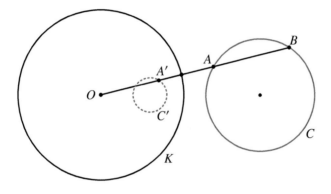
>
> Figure 9.12

Thus far, we have seen that under inversion, lines are inverted to lines or circles and circles are inverted to circles or lines, with the outcome depending on whether the original object passes through the center of the inversion. Clearly, inversion cannot be an isometry since points inside the circle of inversion will become spread out over the rest of the plane (the center of inversion gets mapped to the "point at infinity"). The results we have obtained so far also show that inversion is not a simple dilation because lines can be inverted into circles and vice-versa. In Problem 9 at the end of this section, the reader will be asked to prove that if C_1 and C_2 are circles centered at O of radii r_1 and r_2, respectively, then the composition of inversions $I_{C_2} \circ I_{C_1}$ is a dilation about O by ratio $\frac{r_2^2}{r_1^2}$.

We now turn our attention to angles between curves, are these preserved under inversion? Given that we know that straight lines do not necessarily invert to straight lines, we will need to clarify what we mean by an angle between two curves. Consider two curves β and γ intersecting at a point P. The angle α between them at the point of intersection is the angle between their tangent lines at point P, as in Figure 9.13. Since there are two such angles which are supplementary, we will agree to always choose $0 \leq \alpha \leq \frac{\pi}{2}$ as the angle between the curves. In the next theorem we will prove that inversion preserves the measure of angles between curves.

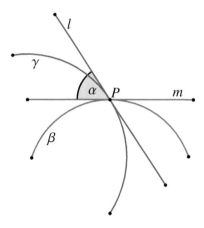

Figure 9.13

Theorem 9.4

The angle measure between any two intersecting curves is preserved under inversion.

Proof. Let C be a circle centered O and let the two curves β and γ intersect at P. Furthermore, let β', γ' be the images of β, γ, respectively, under inversion in C, and let P' be the image of P. To prove that the measure of the angle between the curves β and γ at P is the same as the measure of the angle between β' and γ' at P', we need to realize these angles as the limits of corresponding (hence equal) angles in two similar triangles. The first triangle will have P as a vertex and its two other vertices, M and N, will be distinct points on β and γ, respectively. The second triangle will have the images of the vertices of the first triangle as its vertices. Furthermore, we need the triangles to be such that as the first triangle gets smaller and smaller and degenerates into the point P, the second triangle degenerates into the point P' while remaining similar to the first triangle throughout the limiting process. The sides of the first triangle going through P limit to tangents to the curves β and γ and hence the angles between these sides limit to the angle between the curves. Similarly, the sides of the second triangle going through P' limit to tangents to the curves β' and γ' and hence the angles between these sides limit to the angle between β' and γ'. This will allow us to conclude that the measure of the angle between the curves at P is the same as the measure of the angle between their images at P'.

To do this formally, take a line through O, distinct from $\overleftrightarrow{OP}$, that intersects both β and γ, and let the points of intersection be M and N, respectively. Let M' be the inverse of M and N' be the inverse of N under inversion in C. Note that M' and N' also lie on the line $\overleftrightarrow{OM}$.

9.2 Properties of Inversions

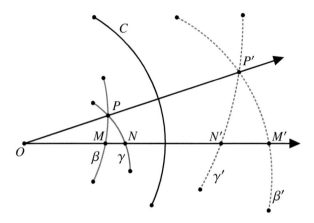

Figure 9.14

Since $\triangle OPM$ and $\triangle OM'P'$ share $\angle POM$ and the ratios of their corresponding sides adjacent to this angle are equal, these triangles are similar. Therefore, $\angle OMP \cong \angle OP'M'$. Similarly, $\triangle OPN$ and $\triangle ON'P''$ are similar, and hence $\angle ONP \cong \angle OP'N'$. Since $\angle OMP$ is an exterior angle for $\triangle MNP$, its measure equals the sum of measures of the two interior angles $\angle MPN$ and $\angle ONP$. Consequently, $m(\angle MPN) = m(\angle OMP) - m(\angle ONP) = m(\angle OP'M') - m(\angle OP'N') = m(\angle M'P'N')$.

As we let $\overleftrightarrow{OM}$ approach the line $\overleftrightarrow{OP}$, M and N will tend to P, and M' and N' will tend to P'. In addition, the secants $\overleftrightarrow{PM}$ and $\overleftrightarrow{PN}$ will limit to the tangents of β and γ, respectively. Likewise, $\overleftrightarrow{P'M'}$ and $\overleftrightarrow{P'N'}$ will limit to the tangents of β' and γ', respectively. The equality of the angle between the secants holds as $\overleftrightarrow{OM}$ approaches $\overleftrightarrow{OP}$. Therefore, the angles between the tangent lines, and hence between the curves β and γ, and β' and γ', are also equal. ∎

Note that although the angle measure is preserved, the direction of the angle is reversed. This is because when measuring the angle from β to γ we move in a counterclockwise direction, while when measuring the angle from β' to γ' we move in a clockwise direction. Therefore, inversion is an orientation-reversing transformation just as reflection in a line. In fact, inversion is sometimes thought of as reflection in the circle of inversion (it's referred to as such in the Transformation Tools menu in GeoGebra). Note that if P' is the image of P under reflection in line l, then $\overline{PP'}$ is perpendicular to l and any circle going through P and P' intersects the line l at right angles (why?). Analogously, if P' is the image of P under inversion in a circle C, then $\overline{PP'}$ is perpendicular to C and any circle going through P and P' is orthogonal to C (see Problem 7 at the end of this section).

The fact that inversion preserves angle measure leads to many important results including one regarding circles that intersect the circle of inversion at right angles. Circles such as C and C_1 in Figure 9.15 that intersect at right angles at points P and Q are called *orthogonal* circles. We will now prove that a circle is orthogonal to the circle of inversion if and only if it inverts onto itself. This will be a useful tool in many of the constructions and applications that will be investigated in the next section.

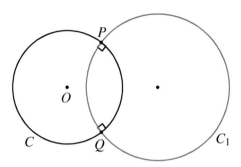

Figure 9.15

> **Theorem 9.5**
>
> Let C be a circle with center O and C_1 a circle not through O. Then, the image of C_1 under inversion in C is C_1 itself if and only if C_1 is orthogonal to C.

Proof. Let C_1 be a circle orthogonal to C and not going through the center of inversion O. Suppose C_1 intersects C in two points P and Q, as in Figure 9.15. What is the image of C_1 under inversion in C? By Theorem 9.3, C_1 inverts onto some circle C_1' (not shown). Since points on C are fixed under inversion, C_1' also intersects C at P and Q. Moreover, since inversion preserves the measure of angles, C_1' intersects C at right angles at P and Q.

We claim that $C_1 = C_1'$. This can be proved by showing that there is a unique circle that intersects C orthogonally at P and Q. To do so, let C' (not shown) be a circle passing through P and Q that is orthogonal to C and show that $C' = C_1$.

The line tangent to C at P is perpendicular to both $\overleftrightarrow{OP}$ and to the line tangent to C' at P. This is true because C and C' are orthogonal and the tangent to a circle at a point P is perpendicular to its radius through P. Therefore, $\overleftrightarrow{OP}$ is the same line as the tangent to C' at P; otherwise, we will have two right angles adding up to less than a straight angle (why?). Similarly, $\overleftrightarrow{OQ}$ is the tangent to C' at Q. Hence, if l is the perpendicular to $\overleftrightarrow{OP}$ through P and k is the perpendicular to $\overleftrightarrow{OQ}$ through Q, then the center of C' is the point of intersection O' of k and l, which is unique. Therefore, there is a unique circle with center O' orthogonal to C at P and Q, which proves that $C_1' = C_1$.

Conversely, suppose C_1 is a circle centered at O_1 and not passing through O such that $I_C(C_1) = C_1$, and show that C_1 is orthogonal to C. To do this, consider the two supplementary angles $\angle \alpha_1$ and $\angle \alpha_2$ formed by $\overleftrightarrow{OP}$ and $\overleftrightarrow{O_1P}$ at a point of intersection of C and C_1, as in Figure 9.16.

9.2 Properties of Inversions

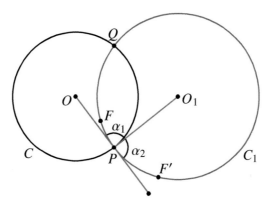

Figure 9.16

To find the image of $\angle\alpha_1$ under I_C, take a point F on C_1 inside C. As F approaches P, the angle formed by $\overleftrightarrow{FP}$ and the line tangent to C at P approaches $\angle\alpha_1$. Furthermore, as F approaches P, $F' = I_C(F)$ approaches P and the angle formed by $\overleftrightarrow{F'P}$ and the line tangent to C at P approaches $\angle\alpha_2$. Thus, the image of $\angle\alpha_1$ under inversion is $\angle\alpha_2$. Since inversion preserves the size of angles, $m(\angle\alpha_1) = m(\angle\alpha_2)$. So it must be that $\angle\alpha_1$ and $\angle\alpha_2$ are right angles. Hence, C and C_1 are orthogonal. ∎

9.2.1 Problem Set

1. Draw the image of an equilateral triangle under inversion in a circle centered at the centroid of the triangle.

2. Let C_1 and C_2 be two circles orthogonal to a circle K. Prove that if C_1 and C_2 intersect in two distinct points, then these points are the images of each other under inversion in K.

3. Given a circle C centered at point O and a point O_1 lying outside C, let P and Q be the points of intersection of C and the circle centered at the midpoint of $\overline{OO_1}$ and going through O and O_1. Prove that the circle C_1 centered at O_1 and going through P and Q is orthogonal to C at P and Q.

4. Let C be a circle with center O and C_1 a circle that doesn't go through O. We have shown that the image of C_1 under inversion in C is a circle C_1' not going through O. Show that the inverse of the center of C_1 is not necessarily the center of C_1'.

5. If the circle C' is the image of the circle C under inversion in a circle K, find a formula relating the radius of C' to the radius of C.

6. In the figure below, semicircles C_2 and C_3, centered at O_2 and O_3 with radii 2 units and 1 unit, respectively, are drawn in the interior of semicircle C_1 centered at O_1 such that C_2 and C_3 are externally tangent to each other and internally tangent to C_1. The circle C_4 centered at O_4 is drawn externally tangent to C_2 and C_3 and internally tangent to C_1. Determine the radius of C_4.

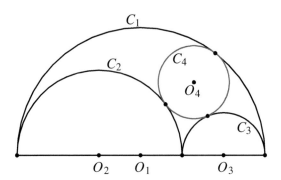

7. Let P' be the image of a point P under inversion in circle C. Prove that any circle through P and P' is orthogonal to C.

8. Given two noncongruent circles C_1 centered at O_1 with radius r_1, and C_2 centered at O_2 with radius r_2, find an inversion that takes C_1 to C_2 and vice versa. *Hint: Take the center of the dilation mapping the smaller circle to the larger one to be the center of inversion. You need to consider three cases: (1) The two circles are external to each other, (2) one of the circles is in the interior of the other, and (3) the two circles intersect in two points.*

9. Prove that if C_1 and C_2 are circles centered at O of radii r_1 and r_2, respectively, then the composition $I_{C_2} \circ I_{C_1}$ is a dilation about O by ratio $\frac{r_2^2}{r_1^2}$.

10. Let $\triangle ABC$ be a right triangle with $m(\angle C) = 90°$, and let X and Y be points on $\overline{AC}$ and $\overline{BC}$, respectively, and distinct from A, B, C. Construct four circles passing through C, centered at A, B, X, and Y. Prove that the resulting four intersection points, each of which is lying on exactly two of these four circles, lie on a circle.

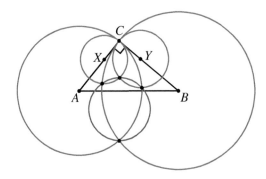

11. Let C_1, C_2, C_3, and C_4 be circles such that C_1 and C_2 are tangent at A, C_2 and C_3 are tangent at B, C_3 and C_4 are tangent at C, and C_4 and C_1 are tangent at D. Prove that the quadrilateral $ABCD$ is cyclic.

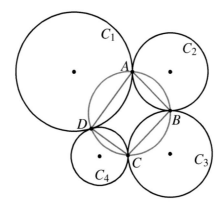

12. Let $\triangle ABC$ be a triangle and let A', B', C' be the respective images of A, B, C under inversion in the incircle of $\triangle ABC$.
    a. Show that the images of $\overleftrightarrow{AB}, \overleftrightarrow{BC}, \overleftrightarrow{AC}$ under inversion in the incircle are congruent circles, each with radius $\frac{r}{2}$, where r is the radius of the incircle.
    b. Show that A', B', C' lie on a circle whose image under inversion in the incircle is the circumcircle.
    c. Show that $\triangle A'B'C'$ is congruent to $\triangle XYZ$, where X, Y, Z are the respective centers of the circles which are the images $\overleftrightarrow{BC}, \overleftrightarrow{AC}, \overleftrightarrow{AB}$ under inversion in the incircle.

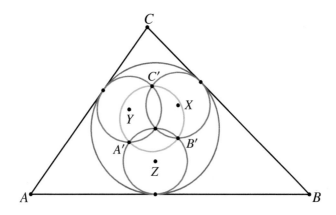

13. Given four distinct, collinear points A, B, C, D (in that order) in the Euclidean plane. The *cross ratio* of these points, denoted by $(A, B; C, D)$, is defined as $(A, B; C, D) = \frac{AC \cdot BD}{BC \cdot AD}$. Prove that the cross ratio of any four distinct points (none of which is the center of inversion) is preserved under inversion. That is, if A', B', C', D' are the respective images of A, B, C, D under inversion in some circle, then $\frac{AC \cdot BD}{BC \cdot AD} = \frac{A'C' \cdot B'D'}{B'C' \cdot A'D'}$. This is true whether the points are collinear or not (see figure below). Hint: Use the result proved in Problem 7 in Section 9.1.

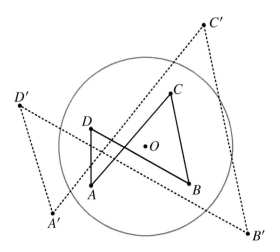

14. Let C_1, C_2, C_3, C_4 be distinct circles such that C_1, C_3 are externally tangent at P, and let C_2, C_4 be externally tangent at the same point P, as shown in the figure below [42, IMO Shortlist 2003/G4]. Suppose that C_1 and C_2, C_2 and C_3, C_3 and C_4, and C_4 and C_1 meet at $A, B, C,$ and D, respectively; and that all these points are distinct from P. Prove that $\frac{AB \cdot BC}{AD \cdot DC} = \frac{(PB)^2}{(PD)^2}$. Hint: Invert the circles and their points of intersection in a circle centered at the point P of some radius r. This results in the parallelogram $A'B'C'D'$. Then use the fact that $AB = \frac{r^2}{PA' \cdot PB'} A'B'$, $BC = \frac{r^2}{PB' \cdot PC'} B'C'$, etc., to finish the proof.

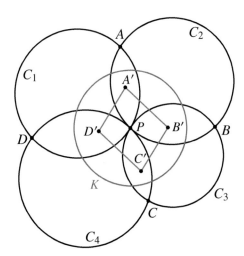

15. Given two fixed points A and B and a real number $r > 0$. Use inversion to prove that the set of points P satisfying $\frac{AP}{BP} = r$ is a circle. This circle is known as the Apollonian circle. Hint: Invert B and P in a circle centered at A of some radius R, and conclude that the inverse images of all points P satisfying $\frac{AP}{BP} = r$ lie on a circle centered at the inverse image of B.

9.2 Properties of Inversions

16. Given two distinct points A and B in the Euclidean plane. Consider the line $\overleftrightarrow{AB}$ and the family of all circles going through the points A and B. Show that there is a second family of circles (with centers on $\overleftrightarrow{AB}$) such that each circle of the second family is orthogonal to each circle of the first family, and each point in the plane, other than A and B, lies on exactly one of the circles of the second family.

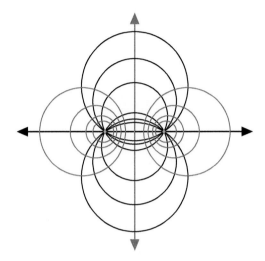

17. Let C_1, C_2 be two nonconcentric circles. Show how to transform these into two concentric circles C_1', C_2'. *Hint: Let K be a circle orthogonal to both C_1 and C_2, then take the circle of inversion to be the circle C centered at one of the points of intersection of K and the line segment connecting the centers of C_1 and C_2.*

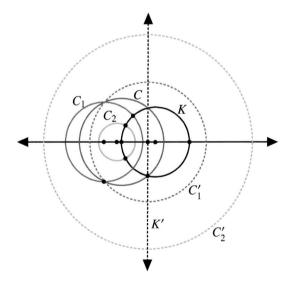

9.3 Applications of Inversions

Thus far, we have studied some key properties of inversions such as circles being inverted into circles or lines and the measure of angles being preserved under inversions. In this section we will discover that inversions are especially useful when dealing with problems involving circles. For example, recall how in Problem 17 of Problem Set 9.2.1 we managed to transform two nonconcentric circles into two concentric circles. Hereinafter, we will show that inversions can provide ingenious ways of looking at some classic geometry problems, often allowing for much easier solutions to complex problems. For example, consider four circles C_1, C_2, C_3, and C_4 such that each pair is internally tangent at a fixed point P. Let C be any circle that passes through P, as in Figure 9.17. How is C related to the other circles?

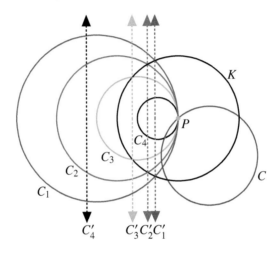

Figure 9.17

From the fact that the four circles are internally tangent, it follows that the line tangent to each of the four circles at point P is the same line. Thus, the angle between the circle C at P and each of the four circles is the same. When we invert all the circles in a circle K centered at P, we get five lines (the fifth line which is the image of C is not shown in Figure 9.17). Since C_1, C_2, C_3, and C_4 only intersect at the center of inversion P, their images under inversion only intersect at "the point at infinity." Hence the images of C_1, C_2, C_3, and C_4 are four parallel lines C'_1, C'_2, C'_3, and C'_4, respectively. Furthermore, since C intersects each of the other four circles, its image C', which is a line, also intersects the four lines C'_1, C'_2, C'_3, and C'_4. Thus, any question about the circles is transformed into a question about parallel lines cut by a transversal, which is more likely than not easier to handle. For instance, if we want to know how the angles formed by the circle C and each of the circles C_i at intersection points other than P are related, the discussion above shows that the four angles have equal measure. This is because the line C' intersects the four parallel lines C'_1, C'_2, C'_3, and C'_4 at four equal corresponding angles.

In the rest of this section, we will investigate a few complex problems and show how inversion, when used to solve these problems, can often lead to solutions that are far less

9.3 Applications of Inversions

complicated than traditional solutions. Among these are the Shoemaker's Knife problem, Apollonius' problem, Peaucellier–Lipkin linkage, Ptolemy's theorem, and Miquel's theorem. Additional interesting problems are investigated in the problem set at the end of this section.

9.3.1 Shoemaker's Knife Problem

Archimedes (287–212 B.C.E.) investigated a figure he called the *arbelos*, which is Greek for a shoemaker's knife. Informally, the figure is a plane region bounded by three semi-circles whose centers lie on a common straight line, as in Figure 9.18. A formal definition of the arbelos is given below.

> **Definition: Arbelos**
>
> Given three collinear points A, B, C, the *arbelos* is the plane figure bounded by the semicircles having diameters $\overline{AB}, \overline{BC}, \overline{AC}$ and lying on the same side of $\overleftrightarrow{AB}$.

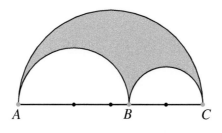

Figure 9.18

In his *Book of Lemmas*, Archimedes proved a number of interesting results about the arbelos. In Proposition 4 he proved that that the area of the arbelos is equal to the area of the circle shown in Figure 9.19 whose diameter is CD, where C is the point of tangency of the two smaller semicircles, and D is where the perpendicular to the base of the arbelos at C intersects the large semicircle.

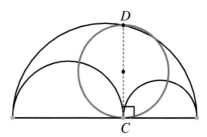

Figure 9.19

> **Now Try This 9.4**
>
> In Proposition 5 of the *Book of Lemmas*, Archimedes proved that the two circles C_1 and C_2, shown in Figure 9.20, are congruent. Each of these circles (known as Archimedes' twin circles) is tangent to $\overline{CD}$ and to a pair of the semicircles bounding the arbelos.
>
> Use dynamic geometry software such as GeoGebra to construct the circles C_1 and C_2, and to demonstrate that the circles are congruent and that the radius of each circle is equal to $\frac{r_1 r_2}{r_1 + r_2}$ (see Figure 9.20). Note that this ratio is equal to one half the harmonic mean of the radii of the two smaller semicircles bounding the arbelos. The reader will be asked to use inversion to formally prove these results in Problem 2 at the end of this section.
>
>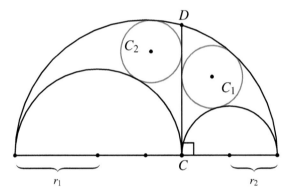
>
> Figure 9.20

Nearly 500 years after Archimedes, Pappus of Alexandria (300–350 C.E.) wrote a treatise known as the *Synagoge* or *Collection*, which is believed to have consisted of 8-12 books. The second section of Book IV dealt with circles inscribed in the arbelos, as in Figure 9.21. Pappus proved that the distance from the center of the nth circle C_n to the base of the arbelos is equal to n times d_n, the diameter of C_n. Pappus' proof is somewhat cumbersome, so instead we present a more elegant proof using inversion, a topic not formally studied until roughly the 19th century.

> **Theorem 9.6**
>
> *The distance from the center of the nth circle C_n (see Figure 9.21) to the base of the arbelos is equal to n times d_n, the diameter of C_n.*

9.3 Applications of Inversions

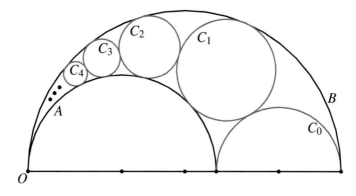

Figure 9.21

Proof. For the sake of brevity, we will hereinafter refer to $C_0, C_1, C_2, \ldots$ and $C'_0, C'_1, C'_2, \ldots$ appearing in Figure 9.22 as chains of circles even though C_0 and C'_0 are clearly semicircles. Also, for the sake of clarity, we will first prove the theorem for the circle C_2. A similar argument can then be applied to prove the theorem for the nth circle C_n. The proof will rely on the following properties of inversion proved earlier in this chapter:

1. *Lines that don't pass through the center of inversion invert into circles that pass through it.*
2. *Circles that pass through the center of inversion invert into lines that don't pass through it.*
3. *Circles that don't pass through the center of inversion invert into circles that don't pass through it.*
4. *A circle orthogonal to the circle of inversion inverts onto itself.*
5. *The absolute value of the angle measure between two arbitrary intersecting curves is preserved under inversion.*

Let the semicircle on the left be A and the largest semicircle be B, as in Figure 9.21. At first glance, the chain of circles $C_0, C_1, C_2, \ldots$ inscribed in the arbelos looks to be somewhat intimidating given how they are sandwiched between the semicircles A and B which are tangent at O. Fortunately, inversion can be used here to make the problem more manageable. But we have to choose the circle of inversion carefully in order to transform the chain of circles into something useful and easier to handle.

As shown in Figure 9.22, choosing O to be the center of inversion allows us to transform the semicircles A and B into the rays A' and B', respectively (why?). This also allows us to invert the circles in the chain $C_0, C_1, C_2, \ldots$ which are tangent to A and B into circles that are tangent to A' and B' (why?). Furthermore, since each circle C_i is tangent to the circle preceding it and the circle succeeding it in the chain, the images of these circles are tangent as well. Now since we are proving the theorem first for the circle C_2, it would be desirable to have this circle invert onto itself (why?). This can be achieved if the circle of inversion is orthogonal to C_2.

Therefore we should choose K_2, the circle of inversion corresponding to C_2, to be centered at O and orthogonal to C_2. To construct K_2 you can simply construct a tangent segment from O to C_2 and then use it as the radius for K_2. Where the tangent segment intersects C_2 is determined by constructing a circle centered at the midpoint of the segment connecting O to the center of C_2 and passing through O (why?).

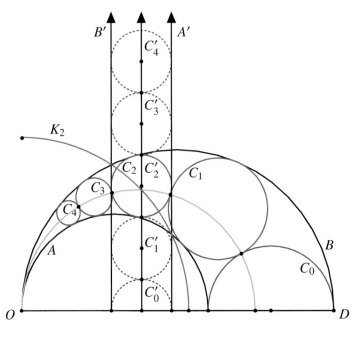

Figure 9.22

Having determined the circle of inversion K_2, we now proceed to determine the images of the semicircles A and B and the chain of circles $C_0, C_1, C_2, \ldots$. By the properties of inversion listed earlier, inverting A and B in K_2 yields the rays A' and B', respectively. Also, the line $\overleftrightarrow{OD}$ inverts onto itself (why?). Now since A and B are orthogonal to $\overline{OD}$ (why?), the rays A' and B' are also perpendicular to the base of the arbelos $\overline{OD}$. Finally, the circles $C_0, C_1, C_2, \ldots$ invert into the circles $C'_0, C'_1, C'_2, \ldots$, as in Figure 9.22. Note that since C_2 is orthogonal to K_2, it inverts onto itself and hence $C'_2 = C_2$. The circles $C'_0, C'_1, C'_2, \ldots$ are tangent to the parallel rays A' and B', and hence are clearly congruent and their diameters are equal in length to $d_2 = \text{diam}(C_2)$, the diameter of C_2. Since the circle $C_2 = C'_2$ is preceded by C'_1 and C'_0 in the vertical chain, we can conclude that the distance between the center of C_2 and $\overline{OD}$ is $\frac{1}{2} \text{diam}(C'_0) + \text{diam}(C'_1) + \frac{1}{2} \text{diam}(C'_2)$ which is equal to $2d_2$. This proves the theorem for C_2.

The argument above can be generalized to prove that for any n, the distance from the center of the circle C_n to the base of the arbelos is nd_n. To achieve this, you can simply use as the circle of inversion the circle K_n centered at O and orthogonal to C_n. Inverting $C_0, C_1, C_2, \ldots$ in the circle K_n produces a vertical chain of congruent circles sandwiched

9.3 Applications of Inversions

between two rays perpendicular to the base of the arbelos and tangent to C_n, which inverts onto itself. Since the diameters of these image circles are equal to d_n, the diameter of C_n, we can conclude that the center of C_n is distance $\frac{1}{2}d_n + \underbrace{d_n + \cdots + d_n}_{n-1 \text{ times}} + \frac{1}{2}d_n = nd_n$ from the base of the arbelos. This completes the proof of Theorem 9.6. ∎

9.3.2 Apollonius' Problem

Inversions in circles can also be a helpful tool in many geometric constructions. One such example is *Apollonius' problem*: *Given three circles, construct a circle tangent to all three circles*. In general, the three circles may or may not intersect, and they may have different radii. In some cases, the problem is impossible, for example, when the circles are concentric with different radii. When a solution does exist, there are generally up to eight different circles that are tangent to all three given circles. The constructed circle may be externally tangent to all three circles, internally tangent to one circle and externally tangent to the other two, and so on. We next demonstrate the construction of a circle that is externally tangent to three nonintersecting circles.

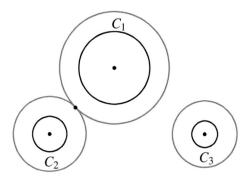

Figure 9.23

Consider three nonintersecting circles C_1, C_2, and C_3. We can increase the radii of all three circles by a fixed amount δ, so that the two closest circles are now tangent, as in Figure 9.23. If we can find a circle that is externally tangent to all three enlarged circles, we can increase the radius of that circle by δ, which will result in a circle tangent to the three original circles.

Therefore, we can simplify the problem to the case of three circles such that two of them, C_1 and C_2, are tangent at a point O. Inverting C_1 and C_2 about a circle K centered at O yields two parallel lines, C_1' and C_2', as in Figure 9.24. If the inversion circle is chosen such that it is orthogonal to C_3, then C_3 inverts onto itself. Now, inverting a circle C tangent to C_1', C_2' and C_3' in K gives the desired circle tangent to C_1, C_2 and C_3.

To determine the circle of inversion K, construct a line through O that is tangent to C_3 (this line is not shown in Figure 9.24). The circle K going through the point where this tangent intersects C_3 and centered at O is orthogonal to C_3 (why?). Since C_1 and C_2

are tangent at O, the center of inversion, they invert onto two parallel lines C'_1 and C'_2, respectively. Also, since C_3 is orthogonal to K, it inverts onto itself.

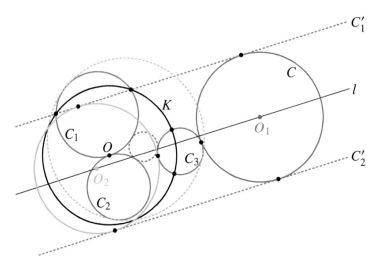

Figure 9.24

We next construct a circle C tangent to C'_1, C'_2 and C_3. Clearly the center of C must lie on l, a line exactly halfway between C'_1 and C'_2, and its radius must be half the distance between these lines. Also, the center of C must be at a distance $r + \frac{d}{2}$ from the center of C_3 (why?), where r is the radius of C_3 and d is the distance between the parallel lines C'_1 and C'_2. Clearly, there are two such points on l, O_1 and O_2. If we take C to be centered at O_2, then its inverse image will be a circle internally tangent to C_1, C_2, and C_3 (why?). Taking C to be centered at O_1 leads to its inverse image being a circle externally tangent to C_1, C_2, and C_3 (why?). Technically, to find the inverse image of C, take any three points on C and construct their inverses under the inversion. Then, construct the circle through these three points. The resulting circle is tangent to the three circles C_1, C_2, and C_3. Having said that, dynamic geometry software such as GeoGebra readily allows for determining the images of circles and other objects. This completes the construction required in Apollonius' problem.

9.3.3 Peaucellier–Lipkin Linkage

Another application of inversion is a solution to a problem that was of great interest in the 19th century. At that time, it was thought that there was no way to construct a mechanical device that could take rotational motion and convert it into linear motion. In 1864, a French engineer and captain in the French army by the name of Charles-Nicolas Peaucellier used properties of inversion in a circle to show that such a device was indeed possible. The linkage was independently discovered in 1871 by a Lithuanian mathematician and inventor named Yom Tov Lipman Lipkin.

9.3 Applications of Inversions

The Peaucellier–Lipkin linkage consists of six bars, as shown in Figure 9.25. The four bars PR, PS, QR, QS are equal in length, and AP, AQ are also equal in length. The bars are free to pivot at each of the points $A, P, Q, R,$ and S, while point A is fixed so it can't move.

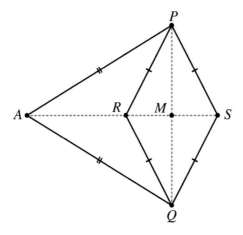

Figure 9.25

Let M be the point of intersection of the diagonals of the rhombus $PSQR$. Recall that the diagonals of a rhombus are the perpendicular bisectors of each other. Also, since $AP = AQ$, the point A is equidistant from the end points of $\overline{PQ}$ and hence it is on the perpendicular bisector of $\overline{PQ}$. Therefore the points $A, R,$ and S are collinear and remain as such as the linkage is moved around to various positions.

We next show that the product $AR \cdot AS$ is fixed and hence R and S can be thought of as inverse points with respect to a circle centered at A of radius $\sqrt{AR \cdot AS}$. Knowing that $A, R,$ and S are collinear, and $\triangle AMP$ and $\triangle RMP$ are right triangles, we have

$$AR \cdot AS = (AM - MR)(AM + MS) = (AM - MR)(AM + MR) = (AM)^2 - (MR)^2 = ((AP)^2 - (PM)^2) - ((PR)^2 - (PM)^2) = (AP)^2 - (PR)^2.$$

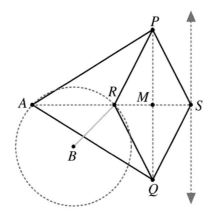

Figure 9.26

Because AP and PR are fixed in length, $AR \cdot AS = (AP)^2 - (PR)^2$ is also fixed. Therefore, S can be considered as the inverse of R under inversion in a circle of radius $\sqrt{(AP)^2 - (PR)^2}$ centered at A and having R inside it (circle not shown in Figure 9.26). We proved earlier that the inverse of a circle through the center of inversion is a straight line. Thus, if we can make R travel in a circle that also passes through A, then S will travel in a straight line. This can easily be accomplished by adding a seventh bar BR and fixing the position of B on the perpendicular bisector of $\overline{AR}$ such that $AB = BR$. Therefore, R traces a circle that passes through A, and S traces a straight line.

In Problem 18 at the end of this section, the reader will be asked to investigate another linkage called Hart's linkage invented around 1874 which can also be used to convert rotational motion into linear motion and vice versa.

9.3.4 Ptolemy's Theorem

We now investigate an important theorem from elementary geometry called Ptolemy's theorem, which states that, *in a cyclic quadrilateral, the product of the diagonals is equal to the sum the products of the opposite sides.* Ptolemy (85–165 C.E.) was an influential Greek astronomer and geographer, hence his interest in the study of circles and cyclic quadrilaterals. He detailed in his writings the mathematical theory of the motion of the sun, moon, and planets. He introduced trigonometrical methods based on the *chord function* (no longer in use), which is related to the modern sine function, and this relationship can be expressed as $Crd(\alpha) = 120 \cdot \sin(\alpha/2)$.

Ptolemy's theorem can be proved by simply using the law of cosines as outlined in Now Try This 9.5.

> **Now Try This 9.5**
>
> The law of cosines states that in a triangle $\triangle ABC$, if we assume that a, b, c are the lengths of the sides opposite the vertices A, B, C, respectively, then without loss of generality, $a^2 = b^2 + c^2 - 2bc \cdot \cos(A)$. Use the law of cosines to prove that $x \cdot y = b \cdot d + a \cdot c$ for the cyclic quadrilateral in Figure 9.27.
>
>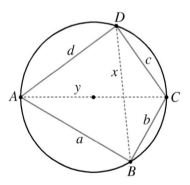
>
> Figure 9.27

9.3 Applications of Inversions

We now proceed to give a simple, yet elegant proof of Ptolemy's theorem which uses inversion. For this, it would be very convenient to fix one of the vertices of the cyclic quadrilateral $ABCD$ and carefully project the rest of the vertices onto a line, since comparing the distances between points on a line is often much easier than comparing distances between points on a circle. This can be achieved by inverting three of the vertices of the quadrilateral about a circle centered at the fourth vertex. We make use of the definition of inversion, the fact that a triangle having the center of inversion as one of its vertices inverts onto a similar triangle, and the fact the images of three of the vertices are collinear to present a formal proof of Ptolemy's theorem.

Theorem 9.7: Ptolemy's Theorem

Let $ABCD$ be a cyclic quadrilateral, that is, the vertices $A, B, C,$ and D lie on a given circle. Then, $AC \cdot BD = AD \cdot BC + AB \cdot CD$.

Proof. Without loss of generality, let us invert the vertices B, C, and D about a circle K centered at vertex A with radius r. Since these points lie on a circle passing through the center of inversion, their images B', C', and D' lie on a line, as in Figure 9.28.

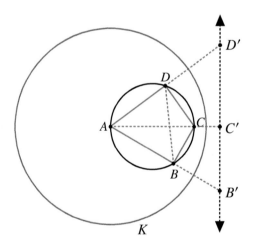

Figure 9.28

Now by the definition of inversion, $AB \cdot AB' = AC \cdot AC' = AD \cdot AD' = r^2$. Hence, $\triangle ABC \sim \triangle AC'B'$, since these triangles share $\angle BAC$ and satisfy $\frac{AB}{AC'} = \frac{AC}{AB'}$. Therefore, $\frac{B'C'}{BC} = \frac{AC'}{AB}$, which gives $B'C' = \frac{BC \cdot AC'}{AB}$. Multiplying the right side of this equation by $\frac{AC}{AC}$ gives $B'C' = \frac{BC \cdot r^2}{AC \cdot AB}$ (why?). Using very similar arguments, we arrive at $B'D' = \frac{BD \cdot r^2}{AD \cdot AB}$, and $C'D' = \frac{CD \cdot r^2}{AC \cdot AD}$.

Since B', C', D' lie on a straight line with C' lying between B' and D', we have $B'D' = B'C' + C'D'$. Substituting the expressions we arrived at for $B'D'$, $B'C'$ and $C'D'$

in this equation gives

$$\frac{BD \cdot r^2}{AD \cdot AB} = \frac{BC \cdot r^2}{AC \cdot AB} + \frac{CD \cdot r^2}{AC \cdot AD}.$$

Finally, multiplying this equation by $\frac{AB \cdot AC \cdot AD}{r^2}$ gives $AC \cdot BD = AD \cdot BC + AB \cdot CD$ which completes the proof. ∎

> **Now Try This 9.6**
>
> Do the following:
>
> 1. What conclusion would you arrive at if the cyclic quadrilateral $ABCD$ in Ptolemy's theorem is a rectangle?
>
> 2. What conclusion would you arrive at if a regular pentagon $ABCDE$ is inscribed in a circle and you take the cyclic quadrilateral $ABCD$ in Ptolemy's theorem to be the isosceles trapezoid $ABCD$?
>
>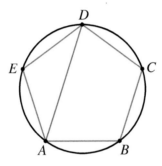
>
> Figure 9.29
>
> 3. What conclusion would you arrive at if the cyclic quadrilateral $ABCD$ in Ptolemy's theorem is such that $\triangle ABC$ is an equilateral triangle?

9.3.5 Miquel's Theorem

In 1838, a French mathematician named Auguste Miquel published an interesting result now known as *Miquel's theorem*, which states that *if three points lie on the three sides of a triangle (one on each side), then the three circles, each passing through a vertex of the triangle and the two points on the sides adjacent to that vertex, are concurrent in a single point.* This point is called the *Miquel point* and its existence will be proved using inversion in Theorem 9.8. Conversely, *given three circles concurrent in a point, a triangle exists having a vertex on each of the three circles and a side passing through each of the intersection points of the circles other than the point of concurrency.* The reader will be asked, in Problem 15 at the end of this section, to use inversion to prove an extension of Miquel's theorem stating that *for any four points on a circle and four circles one passing through each pair of adjacent*

9.3 Applications of Inversions

points, the four intersection points that do not lie on the given circle are also concyclic. This is often referred to as *Miquel's six-circle theorem*.

> **Theorem 9.8: Miquel's Theorem**
>
> Let $\triangle ABC$ be an arbitrary triangle and let D, E, F be points on $\overleftrightarrow{BC}, \overleftrightarrow{AC}, \overleftrightarrow{AB}$, respectively. Then the three circles passing through the points $\{A, F, E\}$, $\{B, D, F\}$, and $\{D, C, E\}$ are concurrent.

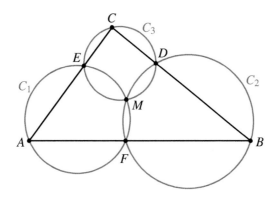

Figure 9.30

Proof. This proof was adapted from [22].

Case I: Let C_1 intersect C_2 in the points F and M, as in Figure 9.31. We will show that M lies also on C_3. To achieve this, choose as the circle of inversion a circle K centered at M. Since F, A, and E lie on circle C_1 that goes through M, the center of inversion, their inverse images F', A', and E' under inversion in circle K lie on a line. Similarly, F', B', and D' lie on a line. Recall that a line not passing through the center of inversion M is inverted onto a circle going through M. Hence the lines $\overleftrightarrow{AB}$, $\overleftrightarrow{BC}$, and $\overleftrightarrow{AC}$ are inverted onto circles going through $\{A', F', B', M\}$, $\{B', D', C', M\}$, and $\{A', E', C', M\}$, respectively (circles not shown). If we can show that D', C', and E' lie on the same line, then inverting this line about K will give us a circle going through the points $\{E, C, D, M\}$. This circle must be the circle C_3, which completes the proof.

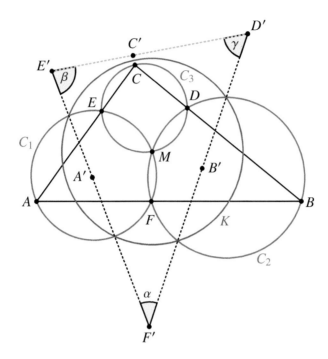

Figure 9.31

For the sake of clarity, let $m(\angle A'F'B') = \alpha$, $m(\angle A'E'C') = \beta$, and $m(\angle B'D'C') = \gamma$. Since $A'F'B'M$, $B'D'C'M$, and $A'E'C'M$ are all cyclic quadrilaterals and since opposite angles in a cyclic quadrilateral add up to $180°$, we have $m(\angle B'MA') = 180° - \alpha$, $m(\angle A'MC') = 180° - \beta$, and $m(\angle B'MC') = 180° - \gamma$. Therefore the sum of the angles around the point M is $(180° - \alpha) + (180° - \beta) + (180° - \gamma) = 360°$, which implies that $\alpha + \beta + \gamma = 180°$. But if C' does not lie on $D'E'$, then $F'D'C'E'$ is a quadrilateral whose angle sum is $m(\angle D'C'E') + \alpha + \beta + \gamma$. Now if C' is not on $\overleftrightarrow{E'D'}$, then the measure of $\angle D'C'E'$ is either less than $180°$, which means $\alpha + \beta + \gamma > 180°$ in quadrilateral $F'D'C'E'$, or is larger than $180°$, which implies that $\alpha + \beta + \gamma < 180°$, which is a contradiction. Therefore, C' must lie on $D'E'$. Hence, the line passing through E', C', and D' inverts into a circle C_3 going through E, C, D, and M. The proof given above is still valid if M lies outside $\triangle ABC$ (why?). This completes the proof of Case I.

Case II: Let C_1 intersect C_2 in a single point F, that is, C_1 and C_2 are tangent at F. We need to show that F lies also on C_3. The reader is encouraged to use Figure 9.32 to prove this using the fact that $C'E'A'F$ and $FB'D'C'$ are both cyclic quadrilaterals (why?), which implies that $m(\angle FB'D') + m(\angle D'C'F) = 180°$ and $m(\angle E'A'F) + m(\angle FC'E') = 180°$. But C_1' is parallel to C_2' (why?), hence $m(\angle D'C'F) + m(\angle FC'E') = 180°$, which implies that E', C', and D' are collinear. Therefore, the image of $\overleftrightarrow{E'D'}$ under inversion in K is C_3 (why?), which means that F lies on C_3, and this completes the proof of Case II.

9.3 Applications of Inversions

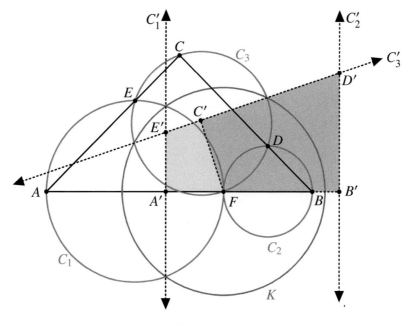

Figure 9.32

This completes the proof of Theorem 9.8.

Generalization: Given n lines each of which intersects the other $n-1$ lines, the n Miquel circles passing through each subset of $n-1$ intersection points are concurrent in a point M, and their centers lie on a circle K. Figure 9.33 shows this result if we start with four lines l_1, l_2, l_3, l_4. The reader will be asked to prove this in Problem 17 at the end of this section.

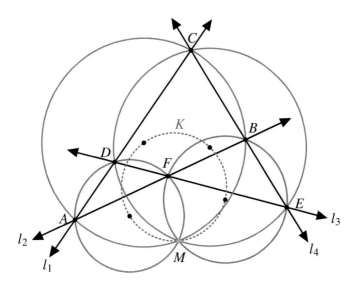

Figure 9.33

9.3.6 Problem Set

1. Given a circle C centered at O and two points A, B outside of C, prove that there exists a circle through A, B and orthogonal to C.

2. Prove that the Archimedes' twin circles, C_1 and C_2, shown in Figure 9.20, are congruent and conclude that the radius of each circle is equal to $\frac{r_1 r_2}{r_1+r_2}$, which is one half the harmonic mean of the radii of the two smaller semicircles bounding the arbelos [47, Proposition 5].

3. Given three noncoaxial concurrent circles (their centers are noncollinear), construct a circle C tangent to all three circles.

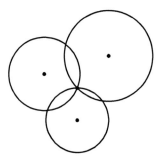

4. In the construction of Apollonius' circle, there were two possibilities for the center of the circle C. What would have happened if we chose the center of C to be the point closer to O? Why did we choose the point farther from O?

5. **Steiner Chains**

    a. Let C_1 and C_2 be two circles centered at O_1 and O_2, respectively. Then, any circle C orthogonal to both C_1 and C_2 intersects the line $O_1 O_2$ in two points P and Q as shown in the figure below. Use this fact to show that given two nonintersecting circles, there exists an inversion such that the image of the two circles is two concentric circles.

    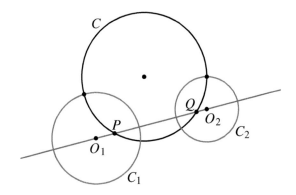

    b. Let C_1 be a circle lying within the interior of a second circle C_2. Suppose that there exists a chain of circles such that each circle is tangent to both C_1 and

9.3 Applications of Inversions

C_2, and such that adjacent circles are tangent, as shown below. This is called a Steiner chain, named after the discoverer of the following property: *if one such chain exists, then no matter where we start the first circle, we will end up with a Steiner chain.* Prove this property.

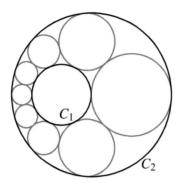

6. Given two circles C_1 and C_2 and a point O as in the figure below, use inversion to construct a circle through O that is tangent to C_1 and C_2. *Hint: Inverting C_1 and C_2 in a circle K centered at O gives two circles C_1' and C_2'. The desired circle is the image (under inversion in K) of a line tangent to C_1' and C_2'.*

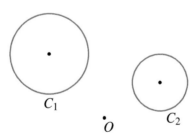

7. Describe how to use Problem 6 to give an alternative method for solving Apollonius' problem. *Hint: Given three circles C_1, C_2, C_3 with radii r_1, r_2, r_3, respectively. Without loss of generality, assume that r_3 is the smallest among r_1, r_2, r_3. Replace each circle C_i with a concentric circle ψ_i whose radius is $r_i - r_3$. This reduces the three circles to two circles ψ_1, ψ_2 and the point O (the center of C_3) (see figure below). Next, use Problem 6 to construct a circle tangent to ψ_1, ψ_2 and going through O.*

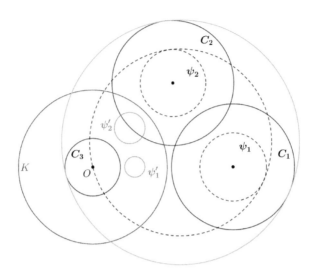

8. A hexagon is inscribed in a circle, as in the figure below [68, 1991 AIME, Problem 14]. Five of the sides have length 81 units and the sixth, denoted by $\overline{AB}$, has length 31 units. Find the sum of the lengths of the three diagonals that can be drawn from A. *Hint: Apply Ptolemy's theorem to $ABCD$ and $ACDF$ to find y. Then apply it to $ADEF$ to find z. Finally, use one of the equations relating x, y, and z to find x.*

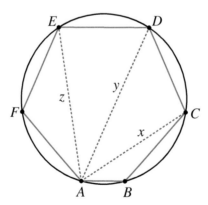

9. Let $\triangle ABC$ be a right triangle with its right angle at C, and let X and Y be points on $\overleftrightarrow{AC}$ and $\overleftrightarrow{BC}$, respectively. Consider the four circles passing through C and centered at A, B, X, and Y (see the figure below). Prove that the four points of intersection (other than C) of these circles lie on a circle.

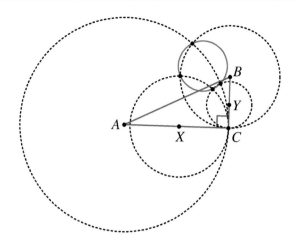

10. Let circles C_1 and C_2 intersect in points O and A. Let B be the point of intersection of C_1 with the diameter $\overline{OE}$ of C_2. Similarly, let D be the point of intersection of C_2 with the diameter $\overline{OF}$ of C_1 (see figure below). Prove that $\overline{AO}$ goes through the center of the circle passing through the points $O, B,$ and D. *Hint: Invert the points A, B, D, E, F about a circle K centered at O of some radius r, and use the fact that the altitudes of a triangle are concurrent.*

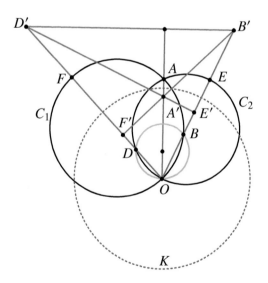

11. Let $ABCD$ be a quadrilateral whose diagonals $\overline{AC}$ and $\overline{BD}$ are perpendicular and intersect at E [67, USAMO 1993/2]. Prove that the reflections of E across $\overline{AB}, \overline{BC}, \overline{CD},$ and $\overline{DA}$ are concyclic. In the figure below, these reflections are $W, X, Y,$ and Z, respectively. *Hint: The points $W, X, Y,$ and Z are the points of intersection of circles centered at the vertices of $ABCD$ and going through E. Inverting these circles and their points of intersection about a circle centered at E of some radius r results in a rectangle $W'X'Y'Z'$, which is cyclic. Hence the points $W, X, Y,$ and Z are cyclic.*

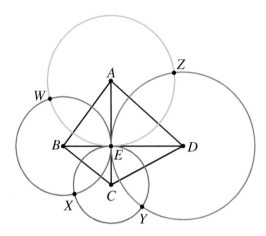

12. Point D lies inside $\triangle ABC$ [5, BAMO 2008/6]. If A_1, B_1, and C_1 are the second intersection points of the lines AD, BD, and CD with the circles circumscribed about $\triangle BDC$, $\triangle CDA$, and $\triangle ADB$, prove that

$$\frac{AD}{AA_1} + \frac{BD}{BB_1} + \frac{CD}{CC_1} = 1.$$

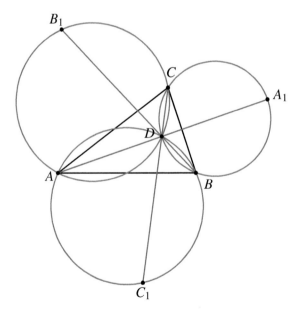

Hint: Invert the circumcircles and the points $A, B, C, A_1, B_1,$ and C_1 about a circle K centered at point D of radius r. This results in a triangle $\triangle A'B'C'$ with $A_1', B_1',$ and C_1' lying on $\overline{B'C'}, \overline{A'C'},$ and $\overline{A'B'}$, respectively, such that $\overline{A'A_1'}, \overline{B'B_1'},$ and $\overline{C'C_1'}$ are concurrent in D. Then use basic properties of inversion to transform $\frac{AD}{AA_1} + \frac{BD}{BB_1} + \frac{CD}{CC_1}$

9.3 Applications of Inversions

into $\frac{A'_1 D}{A' A'_1} + \frac{B'_1 D}{B' B'_1} + \frac{C'_1 D}{C' C'_1}$. Finally, use basic Euclidean results to transform the latter into $\frac{Area(\triangle B'C'D)}{Area(\triangle A'B'C')} + \frac{Area(\triangle C'A'D)}{Area(\triangle A'B'C')} + \frac{Area(\triangle A'B'D)}{Area(\triangle A'B'C')}$.

13. Consider four circles $C_1, C_2, C_3,$ and C_4 such that C_1 is tangent to C_2, C_2 is tangent to C_3, C_3 is tangent to C_4, and C_4 is tangent to C_1. Show that the set of four points of tangency lie either on a circle C or on a straight line. *Hint: Invert the four circles about a circle K (not shown) centered at one of the points of tangency.*

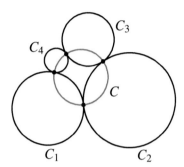

14. Let $\triangle ABC$ be a triangle, D be a point on $\overline{BC}$, and K be the circumcircle of $\triangle ABC$ [78, RMO 1997/4]. Show that the circle tangent to K, $\overline{AD}$, and $\overline{BD}$, and the circle tangent to K, $\overline{AD}$, and $\overline{DC}$ are tangent to each other if and only if $\angle BAD = \angle CAD$.

15. Let C_1, C_2, C_3, C_4 be as in the figure below [74].
    a. Prove that A, B, C, D are concyclic if and only if A', B', C', D' are concyclic. *Hint: Invert the the four circles about a circle centered at one of the points of intersection of two of the circles.*
    b. Let $\theta_1, \theta_2, \theta_3, \theta_4$ be the angles between C_1 and C_2, C_2 and C_3, C_3 and C_4, and C_4 and C_1, respectively. Prove that A, B, C, D are concyclic if and only if $\theta_1 + \theta_3 = \theta_2 + \theta_4$.

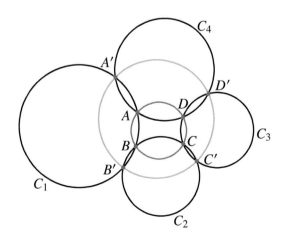

16. Given three mutually tangent circles each of which is tangent to a given straight line, as in the figure below. Prove that the points of tangency of the circles lie on a circle. Also, prove that $\frac{1}{\sqrt{r_2}} = \frac{1}{\sqrt{r_1}} + \frac{1}{\sqrt{r_3}}$, where r_i is the radius of C_i. *Hint: Use the Pythagorean theorem to prove* $AB = 2\sqrt{r_1 r_3}$, $AD = 2\sqrt{r_1 r_2}$, *and* $DB = 2\sqrt{r_2 r_3}$.

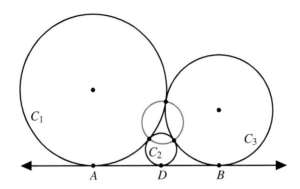

17. Let l_1, l_2, l_3, l_4 be four lines in the plane; no two are parallel [94]. Let C_{ijk} denote the circumcircle of the triangle formed by the lines l_i, l_j, l_k. Prove that $C_{123}, C_{124}, C_{134}, C_{234}$ pass through a common point M, called the Miquel point (see Figure 9.33). *Hint: Apply Theorem 9.8 to* $\triangle ABC$ *and* $\triangle CDE$.

18. *Hart's linkage* is another mechanical device that can be used for tracing the inverse of a curve. The linkage consists of four arms of fixed arbitrary lengths such that $AD = BC$ and $AB = CD$, four joints at points A, B, C, D, and three fixed positions at points O, P, Q on the arms such that O, P, Q are collinear and the line through them is parallel to AC. To trace the inverse of a given curve γ, we fix point O in the plane and move P along γ. The curve that point Q traces is the inverse of γ with respect to O.

   a. Find the radius of inversion for the Hart's linkage. *Hint: Start with* $OP \cdot OQ = r^2$ *and show that* $r = \sqrt{\frac{AO \cdot BO \cdot ((AD)^2 - (AB)^2)}{(AB)^2}}$.

   b. What happens if the curve γ that point P is moving along is a circle passing through O?

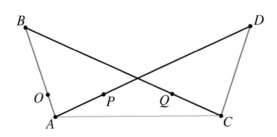

19. Use dynamic geometry software to construct the figure below. Start with three circles tangent to the x-axis and centered at $O_1 = (0, 8)$, $O_2 = (16, 8)$, and $O_3 = (8, 2)$.

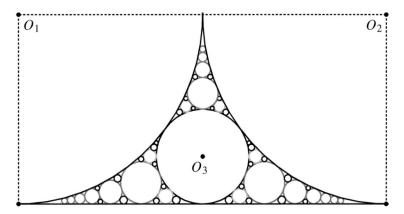

9.4 The Nine-Point Circle and Feuerbach's Theorem

The discovery of the nine-point circle can be traced back to the early 1800s, although it's unclear who actually was the first to discover it. The *nine-point circle* of a $\triangle ABC$ passes through the following nine points: the three midpoints of the sides, the three feet of the altitudes, and the three midpoints of the line segments joining the orthocenter to the vertices of the triangle. There are several proofs of the existence of the nine-point circle including one where the nine points are dilated about the orthocenter by ratio 2, which takes them to points on the circumcircle, and then the circumcircle is dilated about the orthocenter by ratio $\frac{1}{2}$, which takes the circumcircle to a circle passing through the nine points. Using basic properties of dilations, it can be shown that the center of the nine-point circle is the midpoint of the segment connecting the orthocenter and the circumcenter, and the area of the nine-point circle is $\frac{1}{4}$ the area of the circumcircle.

The nine-point circle theorem was proved in Chapter 8, and is listed below for convenience. The interested reader can fill in the details of the dilation proof outlined above, refer to the proof of Theorem 8.1, or prove it using a coordinate approach as outlined in Problem 4.

> **Theorem 9.9: Nine-Point Circle Theorem**
>
> Let $\triangle ABC$ be a triangle. The three midpoints of the sides, the three feet of the altitudes, and the three midpoints of the segments joining the orthocenter to the vertices lie on a circle called the nine-point circle.

We next prove that an inversion can be defined that takes the circumcircle of a triangle onto the nine-point circle and vice versa. The circle of inversion is centered at the centroid of the triangle. Its radius is the square root of the product of the distance from the centroid to one of the vertices of the triangle and the distance from the centroid to the point of intersection of the nine-point circle with the ray from centroid to that vertex.

> **Theorem 9.10**
>
> Let O be the centroid of $\triangle ABC$ and $r = \sqrt{OA \cdot OA'}$, where A' is the point of intersection of $\overrightarrow{OA}$ and the nine-point circle (colored orange in Figure 9.34). Then the circumcircle and the nine-point circle of $\triangle ABC$ are images of each other under inversion in the circle centered at O with radius r.

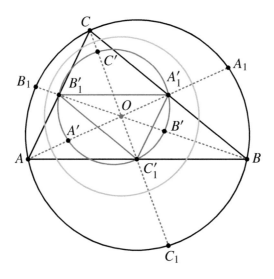

Figure 9.34

Proof. Given $\triangle ABC$, let A_1', B_1', C_1' be the midpoints of the sides $\overline{BC}$, $\overline{AC}$, $\overline{AB}$, respectively, as in Figure 9.34 (the choice of notation will become apparent shortly). Elementary Euclidean results tell us that the medians are concurrent in the centroid O, which divides each median into two segments whose lengths are in the ratio of $2 : 1$. That is, $OA = 2 \cdot OA_1'$, $OB = 2 \cdot OB_1'$, and $OC = 2 \cdot OC_1'$.

Let A_1, B_1, C_1 be the intersection points of $\overrightarrow{AO}$, $\overrightarrow{BO}$, and $\overrightarrow{CO}$ with the circumcircle, respectively. A dilation H centered at O of ratio $-\frac{1}{2}$ takes A to A_1', B to B_1', and C to C_1'. Recall that a dilation maps a circle onto a circle and a straight line through the center of dilation onto itself. Hence, the dilation H takes A_1 to A', B_1 to B', and C_1 to C'. This gives $\frac{OA_1'}{OA} = \frac{OA'}{OA_1} = \frac{1}{2}$, $\frac{OB_1'}{OB} = \frac{OB'}{OB_1} = \frac{1}{2}$, and $\frac{OC_1'}{OC} = \frac{OC'}{OC_1} = \frac{1}{2}$.

Consequently, we have $OA \cdot OA' = OA_1 \cdot OA_1'$, $OB \cdot OB' = OB_1 \cdot OB_1'$, and $OC \cdot OC' = OC_1 \cdot OC_1'$. Now if we can show that $OA \cdot OA' = OB \cdot OB' = OC \cdot OC'$, then we can use $r = \sqrt{OA \cdot OA'}$ as the radius of a circle of inversion centered at O. Inverting points A, B, C

9.4 The Nine-Point Circle and Feuerbach's Theorem

in this circle gives A', B', C', respectively. Since any three noncollinear points determine a unique circle, we can conclude that this inversion takes the circumcircle to the nine-point circle and vice versa.

To show that $OA \cdot OA' = OC \cdot OC'$, we apply the *Intersecting Chords Theorem* to the circumcircle chords $\overline{AA_1}$ and $\overline{CC_1}$, and to the nine-point circle chords $\overline{A'A_1'}$ and $\overline{C'C_1'}$. This leads to $OA \cdot OA_1 = OC \cdot OC_1$ and $OA' \cdot OA_1' = OC' \cdot OC_1'$. Multiplying these equations gives $(OA \cdot OA')(OA_1 \cdot OA_1') = (OC \cdot OC')(OC_1 \cdot OC_1')$, which implies that $(OA \cdot OA')^2 = (OC \cdot OC')^2$ (why?). Hence, $OA \cdot OA' = OC \cdot OC'$ as desired. Similarly, we can show that $OA \cdot OA' = OB \cdot OB'$.

> **Now Try This 9.7**
>
> 1. Show that if $\triangle ABC$ is a right triangle, then the nine-point circle, the circumcircle, and the inversion circle K (centered at the centroid O with radius $\overline{OB}$) are tangent at B, the vertex of the right angle.
>
> 2. Use dynamic geometry software to construct the chains of circles shown in Figure 9.35.
>
> 3. How is the radius of the nine-point circle related to the radii of the shaded circles whose centers lie on the median through B? *Hint: Invert the circles about a circle centered at B.*
>
>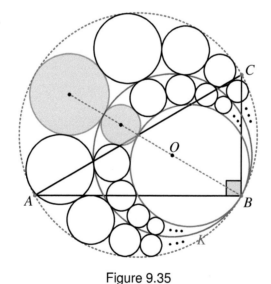
>
> Figure 9.35

Feuerbach's Theorem

One of the most interesting results about the the nine-point circle is *Feuerbach's theorem* published in 1822, which states that *the nine-point circle is internally tangent to the incircle (at a point called the Feuerbach point) and externally tangent to the three excircles of the triangle*. Two circles in the same plane are said to be *internally tangent* if they intersect in exactly one point and the intersection of their interiors is nonempty. On the other hand, two circles in the same plane are said to be *externally tangent* if they intersect in exactly one point and the intersection of their interiors is empty.

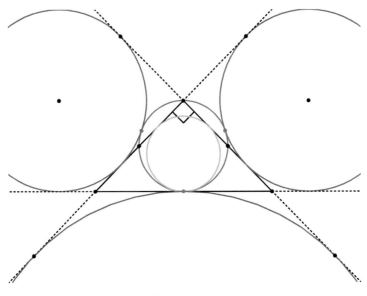

Figure 9.36

Recall that the intersection points of the angle bisectors of the exterior angles of a triangle determine the centers of the excircles (why?). Figure 9.36 shows the incircle, nine-point circle, and excircles for an isosceles right triangle; note that in this special case the Feuerbach point is the midpoint of the hypotenuse.

The results in Lemma 9.1, Lemma 9.2, and Now Prove This 9.2 will be used in the proof of Feuerbach's theorem. The reader is encouraged to prove these results before attempting to analyze the proof of the Theorem 9.11.

> **Lemma 9.1**
>
> Let $\triangle ABC$ be a triangle and I be its incircle. Let a, b, c denote the lengths of the sides opposite the vertices A, B, C, respectively. Finally, let X, Y, Z be the points of tangency of I with the sides $\overline{BC}$, $\overline{AC}$, and $\overline{AB}$, respectively, as in Figure 9.37. Then we have the following:
> 1. $AY = AZ$, $BX = BZ$, and $CX = CY$.
> 2. If $s = \frac{a+b+c}{2} = x + y + z$, then $x = s - a$, $y = s - b$, and $z = s - c$.

Proof. The proof is straightforward. Use similar triangles to prove (1), and then use algebraic manipulations to prove (2). ∎

9.4 The Nine-Point Circle and Feuerbach's Theorem

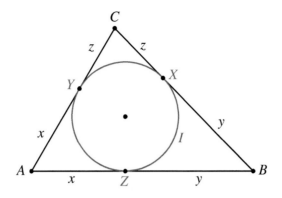

Figure 9.37

> **Lemma 9.2**
>
> Let $\overline{CD}$ be the bisector of $\angle C$ in $\triangle ABC$, and let a, b, and c denote the lengths of the sides opposite the vertices A, B, and C, respectively, as in Figure 9.38. Then we have the following:
> 1. $\frac{AD}{BD} = \frac{b}{a}$.
> 2. $BD = \frac{ac}{a+b}$ and $AD = \frac{bc}{a+b}$.

Proof. The proof is straightforward. Use the *law of sines* to prove (1), and then use algebraic manipulations to prove (2). ∎

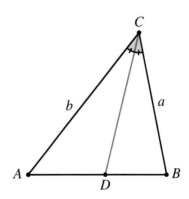

Figure 9.38

Now Prove This 9.2

Let C_1 and C_2 be two circles centered at O_1 and O_2, respectively. Let $\overleftrightarrow{PQ}$ be a common tangent to C_1 and C_2 at P and Q, respectively, as in Figure 9.39.

1. Prove that the circle K centered at M, the midpoint of the segment $\overline{PQ}$, and going through P is orthogonal to C_1 and C_2.
2. Prove that the reflection of $\overleftrightarrow{PQ}$ about the line $\overleftrightarrow{O_1O_2}$ is also tangent to C_1 and C_2.
3. Do (1) and (2) still hold if $\overleftrightarrow{PQ}$ is a common tangent that does not intersect the segment $\overline{O_1O_2}$?

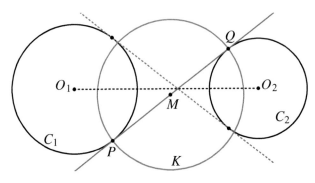

Figure 9.39

We are now ready to prove Feuerbach's theorem. The proof given below was adapted from [19]. For any of the excircles of a triangle, it uses inversion to map the nine-point circle to a common tangent to the incircle and the excircle. But the inversion transformation is its own inverse and it preserves angle measure; hence the same inversion takes the common tangent to the nine-point circle, thus proving that this circle is tangent to the incircle and each of the excircles of the triangle.

Theorem 9.11: Feuerbach's Theorem

The nine-point circle of a triangle is internally tangent to the incircle and externally tangent to each of the three excircles.

Proof. Without loss of generality, we will prove the theorem for the excircle opposite vertex A. The same argument can be used to prove the theorem for the rest of the excircles. To do this, let N and I denote the nine-point circle and the incircle of $\triangle ABC$, respectively, and let E be the excircle opposite A, as in Figure 9.40. Furthermore, let P and Q denote the points of tangency of circles I and E with $\overline{BC}$, respectively, and let A', B', C' denote the respective midpoints of $\overline{BC}, \overline{AC}, \overline{AB}$. Finally, let m be the reflection of $\overleftrightarrow{BC}$ about the angle bisector $\overline{AO}$, and use C_1 and B_1 to label its points of intersection with $\overline{AB}$ and $\overline{AC}$, respectively. We know from Now Prove This 9.2 that line m is also tangent to the circles I and E.

9.4 The Nine-Point Circle and Feuerbach's Theorem

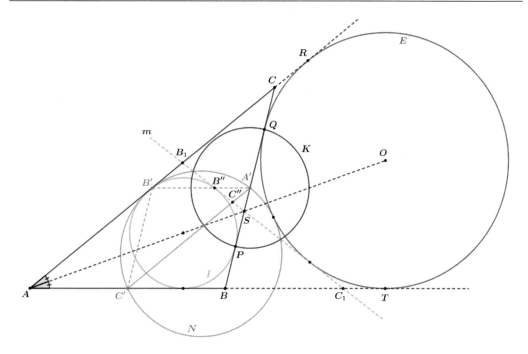

Figure 9.40

Lemma 9.1 tells us that $AR = AT$, $BQ = BT$, and $CR = CQ$. But $AR = b+CR = b+CQ$ and $AT = c + BT = c + BQ = c + a - CQ$, hence $b + CQ = c + a - CQ$, or equivalently $CQ = \frac{a-b+c}{2} = s - b$, where $s = \frac{a+b+c}{2}$ is half the perimeter of $\triangle ABC$. It also tells us that $BP = s - b$ (why?). Now since $A'C = A'B$, we have $A'P = A'Q$. Let B'' and C'' denote the points of intersection of line m with $\overline{A'B'}$ and $\overline{A'C'}$, respectively, and let K be the circle centered at A' and passing through P and Q. Since the nine-point circle N goes through the center of inversion A', its image under inversion in K is a line. If we can prove that this inversion takes B' to B'' and C' to C'', then B'' and C'' lie on the line that is the image of N. This allows us to conclude that N is inverted onto the line m which is tangent to both I and E (why?). Therefore, the nine-point circle N, which is the image of line m under inversion in K, is tangent to both the incircle I and the excircle E.

We now turn our attention to showing that inversion in circle K takes B' to B'' and C' to C''. Assuming that the radius of K is r, we need to show that $A'B' \cdot A'B'' = r^2$ and $A'C' \cdot A'C'' = r^2$. Basic Euclidean results show that the midsegment $\overline{A'B'}$ is parallel to $\overline{AB}$ and $A'B' = \frac{1}{2}AB$. Similarly, for $\overline{B'C'}$ and $\overline{A'C'}$. Hence $\triangle SA'B''$ is similar to $\triangle SBC_1$ and $\triangle SA'C''$ is similar to $\triangle SCB_1$ (why?). This leads to

$$\frac{A'B''}{BC_1} = \frac{SA'}{SB} \quad \text{and} \quad \frac{A'C''}{CB_1} = \frac{SA'}{SC}. \tag{1}$$

Next, we need to find values for BC_1, SA', SB, CB_1, and SC in terms of a, b, c, the lengths of the sides of the triangle. Since C_1 is the reflection of C about $\overline{AO}$, we have $AC = AC_1$. Hence $AC = AC_1 = AB + BC_1$, or equivalently $b = c + BC_1$, which means that

$BC_1 = b - c$. Similarly, $B_1C = b - c$ (why?). Note that $b > c$ in Figure 9.40. If $b < c$, then interchanging the labels of vertices B and C results in $b - c$ having a positive value, so the argument still holds. Furthermore, since S is on the bisector of $\angle A$, Lemma 9.2 gives us that $SB = \frac{ac}{b+c}$ and $SC = \frac{ab}{b+c}$. Now $SA' + SB = A'B = \frac{1}{2}a$; hence $SA' = \frac{1}{2}a - \frac{ac}{b+c} = \frac{a(b-c)}{2(b+c)}$. Note that using $SA' + A'C = SC$ results in the same value for SA'. Substituting these values in (1) and solving for $A'B''$ and $A'C''$ yields

$$A'B'' = \frac{(b-c)^2}{2c} \quad \text{and} \quad A'C'' = \frac{(b-c)^2}{2b}. \tag{2}$$

But $A'B' = \frac{c}{2}$ and $A'C' = \frac{b}{2}$; therefore

$$A'B' \cdot A'B'' = \left(\frac{b-c}{2}\right)^2 \quad \text{and} \quad A'C' \cdot A'C'' = \left(\frac{b-c}{2}\right)^2. \tag{3}$$

Note that the radius of circle K is $r = \frac{1}{2}(BC - BP - CQ) = \frac{b-c}{2}$ (why?). Hence (3) implies that inversion in K takes B' to B'' and C' to C'' as desired. ∎

9.4.1 Problem Set

1. Let $\triangle ABC$ be a triangle with orthocenter O. Let A_1, B_1, and C_1 be the centers of the circles circumscribing $\triangle BOC$, $\triangle AOC$, and $\triangle BOA$, respectively, as in the figure below. Prove that $\triangle ABC \cong \triangle A_1B_1C_1$ and the nine-point circle of $\triangle ABC$ is the same as the nine-point circle $\triangle A_1B_1C_1$.

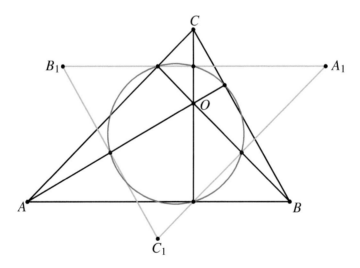

2. Consider $\triangle ABC$ with incircle I and circumcircle K.
    a. Prove that under inversion in I, the circle K inverts onto the nine-point circle of $\triangle XYZ$ determined by the points of contact of I with the sides of $\triangle ABC$.

b. Prove that $d^2 = R^2 - 2Rr$, where d is the distance between the incenter and the circumcenter, R is the radius of the circumcircle, and r is the radius of the incircle.

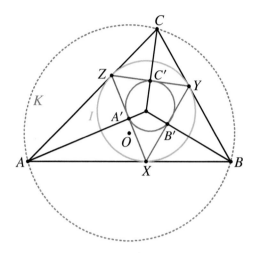

3. Let $\triangle ABC$ be a triangle and let K be the excircle opposite vertex A and tangent to $\overleftrightarrow{AB}$, $\overleftrightarrow{BC}$, and $\overleftrightarrow{AC}$. If the center of K is E and its radius is s, prove that $(OE)^2 = R^2 + 2Rs$, where R is the radius of the circumcircle. The same holds for the rest of the excircles of $\triangle ABC$.

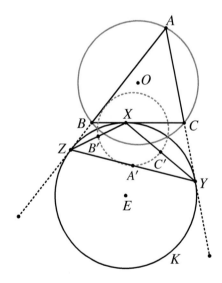

4. Given any triangle ABC, we may assume that $A = (a, 0)$, $B = (b, 0)$, and $C = (0, c)$ (why?). Let A', B', and C' be the midpoint of the sides opposite A, B and C, respectively. Let F_A, F_B, and F_C be the feet of the altitudes from A, B, and C, respectively. Finally, let E_A, E_B, and E_C be the midpoints of the segments connecting the orthocenter to A, B, and C respectively. Follow the steps below to show that the nine points A', B', C', F_A, F_B, F_C, E_A, E_B, and E_C are equidistant from O_N, the midpoint of the segment connecting the orthocenter to the circumcenter. The circle going through these nine points is the nine-point circle and O_N is its center. Consider the figure below and use the steps outlined to prove the existence of the nine-point circle. Some of what follows was adapted from [77].

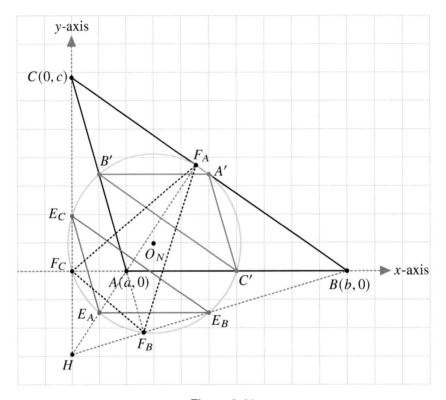

Figure 9.41

a. Find the coordinates of A', B', and C'.

b. Show that $E_A = \left(\dfrac{a}{2}, -\dfrac{ab}{2c}\right)$, $E_B = \left(\dfrac{b}{2}, -\dfrac{ab}{2c}\right)$, and $E_C = \left(0, \dfrac{c^2 - ab}{2c}\right)$.

c. Show that $F_A = \left(\dfrac{ab^2 + bc^2}{b^2 + c^2}, \dfrac{b^2c - abc}{b^2 + c^2}\right)$, $F_B = \left(\dfrac{ba^2 + ac^2}{a^2 + c^2}, \dfrac{a^2c - abc}{a^2 + c^2}\right)$, and $F_C = (0, 0)$.

d. Show that the orthocenter and circumcenter of $\triangle ABC$ have coordinates $\left(0, -\dfrac{ab}{c}\right)$ and $\left(\dfrac{a+b}{2}, \dfrac{c^2 + ab}{2c}\right)$, respectively.

e. Show that the midpoint O_N of the segment connecting the orthocenter to the circumcenter has coordinates $\left(\dfrac{a+b}{4}, \dfrac{c^2 - ab}{4c}\right)$.

f. Show that O_N is equidistant from $A', B, C', E_A, E_B, E_C, F_A, F_B,$ and F_C.

Hint: The common distance is $\sqrt{\left(\dfrac{a+b}{4}\right)^2 + \left(\dfrac{c^2-ab}{4c}\right)^2}$.

g. Note that this also shows that the triangles $\triangle A'B'C'$, $\triangle E_A E_B E_C$, and $\triangle F_A F_B F_C$ have the same circumcenter. These triangles are called the *medial*, *Euler*, and *orthic* triangles, respectively.

h. *Optional*: Find coordinates of the centroid of $\triangle ABC$.

i. *Optional*: Justify that the equations of the circumcircle (1) and the nine-point circle (2) are as follows:

$$\left(x - \frac{a+b}{2}\right)^2 + \left(y - \frac{c^2+ab}{2c}\right)^2 = \left(\frac{a-b}{2}\right)^2 + \left(\frac{c^2+ab}{2c}\right)^2, \quad (1)$$

$$\left(x - \frac{a+b}{4}\right)^2 + \left(y - \frac{c^2-ab}{4c}\right)^2 = \left(\frac{a+b}{4}\right)^2 + \left(\frac{c^2-ab}{4c}\right)^2. \quad (2)$$

9.5 Stereographic Projection and Inversion

We have previously pointed out (see Section 9.1) that in a sense, inversion takes the center of inversion onto the "point at infinity" since as a point P gets closer and closer to the center of inversion, its image gets farther and farther from the circle of inversion. In what follows, we will attempt to give this intuitive idea a more rigorous treatment.

Stereographic projection is a projection of a sphere from one of its points N, called the *center of projection*, onto the plane H tangent to the sphere at a point S diametrically opposite N. The plane H is called the *projection plane*, and in general any plane parallel to H and not going through N can serve as the projection plane. For instance, the projection plane is often taken to be the plane through the equator of the sphere. For convenience, let the sphere be of radius R and the plane of projection be the xy-plane. Think of the sphere as being placed "on top" of the xy-plane so that the plane is tangent to the sphere at S, where $S = (0,0)$. We refer to N and S as the north and south poles of the sphere, respectively.

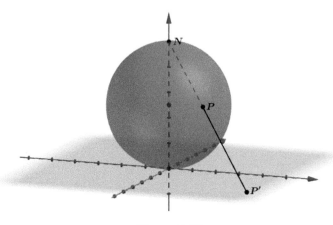

Figure 9.42

Let P be a point on the sphere distinct from N, the north pole of the sphere. We now describe how its image under stereographic projection is determined. Construct the straight line that passes through P and N, as in Figure 9.42. The point P' where $\overleftrightarrow{PN}$ intersects the projection plane is the image of P under stereographic projection. Notice that the image of the south pole is the point S itself, and the image of a point near the north pole is infinitely far from the origin S (why?). Since this mapping from the sphere minus the point N onto the projection plane is one-to-one, it has an inverse. This inverse map takes points in the projection plane and maps them onto the sphere. We denote the stereographic projection by π and write $\pi(P) = P'$.

We now investigate what happens to the center of projection N. Stereographic projection is not defined at the point N, since there is no unique straight line through N that intersects the sphere only at N. As noted in the previous paragraph, as points approach N, their stereographic projections onto the plane get infinitely farther and farther from S, where the sphere touches the plane. Therefore, it is convenient and plausible to agree to map N, under stereographic projection, onto "the point at infinity" in the extended plane. Thus, the sphere provides a geometric representation of the extended plane, with the north pole N corresponding to the point at infinity.

We next investigate what happens to basic geometric objects lying on the sphere under stereographic projection. When we studied the properties of inversion in a circle in Section 9.2, we first investigated the images of straight lines. Since great circles on the sphere play a role similar to that of straight lines on a flat plane (why?), we should start by investigating what happens to circles under stereographic projection. We will do this by considering two cases: Circles passing through N and circles not passing through N.

First, we will show that stereographic projection maps circles passing through N onto straight lines in the projection plane. For the sake of convenience, we will start by considering a great circle C passing through N (similar arguments can be used for other circles passing through N). If C is a great circle passing through N, then C lies on some plane L that cuts the sphere into two congruent halves, as shown in Figure 9.43.

9.5 Stereographic Projection and Inversion

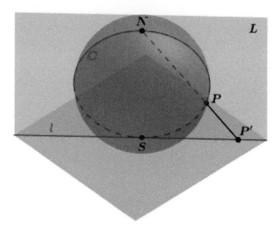

Figure 9.43

In particular, this plane cannot be tangent to the sphere at N, so it is not parallel to the projection plane. Furthermore, for every point $P \neq N$ on C, $\overrightarrow{NP}$ lies on the plane L, since both N and P are on L. Consequently, the stereographic projection P' of P must also lie on L. It follows that P' is on the intersection of plane L and the projection plane. But this intersection is just a straight line l, so the image of $C - \{N\}$ is a subset of the line l in the projection plane.

Moreover, the only line through N on plane L that does not intersect the projection plane is the line tangent to C at N. Hence, for every point P' on l, there exists a point P on $C - \{N\}$ such that $\overrightarrow{NP}$ intersects the projection plane in P'. Thus every point on the line l has a preimage on C under stereographic projection. Therefore, the image of $C - \{N\}$ is precisely l, the line of intersection of plane L and the projection plane. Since we agreed to map N to the "point at infinity," we can conclude that the image of C is the extended line l.

A similar argument can be used to show that if C is any circle passing through N but is not a great circle, then its image is still a straight line in the extended projection plane. The discussion above proves the following theorem.

Theorem 9.12

The image under stereographic projection of a circle passing through the center of projection is a straight line in the extended projection plane.

Second, we now show that stereographic projection maps circles lying in planes that are parallel to the projection plane onto a circle centered at S in the projection plane. We will start by investigating the case where the circle under consideration is the equator. If we take O to be the center of the sphere and P a point on the equator, as in Figure 9.44, then $\triangle NOP$ is an isosceles right triangle whose right angle is $\angle NOP$. If we extend the segment NP to the point $P' = \pi(P)$, then $\overline{SP'} \| \overline{OP}$, hence $\triangle NOP$ and $\triangle NSP'$ are similar.

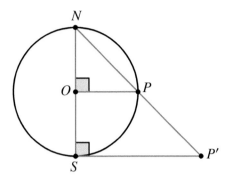

Figure 9.44

Consequently, $\Delta NSP'$ is also an isosceles right triangle, and $SP' = NS$. Since this is true for any point P' in the image of the equator, the image of the equator is a subset of the circle in the projection plane centered at the origin S and of radius $SP' = NS$. By reversing the argument, every point on this circle has a point on the equator that gets mapped to it. Therefore, the image of the equator is the circle centered at S whose radius is twice that of the equator or the sphere.

A similar argument can be used to show that the image of any circle C that is parallel to the projection plane is a circle C' in the projection plane centered at S. Furthermore, if the radius of C is r, the radius of the sphere is R, and the distance between the center of C and N is d, then the radius of C' is $\frac{2R}{d} \cdot r$ (why?).

Before proving that circles, not passing through N and not parallel to the projection plane, are mapped onto circles in the projection plane, we need to review a few facts about the geometry of cones in three-dimensional space. Let C be a circle and A be a point that is not in the plane of the circle. From each point on the circle, draw a line segment connecting that point to A. The set of all points that lie on these segments is called a *circular cone* with *apex* or *vertex* A and *base* C. If the line from A to the center of C is perpendicular to the plane of C, then the cone is a *right circular cone*. Otherwise, it is an *oblique circular cone*. The line passing through A and the center of the circle C is the *axis* of the cone.

One important fact about cones is that the intersection of any plane with a right or oblique circular cone is a conic section (circle, ellipse, hyperbola, or parabola), as shown in Figure 9.45. Can you identify the various conic sections in the figure and specify how a plane needs to intersect the cone to result in each of the sections? Beautiful illustrations and geometric proofs of conic sections can be found in [87]. Note that if a plane intersects a cone in all of the line segments forming the cone through its vertex, but does not contain the vertex itself, then the intersection is an ellipse.

9.5 Stereographic Projection and Inversion

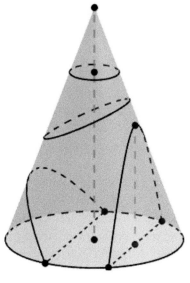

Figure 9.45

In particular, cross sections parallel to the base of a circular cone such as the one whose diameter is $\overline{B'C'}$ (see Figure 9.46) are circles.

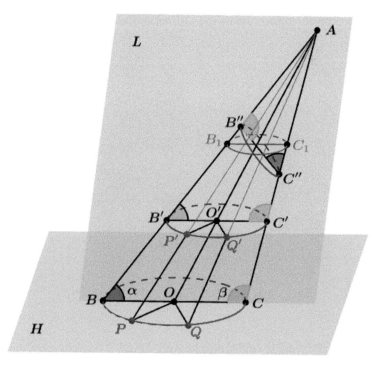

Figure 9.46

To show that this cross section is indeed a circle, take two arbitrary points P' and Q' on it and show that $O'P' = O'Q'$, where O' is the point of intersection of the cross section with the axis of the cone. To do this let P and Q be the points of intersection of the base of the cone with $\overrightarrow{AP'}$ and $\overrightarrow{AQ'}$, respectively. Now since $\triangle P'O'A$ is similar to $\triangle POA$, and $\triangle Q'O'A$ is similar to $\triangle QOA$, then $\frac{P'O'}{PO} = \frac{O'A}{OA} = \frac{Q'O'}{QO}$. But $PO = QO$, hence $P'O' = Q'O'$. This implies that the cross section parallel to the base of the cone and whose diameter is $\overline{B'C'}$ is a circle.

For oblique circular cones, cross sections parallel to the base of the cone are not the only cross sections that are circular. Cross sections such as the one whose diameter is $\overline{B''C''}$ (see Figure 9.46) are also circles. Note that this cross section is such that $\angle AB''C'' = \angle ACB$, or equivalently $\angle AC''B'' = \angle ABC$. Proving that such cross sections are circles requires a little bit more effort and hence will not be presented here. The interested reader can refer to [79] to see how this can be done with the help of the cross section with diameter $\overline{B_1C_1}$, whose plane is parallel to the plane of the base of the cone and hence is a circular cross section.

We now turn our attention to proving that stereographic projection takes circles on the sphere not passing through the center of projection N onto circles lying in the projection plane.

> **Theorem 9.13**
>
> *The image under stereographic projection of a circle on the sphere not passing through N is a circle in the projection plane.*

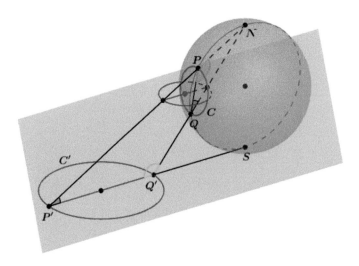

Figure 9.47

Proof. Let C be a circle on the sphere not passing through N. Consider the circular cone going through C and having N as its vertex. Let C' be the intersection of this cone with the projection plane. If the plane of C is parallel to the projection plane, then this cone is a right circular cone, and by the discussion preceding the theorem, C is projected onto a circle in the projection plane.

If the plane of C is not parallel to the projection plane, then the cone going through C and having N as its vertex is an oblique circular cone, as in Figure 9.47. Let L (not shown) be a plane through N and S such that L divides C into two congruent semicircles.

For the sake of clarity, consider the two-dimensional view of the great circle which is the intersection of the sphere with plane L, as shown in Figure 9.48. In this figure, $\overline{PQ}$ represents a diameter of C, and P' and Q' represent the images of P and Q, respectively. By the discussion preceding the theorem, if we can show that $\angle NP'S \cong \angle PQN$ or $\angle NPQ \cong \angle P'Q'N$, then C is a circle if and only if its image in the projection plane is a circle.

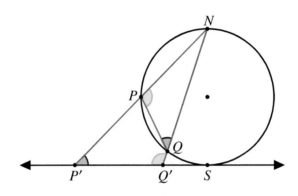

Figure 9.48

Note that $\angle NP'S$ is an exterior angle to the circle, and hence its measure is equal to half the difference of the measure of intercepted arcs, that is, $\frac{1}{2}(m(\widehat{SN}) - m(\widehat{PS})) = \frac{1}{2}m(\widehat{NP})$ (why?). Note further that $\angle PQN$ is an inscribed angle and hence its measure is $\frac{1}{2}m(\widehat{NP})$. Hence $\angle NP'S \cong \angle PQN$, and since $\overline{PQ}$ is the diameter of a cross section that is a circle, then $\overline{P'Q'}$ is the diameter of a cross section in the projection plane that is also a circle. Thus we have shown that a circle not going through the center of projection N is projected onto a circle in the projection plane. ∎

Thus far, we have seen that circles through N map onto lines under stereographic projection and circles not through N map onto circles. Conversely, if we take lines or circles in the plane and map them to the sphere by the inverse of the stereographic projection, they will map to circles through N or circles not through N, respectively. Thus, both lines and circles in the plane correspond to circles on the sphere, and hence we can think of inversions on the extended plane as simply transformations that take circles to circles on the sphere (why?).

However, to make the analogy complete, we must first verify that angle measure is preserved under stereographic projection. Note that if two curves on the sphere intersect at N, then their projections intersect at the point at infinity in the extended plane, so it is tricky to define an angle of intersection between two such curves. In order to avoid any confusion, we will focus our attention on angles between two curves intersecting at points other than N.

> **Theorem 9.14**
>
> *The angle measure between two curves on the sphere intersecting at a point other than N is preserved under stereographic projection.*

Proof. Let α and β be curves on the sphere that intersect at a point P. Denote the line tangent to α at P by α' and the line tangent to β at P by β', as in Figure 9.49. Let K be the plane through the line α' and the point N, and let L be the plane through the line β' and the point N. The planes K and L are not shown in Figure 9.49. The intersection of the plane K with the sphere is a circle C_1 passing through N. Similarly, the intersection of L with the sphere is a circle C_2 passing through N. Consider two points R on C_1 and Q on C_2. Note that as R and Q approach P, $\angle RPQ$ approaches the angle between α' and β'. Thus, to find the image of the angle between α and β at P, we can instead find the angle between the images of C_1 and C_2 under stereographic projection (note that these images are lines that intersect at $P' = \pi(P)$).

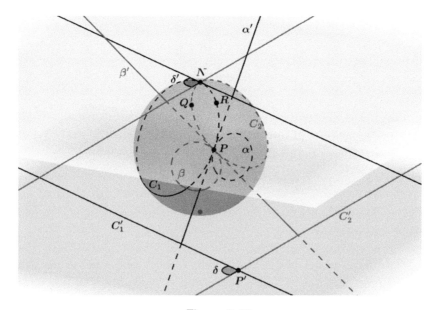

Figure 9.49

The image of C_1 under stereographic projection is C_1', which is precisely the intersection of the projection plane with plane K, and the image of C_2 is C_2', which is the intersection of the projection plane with plane L. Let δ denote the angle between these two lines. Now consider the plane T tangent to the sphere at N. This plane is parallel to the projection plane, and hence if we look at the angle δ' between the line of intersection of K with T and the line of intersection of L with T, we have $m(\angle \delta) = m(\angle \delta')$.

9.5 Stereographic Projection and Inversion

Moreover, since C_1 and C_2 intersect at P and N, it's not hard to see that if two circles intersect in two points such that the angle of intersection at one of the points is γ, then the angle of intersection at the other intersection point is also γ. The proof follows by reflecting in a plane through the centers of both circles, which takes one of the points of intersection of the circles onto the other. This is analogous to how two circles intersecting in two points are related in the Euclidean plane.

Therefore, the angle formed at N by the intersection of C_1 and C_2 is the same as the angle they form at P. But by construction, this angle is the angle between α' and β'. Consequently, $\angle \delta$ (the image of the angle between α and β at P under stereographic projection) has equal measure as the angle between α' and β' at P. ∎

Since stereographic projections and inversions preserve angle measure, we can think of inversions as angle measure-preserving maps on the sphere. When we want to apply an inversion, we first use stereographic projection π to map the sphere to the projection plane, then apply the inversion, and finally use π^{-1} to return to the sphere. Recall that circles on the sphere map to circles or lines on the plane, and inversions map circles and lines in the plane to circles and lines. Therefore, when we return to the sphere via π^{-1}, the circles and lines on the plane are mapped onto circles on the sphere.

We have shown earlier that the image under a stereographic projection of a circle of latitude is a circle centered at S and the image of a circle of longitude is a straight line. The images of the circles of longitude passing through S are lines intersecting at S. These lines are orthogonal to the images of the latitude circles which are concentric circles centered at S.

We next investigate how the images of two points lying on a longitudinal circle through N are related in the projection plane tangent to the sphere at S.

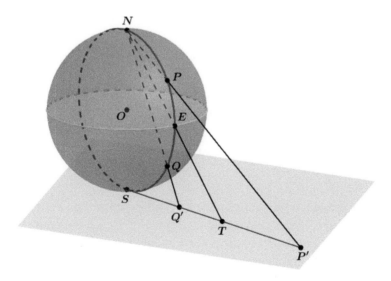

Figure 9.50

Suppose that P and Q are two points on the same longitudinal circle and symmetrical with respect to the plane containing the equator, as in Figure 9.50. Let P' and Q' be the images of P and Q, respectively. How are P' and Q' related? That is, if we know P', how can we find Q' without going through Q? We want a relationship between P' and Q' which does not involve going back to P and Q.

Figure 9.51 shows the circle of longitude that goes through P, Q, N, and S. In the plane of this circle, the points P and Q are symmetrical with respect to the line $\overleftrightarrow{OE}$ (P is the reflection of Q in the line). Let R denote the radius of the sphere. Note that $\triangle OEN \sim \triangle STN$, hence $ST = 2 \cdot OE = 2R$. Since we want a connection between P' and Q' in the projection plane, it is useful to look for a connection between SP' and SQ'. Note that $\overline{SQ'}$ is in $\triangle NSQ'$ and $\overline{SP'}$ is in $\triangle NSP'$. How are these triangles related? Clearly, $\triangle NSQ' \sim \triangle NQS$ and $\triangle SPN \sim \triangle P'SN$ (why?). But $\triangle NQS \cong \triangle SPN$ (reflection of each other in $\overleftrightarrow{OE}$). Hence $\triangle NSQ' \sim \triangle P'SN$ which implies that $\frac{SP'}{NS} = \frac{NS}{SQ'}$. Therefore, $SP' \cdot SQ' = (NS)^2 = (2R)^2 = (ST)^2$.

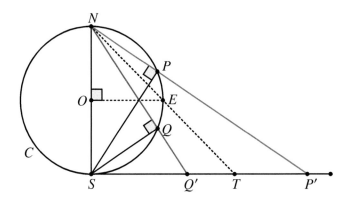

Figure 9.51

Accordingly, if we consider the circle K in the projection plane, centered at S of radius $ST = 2R$, then P' is the image Q' under inversion in K (see Figure 9.52). The circle K is the stereographic image of the equator. Let $k = 2R$ be the radius of K, $x = Q'T$, and $y = P'T$. Using what we know about Q' and P' we have $SP' \cdot SQ' = k^2$. Hence $(k+y)(k-x) = k^2$, which gives $y = \dfrac{kx}{k-x} = \dfrac{x}{1-\frac{x}{k}}$.

Note that as x goes to k (or equivalently, as Q approaches S on the sphere), y goes to infinity (or equivalently, P approaches N on the sphere). Hence the point S on the sphere represents the center of inversion and the point N represents the point at infinity. This gives a geometric interpretation (on the sphere) of the convention adopted in Section 9.1, which declared the point at infinity (in the extended plane) as being the inverse image of the center of inversion.

9.5 Stereographic Projection and Inversion

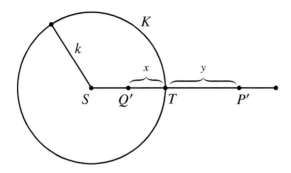

Figure 9.52

9.5.1 Problem Set

1. Suppose x is the real axis of the complex plane, y the imaginary axis, and z an axis perpendicular to the complex plane. Then, the equation of the unit sphere tangent to the complex plane at $(0, 0, 0)$ is $x^2 + y^2 + (z-1)^2 = 1$. The north pole of the sphere N is located at $(0, 0, 2)$. Let $(x, y, 1 + \sqrt{1 - x^2 - y^2})$ be the coordinates of a point P in the northern hemisphere, where $x^2 + y^2 \leq 1$. Compute the coordinates of the image of P under stereographic projection with center N.

2. Using the same setup as in Problem 1, find the coordinates of the stereographic projection of an arbitrary point in the southern hemisphere.

3. The equation of an arbitrary plane is $ax + by + cz + d = 0$, for constants a, b, c, d. Any circle on the sphere through N can be written as the intersection of the sphere $x^2 + y^2 + (z-1)^2 = 1$ and a plane described by the equation $ax + by + z - 2 = 0$. Prove, using coordinates, that stereographic projection takes a circle on the sphere through N to a line on the xy-plane.

4. Prove, using coordinates, that stereographic projection takes a circle on the sphere not through N to a circle on the xy-plane.

10 Hyperbolic Geometry

10.1 Introduction

Some of the material in this chapter was adapted from [44]. For centuries, Euclid's monumental work *The Elements* was regarded as a systematic discussion of absolute geometric truth. However, *The Elements* contains many explicit and implicit unproven assumptions. Euclid stated some of these assumptions as *Postulates* and *Common Notions*, while others, such as the infinitude of a straight line and the continuity of the plane, are merely implied in his proofs. In this chapter, we will see that by eliminating one or more of these assumptions, we may arrive at geometries which are dramatically different from the familiar Euclidean geometry. To derive these geometries, the primary assumption to disregard is that of the historically controversial parallel postulate. Euclid's parallel postulate, Postulate 5 of *The Elements* [37], states:

> *That, if a straight line falling on two straight lines make the interior angles on the same side less than two right angles, the two straight lines, if produced indefinitely, meet on that side on which are the angles less than two right angles.*

This postulate garnered much criticism from early geometers, not because its truth was in doubt (on the contrary, it was generally considered as a logical necessity), but because its complexity left them uneasy about its classification as a postulate rather than a proposition that can be proven using the first four postulates. Geometric results derived using only the first four postulates constitute what is called *absolute* or *neutral* geometry. For instance, the first 28 propositions of *The Elements* are valid in neutral geometry. There were several attempts by early geometers to prove the parallel postulate in absolute geometry, but they all failed because they often unwittingly assumed some "obvious" property that turned out to be equivalent to the fifth postulate. This resulted in the discovery of a host of statements which are equivalent to the fifth postulate. Some of these include:

1. If a line intersects one of two parallel lines, it must also intersect the other (Proclus' axiom).
2. Two lines parallel to the same line are parallel to each other.
3. Parallel lines are everywhere equidistant.
4. Through a point not on a given line there exists a unique line parallel to the given line (Playfair's axiom).
5. The sum of the angles of a triangle is two right angles.
6. If two parallel lines are cut by a transversal, the alternate interior angles are equal.
7. Similar triangles exist which are not congruent.
8. Any three distinct noncollinear points have a circle going through them.

9. The diagonals of a parallelogram bisect each other.

10. Through any point D within any angle $\angle ABC$, a line can be drawn which meets both sides of the angle $\overrightarrow{BA}$ and $\overrightarrow{BC}$.

11. An angle inscribed in a semicircle is a right angle (Thales' theorem).

12. The Pythagorean theorem.

By accepting all of Euclid's assumptions other than the fifth postulate, and substituting one of the statements listed above for the fifth postulate, we arrive at the familiar Euclidean geometry.

In the early 18th century, an attempt was made by the Italian Jesuit mathematician Girolamo Saccheri (1667–1733) to prove the parallel postulate without the use of any additional assumptions. In the process, he derived some of the first results in what is now called *elliptic* and *hyperbolic* geometries. Saccheri considered a quadrilateral $ABCD$ in which the sides $\overline{AD}$ and $\overline{BC}$ are equal in length and perpendicular to the base $\overline{AB}$, as in Figure 10.1. Such a quadrilateral, now known as a Saccheri quadrilateral, was first used for the same purpose by medieval Islamic mathematicians Omar Khayyam (1048–1131) and later by Nasir al-Din al-Tusi (1201–1274) over five centuries before Saccheri [80].

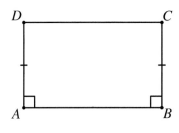

Figure 10.1

Saccheri proved, without using the fifth postulate or any of its consequences, that in such a quadrilateral the summit angles $\angle ADC$ and $\angle BCD$ have equal measure and that the line joining the midpoints of $\overline{AB}$ and $\overline{CD}$ is perpendicular to both segments. Proofs of these results will be given in the section that follows. Proposition 29 of *The Elements*, if invoked at this point, would yield that the summit angles are right angles (*Hypothesis of the Right Angle*). But Saccheri was well aware of the fact that Proposition 29 is a consequence of the fifth postulate, so instead, he attempted to arrive at this result by eliminating its negation, that is, eliminating the possibility of the summit angles being obtuse or acute.

By assuming that the summit angles were obtuse (*Hypothesis of the Obtuse Angle*), as shown in Figure 10.2, Saccheri arrived at the following results:

1. $AB > CD$.

2. The sum of the angles of a triangle is greater than two right angles.

3. An angle inscribed in a semicircle is always obtuse.

10.1 Introduction

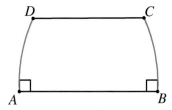

Figure 10.2

In this case, Saccheri was able to arrive at contradictions to Euclid's propositions 16, 17, and 18; however, all these propositions use Euclid's unstated assumption that lines are infinite in extent. Without this unstated assumption, there would be no contradiction. In fact, in Chapter 11, we will show that the three properties listed above hold in *elliptic* geometry, in which lines are finite in extent and Proposition 16 does not hold. Books I–XIII of *The Elements* can be found online at https://mathcs.clarku.edu/~djoyce/elements/toc.htm.

Next, by assuming that the summit angles were acute (*Hypothesis of the Acute Angle*), as shown in Figure 10.3, Saccheri proved the following results:

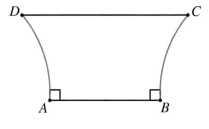

Figure 10.3

1. $AB < CD$.
2. The sum of the angles of a triangle is less than two right angles.
3. An angle inscribed in a semicircle is always acute.
4. If two lines are cut by a transversal so that the sum of the interior angles on the same side of the transversal is less than two right angles, then the lines do not necessarily meet, that is, they are sometimes parallel.
5. Through any point not on a given line, there exists more than one parallel to the given line.
6. Two parallel lines need not have a common perpendicular.
7. Parallel lines are not equidistant. In fact, we have two cases as shown in Figure 10.4.
    a. When the parallel lines have a common perpendicular they diverge away from each other on each side of the perpendicular and are often called *hyper/ultra/super-parallel or divergently parallel* [see Figure 10.4(a)].

b. When the parallel lines have no common perpendicular, they diverge away from each other in one direction and are asymptotic (come arbitrarily close to each other) in the other direction and are often called *asymptotic parallels*, *limiting parallels*, or *horoparallels* [see Figure 10.4(b)].

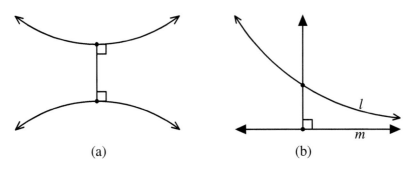

Figure 10.4

To dispose of the possibility that the summit angles are acute, Saccheri showed that two straight lines can either intersect, be divergently parallel, or be asymptotically parallel. In the case of asymptotically parallel lines, Saccheri concluded that these lines must intersect and have a common perpendicular at infinity. To explain this intuitively, if one drops perpendiculars from points on line l to line m, shown in Figure 10.4(b), then as the point from which the perpendicular is being dropped tends to infinity, the length of the perpendicular tends to zero and the angle between l and the perpendicular tends to a right angle. Saccheri considered this point of intersection at infinity and the limit common perpendicular as ordinary objects (even though the idea of "limit" was yet to be formalized in mathematics) and concluded that *asymptotically parallel* lines intersect at infinity. This led him to conclude that: *"The acute angle hypothesis is absolutely false, because it is repugnant to the nature of a straight line"* [36, Proposition 33, page 173]. But is this really a valid contradiction, as it relies on the notion of the "nature of a straight line"? Clearly it is not, as Saccheri himself points out in a book titled *Euclides ab omni naevo vindicatus* (*Euclid Freed of Every Flaw*), which was published shortly after his death in 1733. Toward the end of the book, he admits that he has not completely proven the "acute case" [80, page 98].

> *I do not attain to proving the falsity of the other [acute angle] hypothesis without previously proving that the line, all of whose points are equidistant from an assumed straight line lying in the same plane with it, is equal to this straight line, which itself finally I do not appear to demonstrate from the viscera of the very hypothesis, as must be done for a perfect refutation.... But this is now enough.* [57]

Saccheri's carefully proven results (assuming that the summit angles of a Saccheri quadrilateral are acute) would become the initial results of what was later termed *hyperbolic*

10.1 Introduction

geometry. Surprisingly, the two new geometries (*elliptic and hyperbolic*) stumbled upon by Saccheri would not be actively acknowledged and researched until the early 19th century. Some attribute earlier lack of interest in Saccheri's work to past historians viewing with contempt scholarly works by members of the Jesuit order. That being said, Saccheri's departure from rigorous logical reasoning in the asymptotic parallels case, his failure to realize the true significance of his discoveries, and his belief that Euclid's was the only true geometry may have also contributed to this lack of interest. Regardless of the fact that Saccheri broke away from his rigorous logical reasoning when he claimed in the proof of his 33rd proposition [36] that he arrived at a contradiction due to the "nature of straight lines," it is undeniable that his work paved the way for 19th-century geometers to formally discover and study non-Euclidean geometry.

The credit for first recognizing non-Euclidean geometry for what it was often goes to Carl Friedrich Gauss (1777–1855), despite the fact that Gauss did not publish anything formally on the subject. Like many others, Gauss began by wishing to firmly establish Euclidean geometry free from all ambiguities. His objective was to prove that the angle measures of a triangle must sum to $180°$ (recall, this is equivalent to the parallel postulate). He supposed the contrary, and so was left with two possibilities: either the angle sum is greater than $180°$, or the angle sum is less than $180°$.

Using, as Saccheri had done, Euclid's assumption that lines are infinite in length, Gauss arrived at a contradiction in the case where the angle measures of a triangle sum to more than $180°$. However, the case where the angle sum is less than $180°$ did not lend itself to such a contradiction. In a private letter written in 1824, Gauss asserted:

The assumption that the sum of the three angles is less that $180°$ leads to a curious geometry, quite different from ours, but thoroughly consistent, which I have developed to my entire satisfaction. [25]

While Gauss may have developed this geometry to his own satisfaction, for some reason, he did not see fit to publish any of his findings. As a result, the credit for formally recognizing this "curious geometry" goes to two mathematicians in different parts of the world who, unbeknownst to each other, arrived at the same conclusion around the same time, namely that the geometry in which the angle sum of a triangle is less than $180°$ is entirely valid.

In 1829 a Russian mathematics professor named Nikolai Lobachevsky from the University of Kazan published an article titled "On the Principles of Geometry" in the *Kazan Bulletin*. In this article, he described a geometry in which more than one parallel to a given line may be drawn through a point not on the line. He found that this was tantamount to the angle sum of a triangle being less than $180°$. This was the first publication on non-Euclidean geometry, and so Lobachevsky is recognized as the first to clearly state its properties. Today, hyperbolic geometry is sometimes referred to as Lobachevskian geometry. However, his work was not widely regarded by the mathematical community at the time, and he died in 1856 before his work received wide acceptance. As a side note, Lobachevsky became a high official of the University of Kazan and was instrumental in stopping a cholera epidemic (1830–1831) among the teachers and students by means of a rigid quarantine.

The same year that Nikolai Lobachevsky published his work on non-Euclidean geometry, a Hungarian officer in the Austrian army named Johann Bolyai submitted a manuscript to his father, Wolfgang Bolyai, a mathematics teacher with ties to Gauss. The manuscript contained the younger Bolyai's discovery of non-Euclidean geometry with many of its surprising results. *"Out of nothing, I have created a strange new universe,"* Bolyai is credited with stating in a letter to his father. His work was published in 1832 as an appendix entitled *"The Science of Absolute Space"* to his father's book on elementary mathematics. It was in a letter to the elder Bolyai after reading this appendix that Gauss confessed to having come to the same conclusions thirty to thirty-five years prior. Today, Gauss, Lobachevsky, and Bolyai are given some share of the credit for discovering the non-Euclidean geometry now known as *hyperbolic* geometry.

While Gauss, Lobachevsky, and Bolyai all focused their attention on the geometry formed by assuming the angle sum of a triangle is less than 180°, a mathematician named Georg Friedrich Bernhard Riemann (1826–1866) discovered that by disregarding the assumption that lines have infinite length, one arrives at a valid geometry in which the angle sum of a triangle is greater than 180°. Euclid's second postulate states that a straight line may be continued continuously or indefinitely in a straight line. However, one might imagine a line as being somewhat like a circle, "continuing" forever (unbounded) yet by no means infinite. Riemann was the first to formally make the distinction between *unboundedness* and *infinite extent* when he wrote: *"In the extension of space construction to the infinitely great, we must distinguish between unboundedness and infinite extent; the former belong to the extent relations; the latter to the measure relations"* [9].

Having substituted the more general hypothesis that a straight line is *unbounded* (and hence could be finite) for the hypothesis that it is *infinite*, Riemann embarked on studying geometry free of the parallel postulate. He found that this eliminated any contradiction in the case where the angles of a triangle sum to more than 180°. Interestingly, he found that in such a geometry parallel lines do not exist. This new non-Euclidean geometry came to be known as *elliptic* or *Riemannian* geometry (see Chapter 11).

Thus, by the mid-19th century there were two geometries competing with the Euclidean geometry. Unless the parallel postulate could be proven, both hyperbolic and elliptic geometry seemed logically consistent. But it was not until 1868 that an Italian mathematician named Eugenio Beltrami (1835–1900) proved beyond any doubt that these new geometries were every bit as valid as Euclid's own. He showed through clever analysis that if a contradiction existed in either hyperbolic or elliptic geometry, then a contradiction also existed in Euclidean geometry. Therefore, the mathematical community had to accept these new geometries as valid alternatives, and the quest to prove the parallel postulate finally came to an end.

While elliptic and hyperbolic geometry share most of the spotlight for non-Euclidean geometry, there do exist other geometries which are non-Euclidean. A fairly recent development is *taxicab* geometry (see Chapter 13), the beginnings of which were formulated by the mathematician Hermann Minkowski (1864–1909). Taxicab geometry is derived by taking the regular geometry in the Euclidean coordinate plane and redefining the way distance between points is measured. This means that the assumption that line segments of the same length are congruent must be discarded, and with the loss of that assumption go many of Euclid's most well-known results. Congruence conditions for triangles, for example, do not apply in taxicab geometry. Research continues to be carried out to see what other geometries might be formed by defining distance in still different ways.

A more recently discovered non-Euclidean geometry is *fractal* geometry (see Chapter 14), formally discovered in 1980 by the mathematician Benoit Mandelbrot (1924–2010). Mandelbrot felt that Euclidean geometry was not satisfactory as a model for natural objects.

> *Why is geometry often described as "cold" and "dry"? One reason lies in its inability to describe the shape of a cloud, a mountain, a coastline, or a tree. Clouds are not spheres, mountains are not cones, coastlines are not circles, and bark is not smooth, nor does lightning travel in a straight line.* [59]

The concept of fractal dimension which measures the complexity of fractal objects plays an important role in this geometry. Since many fractals are self-similar or nearly self-similar objects, the role of the concept of size (length, area, volume) is diminished in this geometry. Fractal geometry is currently under development and nothing like a formal axiomatic structure has yet been developed for it.

Today, non-Euclidean geometries are commonly used in mathematics. There are even important applications of these geometries outside the area of pure mathematics. Hyperbolic geometry, for example, is invoked by physicists studying Einstein's general theory of relativity to describe the shape of our universe and is used in the field of computer graphics [23]. Spherical geometry, a simple form of elliptic geometry, is used in navigational calculations for movement on the earth, and taxicab geometry provides a good model of urban geography. Finally, fractal geometry has growing applications in the medical [56] and financial [58] fields as well as in computer graphics.

While Euclidean geometry is still most often used to model objects produced by humans rather than by natural processes, scientists continue to discover surprising and practical new applications for non-Euclidean geometries. As Poincaré once asserted: *"One geometry cannot be more true than another; it can only be more convenient."*

10.2 Hyperbolic Geometry

We begin our discussion of hyperbolic geometry with the quadrilateral construction used by Girolamo Saccheri in the 18th century. Our goal will be to construct a new geometry in which Euclid's parallel postulate does not hold and in which the angle sum of a triangle is less than 180°. We allow ourselves the use of Euclid's first four postulates and all his assumptions that are not equivalent to the parallel postulate, including the assumption that lines are infinite. Since the first 28 propositions of *The Elements* do not require the parallel postulate, these results will be valid in this new geometry. Any time one of these proposition is invoked in our discussion, it will be referenced by number.

Consider the quadrilateral $ABCD$ shown in Figure 10.5, which was used by Saccheri in his attempt to prove the parallel postulate.

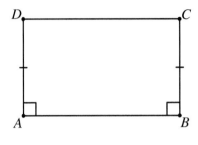

Figure 10.5

$\overline{AB}$ is referred to as the *lower base*, $\overline{CD}$ as the *upper base*, and $\angle C$ and $\angle D$ as the upper base angles or *summit angles*. The *arms* $\overline{AD}$ and $\overline{BC}$ are equal in length and are perpendicular to the lower base $\overline{AB}$. Henceforth, we will refer to such a quadrilateral as a Saccheri quadrilateral with the acknowledgment that Saccheri was not necessarily the first to use this quadrilateral to try to prove the parallel postulate. We next prove a few results about Saccheri quadrilaterals.

Theorem 10.1

In hyperbolic geometry, the summit angles of a Saccheri quadrilateral are congruent.

Proof. By the side-angle-side (SAS) congruency condition, triangles $\triangle ABC$ and $\triangle BAD$ are congruent.

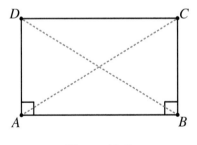

Figure 10.6

Thus, $\overline{AC} \cong \overline{BD}$. Then $\triangle ADC$ and $\triangle BCD$ are congruent by the side-side-side (SSS) congruency condition. Therefore, $\angle C \cong \angle D$. ∎

Propositions 4 and 8 of *The Elements* give the SAS and SSS congruency conditions. Thus, Theorem 10.1 is also valid in neutral geometry, since its proof does not require the parallel postulate. Furthermore, Proposition 27 states that if two lines share a perpendicular, then they are parallel. We are now ready to prove Theorem 10.2.

10.2 Hyperbolic Geometry

> **Theorem 10.2**
>
> The line joining the midpoints of the upper and lower bases of the Saccheri quadrilateral (called the altitude) is perpendicular to both. Therefore, the upper base and lower bases lie on parallel lines sharing a common perpendicular.

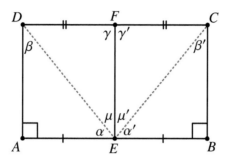

Figure 10.7

Proof. Let E and F be the midpoints of the lower base and upper base, respectively. Also, let α, α', β, β', γ, γ', μ, and μ' be the measures of the angles indicated in Figure 10.7. Note $\triangle DEA \cong \triangle CEB$ by SAS. Thus $\overline{DE} \cong \overline{CE}$, $\alpha = \alpha'$, and $\beta = \beta'$. Invoking SSS, we can conclude that $\triangle DEF \cong \triangle CEF$. Hence $\gamma = \gamma'$, and since their corresponding angles are supplementary, each must be $90°$. Also, $\mu = \mu'$, and so $\alpha + \mu = \alpha' + \mu = 90°$ since $\angle AEF$ and $\angle BEF$ are supplementary. Thus, $\overline{EF} \perp \overline{AB}$ and $\overline{EF} \perp \overline{CD}$, and it follows from Euclid's Proposition 27 that $\overleftrightarrow{AB} \parallel \overleftrightarrow{CD}$ with a common perpendicular $\overleftrightarrow{EF}$. ∎

Note that Theorem 10.2 implies that a Saccheri quadrilateral is a parallelogram with each pair of opposite sides having a common perpendicular.

It is a well-known fact that in hyperbolic geometry the angle sum of a triangle is always less than $180°$, but we are not ready to prove this yet. At this stage, we are able to prove (without any additional postulates) that the angle sum does not exceed $180°$. Later in this chapter, we will require the hyperbolic parallel postulate to show that the angle sum is strictly less than $180°$.

The proof of Theorem 10.3 (*Saccheri–Legendre theorem*) given below uses the fact that a triangle cannot have two angles whose measures add up to more than two right angles. This is proved in Proposition 17 of *The Elements*.

> **Theorem 10.3: Saccheri–Legendre Theorem**
>
> In neutral geometry, the angle sum of a triangle does not exceed $180°$ (sum $\leq 180°$).

Proof. By way of contradiction, suppose that there exists a triangle $\triangle ABC$ whose angle sum is $180° + \epsilon$, for some $\epsilon > 0$. We will construct a sequence of triangles $\triangle ABE_n$, each having the same angle sum as $\triangle ABC$, such that the measure of the smallest angle in $\triangle ABE_n$ goes to zero as n goes to infinity. This will contradict Proposition 17 of *The Elements* (which states that the sum of any two angles in a triangle is less than two right angles) (why?).

Let D_1 be the midpoint of $\overline{BC}$, and construct the line segment $\overline{AE_1}$ through D_1, such that $\overline{AD_1} \cong \overline{D_1E_1}$ (see Figure 10.8). Observe that here we have invoked the assumption that lines may be extended indefinitely.

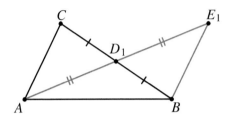

Figure 10.8

Proposition 15 of *The Elements* tells us that vertical angles are congruent, thus we have $\angle AD_1C \cong \angle E_1D_1B$. Hence by SAS, $\triangle AD_1C \cong \triangle E_1D_1B$. This means that the angle sum of $\triangle ABC$ is the same as the angle sum of $\triangle ABE_1$ (why?). Furthermore, $\angle A = \angle CAD_1 + \angle E_1AB$, so either $\angle CAD_1 \leq \frac{1}{2}\angle A$ or $\angle E_1AB \leq \frac{1}{2}\angle A$. Suppose (WLOG) that $\angle E_1AB \leq \frac{1}{2}\angle A$. [1]

Now, repeat the construction above on $\triangle ABE_1$ to produce a second triangle, WLOG $\triangle ABE_2$, having $\angle E_2AB \leq \frac{1}{4}\angle A$, and whose angle sum is equal to that of $\triangle ABC$ (see Figure 10.9). Note that if instead we have $\angle E_1AD_2 \leq \frac{1}{4}\angle A$, then the desired triangle would be $\triangle AE_2E_1$ not $\triangle ABE_2$.

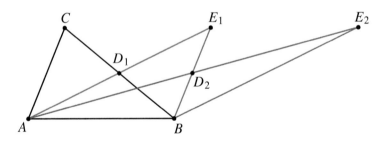

Figure 10.9

Repeating this construction n times, we arrive at a triangle $\triangle ABE_n$, such that $\triangle ABE_n$ has the same angle sum as $\triangle ABC$ and contains (WLOG) the angle $\angle E_nAB \leq \frac{1}{2^n}\angle A$. If we choose n such that $\frac{1}{2^n}\angle A < \epsilon$, then we have $\angle E_nAB \leq \frac{1}{2^n}\angle A < \epsilon$. But since the angle sum of $\triangle ABE_n$ equals $180° + \epsilon$, we conclude that $180° + \epsilon = \angle E_nAB + \angle ABE_n + \angle BE_nA < \epsilon + \angle ABE_n + \angle BE_nA$, and so $180° < \angle ABE_n + \angle BE_nA$. Thus, $\triangle E_nAB$ is a triangle with

---

[1] For sake of brevity, in the proof of Theorem 10.3 we used the $\angle A$ to denote both the angle itself and its measure. It should be clear from the context what is being referred to.

two angles summing to more than two right angles, which contradicts Proposition 17 of *The Elements*.

> **Historical Note:** *Adrien-Marie Legendre (1752–1833)*
> Legendre was a French mathematician who in 1794 published *Eléments de géométrie* in which he rearranged and simplified many of the propositions from Euclid's *Elements* to create a more practical and effective geometry textbook. The *Eléments* was the leading elementary geometry text for nearly a century, first in Europe and then in the United States. Legendre's attempt to prove the parallel postulate extended over 30 years. Using Euclid's first four postulates, he succeeded in proving that the sum cannot be greater than two right angles (using the assumptions that straight lines are infinite). But in his attempts to prove that the angle sum cannot be less than two right angles, he unknowingly used results that turned out to be equivalent to Euclid's fifth postulate. In 1832, when Bolyai published his work on non-Euclidean geometry, Legendre, confirming his absolute belief in the dominance of Euclidean geometry, wrote: "It is nevertheless certain that the theorem on the sum of the three angles of the triangle should be considered one of those fundamental truths that are impossible to contest and that are an enduring example of mathematical certitude." [1]

Notice that Theorem 10.3 implies that the summit angles of the Saccheri quadrilateral are not obtuse (why?). Having investigated Saccheri quadrilaterals, one might ask how such quadrilaterals would behave if the condition that the arms are equal in length is no longer required. In Theorem 10.4 we tackle this question using the same terminology developed earlier. This theorem will turn out to be a useful tool in some of the proofs that follow.

> **Theorem 10.4**
>
> *Consider a quadrilateral with a lower base that makes right angles with its two arms.*
> 1. *If the summit angles are unequal, so are the arms.*
> 2. *If the arms are unequal, so are the summit angles, with the greater summit angle being opposite the greater arm.*

Proof. In order to prove (1) we simply need to note that it is the contrapositive of Theorem 10.1 (why?).

To prove (2), suppose (WLOG) that $BC > AD$ in the quadrilateral $ABCD$, as shown in Figure 10.10.

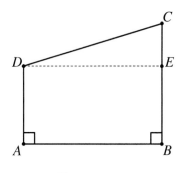

Figure 10.10

Let E be the point on the segment $\overline{BC}$ such that $\overline{AD} \cong \overline{BE}$. Then $ABED$ is a Saccheri quadrilateral, so $\angle ADE \cong \angle BED$ by Theorem 10.1. Proposition 16 of *The Elements* gives us the Exterior Angle Theorem, which states that, for any triangle, an exterior angle is greater than either of the remote interior angles. So we have $\angle BED = \angle ADE > \angle ECD$, and therefore $\angle ECD < \angle ADE + \angle EDC = \angle ADC$. Hence, the summit angles of $ABCD$ are unequal with the greater angle being opposite the greater arm ($\angle ADC$ is opposite $\overline{BC}$ in Figure 10.10).

Theorems 10.1 through 10.4 are part of neutral geometry, because they make no assumptions about parallel lines. This means that they hold in both Euclidean and hyperbolic geometry. In order to examine some results that hold in hyperbolic geometry but not Euclidean geometry we must first state a replacement for the parallel postulate.

Axiom 10.1: Hyperbolic Parallel Postulate

The summit angles of a Saccheri quadrilateral are acute.

Recall that given Euclid's first four postulates, the summit angles of the Saccheri quadrilateral are not obtuse. Therefore, two possibilities remain: either the angles are right angles, or they are acute. The parallel postulate is equivalent to the angles being right angles (why?). Therefore, the hyperbolic parallel postulate is the negation of the parallel postulate (in hyperbolic geometry). As such, the negation of any statement equivalent to the parallel postulate will belong to hyperbolic geometry. This, for example, gives us the following:

1. *There exist parallel lines which are not equidistant from one another.*
   To show that (1) holds in hyperbolic geometry, consider (1) and its negation which is "*any two parallel lines are everywhere equidistant.*" Either (1) or its negation holds in hyperbolic geometry. By way of contradiction, assume the negation of (1) holds in hyperbolic geometry. Let l and m be parallel lines, D and C be arbitrary points on l, and A and B be their respective perpendicular projections on m, as in Figure 10.11. Since l and m are everywhere equidistant, $AD = BC$ and hence $ABCD$ is a Saccheri quadrilateral. Theorem 10.2 implies that $\overline{EF}$, the line segment joining the midpoints of $\overline{AB}$ and $\overline{DC}$, is perpendicular to both. Hence $AD = EF = BC$ and $AEFD$ and $EBCF$

are Saccheri quadrilaterals. But the summit angles of a Saccheri quadrilateral are equal (Theorem 10.1), hence $m(\angle 1) = m(\angle 4)$, $m(\angle 2) = m(\angle 3)$, and $m(\angle 1) = m(\angle 2)$. This implies that $\angle 1$ and $\angle 2$, the summit angles of the Saccheri quadrilateral $ABCD$, are right angles (why?), which contradicts the hyperbolic parallel postulate. Hence (1) holds in hyperbolic geometry.

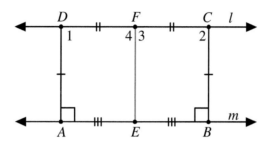

Figure 10.11

2. *There exist a line and a point not on the line through which we have more than one parallel to the line.*

   To show that (2) holds in hyperbolic geometry, consider (2) and its negation which is "*given a line and a point not on it, exactly one line parallel to the given line goes through this point*" (recall that Propositions 23 and 27 guarantee the existence of at least one parallel). Either (2) or its negation holds in hyperbolic geometry. By way of contradiction, assume the negation of (2) holds in hyperbolic geometry. Let l and m be two lines with a transversal $\overleftrightarrow{PQ}$ such that the interior angles on one side of the transversal satisfy $\alpha + \beta < 180°$, as shown in Figure 10.12. We will show that l must intersect m on the right, which proves that Playfair's postulate [the negation of (2)] implies Euclid's fifth postulate. Using Proposition 23, we can construct a line k through P such that $\alpha = \gamma$. Proposition 27 implies that k is parallel to m. Now, $\gamma + \beta = \alpha + \beta < 180°$, hence l and k are distinct lines through P. Therefore, by the negation of (2), l must intersect m. Furthermore, since l is below k on the right, it must intersects it there. Otherwise, l (going through P) would have to intersect m on the left, which would mean that l would have to intersect k in a second point P' to the left of P. This contradicts Euclid's Postulate 1, which states that exactly one line goes through two distinct points. Hence Euclid's fifth postulate holds in hyperbolic geometry, which implies that the summit angles of a Saccheri quadrilateral are right angles (why?), and this contradicts the hyperbolic parallel postulate. Therefore, statement (2) holds in hyperbolic geometry.

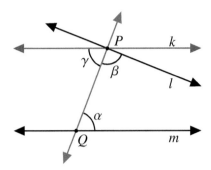

Figure 10.12

3. *Similar triangles are always congruent.*

Showing that (3) holds in hyperbolic geometry is the subject of Theorem 10.13.

Now we will examine some results unique to hyperbolic geometry. Most of our figures henceforth will be drawn in such a way as to approximate the nature of hyperbolic lines in models of the hyperbolic plane discussed in Section 10.3.

Theorem 10.5

In hyperbolic geometry, a Saccheri quadrilateral satisfies:

1. *The altitude is shorter than the arms.*

2. *The upper base is longer than the lower base.*

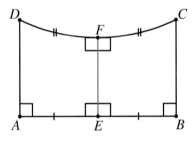

Figure 10.13

Proof. On the Saccheri quadrilateral $ABCD$ shown in Figure 10.13, let E and F be the midpoints of the lower base and upper base respectively. Theorem 10.2 implies that $\overline{EF}$ is perpendicular to both the upper and lower bases of $ABCD$. By the Hyperbolic Parallel Postulate, $\angle C$ and $\angle D$ are acute and by Theorem 10.4, $AD > EF$ in $AEFD$ and $CB > EF$ in $EBCF$. This completes the proof of (1).

To prove (2), consider $EFDA$ as having lower base $\overline{EF}$ and arms $\overline{AE}$ and $\overline{DF}$. Theorem 10.4 implies that $DF > AE$ and similarly we can conclude that $FC > EB$. Therefore, $DF + FC > AE + EB$ or $DC > AB$. Thus the upper base is longer than the lower base. ■

10.2 Hyperbolic Geometry

Consider the Saccheri quadrilateral shown in Figure 10.13. In Theorem 10.2 we proved that lines $\overleftrightarrow{DC}$ and $\overleftrightarrow{AB}$ are parallel with a common perpendicular $\overline{EF}$. Theorem 10.5 gave us that $AD > EF$ and $BC > EF$. Thus the parallel lines $\overleftrightarrow{DC}$ and $\overleftrightarrow{AB}$ are not equidistant. Note that each of the two quadrilaterals $AEFD$ and $EBCF$ has three right angles. A quadrilateral with three right angles is commonly known as a *Lambert quadrilateral*, named after Johann Lambert (1728–1777), who used its construction in an attempt to prove Euclid's fifth postulate, much like Saccheri had done. However, it's worth pointing out here that a quadrilateral with three right angles was first used by the medieval Islamic mathematician Ibn al-Haytham (956–1039) for the same purpose.

> **Theorem 10.6**
>
> *In hyperbolic geometry, the fourth angle of a Lambert quadrilateral is acute, and each side adjacent to the acute angle is longer than its opposite side.*

The proof of this theorem is the subject of Problem 4 at the end of this section.

We now turn our attention to the subject of common perpendiculars between parallel lines. Remember that in Euclidean geometry any two parallel lines always have a common perpendicular. In fact, they have infinitely many common perpendiculars. This often conjures the image of a pair of railroad tracks, with their ties, equal in length and perpendicular to both tracks, representing the common perpendiculars. However, in hyperbolic geometry the situation is very different, as any two parallel lines either have no common perpendiculars or exactly one.

> **Theorem 10.7: Uniqueness of a common perpendicular to parallel lines**
>
> *In hyperbolic geometry, if two parallel lines have a common perpendicular, then they cannot have a second common perpendicular.*

Proof. By way of contradiction, assume l and m are parallel lines with two common perpendiculars p_1 and p_2, as in Figure 10.14.

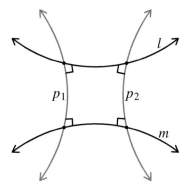

Figure 10.14

Then l, m, p_1, and p_2 define a Lambert quadrilateral with four right angles, which contradicts Theorem 10.6. ∎

The following three theorems give us an idea of what certain parallel lines "look" like in hyperbolic geometry. They will also be useful in proving that the angle sum of a hyperbolic triangle is always less than 180°.

> **Theorem 10.8**
>
> *Given two lines, if there exists a transversal which cuts the lines so as to form equal alternate interior angles (or equal corresponding angles), then the lines are parallel with a common perpendicular.*

Proof. It is sufficient to prove the case when the transversal makes equal alternate interior angles (why?). Let l and m be two lines cut by a transversal $\overleftrightarrow{AB}$, where A is a point on m and B is a point on l, such that $\overleftrightarrow{AB}$ makes equal alternate interior angles with respect to l and m (see Figure 10.15). Proposition 27 of *The Elements* tells us that $l \parallel m$. It remains to show that l and m have a common perpendicular. To do this, let P be the midpoint of $\overline{AB}$ and construct the line perpendicular to l which passes through P. Let C be the point where this perpendicular meets l, hence $\overleftrightarrow{CP} \perp l$. Now construct the line perpendicular to m and passing through point P. Let D be the point where this perpendicular meets m, hence $\overleftrightarrow{DP} \perp m$.

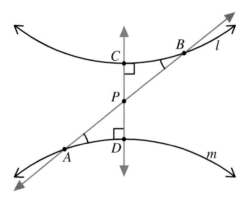

Figure 10.15

We will show that $\overleftrightarrow{CP}$ is the same line as $\overleftrightarrow{DP}$, which, based on Euclid's Proposition 14, is true if the sum of the adjacent angles $\angle CPB$ and $\angle BPD$ equals two right angles. Now, $\angle APD$ and $\angle DPB$ are supplementary, so by Proposition 13 their sum is two right angles. Thus, if we show that $\angle BPC \cong \angle APD$, we will also have that the sum of $\angle CPB$ and $\angle BPD$ is two right angles. By the hypothesis of the theorem, we have $\angle CBP \cong \angle DAP$, and by the construction of the perpendiculars to l and m through P, we know $\angle BCP$ and $\angle ADP$ are both right angles. Furthermore, since P is the midpoint of $\overline{AB}$, we have $AP = PB$.

10.2 Hyperbolic Geometry

Thus, by Proposition 26 (AAS), $\triangle BCP \cong \triangle ADP$, and hence $\angle APD \cong \angle BPC$. Therefore, $\overleftrightarrow{CP}, \overleftrightarrow{DP}, \overleftrightarrow{CD}$ all represent the same line, and as a result $\overleftrightarrow{CD}$ is a common perpendicular to the parallel lines l and m. ∎

> **Corollary 10.1**
>
> *If two lines are perpendicular to the same line, then they are parallel.*

Corollary 10.1 follows directly from Theorem 10.8 and shows that this intuitive property of Euclidean geometry also holds in hyperbolic geometry.

We already mentioned that Euclid often relied on unstated assumptions to prove his propositions in *The Elements*. One such assumption was the infinite extent of a straight line. Another was what is now called the *Plane Separation Axiom*. This axiom states that *any straight line in a plane splits the plane into two disjoint (nonempty) sets, and if any two points lie in separate halves of the split plane, they determine a line which intersects the original straight line*. This ensures that if a line has two points on different sides of another line, the lines intersect somewhere. The *plane separation axiom* is required for the proof of the following theorem, which is the subject of Problem 7 at the end of this section.

> **Theorem 10.9**
>
> *If two lines have a common perpendicular, then there exist transversals, other than the perpendicular, which cut the lines so as to form equal alternate interior angles (or equal corresponding angles). Moreover, the only transversals with this property are those which go through the midpoint of the common perpendicular.*

If in hyperbolic geometry parallel lines are not everywhere equidistant (see Theorem 10.14), how is the distance between two parallel lines defined? The answer to this question is the subject of Theorem 10.10.

> **Theorem 10.10**
>
> *The distance between two parallels with a common perpendicular is least when measured along that perpendicular. The distance from a point on either parallel to the other increases as the point recedes from the perpendicular in either direction.*

Proof. Let l and m be parallel lines with a common perpendicular $\overleftrightarrow{AB}$, which intersects m at A and l at B. Let C be any point on l other than B and construct a perpendicular to m through C. Let D be the point of intersection of this perpendicular with m (or equivalently, the perpendicular projection of C on m).

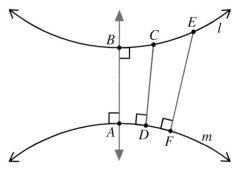

Figure 10.16

Then $ABCD$ is a Lambert quadrilateral, so $\angle BCD$ is acute and $CD > BA$ by Theorem 10.6. We now show that the distance measured from a point on l to the line m continues to increase as the point recedes farther and farther from B. We do this by choosing a point E on l such that C is between E and B and showing that the distance from E to m is larger than CD. As before, construct the perpendicular to m through point E and assume it intersects m at F. Since $\angle BCD$ is acute, $\angle DCE$ is obtuse. Also, $AFEB$ is a Lambert quadrilateral, so $\angle CEF$ is acute. Thus $\angle DCE > \angle CEF$. Invoking Theorem 10.4 allows us to conclude that $EF > DC$, and it follows that the distance between the lines continues to increase as we recede from the common perpendicular. ∎

Playfair's postulate is often used as an equivalent to the parallel postulate. It states that *given a line and a point not on the line, there exists a unique parallel to the given line through the given point*. The negation of Playfair's postulate says that there might be no parallels (as is the case in spherical geometry) or more than one parallel (as is the case in hyperbolic geometry) through a point not on a given line. In fact, in hyperbolic geometry we can prove a stronger statement implying that the number of parallels is infinite, as stated in Theorem 10.11.

Theorem 10.11

In hyperbolic geometry, given a line and a point not on the line there exist infinitely many parallels to the given line through the given point.

Proof. Let l be a line and E a point not on l. Let F be the perpendicular projection of E on l, hence $\overleftrightarrow{EF} \perp l$. Proposition 11 of *The Elements* allows us to construct a line k through E which is perpendicular to $\overleftrightarrow{EF}$. Then k is a line parallel to l and sharing the common perpendicular $\overleftrightarrow{EF}$. Let E' be any point on k (WLOG) to the right of E. Let F' be the perpendicular projection of E' on l. Since Theorem 10.2 implies that $E'F' > EF$, we can let P' be the point on $\overline{E'F'}$ such that $\overline{P'F'} \cong \overline{EF}$.

10.2 Hyperbolic Geometry

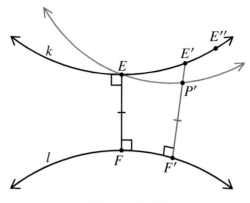

Figure 10.17

Then $FF'P'E$ is a Saccheri quadrilateral, and hence by Theorem 10.2 we know that the upper base $\overline{EP'}$ is on a line parallel to the lower base $\overline{FF'}$ and sharing a common perpendicular with it (where is this common perpendicular?). Therefore, k and $\overleftrightarrow{EP'}$ are two lines parallel to l and passing through E. Note that the line $\overleftrightarrow{EP'}$ was determined by our choice of E' to the right of E on k. Since E' was an arbitrary point, we would like to conclude that if we choose a different point on k, E'' for example, then this will give us another parallel to l through E distinct from both k and $\overleftrightarrow{EP'}$. This would imply that there exist an infinite number of lines through E that are parallel to l. The rest of the proof (proving that these lines are distinct) is left as an exercise in the problem set at the end of this section (see Problem 8). ■

So far, we have only discussed parallel lines that have a common perpendicular. Recall that two lines that are parallel with a common perpendicular are said to be hyperparallel. There are, however, parallel lines in hyperbolic geometry that are not hyperparallel, namely, horoparallel lines, which are asymptotic in one direction. It can be proved that for any line l and any point P not on l, there are exactly two lines through P horoparallel to l. The discussion below will help shed some light on this claim. Figure 10.18 shows the two lines k_1 and k_2 passing through P and horoparallel to l.

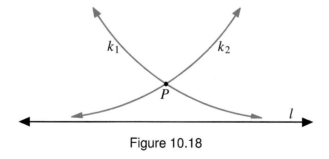

Figure 10.18

Informally, let l be a line, P be a point not on l, A be the perpendicular projection of P onto l, and B be a point on l to the right of A, as in Figure 10.19. For the sake of clarity in this informal discussion, rays and lines are depicted as Euclidean rays and lines (for a more formal discussion, see [82, Chapter 7]). For each $0 \le r \le 90$, let $\overrightarrow{PQ_r}$ be the ray emanating from P such that $m(\angle APQ_r) = r°$. As r increases from 0 to 90, the rays fan out from $\overrightarrow{PA}$ to $\overrightarrow{PQ_{90}}$. Some of these rays such as $\overrightarrow{PB}$ intersect l, while others such as $\overrightarrow{PQ_{90}}$ do not (why?). There must be a "first" ray $\overrightarrow{PQ_{r_0}}$, where $r_0 < 90$, that does not intersect l. Otherwise, if $\overrightarrow{PQ_{90}}$ is the "first" ray that does not intersect l, then $\overleftrightarrow{PQ_{90}}$ is the only line through P that is parallel to l, which contradicts Theorem 10.11. The $\angle APQ_{r_0}$ is called the *angle of parallelism* of P and l, and r_0 is called the *critical value* for P. Note that since there are two sides of $\overleftrightarrow{AP}$, we can talk about two angles of parallelism for P and $\overleftrightarrow{AB}$: A left angle determined by the "first" left parallel k_2 and a right angle determined by the "first" right parallel k_1, as in Figure 10.19. Fortunately, we can prove that these two angles are congruent; see Problem 18 at the end of this section.

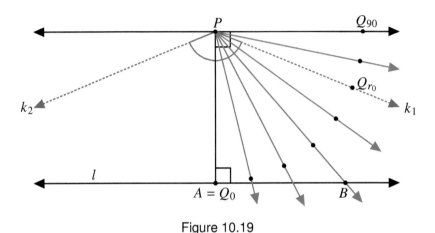

Figure 10.19

> **Definition: Angle of Parallelism**
>
> Given a hyperbolic line l and a point P not on l, let Q be the perpendicular projection of P on l, and consider hyperbolic rays emanating from P. Some of these rays intersect l, while others do not. Let θ be the angle with the smallest positive measure such that a ray k emanating from P and making an angle θ with $\overrightarrow{PQ}$ does not intersect l. This angle θ is the angle of parallelism of P and l.

In Euclidean geometry the angle of parallelism is uninterestingly always a right angle, whereas in hyperbolic geometry the angle of parallelism is an acute angle (see Problem 19 at the end of this section) and every acute angle is the angle of parallelism for some line l and point P not on l. The angle of parallelism is a fundamental concept in hyperbolic geometry as it relates length to angles. In fact, we will prove in Theorem 10.16 in the next section that the angle of parallelism for a line l and a point P not on l depends only on the length of the perpendicular segment from P to l.

10.2 Hyperbolic Geometry

We now describe a simple straightedge and compass construction discovered by J. Bolyai for the ray determining the angle of parallelism for a point P and a line l [35, page 198]. Let Q be the perpendicular projection of P on l, and let m be the line through P perpendicular to $\overleftrightarrow{PQ}$. Furthermore, let R be a point on l distinct from Q, and let S be the perpendicular projection of R onto m, as in Figure 10.20.

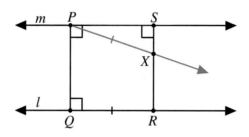

Figure 10.20

Note that $QRSP$ is a Lambert quadrilateral, hence, by Proposition 19 (*if one angle in a triangle is greater than another, then the side opposite the greater angle is greater than the side opposite the smaller angle*) and Theorem 10.6, we have $PR > QR$ and $PS < QR$. Now we construct a circle C (not shown in the figure) with center P and radius QR. By the *Elementary Continuity Principle* (*if one endpoint of a segment is inside a circle and the other is outside, then the segment intersects the circle*), since S lies inside C and R lies outside C, the segment $\overline{RS}$ intersects C in exactly one point X lying between S and R, as shown in Figure 10.20. It can be shown that $\overrightarrow{PX}$ is the ray determining the angle of parallelism of P and l (see [35]). Figure 10.21 shows Bolyai's construction carried out in the Poincaré disc, one of the hyperbolic models discussed later in this chapter.

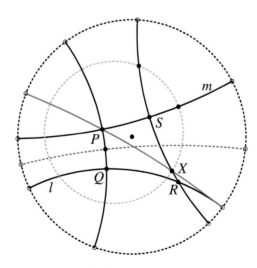

Figure 10.21

An important, well-known result from hyperbolic geometry is that the angle sum of a triangle is always less than 180°. We now present a proof of this result, and then state some other interesting theorems dealing with hyperbolic triangles.

> **Theorem 10.12**
>
> *In hyperbolic geometry, the angle sum of a triangle is always less than 180°.*

Proof. We first show that the theorem holds for right triangles and then prove the general case. Let $\triangle ABC$ be a right triangle with its right angle at A, as in Figure 10.22. Using Proposition 23 of *The Elements*, construct $\overleftrightarrow{CD}$ through C such that $\angle ABC \cong \angle BCD$.

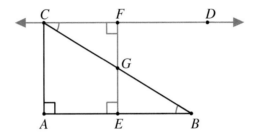

Figure 10.22

Hence, $\overleftrightarrow{BC}$ is a transversal making equal alternate interior angles with $\overleftrightarrow{CD}$ and $\overleftrightarrow{AB}$. Then by Theorem 10.8, $\overleftrightarrow{CD} \parallel \overleftrightarrow{AB}$ with a common perpendicular. Let E and F be the intersection points of this common perpendicular with $\overleftrightarrow{AB}$ and $\overleftrightarrow{CD}$, respectively. Let G be the midpoint of $\overline{EF}$. Invoking Theorem 10.9 allows us to conclude that $\overleftrightarrow{CB}$ passes through G, and $\overline{CG} \cong \overline{GB}$ (why?). Notice that $AEFC$ is a Lambert quadrilateral, so $\angle ACF$ must be acute. Therefore, $m(\angle ACB) + m(\angle CBA) = m(\angle ACB) + m(\angle BCF) = m(\angle ACF) < 90°$, and this implies that the angle sum of $\triangle ABC$ is less than 180°. Hence, for any right triangle in hyperbolic geometry the angle sum is always less than 180°.

Now consider any nonright triangle $\triangle PQR$. Since Theorem 10.3 implies that the angle sum of a triangle is at most 180°, $\triangle PQR$ can have at most one obtuse angle, that is, it must have at least two acute angles; assume (WLOG) these are the angles at P and Q.

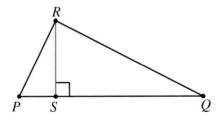

Figure 10.23

Let S be the perpendicular projection of R onto $\overleftrightarrow{PQ}$. Then S lies between P and Q (why?), and hence $\overline{RS}$ divides $\triangle PQR$ into two right triangles. Having proved that each right triangle has angle sum less than 180°, and knowing that $\angle PSR$ and $\angle QSR$ form a straight angle, we can conclude that the angle sum of $\triangle PQR$ is the sum of the angle sums of $\triangle PSR$ and $\triangle SQR$ minus 180°. Hence, the angle sum of $\triangle PQR$ is less than 360° − 180° = 180°. This completes the proof of Theorem 10.12. ∎

> **Theorem 10.13: AAA Congruence Condition**
>
> *In hyperbolic geometry, if two triangles $\triangle ABC$ and $\triangle A'B'C'$ satisfy $\angle A \cong \angle A'$, $\angle B \cong \angle B'$, and $\angle C \cong \angle C'$, then the triangles are congruent.*

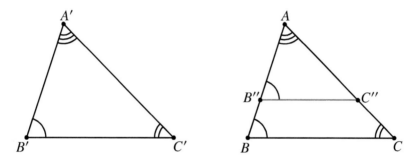

Figure 10.24

Proof. Let $\triangle ABC$ and $\triangle A'B'C'$ be triangles such that $\angle A \cong \angle A'$, $\angle B \cong \angle B'$, and $\angle C \cong \angle C'$, as in Figure 10.24. By way of contradiction, suppose that the triangles are not congruent. Then there is a side of one triangle that is longer than the corresponding side of the other triangle (otherwise the triangles would be congruent by ASA). Suppose (WLOG) that $AB > A'B'$, and construct B'' on $\overline{AB}$ such that $\overline{AB''} \cong \overline{A'B'}$. Using Proposition 23, construct $\angle AB''C'' \cong \angle A'B'C'$. Proposition 28 implies that $\overline{B''C''}$ is parallel to $\overline{BC}$, and this together with Pasch's axiom, or equivalently, the plane separation axiom [35] allow us to conclude that C'' is between A and C. Therefore, the quadrilateral $BCC''B''$ is convex, and by ASA, $\triangle AB''C'' \cong \triangle A'B'C'$. Hence $\angle AB''C'' \cong \angle B$ and $\angle AC''B'' \cong \angle C$. This leads to the conclusion that $m(\angle BB''C'') = 180° - m(\angle B)$ and $m(\angle CC''B'') = 180° - m(\angle C)$. Hence the angle sum of the quadrilateral $BCC''B''$ is 360°, which means that either $\triangle BCC''$ or $\triangle BC''B''$ has an angle sum greater or equal to 180° (why?), and this clearly contradicts Theorem 10.12. ∎

Note that Theorem 10.13 implies that the length of a segment in hyperbolic geometry can be related to the measure of an angle. More precisely, given a segment, the angle of the equilateral triangle constructed on this segment uniquely determines the length of the segment. We will show in Chapter 11 that Theorem 10.13 also holds in elliptic geometry, so here again the length of a segment can be related to the measure of an angle. Compare this to Euclidean geometry, where the side length of an equilateral triangle is clearly not uniquely determined by the angle of the triangle, which is 60°.

Having already proved that the angle sum of a hyperbolic triangle is always less than 180° (see Theorem 10.12), we next show that this angle sum can vary. In fact, for any positive real number $x < 180$, it can be shown that there exists a hyperbolic triangle with angle sum x degrees. In Problem 12 at the end of this section, the reader will demonstrate that there are triangles in hyperbolic geometry with angle sums arbitrarily close to 180°.

> **Corollary 10.2**
>
> *In hyperbolic geometry, not all triangles have the same angle sum.*

> **Now Prove This 10.1**
>
> Prove Corollary 10.2 by assuming that all hyperbolic triangles have the same angle sum and arriving at a contradiction to Theorem 10.12. *Hint: Let D be an arbitrary point on $\overline{AB}$ (other than A and B), and conclude that the angle sum of $\triangle ABC$ in Figure 10.25 is 180°.*
>
>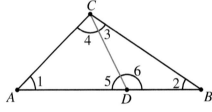
>
> Figure 10.25

10.2.1 Problem Set

1. List the axioms and assumptions we used in our development of hyperbolic geometry.

2. Find the flaw in the proof of the following:

    Claim: There is a triangle in hyperbolic geometry whose angle sum is 180°.

    Proof. Let l be a line, and let m be a parallel to l through point P not on l. Let Q and R be distinct points on l. Then lines $\overleftrightarrow{QP}$ and $\overleftrightarrow{RP}$ are transversals that cut the parallel lines l and m.

    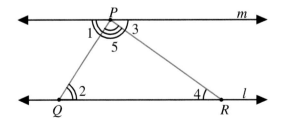

10.2 Hyperbolic Geometry

Next, referring to the preceding figure, we have $\angle 1 = \angle 2$ and $\angle 3 = \angle 4$. Thus, the angle sum of $\triangle QRP$ is given by $\angle 2 + \angle 5 + \angle 4 = \angle 1 + \angle 5 + \angle 3 = 180°$.

3. Complete the following sentences:
   a. The angle sum of a quadrilateral in neutral geometry is _____ 360°.
   b. The angle sum of a quadrilateral in hyperbolic geometry is _____ 360°.
   c. Write a proof for your statement in part (b).

4. Write a proof for Theorem 10.6.

5. Do squares or rectangles exist in hyperbolic geometry? Justify your answer.

6. Show that in hyperbolic geometry there exist rhombuses with equal angles. Do rhombuses with equal angles exist in Euclidean geometry other than the square?

7. Proof of Theorem 10.9: Let l and m be lines with a common perpendicular $\overline{AB}$, where A is on l and B is on m. Theorem 10.8 tells us that $l \parallel m$.
   a. Show that there exists a transversal (other than $\overline{AB}$) cutting l and m that makes equal alternate interior angles (equivalently, equal corresponding angles). *Hint: If P is the midpoint of $\overline{AB}$ construct a transversal through P.*
   b. Where, if anywhere, did you use the plane separation axiom in part (a)?
   c. Show that if there exists a transversal of l and m which makes equal alternate interior angles, and which does not intersect $\overline{AB}$ at the point midway between l and m, then one can construct a second common perpendicular to l and m. *Hint: You may need to consider two cases.*

8. Completing the proof of Theorem 10.11. Using the construction from the proof of Theorem 10.11 as depicted in the figure below, choose a point E'' to the right of E' on k. Then let F'' and P'' be constructed in the same way as F' and P' in the proof of Theorem 10.11. Next, show that the lines $\overleftrightarrow{EP'}$ and $\overleftrightarrow{EP''}$ are not the same line. Explain why this means there are infinitely many lines parallel to a given line through a point not on it.

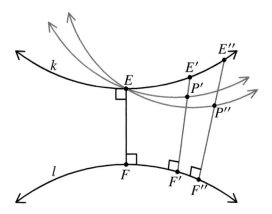

9. Prove that an angle inscribed in a semicircle in hyperbolic geometry is always acute.

10. Exhibit a pair of parallel lines and a transversal so that on one side of the transversal, the sum of the interior angles is less than 180°.

11. Show that no two hyperbolic lines are equidistant. *Hint: Show that the distance from one line to another cannot be the same at more than two points.*

12. Consider a triangle $\triangle ABC$, and let D be a point between A and B. Extend $\overline{AB}$ slightly to create the exterior angle $\angle EAC$, as in the figure below.
    a. Describe what happens to the angles $\angle EDC$ and $\angle ACD$ as D approaches A along segment $\overline{AB}$.
    b. What happens to the angle sum of $\triangle ADC$?
    c. How might you use this construction to create a triangle with angle measure arbitrarily close to 180°?

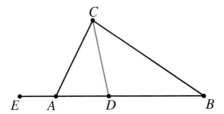

13. The *defect* of a hyperbolic triangle is defined to be the amount by which the angle sum of the triangle differs from 180°. What do you think should be the definition of the defect of an *n*-sided polygon? Give an upper bound for the defect of an *n*-sided hyperbolic polygon.

14. Given $\triangle ABC$ and D on $\overline{BC}$, if the defect of $\triangle ABD$ is δ_1 and the defect of $\triangle ADC$ is δ_2, prove that defect of $\triangle ABC$ is $\delta_1 + \delta_2$.

15. It can be shown that the area of a hyperbolic triangle T_d with defect d is given by the formula $A(T_d) = k \cdot d$ for some constant k which relates the units of length to the units of area. This means that the area of a triangle is determined by its defect.
    a. Give an upper bound for the area of a triangle in hyperbolic geometry. This means that no matter how "big" we make our triangles, the area is always less than this upper bound. Thus, unlike in Euclidean geometry, we cannot construct triangles of arbitrarily large area. Shown below are *ideal* hyperbolic triangles each having the largest possible area.

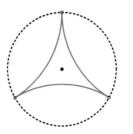 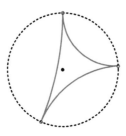 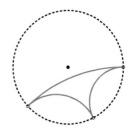

Ideal triangles are triangles whose vertices are ideal points, that is, points lying on the boundary of the Poincaré disc which will be introduced in the next section.

  b. Give a formula for the area of an *n*-sided hyperbolic polygon.
  c. Is there an upper-bound for the area of an *n*-sided hyperbolic polygon?
  d. Can we construct hyperbolic polygons of arbitrarily large area? Why or why not?

16. In hyperbolic geometry, prove that if $ABCD$ and $A'B'C'D'$ are two Saccheri quadrilaterals having the same defect and $\overline{CD} \cong \overline{C'D'}$, then the quadrilaterals are congruent. *Hint:* Let A'' and B'' be on $\overrightarrow{DA}$ and $\overrightarrow{CB}$, respectively, such that $\overline{DA''} \cong \overline{D'A'}$, and $\overline{CB''} \cong \overline{C'B'}$ and prove that $A'' = A$ and $B'' = B$.

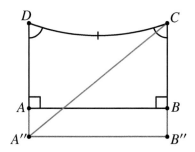
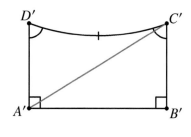

17. Let $\overleftrightarrow{AB}$ be a hyperbolic line and P a point not on it. Let $\overleftrightarrow{PM}$ be a line horoparallel to $\overleftrightarrow{AB}$ as in the figure below. Let $\overline{CP}$ be the segment perpendicular to $\overline{AB}$ through P, and let D be any point on $\overline{AB}$ between C and B. What happens to $\angle CPD$ as D moves to the right along $\overleftrightarrow{AB}$? What angle does it approach? This angle is called the *angle of parallelism* for segment $\overline{CP}$. What range of values can the measure of the angle of parallelism take?

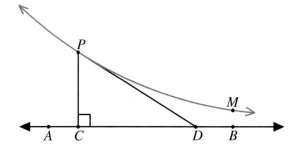

18. Given a hyperbolic line l and a point P not on l, prove that the left and right angles of parallelism for P and l are congruent. *Hint:* WLOG, show that if the right angle is greater than the left angle, then the "first" left parallel intersects l, which is absurd.

19. Given a hyperbolic line l and a point P not on l, prove that the angle of parallelism for P and l is acute.

20. Prove that if two Saccheri quadrilaterals have congruent lower bases and summit angles then they must be congruent. Does the same hold if the quadrilaterals have congruent upper bases and summit angles? What if all you know is the quadrilaterals have congruent arms?

21. Prove that if there exists a triangle whose defect is 0 then:
   a. There is a right triangle of defect 0.
   b. There is an arbitrarily large rectangle of defect 0.
   c. Any right triangle has defect 0.
   d. Any triangle has defect 0.

10.3 Models of Hyperbolic Geometry

There are several models of hyperbolic geometry including the Poincaré half-plane and disc models and the Klein disc model. We will briefly discuss these three models with the disclaimer that a more thorough understanding of them requires a more rigorous study than is presented here. We will focus our attention on the Poincaré disc model in many of the investigations presented in this and in subsequent sections. Various geometric constructions in this model will be emphasized given the availability of dynamic geometry software that can be used for this purpose. This should lead to a better understanding of the material being studied and a deeper appreciation for it.

> **Historical Note:** *Beltrami–Klein–Poincaré*
> In two articles published in 1868 and 1869, Eugenio Beltrami (1835–1900), at the time a professor at the University of Bologna, produced various models of the non-Euclidean geometry developed by Gauss, Lobachevsky and Bolyai. The first model is often referred to as the Klein model, and the other two are often credited to Poincaré. It would be appropriate and more accurate to give credit to Beltrami when referring to these models as he proposed these before Felix Klein (1849–1925) and Henri Poincaré (1854–1912), although it should be mentioned that Klein made the connections between the disc model and projective geometry more explicit, and Poincaré was a pioneer in introducing the use of hyperbolic geometry in areas far from geometry such as complex analysis, mechanics, etc. For the sake of brevity, we will refer to these models as the Klein and Poincaré models with the understanding that Beltrami proposed these models before Klein and Poincaré. [2]

The Poincaré Half-Plane Model $\mathbb{H}^2$

The Poincaré half-plane model, denoted by $\mathbb{H}^2$, is the upper half of the Euclidean plane (all points (x,y) such that $y > 0$) together with a hyperbolic metric, that is, a way for measuring

10.3 Models of Hyperbolic Geometry

distance in hyperbolic geometry. Lines in this model are represented by the arcs of circles in the upper half-plane whose centers lie on the x-axis (hence these arcs are semicircles missing the end points), and by straight Euclidean open rays (missing their end points) which are perpendicular to the x-axis. One might think of these straight Euclidean open rays as being arcs of circles of infinite radii centered on the x-axis (hence they are perpendicular to the x-axis at a finite point and at a point infinitely far from this finite point).

Given two points P and Q in $\mathbb{H}^2$, the unique hyperbolic line passing through them is determined as follows:

1. If the x-coordinates of P and Q are equal, then the hyperbolic line is the open ray through the points that is perpendicular to the x-axis.

2. If the x-coordinates of P and Q are not equal, the hyperbolic line is the semicircle centered at the point of intersection of the perpendicular bisector of $\overline{PQ}$ and the x-axis (again, the endpoints of the semicircle do not belong to the line).

Lines in $\mathbb{H}^2$ that appear to intersect on the x-axis represent horoparallel lines (the x-axis is not part of the Poincaré half-plane model), while lines which do not intersect in the upper half-plane represent hyperparallel lines. Lines in $\mathbb{H}^2$ which intersect at right angles represent perpendicular lines. As in Euclidean geometry, the angle between two intersecting arcs is determined by the measure of the angle between the tangent rays to the arcs at the point of intersection. Figure 10.26 shows some lines in the Poincaré half-plane. Observe that lines l and k are horoparallel, lines n and k are hyperparallel with a common perpendicular p, and lines l, m, and q determine a hyperbolic triangle.

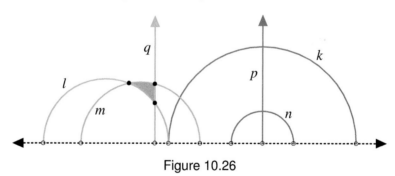

Figure 10.26

We now investigate how hyperbolic circles appear in $\mathbb{H}^2$. Interestingly enough, hyperbolic circles turn out to be Euclidean circles with their hyperbolic centers being distinct from their Euclidean centers. To show that a hyperbolic circle K is in indeed a Euclidean circle, let O be the hyperbolic center of K and A be a point on K, as in Figure 10.27. The set of hyperbolic lines in $\mathbb{H}^2$ going through O is the set of semicircles centered on the x-axis and going through O (and the open ray perpendicular to the x-axis and going through O). Note that the set of circles determining the hyperbolic lines through O also intersect in the point P (the reflection of O about the x-axis). Since these hyperbolic lines go through O, the center of K, they determine diameters of K, and hence they are orthogonal to K (see Theorem 10.15). Inverting A which lies on K in these hyperbolic lines results in points whose hyperbolic distance from O is the same as the distance from A to O (see Theorem 10.14). Hence K, the set of images of A under inversion in these hyperbolic lines, is the same as the unique Euclidean circle through A and orthogonal to the hyperbolic lines through O [84]. But does such a Euclidean circle exist and how can it be constructed?

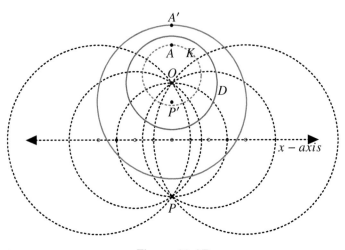

Figure 10.27

Consider a Euclidean circle D centered at point O and let P' be the image of P under inversion in D. Inverting the circles determining the hyperbolic lines through O in the circle D results in a set of Euclidean lines (not shown in the figure) that are concurrent in P'. Let A' be the image of A under inversion in D. This inversion maps the circle centered at P' and passing through A' onto a Euclidean circle K through A that is orthogonal to all the hyperbolic lines through O (why?).

We now investigate how to determine the hyperbolic center of a circle in $\mathbb{H}^2$. To do this we will use the notion that diameters of a circle are line segments that are orthogonal to the circle and go through its center. Given a Euclidean circle C in $\mathbb{H}^2$ with Euclidean center O_1, let B be the perpendicular projection of O_1 onto the x-axis, and P and Q be the points where the lines tangents to C from point B meet C (see Figure 10.28). Note that P and Q can be realized as the points of the intersection of C and the circle centered at the midpoint of $\overline{BO_1}$ (why?). One can easily prove that the semicircle γ centered at B and going through P and Q is orthogonal to C (why?); hence the arc $\widehat{PQ}$ of γ is a diameter of C. But $\overrightarrow{BO_1}$ is also orthogonal to C; hence O_2, the intersection of γ and $\overrightarrow{BO_1}$, is the hyperbolic center of C in $\mathbb{H}^2$.

10.3 Models of Hyperbolic Geometry

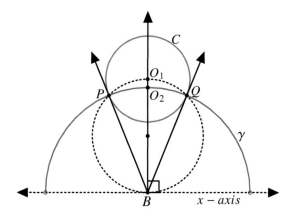

Figure 10.28

> **Now Try This 10.1**
>
> In $\mathbb{H}^2$, assume the Euclidean center O_1 of the circle C is located at $(0, b)$ (see Figure 10.28). Let r be the radius of C and note that $b > r$.
>
> 1. Show that $P = \left(-\sqrt{r^2 - \frac{r^4}{b^2}}, \frac{b^2 - r^2}{b}\right)$ and $Q = \left(\sqrt{r^2 - \frac{r^4}{b^2}}, \frac{b^2 - r^2}{b}\right)$.
> 2. Show that the equation of the circle centered at B and going through P and Q is $x^2 + y^2 = b^2 - r^2$.
> 3. Show that the hyperbolic center of C is located at $O_2 = (0, \sqrt{b^2 - r^2})$.
> 4. Conclude that the Euclidean distance between the points O_1 and O_2 is equal to $b - \sqrt{b^2 - r^2}$.

Note that for any circle of radius $r \neq 0$ in $\mathbb{H}^2$, the hyperbolic and Euclidean centers are distinct, with the hyperbolic center being offset towards the x-axis. The reason is that distances are larger the closer we get to the x-axis, as we will see when we define a hyperbolic metric on $\mathbb{H}^2$ later in this section.

The Poincaré Disc Model $\mathbb{D}^2$

Now suppose one could pick up the "ends" of the x-axis which lie at infinity and glue them together at a point. The result is something like a disc. What happens to our hyperbolic lines in the half-plane when we mold the half-plane into a disc? The open rays perpendicular to the x-axis become diameters of the disc, and the arcs of circles with centers on the x-axis become arcs of circles that intersect the disc's boundary at right angles. This is essentially the Poincaré disc model.

More formally, the Poincaré hyperbolic disc $\mathbb{D}^2$ is defined to be the interior of the disc of radius one unit centered at the origin, together with a hyperbolic metric. A hyperbolic point is thus a point that lies in the interior of the disc. A hyperbolic line is represented by either an arc of a circle that intersects the boundary of the disc at right angles or a diameter of the disc (the endpoints of either type of line do not belong to the hyperbolic line). Diameters of the disc can be thought of as arcs of circles of infinite radii, that is, circles centered at infinity in the extended plane containing the disc. The reader can think of the boundary of the disc as representing points at infinity. Euclidean points on the boundary of the disc are called *ideal points, omega points, or points at infinity*. Lines in $\mathbb{D}^2$ that appear to intersect on the boundary of the disc represent horoparallel lines, which are asymptotic and "meet" at infinity, while lines that do not intersect in the interior of the disc or on the boundary represent hyperparallel lines. Lines that intersect at right angles in the interior of the disc represent perpendicular lines. As in $\mathbb{H}^2$, the angle between two intersecting arcs is determined by the measure of the angle between the tangent rays to the arcs at their point of intersection.

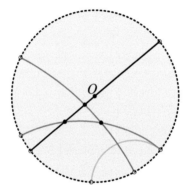

Figure 10.29

Examine Figure 10.29, which represents the Poincaré disc, and identify a pair of horoparallel lines, a pair of hyperparallel lines, a pair of perpendicular lines, and a hyperbolic triangle.

Now the question arises, if P and Q are two distinct points in $\mathbb{D}^2$, how do we construct a hyperbolic line through these points? Obviously, if the points lie on a diameter of $\mathbb{D}^2$, then the construction is trivial. Otherwise, the hyperbolic line through P and Q is determined by constructing the Euclidean circle orthogonal to the boundary of $\mathbb{D}^2$ and passing through these two points. This construction, using a straightedge and a compass, is shown in Figure 10.30. Here P' is the intersection of $\overrightarrow{OP}$ with the tangent to the boundary of $\mathbb{D}^2$ at S, where S is one of the points of intersection of boundary of $\mathbb{D}^2$ with the perpendicular to $\overrightarrow{OP}$ at P. Those who studied inversion (see Chapter 9) should recognize that P' is the inverse image of P under inversion in the boundary of $\mathbb{D}^2$. Therefore the circle γ going through P', P, and Q is orthogonal to the boundary of $\mathbb{D}^2$ (why?), and hence its intersection with $\mathbb{D}^2$ is the desired hyperbolic line through P and Q. To summarize, if P and Q are points in $\mathbb{D}^2$ (that don't lie on a diameter of the disc), then inverting P in the boundary of $\mathbb{D}^2$ gives P', which together with P and Q determine a unique circle orthogonal to boundary of $\mathbb{D}^2$, and this circle determines the unique hyperbolic line going through P and Q.

10.3 Models of Hyperbolic Geometry

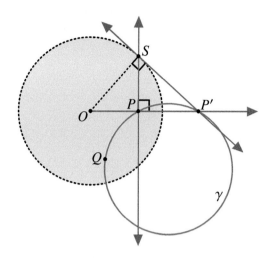

Figure 10.30

Just as in $\mathbb{H}^2$, a hyperbolic circle in $\mathbb{D}^2$ is a Euclidean circle whose center is offset towards the boundary of $\mathbb{D}^2$. Again, the reason is that distances are larger the closer we get to the boundary as we will see when we define a hyperbolic metric on $\mathbb{D}^2$ later in this section. Note that for circles in $\mathbb{D}^2$ centered at the origin (the center of the disc), the Euclidean and hyperbolic centers coincide.

The Klein Disc Model $\mathbb{K}^2$

For the sake of completeness, we now give a brief description of the Klein disc model and discuss some of its surprising features. Just like the Poincaré disc, the Klein disc is defined to be the interior of the unit disc centered at the origin, together with a hyperbolic metric. A hyperbolic point is thus a point inside the disc, but unlike in the Poincaré disc model, hyperbolic lines in the Klein disc model are chords of the unit circle without their endpoints. Since diameters are chords, these are considered hyperbolic lines in this model as well. Points on the boundary of the disc are called ideal points and do not belong to the hyperbolic model. Chords that intersect on the boundary of the disc represent horoparallel lines and chords that do not intersect in the interior of the disc or on the boundary represent hyperparallel lines.

While the notion of lines in the Klein disc model is less complicated than in the Poincaré disc model, the notion of angle is more complicated as evidenced, for example, by how perpendicular lines are defined. More specifically, given a chord in the Klein disc, the point outside the disc where the tangents to the disc at the endpoints of the chord meet is called the *pole* of the chord. Two chords are said to be *perpendicular* in the Klein disc model if when extended outside the disc, each goes through the pole of the other. For example, l_1 and l_2 in Figure 10.31 are perpendicular since l_1 when extended goes through P_2, the pole of l_2, and l_2 when extended goes through P_1, the pole of l_1.

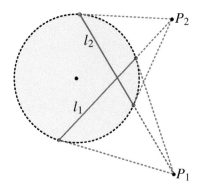

Figure 10.31

If l is a hyperbolic line in the Klein disc that is not a diameter, how do we construct lines perpendicular to l? To do this, we construct the tangents to the unit circle at the points where l intersects the circle. The point of intersection P of these tangents is the pole of l. Any chord whose extension passes through P turns out to be perpendicular to l (see Figure 10.32). This figure shows a few lines perpendicular to l, which in turn shows that the notion of right angle in the Klein disc is not as clear as that in the Poincaré disc. In Problem 16 at the end of this section, the reader will be asked to prove that the line l indeed goes through the poles of each of these chords whose extensions go through P.

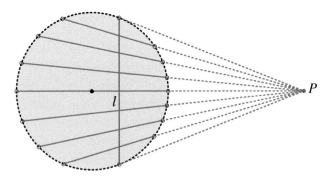

Figure 10.32

We have seen that in the Poincaré disc model the angle concept is inherited from the Euclidean plane, while the concept of a straight line is dramatically different from that in the Euclidean plane. In the Klein disc model, the concept of straight line is inherited from the Euclidean plane, but the concept of angle is more complicated. But how about the concept of circle? As mentioned earlier, circles in the Poincaré disc are Euclidean circles whose hyperbolic centers do not necessarily coincide with their Euclidean centers. In contrast, Figure 10.33 shows (without proof) how circles can be distorted in the Klein disc model. Note that when the center of the circle Q coincides with the center of the disc, the circle is the same as a Euclidean circle; otherwise the circles are transformed into ellipses.

10.3 Models of Hyperbolic Geometry

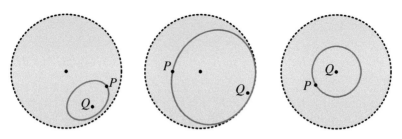

Figure 10.33

Defining a Hyperbolic Metric on $\mathbb{D}^2$

Having defined hyperbolic lines and described how they are constructed and how angles are measured in $\mathbb{D}^2$, we now proceed to introduce a hyperbolic metric d_h on $\mathbb{D}^2$. This metric needs to ensure that lines have infinite extent even though they lie within a unit disc. To achieve this, let A and B be two points in $\mathbb{D}^2$ and let M and N be the points of intersection of the line $\overleftrightarrow{AB}$ with the boundary of $\mathbb{D}^2$, as in Figure 10.34. Define the hyperbolic distance between A and B by

$$d_h(A, B) := \left| \ln \left(\frac{d(A, M) \cdot d(B, N)}{d(A, N) \cdot d(B, M)} \right) \right|,$$

where d is the Euclidean distance. It's worth noting that in Euclidean geometry the ratio $\frac{d(A,M) \cdot d(B,N)}{d(A,N) \cdot d(B,M)}$ is referred to as the *cross ratio* of the four collinear points $M, A, B,$ and N, where B is between A and N, and A is between B and M. It is also worth noting that the cross ratio of four points is invariant under inversion (see Problem 13 - Section 9.2 in Chapter 9) and under central projection (see Section 12.1 in Chapter 12). Justifying that d_h is a metric is the subject of Problem 7 at the end of this section.

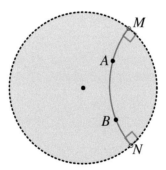

Figure 10.34

At first glance, the definition of d_h looks arbitrary and bizarre, but upon further examination, d_h proves to be a good choice for a way to measure the distance between hyperbolic points. More precisely, let $x = \frac{d(A,M) \cdot d(B,N)}{d(A,N) \cdot d(B,M)}$ and note that:

1. If A and B are very close to each other (in the Euclidean sense), then $d(A, M)$ and $d(B, M)$ are very close, and so are $d(A, N)$ and $d(B, N)$. Hence, x goes to 1 as the Euclidean distance between A and B goes to zero; therefore $d_h(A, B) \approx |\ln(1)| = 0$.

If B gets arbitrarily close to the boundary of the disc and A does not, then $d(B, N)$ goes to 0 while $d(A, M)$, $d(B, M)$, and $d(A, N)$ remain finite, hence x goes to 0 and $d_h(A, B)$ goes to infinity. This means that a hyperbolic ray has the desired infinite length.

2. If A and B are arbitrarily close to M and N respectively, then $d(A, N)$ and $d(B, M)$ are very close but finite and $d(A, M)$ and $d(B, N)$ are arbitrarily close to 0. Hence x goes to 0 as A approaches M and B approaches N, and therefore $d_h(A, B)$ goes to infinity. This implies that a hyperbolic line has infinite length as desired.

3. Note that if A and B are near the center of the disc, then the hyperbolic distance $d_h(A, B)$ is close to Euclidean distance $d(A, B)$, while if A and B are far from the center then $d_h(A, B)$ is much larger than $d(A, B)$. In fact, if A or B approaches the boundary of the disc then $d_h(A, B)$ approaches infinity (why?). This means that a finite small length near the center of the disc is simply that, while the same finite small length near the boundary of the disc represents an arbitrarily large distance.

As we have defined the distance between two points in $\mathbb{D}^2$, the distance between a point P and a line l is defined to be the hyperbolic distance between P and the point of the intersection of l and the hyperbolic line orthogonal to l and going through P. This is consistent with how distance between a point and a line is defined in the Euclidean geometry. One might now wonder about the set of points in $\mathbb{D}^2$ that are at a fixed distance r from a line l. This is formally referred to as the *hypercycle* determined by P and l and will be explored in the next section. For now, the reader should expect that the hypercycle is not going to be a pair of hyperbolic lines parallel to l (one on each side of l) as is the case in Euclidean geometry (see Section 10.4, page 490).

> **Now Try This 10.2**
>
> Show that, as expected, reversing the roles of M and N in the definition of the hyperbolic distance results in the same distance. That is, show that
>
> $$d_h(A, B) = \left| \ln \left(\frac{d(A, M) \cdot d(B, N)}{d(A, N) \cdot d(B, M)} \right) \right| = \left| \ln \left(\frac{d(A, N) \cdot d(B, M)}{d(A, M) \cdot d(B, N)} \right) \right|$$

The same metric defined in the Poincaré disc can be used in the Poincaré half-plane model. Recall that hyperbolic lines in the half-plane model are either (1) arcs of circles in the upper half-plane whose centers lie on the x-axis or (2) Euclidean rays in the upper half-plane which are perpendicular to the x-axis. If A and B lie on a line of the first type then $d_h(A, B)$ is defined the same way as in the Poincaré disc, where M and N are the intersection points of $\overleftrightarrow{AB}$ and the x-axis. If A and B lie on a line of the second type, assume without loss of generality, that $A = (x_1, y_1)$ and $B = (x_1, y_2)$ are such that $y_2 > y_1$. Let $M = (x_1, 0)$ be the point where $\overleftrightarrow{AB}$ intersects the x-axis, and let $N = (x_1, w)$ be a point on

10.3 Models of Hyperbolic Geometry

$\overrightarrow{AB}$ such that $w > y_2$. Note that

$$\frac{d(B,N)}{d(A,N)} = \frac{d(B,N)}{d(A,B) + d(B,N)} = \frac{1}{\frac{d(A,B)}{d(B,N)} + 1}.$$

Hence as w goes to infinity, $\frac{d(B,N)}{d(A,N)}$ goes to 1 (why?). Therefore,

$$d_h(A,B) = \left|\ln\left(\frac{d(A,M) \cdot d(B,N)}{d(A,N) \cdot d(B,M)}\right)\right| = \left|\ln\left(\frac{y_1}{y_2}\right)\right|.$$

> **Now Try This 10.3**
>
> Let A or B lie at the center of the Poincaré disc (the origin) and assume the Euclidean distance $d(A,B) = r$. Let M and N be the points where $\overleftrightarrow{AB}$ meets the boundary of the disc. Show that if A is between M and B, and B is between A and N, then
>
> $$d_h(A,B) = \ln\left(\frac{1+r}{1-r}\right).$$
>
> How is this related to the inverse hyperbolic tangent of r?

The Hyperbolic Models $\mathbb{H}^2$ and $\mathbb{D}^2$ Are Equivalent

To formally relate the Poincaré half-plane model $\mathbb{H}^2$ to the Poincaré disc model $\mathbb{D}^2$, it's convenient to think of the half-plane model as $\mathbb{H}^2 = \{z \in \mathbb{C} : \text{Im}(z) > 0\}$ and the disc model as $\mathbb{D}^2 = \{z \in \mathbb{C} : |z| < 1\}$, where $\mathbb{C}$ denotes the complex plane. Recall that if $z = x + iy \in \mathbb{C}$, then $\text{Im}(z) = y$ (the imaginary part of z), and $|z| = \sqrt{x^2 + y^2}$ (the length of the vector from $(0,0)$ to (x,y)). We will establish a one-to-one correspondence between the points and lines in $\mathbb{H}^2$ and those in $\mathbb{D}^2$. To do so, consider $T : \overline{\mathbb{C}} \to \overline{\mathbb{C}}$ defined on the extended complex plane, $\overline{\mathbb{C}} = \mathbb{C} \cup \{\infty\}$, as follows:

$$T(z) = \begin{cases} \frac{z-i}{z+i}, & \text{if } z \in \mathbb{C} \text{ and } z \neq -i, \\ \infty, & \text{if } z = -i, \\ 1, & \text{if } z = \infty. \end{cases}$$

Note that $\frac{z-i}{z+i} = \frac{z+i-2i}{z+i} = 1 - \frac{2i}{z+i} = 1 + (-2i)\frac{1}{z+i}$ for $z \in \mathbb{C}$, where $z \neq -i$.

The transformation T is represented geometrically as shown in Figure 10.35. Note that $T(-1) = \frac{-1-i}{-1+i} \cdot \frac{-1-i}{-1-i} = i$, $T(0) = -1$, $T(1) = -i$, $T(i) = 0$, and $T(\infty) = 1$. Note also that T maps the x-axis onto the unit circle such that as z travels along the x-axis from left to right, $T(z)$ travels counterclockwise along the unit circle such that $T(0) = -1$ and $T(\infty) = 1$. Figure 10.35 implies that T maps any horizontal line in $\mathbb{H}^2$ onto a circle centered on the x-axis and passing through the point $(1,0)$. Some of the material discussed in this section is based on material from [69].

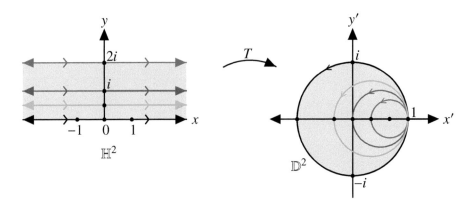

Figure 10.35

We will next show that $T : \overline{\mathbb{C}} \to \overline{\mathbb{C}}$ defined above is a bijection taking the Poincaré half-plane $\mathbb{H}^2$ onto the Poincaré disc $\mathbb{D}^2$. The inverse of T is given by (why?)

$$T^{-1}(z) = \begin{cases} \dfrac{i(z+1)}{1-z}, & \text{if } z \neq 1 \text{ or } \infty, \\ -i, & \text{if } z = \infty, \\ \infty, & \text{if } z = 1. \end{cases}$$

Note that T^{-1} maps $\mathbb{D}^2$ onto $\mathbb{H}^2$. Proving that T is a bijection is the subject of Now Prove This 10.2.

Now Prove This 10.2

Prove that $T : \overline{\mathbb{C}} \to \overline{\mathbb{C}}$ defined by $T(z) = \dfrac{z-i}{z+i}$ if $z \in \mathbb{C}$ and $z \neq -i$, $T(-i) = \infty$, and $T(\infty) = 1$ is a bijection.

1. To prove that T is one-to-one, start with $T(z_1) = T(z_2)$ and show that this leads to $z_1 = z_2$. Hint: Consider the three cases: $T(z_1) = T(z_2) \neq 1$ or ∞, $T(z_1) = T(z_2) = 1$, and $T(z_1) = T(z_2) = \infty$.
2. To prove that T is onto, for every $w \in \overline{\mathbb{C}}$ (the co-domain of T), find a $z \in \overline{\mathbb{C}}$ (the domain of T) such that $T(z) = w$. Hint: Consider the three cases: $w \neq 1$ or ∞, $w = 1$, and $w = \infty$.

We now formally show that T maps the Poincaré half-plane $\mathbb{H}^2 = \{z \in \mathbb{C} : Im(z) > 0\}$ onto the Poincaré disc $\mathbb{D}^2 = \{z \in \mathbb{C} : |z| < 1\}$.
- First, note that T maps the x-axis onto the unit circle. This is because if $z = x + iy$ is on the x-axis, then $y = 0$ and

$$|T(z)| = \frac{|x-i|}{|x+i|} = \frac{\sqrt{x^2+1}}{\sqrt{x^2+1}} = 1.$$

10.3 Models of Hyperbolic Geometry

Here we used the fact that $\left|\frac{z_1}{z_2}\right| = \frac{|z_1|}{|z_2|}$ which can be easily justified using the polar representation of complex numbers ($z = re^{i\theta}$, where $r = |z|$ and θ, measured in radians in the counterclockwise direction, is the angle between the positive x-axis and the vector corresponding to z). Hence T maps the x-axis to the unit circle.

- Second, if $y > 0$, then using $|z - i| = |x + i(y - 1)|$ and $|z + i| = |x + i(y + 1)|$, we arrive at

$$|T(z)| = \frac{\sqrt{x^2 + (y-1)^2}}{\sqrt{x^2 + (y+1)^2}} < 1, \text{ since } (y-1)^2 < (y+1)^2.$$

Hence $T(z)$ lies in the interior of the unit disc.

- Third, to justify that each point $w = u + iv$ in $\mathbb{D}^2$ is $T(z)$ for some z in $\mathbb{H}^2$, we will show that $z = T^{-1}(w)$ lies in $\mathbb{H}^2$. Note that

$$T^{-1}(w) = \frac{i(w+1)}{1-w} = \frac{i(u+iv+1)}{1-(u+iv)}$$
$$= \frac{-2v + i(1 - u^2 - v^2)}{(1-u)^2 + v^2}.$$

If we can show that $(1 - u^2 - v^2) > 0$, then the imaginary part of $T^{-1}(w)$ is positive, which implies that $z = T^{-1}(w)$ lies in $\mathbb{H}^2$. This follows from the fact that $w = u + iv$ lies in $\mathbb{D}^2$, and hence $\sqrt{u^2 + v^2} < 1$, which implies that $1 - u^2 - v^2 > 0$.

Having established a one-to-one correspondence between points in $\mathbb{H}^2$ and points in $\mathbb{D}^2$, we now turn our attention to establishing a one-to-one correspondence between lines in $\mathbb{H}^2$ and lines in $\mathbb{D}^2$. To do this, note that $T(z) = 1 + (-2i)\frac{1}{z+i}$ can be written as the composition of the following functions:

1. $f_1(z) = z + i$, which is a translation by i (a translation by one unit in the positive y-direction).

2. $f_2(z) = \frac{1}{z}$ which is an inversion about the unit circle followed by a reflection about the x-axis, since if $z = x + iy$ then $\frac{1}{x+iy} = \frac{x}{x^2+y^2} - i\frac{y}{x^2+y^2}$ (why?). Figure 10.36 shows z, its image $\frac{z}{|z|^2}$ under inversion in the unit circle, and $\frac{1}{z}$, the reflection of $\frac{z}{|z|^2}$ in the x-axis. Note that the reflection of $\frac{z}{|z|^2}$ about the x-axis can be written as $\frac{\bar{z}}{|z|^2}$, where $\bar{z}$ is the complex conjugate of z (if $z = x + iy$, then $\bar{z} = x - iy$).

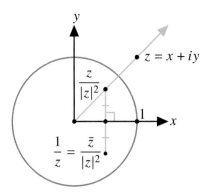

Figure 10.36

3. $f_3(z) = (-2i)z$, which is the composition of a dilation (by a factor of -2) and a rotation (by 90°) about the origin (why?). Equivalently, f_3 is the composition of a dilation (by a factor of 2) and a rotation (by $-90°$) about the origin.

4. $f_4(z) = z + 1$, which is a translation 1 unit in the positive x-direction.

Those who studied inversion (see Chapter 9) know that inversion takes lines to lines or circles and takes circles to lines or circles, and it preserves angle measure. Also, we know that translations, reflections, rotations, and dilations preserve lines, circles, and angle measure. Hence $T(z) = f_4 \circ f_3 \circ f_2 \circ f_1(z)$ takes lines to lines or circles and takes circles to lines or circles and preserves angle measure. Therefore, T takes lines in the Poincaré upper half-plane, which are vertical lines or semicircles centered on the x-axis, onto lines in the Poincaré disc, which are diameters of the disc or arcs of circles intersecting the boundary of the disc at right angles.

Figure 10.37 shows how a few hyperbolic lines in $\mathbb{H}^2$ are mapped onto hyperbolic lines in $\mathbb{D}^2$. It's worth noting that functions $f : \overline{\mathbb{C}} \to \overline{\mathbb{C}}$ of the form $f(z) = \dfrac{az+b}{cz+d}$, where a, b, c, d are complex constants satisfying $ad - bc \neq 0$, are called *Möbius transformations*.

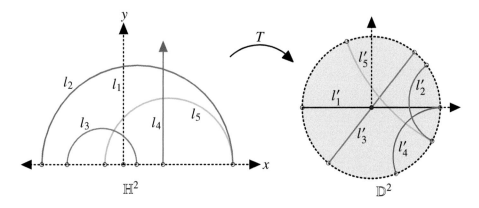

Figure 10.37

10.3 Models of Hyperbolic Geometry

Inversion in a Hyperbolic Line in $\mathbb{D}^2$

Inverting a circle C in an orthogonal circle K takes C onto C and maps the disc bounded by C onto itself (why?). Since a hyperbolic line in the Poincaré disc $\mathbb{D}^2$ is either a diameter of $\mathbb{D}^2$ or an arc of a circle orthogonal to the boundary of $\mathbb{D}^2$, inversion about a hyperbolic line maps $\mathbb{D}^2$ onto itself. The following theorem implies that such inversion preserves hyperbolic distance between points and their images, thus defining an isometry from $\mathbb{D}^2$ onto itself.

> **Theorem 10.14**
>
> *If γ is a hyperbolic line in $\mathbb{D}^2$, then inversion about γ defines an isometry on $\mathbb{D}^2$.*

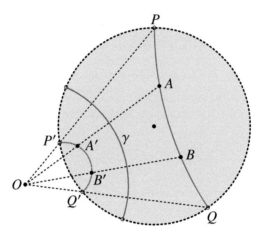

Figure 10.38

Proof. Let A and B be two distinct points in $\mathbb{D}^2$, and A' and B' be their respective images under inversion in the hyperbolic line γ. Furthermore, let P and Q be the points of intersection of $\overleftrightarrow{AB}$ and the boundary of $\mathbb{D}^2$, and P' and Q' be their respective images under inversion in γ, as shown in Figure 10.38. We wish to show that $d_h(A, B)$, the hyperbolic distance between A and B, is equal to $d_h(A', B')$. More specifically, we need to show that

$$\left|\ln\left(\frac{B'P' \cdot A'Q'}{A'P' \cdot B'Q'}\right)\right| = \left|\ln\left(\frac{BP \cdot AQ}{AP \cdot BQ}\right)\right|.$$

If γ is a diameter of $\mathbb{D}^2$, then we can think of it as part of a circle centered at a point at infinity that "lies" on the perpendicular bisector of γ. Therefore, inversion in γ is just reflection about γ, which is clearly an isometry in the Euclidean plane, and hence also in $\mathbb{D}^2$ (see Chapter 9, page 367).

If γ is an arc of a circle orthogonal to the boundary of $\mathbb{D}^2$, let O be its center. Assume A and B are located relative to γ as in Figure 10.38 (other special case configurations are left to the reader to investigate). By properties of inversion in the Euclidean plane, we have

$\triangle OBQ \sim \triangle OQ'B'$, $\triangle OAP \sim \triangle OP'A'$, $\triangle OBP \sim \triangle OP'B'$, and $\triangle OAQ \sim \triangle OQ'A'$ (why?). This gives

$$\frac{BQ}{B'Q'} = \frac{OQ}{OB'}, \quad \frac{AP}{A'P'} = \frac{OP}{OA'}, \quad \frac{BP}{B'P'} = \frac{OP}{OB'}, \quad \text{and} \quad \frac{AQ}{A'Q'} = \frac{OQ}{OA'}.$$

Solving for $B'Q'$, $A'P'$, $B'P'$, and $A'Q'$, and substituting the values obtained into $\frac{B'P' \cdot A'Q'}{A'P' \cdot B'Q'}$ gives $\frac{BP \cdot AQ}{AP \cdot BQ}$, and hence $d_h(A, B) = d_h(A', B')$. ∎

Since inversion about a hyperbolic line preserves distance, it can be used to define hyperbolic isometries of the Poincaré disc $\mathbb{D}^2$. More specifically, if C is the boundary of $\mathbb{D}^2$, let P be a point in $\mathbb{D}^2$ and E be its image under inversion in C. Then inversion about the circle K orthogonal to C and centered at E defines an isometry taking P to O, the center of $\mathbb{D}^2$ (see Now Prove This 10.3, page 480). As a consequence, if P and Q are any points in $\mathbb{D}^2$, we can define a hyperbolic isometry taking P onto Q by first taking P to O and then taking O to Q [11]. Such hyperbolic isometries are useful because they allow us, for example, to map $\triangle PQR$ shown in Figure 10.39 onto $\triangle P'Q'R'$, so that one of its vertices P' now lies at the center of $\mathbb{D}^2$. This makes investigating results about $\triangle PQR$ much easier, since two of the hyperbolic lines determining the triangle are transformed into straight lines lying on two diameters of $\mathbb{D}^2$.

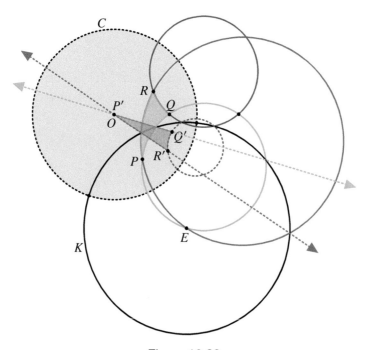

Figure 10.39

In the following theorem we use inversion about a hyperbolic line to show that any hyperbolic line through the center of a given hyperbolic circle in $\mathbb{D}^2$ is orthogonal to the

10.3 Models of Hyperbolic Geometry

given circle. Hence the portion of this line lying on or within the circle defines a diameter of the circle. This is consistent with how diameters of Euclidean circles are defined.

> **Theorem 10.15**
>
> *Consider a hyperbolic circle ω in $\mathbb{D}^2$ centered at Q, and a point P lying on ω. Then*
> 1. *there is a unique hyperbolic tangent γ to ω at P, and*
> 2. *γ is perpendicular to the hyperbolic line $\overleftrightarrow{PQ}$.*

Proof. To prove (1), let ω be a hyperbolic circle in $\mathbb{D}^2$ centered at Q and going through P, as in Figure 10.40. Construct the circle K orthogonal to the boundary of $\mathbb{D}^2$ and centered at P', the image of P under inversion in the boundary of $\mathbb{D}^2$. The intersection points of the Euclidean circle, centered at the midpoint of $\overline{OP'}$ and going through O, and the boundary of $\mathbb{D}^2$ determine the radius of K (why?). As mentioned in the discussion preceding this theorem, inversion in K takes P onto O, the center of $\mathbb{D}^2$, and takes ω onto a circle going through O. This circle has a unique tangent l at O which happens to be a diameter of $\mathbb{D}^2$, and hence is orthogonal to its boundary. Since the boundary of $\mathbb{D}^2$ is also orthogonal to K, the image of l under inversion in K is the hyperbolic line γ tangent to ω at P (why?).

To prove (2), note that inversion about the circle determining the hyperbolic line $\overleftrightarrow{PQ}$ takes ω onto ω and P onto P (why?). But since the tangent γ is unique, this inversion must also take γ onto γ. Hence γ is orthogonal to $\overleftrightarrow{PQ}$.

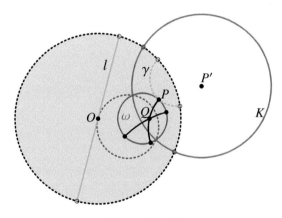

Figure 10.40

We conclude this section with an important theorem which implies that the angle of parallelism (see definition on page 450) associated with a point P and a hyperbolic line l is determined by the hyperbolic distance from P to l. Note that this is not the case in Euclidean geometry, where the angle of parallelism is always 90° no matter the distance from P to l.

> **Theorem 10.16: Bolyai-Lobachevsky Theorem**
>
> Given a point P at a hyperbolic distance d from a hyperbolic line $\overleftrightarrow{AB}$, the angle of parallelism θ associated with P and $\overleftrightarrow{AB}$ satisfies $e^{-d} = \tan(\frac{\theta}{2})$ [82, Chapter 9].

Proof. Given the hyperbolic line $\overleftrightarrow{AB}$ and a point P not on it, construct the line l through P that is perpendicular to $\overleftrightarrow{AB}$, as in Figure 10.41-(a) (see Construction II in the next section). If R is the intersection point of these two lines, then the hyperbolic distance d from P to $\overleftrightarrow{AB}$ is $d_h(P, R)$. The angle of parallelism θ of P and $\overleftrightarrow{AB}$ is $\angle RPB$ (why?). To determine how θ is related to $d_h(P, R)$, we first apply an isometry taking P to O, the center of the disc. Let K (not shown) be a circle orthogonal to the boundary of the Poincaré disc and centered at the image of P under inversion in the boundary of the disc. Inversion in K maps P to P' = O (the center of the disc), l to l', R to R', and $\overleftrightarrow{AB}$ to $\overleftrightarrow{A'B'}$, as shown in Figure 10.41(b). Since inversion is angle preserving, the measure of the angle of parallelism of P and $\overleftrightarrow{AB}$ is the same as the measure of the angle of parallelism of P' and $\overleftrightarrow{A'B'}$, which is easier to determine.

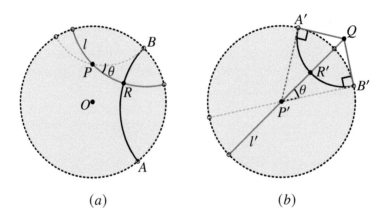

Figure 10.41

Construct the radii from P' to the ideal points A' and B', then construct tangents to the boundary of the disc at A' and B'. Let Q be the point of intersection of these tangents. Clearly Q lies on $\overrightarrow{P'R'}$ and is the center of the circle determining $\overleftrightarrow{A'B'}$ (why?). Let the Euclidean distance from P' to R' be r; then the hyperbolic distance from P' to R' is $d = \ln\left(\frac{1+r}{1-r}\right)$ (see Now Try This 10.3). Note that this is the natural logarithm of the cross ratio of P', R', and the points of intersection of $\overleftrightarrow{P'Q}$ with the boundary of the disc (why?). Equivalently, $e^d = \frac{1+r}{1-r}$ or $e^{-d} = \frac{1-r}{1+r}$. Now using the fact that $\Delta P'B'Q$ is a Euclidean right triangle we can conclude that $r = P'Q - R'Q = P'Q - B'Q = \sec(\angle B'P'Q) - \tan(\angle B'P'Q)$ (why?). But $\angle B'P'Q$ is the angle of parallelism θ of P' and $\overleftrightarrow{A'B'}$. Hence $r = \frac{1-\sin(\theta)}{\cos(\theta)}$ (how?).

10.3 Models of Hyperbolic Geometry

Substituting this for r in the equation $e^{-d} = \frac{1-r}{1+r}$ and using some algebra and trigonometric identities (namely, $\cos^2\theta + \sin^2\theta = 1$ and $\tan\frac{\theta}{2} = \frac{\sin\theta}{1+\cos\theta}$) gives $e^{-d} = \tan(\frac{\theta}{2})$ (how?). ∎

Regarding the range of values θ can take, note that Theorem 10.16 implies that as d goes to 0, θ goes to $\frac{\pi}{2}$, and as d goes to infinity, θ goes to 0.

10.3.1 Problem Set

1. Recall that if P' is the image of P under inversion in a circle C, then any circle going through P and P' is orthogonal to C. Use this result to construct a unique hyperbolic line through any two points in the Poincaré disc.

2. On your own paper, draw in the Poincaré half-plane:
   a. A pair of hyperparallel lines.
   b. The common perpendicular of your hyperparallel lines.
   c. A pair of horoparallel lines.
   d. A hyperbolic triangle.
   e. A Saccheri quadrilateral.

3. Let P and Q be two points in the Poincaré disc and X be a point not lying on the hyperbolic line $\overleftrightarrow{PQ}$. Construct the two lines through X horoparallel to $\overleftrightarrow{PQ}$.

4. Can you draw any other lines in the Poincaré disc, besides those you drew in Problem 3, which are horoparallel to $\overleftrightarrow{PQ}$ through X?

5. Suppose that the radius of the circle C bounding the Poincaré disc is 1, and let A be a hyperbolic point whose Euclidean distance from the origin O is r. The diameter of C passing through A is an hyperbolic line. Show that $d_h(O, A)$ is $|\ln(\frac{1+r}{1-r})|$. Find r when $d_h(A, O) = 10$. This exercise demonstrates the difficulty of constructing hyperbolic segments of large finite lengths.

6. Use the previous problem to show that there is a one-to-one correspondence between points on a diameter of the Poincaré disc and the real numbers in the open interval $(-1, 1)$.

7. Prove that d_h is a metric on the Poincaré disc. That is, prove that for any points A, B, C in the Poincaré disc:
   a. $d_h(A, B) \geq 0$ and $d_h(A, B) = 0$ if and only if $A = B$.
   b. $d(A, B) = d(B, A)$.
   c. $d_h(A, C) + d_h(C, B) \geq d_h(A, B)$.

8. Prove that for any points A, B, C in the Poincaré disc:
   a. If A tends to the boundary of the disc, then $d_h(A, B)$ tends to infinity.
   b. If C lies between A and B on a hyperbolic line through A and B, then $d_h(A, C) + d_h(C, B) = d_h(A, B)$.

9. In hyperbolic geometry, prove that if $ABCD$ is a quadrilateral with $AD = BC$ and $\angle BAD + \angle CBA = 180°$, then $AB < CD$.

10. Prove that in hyperbolic geometry, the line segment joining the midpoints of two sides of a triangle is less than half the length of the third side.

11. Given a unit circle C centered at the origin (0,0), show that the equation of a circle C_1 orthogonal to C and centered at (h, k) is given by $x^2 - 2hx + y^2 - 2ky + 1 = 0$.

12. Use the result of the previous problem to show that if C is the boundary of the Poincaré disc and if $A(x_1, y_1)$ and $B(x_2, y_2)$ are two points in the interior of the disc then there is a unique hyperbolic line through A and B. Do this by showing that there is one and only one choice of (h, k) for which the circle centered at (h, k) is orthogonal to C and passes through A and B.

13. Show that the Pythagorean theorem does not hold in the Poincaré disc model.

14. In the Poincaré disc, draw a right isosceles triangle whose angle sum is 100°. *Hint: Place the right angle at the center of the Poincaré disc.*

15. Let H be a regular hexagon in hyperbolic geometry whose defect is 12.
    a. Find the measure of each angle of H.
    b. If O is the center of H, triangulate H about O by connecting O to each vertex, thus dividing the hexagon into six triangles. Find the measure of each interior angle of each of the six triangles created.
    c. Are each of these six triangles equilateral, as is the case if H were a regular hexagon in Euclidean geometry?

16. Show that all chords whose extensions outside the Klein disc go through the pole of a given chord are perpendicular to that chord. That is, show that given a chord l whose pole is P, any chord m whose extension goes through P has its pole lying on the extension of l (see Figure 10.32).

10.4 Compass and Straightedge Constructions in the Poincaré Disc Model $\mathbb{D}^2$

In this section the reader is presumed to be familiar with elementary Euclidean constructions dealing with bisecting a line segment, constructing a perpendicular bisector, constructing parallel and perpendicular lines, bisecting an angle, constructing the incircle and circumcircle of a triangle, constructing a circle through three noncollinear points, etc. Basic knowledge of circle inversion is also assumed including knowing that a circle orthogonal to the circle of inversion inverts onto itself and a circle going through a point P and its inverse P' is orthogonal to the circle of inversion. Also, if circle C_1 centered at O_1 is orthogonal to circle C_2 centered at O_2, then O_1 lies outside C_2 and O_2 lies outside C_1 (why?). Finally, if P is a point outside C, then there is a unique circle with center P orthogonal to C. These

10.4 Compass and Straightedge Constructions in the Poincaré Disc Model $\mathbb{D}^2$

results were proved in Chapter 9. Unlike in the previous section, $\mathbb{D}^2$ will not be shaded in figures in this section.

Recall that a hyperbolic line in $\mathbb{D}^2$ is either a diameter of $\mathbb{D}^2$ or an arc of a circle orthogonal to the boundary of $\mathbb{D}^2$. Therefore, to construct a hyperbolic line through points P and Q (if they don't lie on a diameter of $\mathbb{D}^2$) we can simply invert P in the boundary of $\mathbb{D}^2$ to get P' and then construct a circle through the points P, P', and Q. Figure 10.42 shows the construction of the circle determining the hyperbolic line through the two points P and Q, where

1. both P and Q are in the interior of $\mathbb{D}^2$,
2. exactly one of the points is an ideal point (lying on the boundary of $\mathbb{D}^2$), and
3. both points are ideal points.

Note that if P and Q are ideal points then the perpendiculars to $\overrightarrow{OP}$ and $\overrightarrow{OQ}$ at P and Q, respectively, intersect at the center of the desired orthogonal circle (why?). Alternatively, one can find the midpoint of the Euclidean segment $\overline{PQ}$ and invert that about the boundary of $\mathbb{D}^2$ to determine the center of the circle determining the hyperbolic line through P and Q (why?).

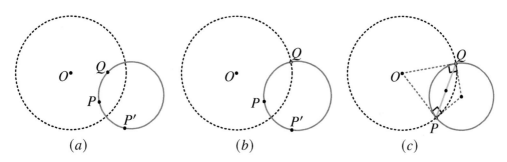

Figure 10.42

Inversion will be particularly useful in various hyperbolic constructions in $\mathbb{D}^2$, since we are often dealing with orthogonal circles, due to the nature of hyperbolic lines in $\mathbb{D}^2$. Therefore, the following two Euclidean constructions will prove to be very useful when carrying out constructions in the Poincaré disc.

Construction I: *Given a circle C_1 and a point Q not on C_1, construct a circle C_2 orthogonal to C_1 and centered at Q.*

Investigation: To discover the steps needed for the construction, suppose the required circle C_2 with center Q has been constructed as in Figure 10.43. Let P be one of the points of intersection of C_1 and C_2. Since the circles are orthogonal, $\angle OPQ$ is a right angle. Hence knowing the location of P determines the required circle C_2, but how can this location be determined? By Thales' theorem we know that the locus of all the points P for which $\angle OPQ$ is a right angle is the circle whose diameter is $\overline{OQ}$. Thus P is on the circle K with center M (the midpoint of $\overline{OQ}$) and radius $\overline{OM}$ and this determines the location of P (which also lies on C_1 and C_2). Therefore, the circle C_2 with center Q and radius $\overline{PQ}$ is the required circle centered at Q and orthogonal to C_1. Notice that C_2 is unique (why?).

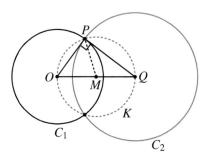

Figure 10.43

Construction Steps and Justification: Consider C_1 and Q shown in Figure 10.43. Let M be the midpoint of $\overline{OQ}$ and P be the point of intersection of C_1 and the circle K centered at M and going through O and Q. The circle C_2 centered at Q and going through P is orthogonal to C_1. This follows from the fact that $\angle OPQ$ is a right angle by Thales' theorem. Alternatively, note that $\angle MOP \cong \angle MPO$ and $\angle MQP \cong \angle MPQ$ since $\triangle OMP$ and $\triangle PMQ$ are isosceles triangles. Furthermore, note that in Euclidean geometry, $\angle MQP + \angle MPQ + \angle MOP + \angle MPO = 180°$. Hence $\angle OPQ$ is a right angle and the circles are orthogonal.

Construction II: *Given two orthogonal circles C_1 and C_2 and a point P, construct a circle C_3 through P and orthogonal to both C_1 and C_2.*

Investigation and Construction Steps: To discover the steps of the construction assume first that P is inside the circle C_1 but not on C_2 and let P' be its image under inversion in C_1. Then any circle through P and P' is orthogonal to C_1. Now let P'' be the image of P' under inversion in C_2; then any circle through P' and P'' is orthogonal to C_2. But we want a circle orthogonal to both C_1 and C_2; hence the required circle needs to go through the three points P, P', and P'', which means that $\overline{PP'}$ and $\overline{PP''}$ will be chords of this circle. Since the center of C_3 is equidistant from P, P', and P'', it must be on the perpendicular bisector of each of the chords $\overline{PP'}$ and $\overline{PP''}$. Since these chords are not parallel, their perpendicular bisectors must meet at a point, call it O_3 (why?). Now take C_3 to be the circle centered at O_3 and whose radius is $\overline{O_3 P}$, as shown in Figure 10.44.

10.4 Compass and Straightedge Constructions in the Poincaré Disc Model $\mathbb{D}^2$

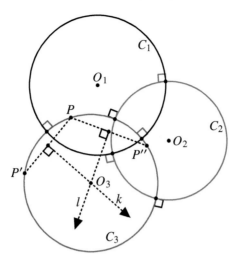

Figure 10.44

If P happens to lie on C_2 as in Figure 10.45, then P', the image of P under inversion in C_1, would also lie on C_2 since C_2 is orthogonal to C_1. If we now invert P' in C_2, we do not obtain P'' distinct from P', and hence the construction outlined above doesn't work. One way to remedy this is to notice that the center of the desired circle C_3 must lie on the tangent to C_2 at P' (or the tangent to C_2 at P) (why?). Furthermore, since C_3 goes through P and P', its center must also lie on the perpendicular bisector of the the common chord $\overline{PP'}$. This determines the center of C_3 orthogonal to C_1 and C_2 and going through P.

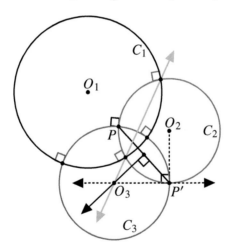

Figure 10.45

Alternatively, if P happens to lie on C_2 and P' is the image of P under inversion in C_1, then we can use the fact that the center of C_3 must lie on the *radical axis* of C_1 and C_2 (which in this case is the line containing the common chord of C_1 and C_2) to determine C_3. Proving this will be the subject of Problem 18 at the end of this section.

Note that *Construction II* is also valid if P is a point on or outside C_1. The reader is encouraged to verify this claim and determine whether circle C_3 is unique or not.

When working in the Poincaré disc model, *Construction II* allows us to draw a hyperbolic line perpendicular to a given hyperbolic line through a given point. This enables us to carry out important constructions such as constructing a hyperbolic circle centered at a given point and going through another point, etc. A number of these hyperbolic constructions will be explored later in this section.

> **Now Prove This 10.3: Inverting a Point in $\mathbb{D}^2$ onto the Center of $\mathbb{D}^2$**
>
> Let C be the boundary of the Poincaré disc $\mathbb{D}^2$, O be its center, and P be a point in $\mathbb{D}^2$ other than O. Let P' be the image of P under inversion in C, and K be the circle orthogonal to C and centered at P' (see *Construction I*). Use Figure 10.46 to prove that the image of P under inversion in K is O, the center of $\mathbb{D}^2$.
>
>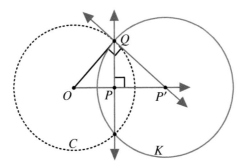
>
> Figure 10.46

Now Prove This 10.3 implies that we can transform a hyperbolic figure in $\mathbb{D}^2$ so that one of its vertices P is located at the origin. This can make investigating properties of the figure easier and more convenient (see the discussion accompanying Figure 10.39).

> **Remark 10.1**
>
> The previous discussion allows us to conclude that if orthogonal circles C_1 and C_2 centered respectively at O_1 and O_2 intersect in Q_1 and Q_2, and if P is the point of intersection of $\overline{O_1 O_2}$ and $\overline{Q_1 Q_2}$, then the inverse of P in C_1 is O_2 and the inverse of P in C_2 is O_1.

Next we explore constructions in the Poincaré disc such as constructing a perpendicular line through a point, a circle centered at a point and going through another point, the perpendicular bisector of a segment, the image of a point under reflection in a line, the angle bisector, regular polygons, and regular polygons with right angles. Many of these constructions are the same as their corresponding Euclidean constructions, but the reader is encouraged to try these constructions for themselves to gain a better understanding of how they can be carried out in the Poincaré disc. It is worth mentioning that there are

10.4 Compass and Straightedge Constructions in the Poincaré Disc Model $\mathbb{D}^2$

currently applets such as *NonEuclid* [12] that allow the user to easily and interactively create straightedge and compass constructions in both the Poincaré disc and the upper half-plane models of hyperbolic geometry. Unfortunately, the degree of support for such applets is often not certain given the rapid advances in technology. Therefore, using well-supported, free dynamic geometry software such as GeoGebra may be preferable, although in its current state it may not allow for easy implementation of hyperbolic constructions. Nonetheless, there are many dynamic GeoGebra worksheets designed by expert users and shared freely online that allow the average user to easily explore hyperbolic constructions.

Many of the constructions we will be exploring can be done more easily using inversion, and therefore this approach will often be given as an alternative approach. This is especially helpful since software such as GeoGebra includes circle inversion as a construction tool. In the problem set at the end of this section, the reader will be asked to apply these construction techniques to test whether basic concepts from Euclidean geometry hold in hyperbolic geometry or not. For example, how do the medians, altitudes, perpendicular bisectors, and angle bisectors of a triangle behave in hyperbolic geometry? Does a triangle always have an incircle or a circumcircle? What properties do the diagonals of a rhombus, kite, or parallelogram have in hyperbolic geometry? Are the base angles of an isosceles triangle congruent? Is the bisector of the angle opposite the base of an isosceles triangle perpendicular to the base? Investigating these and other concepts will help the reader develop a deeper understanding of hyperbolic geometry and the important role the Euclidean parallel postulate plays in Euclidean geometry.

Constructing a Perpendicular Line Through a Point

Given a hyperbolic line γ in the Poincaré disc $\mathbb{D}^2$, how do we construct a perpendicular to γ through a given point P? Let C denote the boundary of $\mathbb{D}^2$ and recall that a hyperbolic line is either a diameter of C or an arc of a circle that intersects C at right angles. Consequently, the desired construction is equivalent to the Euclidean construction of a circle passing through a given point and orthogonal to a given circle and one of its diameters, or orthogonal to two circles that are themselves orthogonal to each other. Figure 10.47 shows line l through P which is perpendicular to the hyperbolic line γ.

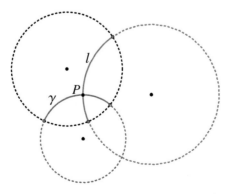

Figure 10.47

The same Euclidean construction works in both cases, regardless of whether P is on γ or not, and can be carried out as outlined in *Construction II* described earlier.

Constructing a Hyperbolic Circle Centered at A and Going Through B

To do this, we will use the fact that the diameters of a hyperbolic circle are hyperbolic lines orthogonal to the circle. Let A and B be two points in the Poincaré disc $\mathbb{D}^2$.

1. Construct the hyperbolic line γ going through the points A and B. This can be a diameter of $\mathbb{D}^2$ or an arc of a circle orthogonal to the boundary of $\mathbb{D}^2$.

2. If γ is a not a diameter of $\mathbb{D}^2$ as in Figure 10.48(a), then the desired hyperbolic circle centered at A and going through B is the Euclidean circle K orthogonal to γ, going through B, and centered at E. The point E is the intersection of the tangent to γ at B and the diameter of $\mathbb{D}^2$ through A (why?).

3. If γ is a diameter of $\mathbb{D}^2$ as in Figure 10.48(b), construct two hyperbolic lines α and β through A. Let B' be the image of B under inversion in the circle determining α, and B'' be the inverse image of B' under inversion in the circle determining β. Then the desired hyperbolic circle centered at A and going through B is the Euclidean circle K determined by B, B', and B'' (centered at E) (see *Construction II* discussed earlier). Note that K is orthogonal to α, β, γ, and all other hyperbolic lines through A. The parts of these lines lying on or inside K represent diameters of the hyperbolic circle K.

Do you notice anything peculiar about the hyperbolic circles in Figure 10.48?

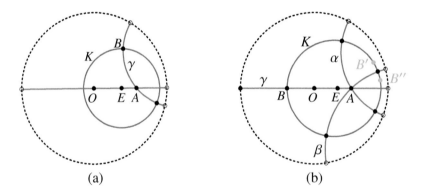

Figure 10.48

In Problem 15 at the end of this section, the reader will be asked to investigate constructing a hyperbolic circle centered at a point A and and whose radius is the length of a given hyperbolic segment $\overline{BC}$.

Constructing the Perpendicular Bisector of a Hyperbolic Segment $\overline{AB}$

The construction of the perpendicular bisector of a hyperbolic segment $\overline{AB}$ is the same as the respective construction in the Euclidean plane. Figure 10.49 shows the hyperbolic circle

10.4 Compass and Straightedge Constructions in the Poincaré Disc Model $\mathbb{D}^2$

C_1 centered at A and going through B and the hyperbolic circle C_2 centered at B and going through A. The hyperbolic line l through the points of intersection of C_1 and C_2 is the perpendicular bisector of the segment $\overline{AB}$ (why?).

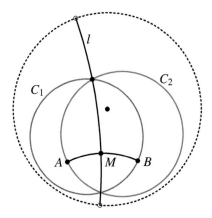

Figure 10.49

Now Try This 10.4

Alternatively, we can construct the hyperbolic perpendicular bisector of a segment $\overline{AB}$ using inversion as described below [34].

1. Let A', B' be the inverse images of the points A, B, respectively, under inversion in C, the boundary of the Poincaré disc, as shown in Figure 10.50.

2. Since l, the desired hyperbolic perpendicular bisector of $\overline{AB}$, is part of a circle orthogonal to the circle that $\overline{AB}$ lies on, inversion about l would take A to B (see Problem 17 at the end of this section). Those who studied inversion know that the composition of inversions in orthogonal circles is commutative. Therefore, if I_C denotes inversion in circle C and I_l denotes inversion in the circle determining l, then $I_C \circ I_l(A) = I_l \circ I_C(A)$, which implies that $I_C(B) = I_l(A')$, and hence $B' = I_l(A')$. This means that inversion about l also takes A' to B'. Hence the center of the circle determining l is E, the point of intersection of $\overleftrightarrow{AB}$ and $\overleftrightarrow{A'B'}$.

3. The tangents from E to the boundary of the Poincaré disc determine the endpoints of l (see *Construction I* discussed earlier).

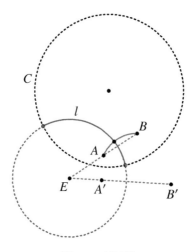

Figure 10.50

Constructing the Reflection of a Point About a Line γ

Let P be a point in $\mathbb{D}^2$ and γ be a hyperbolic line. The point P', the reflection of P about γ, can be constructed using the usual Euclidean construction as described below:

1. Construct the hyperbolic line γ' orthogonal to γ and passing through P.
2. Construct a hyperbolic circle through P centered at the point of intersection of the two lines γ and γ', and take P' to be the second point of intersection of this circle with γ'.

Alternatively, P' can be constructed as the image of P under inversion in the circle determining the line γ. In fact, inversion in a circle is often referred to as reflection about a circle as is the case in the constructions menu in GeoGebra.

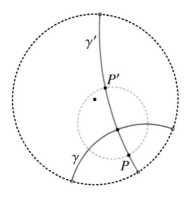

Figure 10.51

10.4 Compass and Straightedge Constructions in the Poincaré Disc Model $\mathbb{D}^2$

> **Now Try This 10.5**
>
> Figure 10.52 shows the image of $\triangle ABC$ under reflection about line l. Carry out this construction and note that reflection about a hyperbolic line is orientation reversing, as is the case in Euclidean geometry.
>
>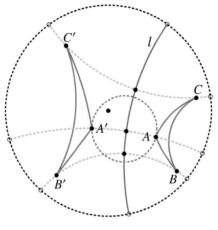
>
> Figure 10.52

Constructing the Angle Bisector

The bisector of a hyperbolic angle can be determined using the corresponding angle bisector construction used in the Euclidean plane. Let l_1 and l_2 be two hyperbolic lines in the Poincaré disc intersecting at B. Figure 10.53 shows a hyperbolic circle centered at B and intersecting l_1 in C and l_2 in A. It also shows two congruent hyperbolic circles centered at C and A, and going through B. The points of intersection of these two circles determine the angle bisector. Hence $\overleftrightarrow{BD}$ is the bisector of $\angle ABC$ formed by l_1 and l_2.

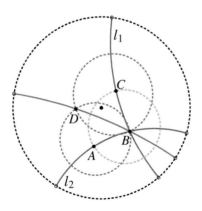

Figure 10.53

Constructing Regular Hyperbolic Polygons

The reader should be aware of the fact that some regular polygons are *constructible* (can be constructed using a straightedge and a compass), while others are not. For instance, a regular pentagon is constructible, while a regular heptagon (7-gon) is not (see Gauss–Wantzel theorem below). The ancient Greeks knew how to construct regular polygons with 3, 4, or 5 sides, and how to construct regular polygons with double the number of sides of a given regular polygon (how?) [8, page 49]. They also proved that if a regular polygon with m sides and one with n sides can be constructed and m and n are relatively prime, then a regular polygon with $m \cdot n$ sides can be constructed. This allows the construction of, for example, a regular polygon with 15 sides. The question "Which regular n-polygons are constructible and which are not?" was finally answered by what is now known as the Gauss–Wantzel theorem (proved in 1837), which states: *A regular n-gon (polygon with n sides) can be constructed with compass and straightedge if and only if n is the product of a power of 2 and any number of distinct Fermat primes (including none).* A *Fermat prime* is a prime number of the form $2^{(2^m)} + 1$, where m is positive integer. We will not try to give a comprehensive treatment of constructibility of regular n-gons here; rather we will show how Euclidean constructions can be used to construct certain regular hyperbolic polygons.

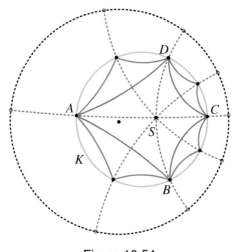

Figure 10.54

Figure 10.54 shows a regular quadrilateral and a regular octagon constructed in the Poincaré disc. Here we made use of the fact that any hyperbolic line that goes through the center of a hyperbolic circle is orthogonal to the circle. To construct the equilateral quadrilateral $ABCD$ we started with two orthogonal lines $\overleftrightarrow{AC}$ and $\overleftrightarrow{BD}$, and then constructed a circle K of some radius centered at S, the point of intersection of $\overleftrightarrow{AC}$ and $\overleftrightarrow{BD}$. The quadrilateral $ABCD$ is equilateral since $\triangle ASD$, $\triangle DSC$, $\triangle CSB$, and $\triangle BSA$ are congruent by SAS. Bisecting the angles at S formed by $\overleftrightarrow{AC}$ and $\overleftrightarrow{BD}$ allows us to construct an octagon whose vertices lie on K. The same reasoning shows that the octagon is equilateral (each of

10.4 Compass and Straightedge Constructions in the Poincaré Disc Model $\mathbb{D}^2$

the angles facing the sides is 45°). Furthermore, the interior angles of $ABCD$ are congruent and so are the interior angles of the octagon (why?); hence these are regular polygons.

Continuing the process of bisecting the angles at the center of the circumscribing circle K allows us to construct hyperbolic regular n-gons, where $n = 2^m$ and $m \geq 2$. Note that this construction will not work for some values of n as it depends on being able to divide an angle into n congruent parts, which is not always possible.

> **Now Try This 10.6**
>
> Figure 10.55 shows a hyperbolic equilateral triangle, its incircle, and its circumcircle. It also shows a hyperbolic regular hexagon. Continuing the process of bisecting the angles at the center of the circumscribing circle, we can construct hyperbolic regular n-gons for $n = 3(2^m)$ where m is a positive integer. The reader is encouraged to carry out this construction to gain experience and confidence in doing such constructions. *Hint: Given $\overline{AB}$, construct the equilateral $\triangle ABC$ using the usual Euclidean construction. The intersection of the angle bisectors determines the incenter of $\triangle ABC$. Note that this is also the circumcenter of $\triangle ABC$ (why?).*

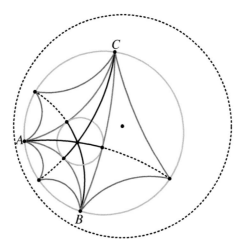

Figure 10.55

Constructing Regular Hyperbolic Polygons with Right Angles [64]

The construction described below works for all constructible n-gons, where $n \geq 5$. For the sake of clarity, we will investigate the case where $n = 5$; the same techniques can be used in the general case. So, let $ABCDE$ be a regular Euclidean pentagon, as in Figure 10.56. Without loss of generality, consider the vertex C and let M be the midpoint of $\overline{OC}$, where O is the center of the Euclidean circle circumscribing $ABCDE$ (circle not shown). Let H be the circle centered at M and going through O and C. Consider $\overline{BD}$, the diagonal connecting the vertices of $ABCDE$ adjacent to C, and let I and J denote its points of intersection with the circle H.

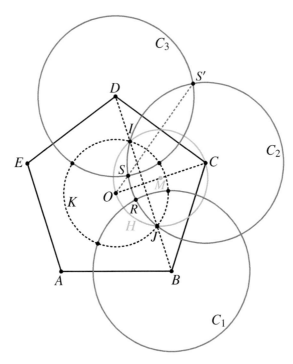

Figure 10.56

Let K and C_2 denote the circles centered at O and C, respectively, and going through the points I and J. The angles $\angle OIC$ and $\angle OJC$ are right angles (why?). Hence K and C_2 are orthogonal. Alternatively, we can say that K and C_2 are orthogonal by *Construction I* described earlier. Let C_3 be the circle centered at D and congruent to C_2, and let S and S' be the points of intersection of C_2 and C_3. The points S and S' lie on $\overrightarrow{OS}$, since $\overrightarrow{OS}$ is the perpendicular bisector of $\overline{DC}$ (the side of the pentagon joining the centers of C_2 and C_3) (why?). They also lie on C_2, which is orthogonal to K. Since inversion about K takes C_2 onto itself, it must take S to S'. But S and S' also lie on C_3, and hence C_3 is orthogonal to K. The circle C_3 is also orthogonal to C_2, since the quadrilateral $SCS'D$ is a square (why?). This conclusion can also be reached using *Construction II* described earlier (how?). Similarly, the circle C_1 centered at B and congruent of C_2 is orthogonal to both K and C_2. Therefore, $\angle DSC$ and $\angle BRC$ are right angles. Thus $\triangle DSC$ and $\triangle BRC$ are isosceles right triangles, which means that $\angle SCD$ and $\angle BCR$ are $45°$ each, since the sum of the angles of a Euclidean triangle is $180°$. Hence the measure of $\angle RCS$ is $108° - 90° = 18°$ (the measure of the interior angle of the regular pentagon in Euclidean geometry is $108°$).

Taking K to be the boundary of the Poincaré disc, we realize that we have just constructed one side, $\overline{RS}$, of the desired hyperbolic regular pentagon with right angles. Continuing this process, we can generate the rest of the pentagon $QRSTU$ shown in Figure 10.57. Each arc which forms a side of $QRSTU$ has the same length when measured as an arc in the Euclidean plane (each arc subtends a central angle of $18°$ in congruent circles). Furthermore, each arc is the same distance from O and is symmetric about the ray from O through the vertex

10.4 Compass and Straightedge Constructions in the Poincaré Disc Model $\mathbb{D}^2$

determining the center of the circle on which the arc lies. Thus, the hyperbolic length of each side of $QRSTU$ is the same (why?). Moreover, since the circles determining adjacent sides are orthogonal, the angles of $QRSTU$ are right angles. Hence $QRSTU$ is the desired hyperbolic regular pentagon with right angles.

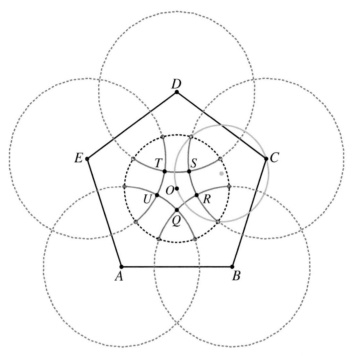

Figure 10.57

> **Remark 10.2**
>
> The measure of $\angle RCS$ in Figure 10.56 in the general case (when constructing a hyperbolic regular n-gon with right angles) would be $\frac{(n-2)\cdot 180°}{n} - 90 = \frac{(n-4)\cdot 90°}{n}$.
> Recall that the measure of the interior angle of a regular Euclidean n-gon is $\frac{(n-2)\cdot 180°}{n}$.

In general, let $V_1 V_2 \ldots V_n$ be a constructible Euclidean regular n-gon. To construct the corresponding hyperbolic regular n-gon with right angles, where $n \geq 5$ (why?), follow the steps listed below.

1. Start with a constructible regular Euclidean n-gon and construct its diagonals.
2. Without loss of generality, construct a circle whose diameter is the segment joining the center of the n-gon and one of its vertices, for example, V_1.
3. Construct the circle centered at V_1 and passing through the points of intersection of the circle in (2) with the diagonal joining V_2 and V_n, the vertices adjacent to V_1.

4. Construct the circle K centered at the center of the n-gon and passing through the points of intersection of the circle in (3) with the diagonal joining V_2 and V_n, the vertices adjacent to V_1. Take K to be the boundary of the Poincaré disc $\mathbb{D}^2$.

5. For each vertex V_i, construct a circle centered at V_i and going through the points of intersection of the circle in (4) with the diagonal joining the two vertices adjacent to V_i. Each of these circles is orthogonal to its two adjacent circles and to K, the boundary of $\mathbb{D}^2$.

6. The points of intersection of the circles in (5) are the vertices of a hyperbolic regular n-gon whose angles are right angles.

Constructing Hypercycles

In the Euclidean plane, the set of points that are at distance r from a given line consists of two lines parallel to the given line, one on each side. The situation is different in hyperbolic geometry since a straight line cannot be equidistant from another straight line (why?). In fact, given a hyperbolic line l and a distance r, the locus of hyperbolic points that are at orthogonal distance r from l is a curve called a *hypercycle, hypercircle,* or *equidistant curve* (see construction below). The distance r is referred to as the *radius* of the hypercycle and the line l as its *center* or *axis*.

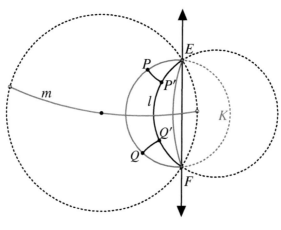

Figure 10.58

Let l be a hyperbolic line in the Poincaré disc $\mathbb{D}^2$ with ideal points E and F and let P be a point not on l, as in Figure 10.58. To construct the hypercycle determined by the line l and the point P we need to determine all points Q whose perpendicular distance from l is the same as that of P. To accomplish this, we construct the Euclidean circle K going through the points P, E and F. Note that K does not determine a hyperbolic line in $\mathbb{D}^2$ (why?). The boundary of $\mathbb{D}^2$, the circle K, and the circle determining l all share the common chord $\overline{EF}$. Hence $\overleftrightarrow{EF}$ is the *radical axis* of any two of these circles, and therefore any circle centered at a point on $\overleftrightarrow{EF}$ and orthogonal to one of the circles will be orthogonal to all three circles (why?).

10.4 Compass and Straightedge Constructions in the Poincaré Disc Model $\mathbb{D}^2$

Let the line orthogonal to l and going through P intersect l in P'; hence the distance from P to l is $d_h(P,P')$. Let Q be an arbitrary point on K lying in $\mathbb{D}^2$ and distinct from P. If the line orthogonal to l through Q meets l at Q', then the distance between Q and l is $d_h(Q,Q')$. To show that $d_h(P,P') = d_h(Q,Q')$ we consider the perpendicular bisector m of $\overline{P'Q'}$. Since m is orthogonal to both l and K, and since reflection preserves angle measure, reflection about m takes P to Q and $\overleftrightarrow{PP'}$ to $\overleftrightarrow{QQ'}$ (why?). Since l is fixed under reflection about m, the point P' is reflected onto Q', and hence $\overline{PP'}$ is reflected onto $\overline{QQ'}$. Therefore, P and Q are equidistant from l, and since Q was chosen randomly, all points on $K \cap \mathbb{D}^2$ are equidistant from l [81]. This means that the part of K that lies within $\mathbb{D}^2$ is part of the hypercycle determined by P and l (the part that lies on the same side of l as P). How about points lying on the other side of l; are any of them part of this hypercycle? Since reflection preserves distance, the reflection of $K \cap \mathbb{D}^2$ about l is also part of the hypercycle determined by P and l. This completes the construction of the hypercycle determined by l and P. Note that these two components of the hypercycle are not hyperbolic lines, since l is the unique hyperbolic line with ideal points E and F (why?).

> **Now Try This 10.7**
>
> Use a similar construction to the one discussed above to find the hypercycle of l when l is a diameter of $\mathbb{D}^2$. What happens to the hypercycle when P approaches the boundary of $\mathbb{D}^2$? What happens when P approaches l?

10.4.1 Problem Set

1. The following results hold in Euclidean geometry. Using dynamic geometry software, verify whether or not each also holds in hyperbolic geometry.
   a. The three altitudes of a triangle are concurrent in a point called the orthocenter.
   b. The three medians of a triangle are concurrent in a point called the centroid.
   c. The three angle bisectors of a triangle are concurrent in a point called the incenter.
   d. The three perpendicular bisectors of the sides of a triangle are concurrent in a point called the circumcenter.

2. The following results hold in Euclidean geometry. Which (if any) hold in hyperbolic geometry? Construct an example or counterexample of each.
   a. If a triangle is equilateral, then it is equiangular.
   b. If a triangle is equiangular, then it is equilateral.
   c. Each angle of an equilateral triangle measures 60 degrees.

3. In this problem the reader investigates the orthocenter of a hyperbolic triangle.
   a. Construct a hyperbolic triangle in the Poincaré disc and its three hyperbolic altitudes. Move the vertices of the triangle around to show that the altitudes are concurrent for some triangles and not for others.
   b. Can you find a triangle for which exactly two of the altitudes intersect?

- **c.** Can you find a triangle for which all three altitudes are asymptotically parallel?
- **d.** Can you find a triangle for which all three of the altitudes admit a common perpendicular?

4. Prove that if two of the perpendicular bisectors of the sides of a hyperbolic triangle intersect, then the third perpendicular bisector also passes through the point of intersection, and hence the triangle has a hyperbolic circumcenter.

5. If the hyperbolic circumcenter, orthocenter, and centroid exist for a hyperbolic triangle, are they collinear as is the case in Euclidean geometry?

6. A rhombus is defined as a quadrilateral whose sides are congruent. The following results hold for rhombuses in Euclidean geometry. Which (if any) hold in hyperbolic geometry? Construct an example or counterexample of each.
   - **a.** The opposite angles of a rhombus are congruent.
   - **b.** The diagonals of a rhombus bisect each other.
   - **c.** The diagonals of a rhombus are perpendicular.
   - **d.** The diagonals of a rhombus bisect the angles they join.

7. Construct a hyperbolic equilateral triangle and its incircle and circumcircle.

8. Investigate how to construct regular polygons in hyperbolic geometry and prove that any regular polygon in hyperbolic geometry is cyclic.

9. Construct a hyperbolic regular hexagon with right angles.

10. Construct a hyperbolic regular quadrilateral, octagon, 16-gon, etc.

11. Construct a hyperbolic isosceles triangle. Are angles opposite the congruent sides congruent? Does the ray bisecting the angle included by the congruent sides bisect the opposite side? Is it also perpendicular to the opposite side?

12. In hyperbolic geometry, can any triangle be circumscribed by a circle? Does every triangle have an incircle? Justify your answers.

13. Given a point P and a diameter l of the Poincaré disc not through P, construct a hyperbolic line through P that is perpendicular to l.

14. Given two points A and B in the Poincaré disc as in the figure below, construct a circle centered at A and going through B.

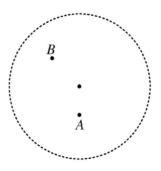

15. Given a point A and a segment $\overline{BC}$, construct a hyperbolic circle centered at A with radius $\overline{BC}$.

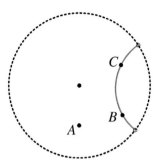

16. Given a hyperbolic line l and a triangle $\triangle ABC$, construct $\triangle A'B'C'$, the image of $\triangle ABC$ under reflection about l.

17. In the Poincaré disc $\mathbb{D}^2$, prove that if the circle S determines the perpendicular bisector of $\overline{AB}$, then inversion in S maps A onto B. Hint: Let M be the midpoint of $\overline{AB}$ and O_W be its image under inversion in C (the boundary of $\mathbb{D}^2$), as shown in the figure below. Consider the circle W centered at O_W and orthogonal to C. Inverting the circles K (which determines $\overleftrightarrow{AB}$) and S in the circle W gives the lines K' and S'. By Now Prove This 10.3, inversion about W takes M to O; hence K' and S' go through the center of $\mathbb{D}^2$. Inverting the circle H, which goes through A', B', and O_W, in W gives the line H', going through A, B, and O_S (the center of S) and orthogonal to S (why?). Since K and S are orthogonal, inversion in S takes A onto B.

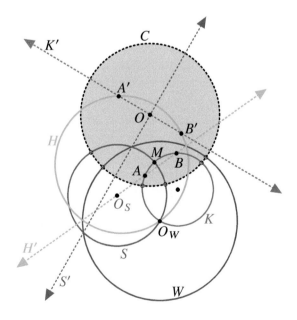

18. Prove that if two circles C_1 and C_2 intersect, then the center of any circle C_3 orthogonal to both of them must lie on their *radical axis*, which is the line containing their common chord (or common tangent if C_1 and C_2 are tangent).

10.5 Hyperbolic Tessellations

Some of the material in this section was adapted from [41, Chapter 7]. A regular *tessellation* or *tiling* of the plane is a covering of the plane (with no gaps and no overlap) by congruent regular polygons so that the same number of polygons meet at each vertex. Recall that the only regular polygons that tessellate the Euclidean plane are the equilateral triangle, square, and regular hexagon. How about the hyperbolic plane, which hyperbolic regular polygons can be used to tessellate it? If k hyperbolic regular n-gons meet at a common vertex of the tiling, then the interior angle of the regular n-gon would be $\frac{360}{k}$. The n-gon can be triangulated around its center producing n congruent triangles. The angles around this center clearly sum up to $360°$, while the angle sum of the n triangles would have to be less than $n(180°)$, since we are in the hyperbolic plane. Hence $n(180) > 360 + n(2\alpha)$ where 2α is the interior angle of the regular n-gon. Therefore, $2\alpha < 180 - \frac{360}{n}$. But $k(2\alpha) = 360$ (why?), and hence $\frac{2\alpha}{360} = \frac{1}{k}$. Dividing both sides of $2\alpha < 180 - \frac{360}{n}$ by 360 gives $\frac{1}{n} + \frac{1}{k} < \frac{1}{2}$. Therefore, a regular tessellation of the hyperbolic plane by n-gons, k of which meet at a vertex, must satisfy the relationship $\frac{1}{n} + \frac{1}{k} < \frac{1}{2}$. Multiplying this inequality by $2nk$ and adding 4 to each side gives $kn - 2k - 2n + 4 > 4$, hence $(n-2)(k-2) > 4$. A regular tessellation by n-gons, k of which meet at a common vertex, is often called an $\{n, k\}$ tiling. This is the *Schläfli* symbol for the tiling, named after the Swiss mathematician Ludwig Schläfli (1814–1895). Note that an $\{n, k\}$ tiling of the Euclidean plane would satisfy $\frac{1}{n} + \frac{1}{k} = \frac{1}{2}$, while an $\{n, k\}$ tiling of the elliptic plane would satisfy $\frac{1}{n} + \frac{1}{k} > \frac{1}{2}$ (why?).

Consider for example the $\{5, 4\}$ tiling of the Poincaré disc $\mathbb{D}^2$ shown in Figure 10.59. Since we have four pentagons meeting at a vertex, the pentagons would have to be right angled. Hence the center regular pentagon (centered at O, where O is the center of $\mathbb{D}^2$) can be constructed as described in the previous section. Recall that inversion about a hyperbolic line in $\mathbb{D}^2$ defines an isometry on $\mathbb{D}^2$ (see Theorem 10.14), and inversion defined in the Euclidean plane preserves angle measure between intersecting curves (see Theorem 9.4). Furthermore, since angles in $\mathbb{D}^2$ are measured the same way as in the Euclidean plane, inversion about a hyperbolic line in $\mathbb{D}^2$ also preserves angle measure.

Inverting the center pentagon (and its center O) about each of the hyperbolic lines determining its sides gives five regular pentagons (and their centers), where each of these pentagons is congruent to the center pentagon and shares a side with it. Can you tell how the five pentagons which intersect the center pentagon in only one vertex were constructed? How about their centers? Continuing this process yields the desired $\{5, 4\}$ tiling.

Note that due to nature of the hyperbolic metric defined on $\mathbb{D}^2$, the congruent pentagons in Figure 10.59 appear deceivingly to be decreasing in size (from a Euclidean point of view) as they approach the boundary of $\mathbb{D}^2$. Triangulating each pentagon around its center

10.5 Hyperbolic Tessellations

produces five congruent triangles each having three angles measuring 45°, 45°, and 72° (why?). This gives a tessellation of $\mathbb{D}^2$ by congruent, nonregular triangles where some vertices are surrounded by five triangles while others are surrounded by eight triangles.

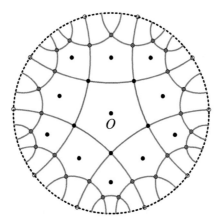

Figure 10.59

It is readily clear that any right-angled regular hyperbolic n-gon, where $n \geq 5$, can be used to tessellate the hyperbolic plane resulting in an $\{n, 4\}$ tiling. So there are infinitely many regular tessellations of the hyperbolic plane.

The dual of an $\{n, k\}$ tessellation is the $\{k, n\}$ tessellation created by reversing the roles of n, the number of sides, and k, the number of polygons meeting at a vertex. For instance, the dual of the $\{5, 4\}$ tessellation shown in Figure 10.59 is the $\{4, 5\}$ partially shown in Figure 10.60, where the hyperbolic plane is tiled using equilateral quadrilaterals having angle measure 72°.

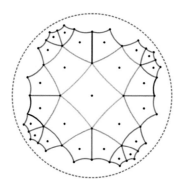

Figure 10.60

When tiling the Euclidean plane with equilateral triangles, squares, or regular hexagons, there is no restriction on the size of the tile used. However, the values of n and k in a hyperbolic $\{n, k\}$ tessellation uniquely determine the size of the tile centered at the center of the $\mathbb{D}^2$. Proving this claim is the subject of Problem 6 at the end of this section.

> **Historical Note:** *M. C. Escher (1898–1972)*
> The Dutch artist M. C. Escher is known for his tessellations of the Euclidean and hyperbolic plane. Despite having limited formal mathematical background, his remarkable visual mathematical intuition led him to represent many mathematical concepts from non-Euclidean geometry with a great degree of accuracy. He created five wood cut works depicting the concept of infinity in a finite space inspired by hyperbolic plane tessellations: Circle Limits I–IV and Snakes. Circle Limit II and Snakes are more artistic than mathematical, while Circle Limits I, III, and IV are mathematically more correct. In 1954, he met R. Penrose (1939–) and H. Coxeter (1907–2003) at the International Congress of Mathematics held in Amsterdam, which led to fruitful collaborations over the following years. Escher was especially influenced by hyperbolic tessellations of the plane contained in a paper published by Coxeter in 1957 and this led to Circle Limit I–IV.

10.5.1 Problem Set

1. Reproduce the tiling of Poincaré disc having the Schläfli symbol $\{5, 4\}$. This tessellation has four pentagons at each vertex, so each pentagon has five 90° angles.

2. Construct the tiling of Poincaré disc having the Schläfli symbol $\{3, 7\}$. This tessellation has seven triangles at each vertex, so each triangle has three congruent angles each measuring $\frac{360°}{7} \approx 51.4°$.

3. Construct the tiling of Poincaré disc having the Schläfli symbol $\{4, 6\}$.

4. What is the Schläfli symbol for the hyperbolic tessellation shown below [62]? What are the measures of the angles of the triangle used in the tessellation? Construct this tessellation.

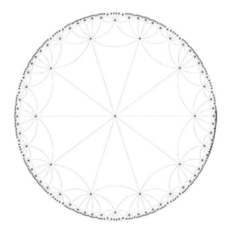

5. What does the dual of the tessellation below [62] look like?

10.5 Hyperbolic Tessellations

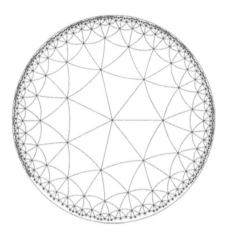

6. [15] Use the figure below to help you prove that in an $\{n, k\}$ hyperbolic tessellation, the Euclidean distance d from the center of the Poincaré disc O to the vertex A of the polygon centered at O is given by $d = \sqrt{\frac{\tan(\pi/2-\pi/k)-\tan(\pi/n)}{\tan(\pi/2-\pi/k)+\tan(\pi/n)}}$.

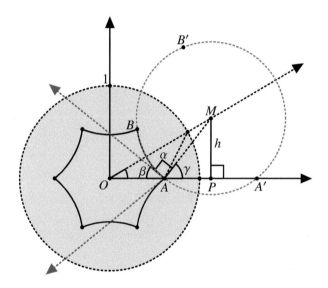

11 Elliptic Geometries

11.1 Introduction and Basic Results

In Chapter 10, we saw that hyperbolic geometry was the result of Saccheri's investigating the *hypothesis of the acute angle* (summit angles of a Saccheri quadrilateral are acute). We now investigate geometric results attained as a result of assuming the *hypothesis of the obtuse angle* (summit angles of a Saccheri quadrilateral are obtuse). As mentioned in the introduction to Chapter 10, Saccheri discovered that the hypothesis of the obtuse angle implied that the upper base is shorter than the lower base in a Saccheri quadrilateral, and the angle sum of any triangle is more than 180°. Saccheri reached a contradiction (to Proposition 17) in this case based on the assumption that lines have infinite length (see the proof of Theorem 10.3). Assuming the hypothesis of the obtuse angle (which implies that parallel lines do not exist, as shown in Theorem 11.2), and substituting the more general hypothesis that a straight line is *unbounded* (having no endpoints and hence could be finite like a circle) for the hypothesis that it is *infinite*, led to the development of two new non-Euclidean geometries now known as *elliptic* geometries. These were first developed by Riemann (1826–1866) and later generalized by Felix Klein (1849–1925), who showed that Euclidean, hyperbolic, and elliptic geometries can all be derived as special cases of a larger system called *projective geometry* [85]. Some of the material in this chapter was adapted from [44].

Note that as a result of allowing lines to behave like finite circles, the usual notion of *betweenness* of points on a line can no longer hold. For instance, given any three points on a circle, each of the points is between the other two. Furthermore, since lines are now finite, we may only construct circles of finite radii, and hence, Euclid's third postulate which implies the existence of arbitrarily large circles, needs to be modified accordingly. Given that straight lines are now considered to be unbounded and not necessarily infinite in extent, we are allowed to use the first 15 propositions of *The Elements*. We can also use certain propositions such as the ASA congruence condition for triangles (Proposition 26) and the ability to construct an angle at a point which is congruent to a given angle (Proposition 23), as these do not depend on any of the changes we have made to Euclid's original assumptions. Finally, we need to allow the use of *Pasch's Axiom* which states: *If a line intersects a side of a triangle, and does not intersect any of the vertices, it also intersects another side of the triangle.* Pasch's axiom, named after the German mathematician Moritz Pasch (1843–1930), can be derived from the *plane separation axiom* as mentioned in Chapter 10. However, in elliptic geometry, the plane separation axiom may not hold, so the assumption of Pasch's axiom needs to be explicitly made. We will discuss this further when we investigate models of elliptic geometry.

Under Euclid's original assumptions, stated and unstated, the existence of parallel lines was certain and was a consequence of Proposition 16 which relied on the infinitude of straight lines. Now that we have altered some of Euclid's fundamental assumptions, the existence of parallels is by no means certain. In fact, the contrary turns out to be true, that is, in elliptic geometry there are no parallel lines! Some other surprising results include:

1. All lines have the same finite length.
2. The area of a triangle is determined by its angle sum.
3. All lines that are perpendicular to a given line meet at a point. This point is called the *pole* of the line.
4. If the three corresponding angles of two triangles are congruent, then the triangles are congruent.

We now give our substitution for the parallel postulate.

Axiom 11.1: Elliptic Parallel Postulate

The summit angles of a Saccheri quadrilateral are obtuse.

Next we prove some results that hold in elliptic geometry. To help visualize some of these results, the reader is encouraged to view the elliptic plane as the surface of a sphere or the surface of a hemisphere. These and other models of the elliptic plane will be discussed further in Section 11.2.

The following two results deal with the Lambert quadrilateral which is a quadrilateral having three right angles.

Theorem 11.1

In elliptic geometry, if $ABCD$ is a Lambert quadrilateral with its right angles at A, B, and C, then $AD < BC$ and $DC < AB$.

Proof. If $AD = BC$, then $ABCD$ is a Saccheri quadrilateral and the elliptic parallel postulate implies that $\angle D$ and $\angle C$ are obtuse. This contradicts the fact that $\angle C$ is a right angle.

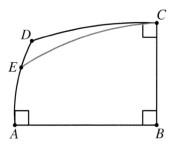

Figure 11.1

Furthermore, if $AD > BC$, let E be a point on AD such that $AE \cong BC$. Hence $ABCE$ is a Saccheri quadrilateral, and $\angle BCE$ is obtuse, which contradict the fact that it is in the interior of $\angle C$ which is a right angle. Therefore, $AD < BC$. Using similar arguments we can prove that $DC < AB$. ∎

11.1 Introduction and Basic Results

Corollary 11.1 follows from Theorem 11.1. The reader will be asked to prove it in the problem set at end of this chapter.

Corollary 11.1

In elliptic geometry, the fourth angle of a Lambert quadrilateral is obtuse.

The following theorem shows that elliptic geometry differs in an important way from both Euclidean geometry and hyperbolic geometry. Recall that Euclidean geometry and hyperbolic geometry are neutral geometries (both assume the first four Euclidean postulates). Euclid's Proposition 27, which is part of neutral geometry, guarantees the existence of at least one line parallel to a given line through any point not on the given line. Therefore, elliptic geometry cannot be a neutral geometry since parallel lines do not exist in elliptic geometry as Theorem 11.2 will show.

Theorem 11.2

In elliptic geometry, any two lines intersect.

Proof. By way of contradiction, let l and m be two lines that do not intersect, as in Figure 11.2. These lines are not of infinite extent, so there exists a point on l which is of least distance to m; call this point A. This implies that the length of the line segment perpendicular to m through A is less than or equal to the length of any other segment perpendicular to m from any other point on l. Let B be the perpendicular projection of A on m. We will now show that:

1. $\overline{BA}$ meets l at right angles.
2. There is a point E on l which is closer to m than A.

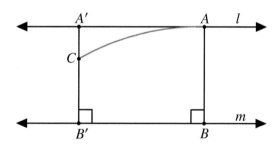

Figure 11.2

To prove (1), suppose $\overline{BA}$ does not meet l at right angles. Then one of the angles at A must be acute. Let A' be a point on l on the "side" of the acute angle; hence $\angle A'AB$ is acute (see Figure 11.2). Remember that in elliptic geometry, there are no "sides" of a line, because "betweenness" no longer applies, but we use the term here to help the reader visualize the argument. Let B' be the perpendicular projection of A' on m. Since $BA \leq B'A'$, let C be the point on $\overline{B'A'}$ such that $B'C = BA$. Then $BACB'$ is a Saccheri quadrilateral, and $\angle CAB$ is obtuse. However, $\angle CAB$ lies in the interior of $\angle A'AB$, which is acute, and this gives us a contradiction. Hence neither of the angles at A is acute, which means that $\overline{BA}$ meets l at right angles.

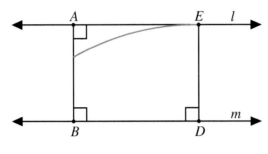

Figure 11.3

To prove (2), let D be any point on m distinct from B, and let E be where the perpendicular to m at D meets l, as in Figure 11.3. Then $BDEA$ is a Lambert quadrilateral, so by Theorem 11.1, $ED < AB$. However, this contradicts our assumption that $\overline{BA}$ was the shortest line segment perpendicular to m from a point on l. Therefore, our assumption at the beginning of the proof that l and m do not intersect is false, and hence any two lines in elliptic geometry must intersect. ∎

The following theorem shows that in elliptic geometry all perpendiculars to a given line are concurrent in a point called the *pole* of the line. It also shows that the distance from the pole to the line (measured along any of the perpendiculars) is constant. Finally, it shows that if lines are permitted to intersect in more than one point, then every line has two poles.

Theorem 11.3

In elliptic geometry, let P and Q be two points on a line l, and let lines α and γ be perpendicular to l at P and Q, respectively.

1. *If α and γ intersect at a point O, then all perpendiculars to l are concurrent at the point O. Furthermore, the distance from O to any point on l (measured along any of the perpendiculars) is constant [41].*

2. *All perpendiculars to l intersect in another point O'.*

Proof. To prove (1), consider Figure 11.4 in which $\triangle POQ$ is congruent to $\triangle QOP$ by ASA (Euclid's Proposition 26). Hence $\overline{OP} \cong \overline{OQ}$. First, let R_1 be the midpoint of $\overline{PQ}$. Then $\triangle POR_1$ is congruent to $\triangle QOR_1$ by SSS (Euclid's Proposition 8), which implies that $\angle PR_1O$ is a right angle. Since $\triangle OPR_1 \cong \triangle OR_1P$ by ASA, we have $\overline{OR_1} \cong \overline{OP}$. Second, similarly let R_2 and R_3 be the midpoints of $\overline{PR_1}$ and $\overline{QR_1}$, respectively, then $\overline{OP} \cong \overline{OR_2} \cong \overline{OR_3}$. Continuing this process, we can construct a sequence of points R_k that, at the nth stage,

11.1 Introduction and Basic Results

divide $\overline{PQ}$ into 2^n segments each of length $\frac{1}{2^n}$ times the length of $\overline{PQ}$, and such that $\overline{OP} \cong \overline{OR_k}$ for $k = 1, 2, 3, \ldots, 2^n - 1$. Using a limiting process when necessary, that is, if $R \neq R_k$ for any k, we can conclude that for any point R on $\overline{PQ}$ (R not shown), $\overline{OR}$ is perpendicular to l and $\overline{OR} \cong \overline{OP}$.

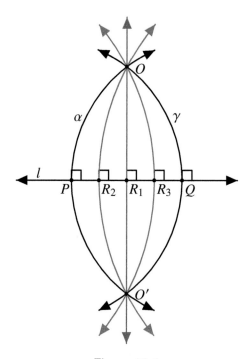

Figure 11.4

Next, let X (not shown in Figure 11.4) be any point on l not on $\overline{PQ}$. Assume WLOG that X is to the right of Q. Mark distinct points $Q_1, Q_2, \ldots, Q_n$ on l (to the right of Q) such that $\overline{PQ} \cong \overline{QQ_1} \cong \overline{Q_1 Q_2} \cong \cdots \cong \overline{Q_{n-1} Q_n}$ and $\overline{Q_{n-1} Q_n}$ contains X (these points are not shown in Figure 11.4). By SAS, $\triangle POQ \cong \triangle QOQ_1$, hence $\angle OQ_1 Q$ is a right angle. Similarly, $\triangle QOQ_1 \cong \triangle Q_1 OQ_2$, hence $\angle OQ_2 Q_1$ is a right angle. Continuing this process we have $\triangle POQ \cong \triangle QOQ_1 \cong \cdots \cong \triangle Q_{n-1} OQ_n$, and hence $\overline{OQ_i}$ is perpendicular to l for all $i = 1, 2, \ldots, n$. Applying the argument in the previous paragraph to $\triangle Q_{n-1} OQ_n$ yields that $\overline{OX}$ is perpendicular to l and $\overline{OP} \cong \overline{OX}$.

To prove (2), extend OP through P to a point O' such that $\overline{O'P} \cong \overline{OP}$, then connect O' to Q. Since $\angle OPQ \cong \angle O'PQ$, SAS gives $\triangle OPQ \cong \triangle O'PQ$, and hence $\overline{O'Q}$ is perpendicular to l. Consequently, O', Q and O lie on the line γ (why?). Repeating this argument we can show that any line perpendicular to l goes through O'. ∎

The conclusions of Theorem 11.3 are consistent with the behavior of great circles on the surface of a sphere. For example, all great circles perpendicular to the equator intersect in the north and south poles and the distance from one of the poles to the equator is one fourth the length of the equator.

Recall that in hyperbolic geometry, the upper base of a Saccheri quadrilateral is longer than the lower base and the altitude (the line segment joining the midpoints of the lower base and the upper base) is perpendicular to each of the bases and is shorter than the arms. Theorem 11.4 gives us the corresponding results in elliptic geometry.

Theorem 11.4

In elliptic geometry, the upper base of the Saccheri quadrilateral is shorter than the lower base, and the altitude is longer than both of the arms.

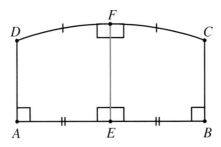

Figure 11.5

Proof. Let $ABCD$ be a Saccheri quadrilateral, and let E and F be the midpoints of the lower base and upper base, respectively, as in Figure 11.5. By Theorem 10.2, $\overline{EF}$ (the altitude of the Saccheri quadrilateral) is perpendicular to each of the bases. Therefore, $AEFD$ and $EBCF$ are both Lambert quadrilaterals. Theorem 11.1 implies that $DF < AE$ and $FC < EB$. Hence, $DC = DF + FC < AE + EB = AB$. Furthermore, Theorem 11.1 implies that $AD < EF$ and $BC < EF$. This completes the proof of Theorem 11.4. ∎

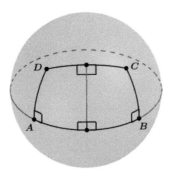

Figure 11.6

Figure 11.6 shows a more precise and realistic depiction of the Saccheri quadrilateral shown in Figure 11.5, now drawn on the surface of a sphere.

11.1 Introduction and Basic Results

An interesting property of both hyperbolic and elliptic geometries is that they locally behave almost like Euclidean geometry on small, restricted areas. For example, the smaller the area of a triangle in either elliptic or hyperbolic geometry is, the closer its angle sum is to 180°. This also means that many of the propositions and results of regular Euclidean plane geometry do hold in small portions of the elliptic plane. Recall that we are not allowed to use any of Euclid's propositions (past Proposition 15) that rely on the assumption that lines are infinite and may be extended indefinitely. However, if we restrict the area we are working in, and make our geometric figures sufficiently small, then we are able to extend lines "enough" to be able to use, at least locally, some propositions that may fail to hold for large figures. In Theorem 11.5, we restrict the area we are allowed to work in and use Euclid's Proposition 25 to prove that, locally, the angle sum of a triangle is greater than 180°.

> **Theorem 11.5**
>
> *Locally, the angle sum of an elliptic triangle is greater than 180°.*

Proof. Let $\triangle ABC$ be a right triangle with its right angle at B. Let D be the point so that $\overline{AD}$ is perpendicular to $\overline{AB}$ and $\overline{AD} \cong \overline{BC}$, as in Figure 11.7. Note that D in this proof exists for small triangles, but not necessarily for large ones (why?), hence the reason why we are working locally.

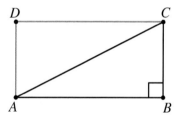

Figure 11.7

Then $ABCD$ is a Saccheri quadrilateral, and so $DC < AB$. Note that $\triangle ADC$ and $\triangle CBA$ are such that $AD = BC$, $\overline{AC}$ is a shared edge, and $DC < AB$. Euclid's Proposition 25 implies that $m(\angle DAC) < m(\angle BCA)$. Hence, $m(\angle BCA) + m(\angle BAC) > m(\angle DAC) + m(\angle BAC) = 90°$. Therefore, the angle sum of the right triangle $\triangle ABC$ is greater than 180°. Now since we may divide any triangle into two right triangles by constructing the altitude from the vertex opposite the longest side, we can conclude that for any small triangle in our restricted area, the angle sum is greater than 180°. Note that if the triangle has two or more equal longest sides, choosing any of these for the construction of the altitude would suffice. ∎

> **Theorem 11.6**
>
> *In general, the angle sum of an elliptic triangle is greater than 180°.*

Proof. Intuitively, the idea is to break up any triangle into smaller and smaller triangles until we are able to apply Theorem 11.5 to each of these smaller triangles, which implies that each of these triangles has angle sum greater than 180°. This in turn implies that the angle sum of the arbitrary triangle we started with is greater than 180° (why?).

More precisely, let ABC be an arbitrary triangle and O be point in its interior, as in Figure 11.8. Let A_1, B_1, and C_1 be the midpoints of the sides $\overline{BC}$, $\overline{AC}$, and $\overline{AB}$, respectively. Partition $\triangle ABC$ into six smaller triangles by connecting O to the midpoints of the sides and to the vertices. If the resulting triangles $\triangle OAC_1$, $\triangle OC_1B$, $\triangle OBA_1$, $\triangle OA_1C$, $\triangle OCB_1$, and $\triangle OB_1A$ are small enough to apply Theorem 11.5, then the sum of their angles is greater than 6π. This sum consists of the angles at A, B, C, A_1, B_1, C_1, and O. Note that the sum of the angles at A_1, B_1, and C_1 is $\pi + \pi + \pi = 3\pi$, and the sum of the angles at O is 2π. Hence the sum of the measures of the angles in the six nonoverlapping triangles in Figure 11.8 is $m(\angle A) + m(\angle B) + m(\angle C) + 3\pi + 2\pi = m(\angle A) + m(\angle B) + m(\angle C) + 5\pi > 6\pi$. Therefore, the sum of the angles of $\triangle ABC$ is greater than π.

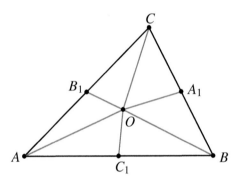

Figure 11.8

If necessary, repeat the partition process described earlier on each of the six smaller triangles discussed above producing 36 smaller triangles, as in Figure 11.9. Note for example that partitioning $\triangle OAC_1$ results in two new angles at X, two new angles at Y, two new angles at Z, and six new angles at O_1. The sum of the new angles in $\triangle OAC_1$ at X, Y, Z, and O_1 is $\pi + \pi + \pi + 2\pi = 5\pi$. If the resulting 36 smaller triangles are small enough to apply Theorem 11.5, then the sum of their angles is greater than 36π. This sum consists of the angles at A, B, C, A_1, B_1, C_1, O, and the angles in each of the 36 triangles corresponding to the angles at X, Y, Z and O_1 in $\triangle OAC_1$. Hence the sum of the measures of the angles in the 36 nonoverlapping triangles in Figure 11.9 is $m(\angle A) + m(\angle B) + m(\angle C) + 6(5\pi) + 5\pi > 36\pi$ (why?). Therefore, the sum of the angles of $\triangle ABC$ is greater that π.

Since the area of $\triangle ABC$ is finite, then for some finite number n, repeating the partition process described earlier n times is guaranteed to produce 6^n smaller triangles satisfying Theorem 11.5. Then $m(\angle A) + m(\angle B) + m(\angle C) + 6^{n-1}(5\pi) + \cdots + 6^2(5\pi) + 6(5\pi) + 5\pi = m(\angle A) + m(\angle B) + m(\angle C) + (6^n - 1)\pi > 6^n\pi$ (why?). Hence the sum of the angles of $\triangle ABC$ is greater than π.

11.1 Introduction and Basic Results

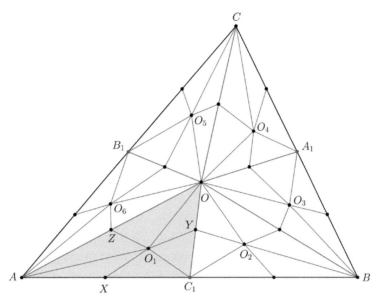

Figure 11.9

Theorem 11.7

In elliptic geometry, the angles of a quadrilateral sum to more than 360°.

Proof. Divide the quadrilateral into two triangles and apply Theorem 11.6. ∎

We now define the *excess* of a triangle in elliptic geometry to be the amount by which the angle sum of the triangle differs from 180°, and the *excess* of a quadrilateral to be the amount by which the angle sum of the quadrilateral differs from 360°.

Theorem 11.8

If a line cuts a triangle so as to form one quadrilateral and one triangle, or two triangles, then the excess of the original triangle is the sum of the excesses of the smaller triangle and quadrilateral, or the sum of excesses of the two smaller triangles.

The proof of Theorem 11.8 is the subject of Problem 7 at the end of this chapter. We next show that if the angles of one triangle are congruent to the angles of another triangle, then the triangles are congruent.

Theorem 11.9: AAA Congruence Condition

In elliptic geometry, if two triangles $\triangle ABC$ and $\triangle A'B'C'$ satisfy $\angle A \cong \angle A'$, $\angle B \cong \angle B'$, and $\angle C \cong \angle C'$, then the triangles are congruent.

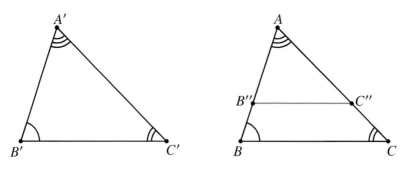

Figure 11.10

Proof. Let $\triangle ABC$ and $\triangle A'B'C'$ be triangles such that $\angle A \cong \angle A'$, $\angle B \cong \angle B'$, and $\angle C \cong \angle C'$, as in Figure 11.10. Since $\triangle ABC$ and $\triangle A'B'C'$ have equal angle sums, it follows that they have equal excesses. By way of contradiction, suppose that the triangles are not congruent. Then at least two sides of one of the triangles are longer than the corresponding sides in the other triangle (why?). WLOG assume $AB > A'B'$ and $AC > A'C'$. Construct B'' on $\overline{AB}$ and C'' on $\overline{AC}$ such that $\overline{AB''} \cong \overline{A'B'}$ and $\overline{AC''} \cong \overline{A'C'}$. Then by SAS, $\triangle AB''C'' \cong \triangle A'B'C'$, and hence they have equal excesses.

If F is a triangle or a quadrilateral, let $\text{EX}(F)$ denote the excess of F. Note that $\overline{B''C''}$ divides $\triangle ABC$ into a triangle and a quadrilateral. Then applying Theorem 11.8, we have

$$\text{EX}(\triangle ABC) = \text{EX}(BCC''B'') + \text{EX}(\triangle AB''C'')$$
$$= \text{EX}(BCC''B'') + \text{EX}(\triangle A'B'C')$$
$$= \text{EX}(BCC''B'') + \text{EX}(\triangle ABC).$$

Thus $\text{EX}(BCC''B'') = 0$, which contradicts Theorem 11.7. Therefore, $\triangle ABC$ and $\triangle A'B'C'$ are congruent. ∎

> **Corollary 11.2**
>
> *In elliptic geometry, similar triangles are always congruent.*

11.2 Models of Elliptic Geometry

Recall that using Euclid's first four postulates (the axioms of neutral geometry) and the right angle hypothesis (respectively, the acute angle hypothesis) resulted in Euclidean (respectively, hyperbolic) geometry. Would using Euclid's first four postulates and the obtuse angle hypothesis lead to a consistent geometry, that is, a geometry with no contradictory theorems? Since the obtuse angle hypothesis implies that parallel lines do not exist, this contradicts Euclid's Proposition 27, which guarantees (in neutral geometry) the existence of at least one line parallel to a given line through a point not on the given line. Therefore, using

11.2 Models of Elliptic Geometry

Euclid's first four postulates and the obtuse angle hypothesis does not lead to a consistent geometry.

Fixing this requires making some modifications to Euclid's first four postulates to remove the cause of this inconsistency, which is the assumption that straight lines are infinite. Assuming instead that straight lines are unbounded leads to straight lines being simple closed curves like circles, as in Figure 11.11. This in turn requires the abandonment of the notion of betweenness as we know it, since given three points A, B, and C on a circle, it is not clear which point is "between" the other two. *Betweenness* is replaced by *separation* in elliptic geometry and requires considering a fourth point D, as in Figure 11.11. Here we say that A is separated from B by D and C, or C is "between" A and B relative to point D.

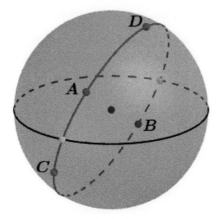

Figure 11.11

Double Elliptic Geometry

Allowing two lines in elliptic geometry to intersect in more than one point, as in Figure 11.11, leads to what is known as *double elliptic* geometry. This geometry is closely related to *spherical geometry*, which was studied by early Greek mathematicians and astronomers as part of solid geometry. However, double elliptic geometry is a non-Euclidean plane geometry (that uses the surface of the sphere as a model, see below) and should not be confused with Euclidean spherical geometry (the study of figures on the surface of the sphere), which is independent of any parallel postulate [90]. This subtle difference is occasionally overlooked in some resources on the subject.

The Beltrami–Riemann Sphere Model

To produce a model for double elliptic geometry, we define points to be points on the surface of a sphere and lines to be *great circles* (ones that cut the sphere in two congruent hemispheres). Great circles may be described geometrically as the intersections of the sphere with planes going through the center of the sphere.

The Beltrami–Riemann sphere model for double elliptic geometry satisfies the following properties [33]:

1. If P and Q are antipodal points, then infinitely many lines pass through them. However, if P and Q are not antipodal, then exactly one line goes through them.

2. Any two distinct lines intersect in two distinct antipodal points.

3. If r is the radius of the sphere, then the length of any line is $2\pi r$, which is the circumference of a circle of radius r.

4. If P and Q are nonantipodal points on the surface of the sphere, then the great circle going through P and Q defines two arcs connecting P and Q. We define the distance between P and Q to be the length of the shorter arc. If P and Q are antipodal points, then the two arcs defined by any great circle going through P and Q have the same length, and this length is taken to be the distance between P and Q. Hence the maximum distance between any two points in this model is πr.

5. If two distinct lines intersect at P and P', then the measure of the angle between the two lines is the measure of the angle between the tangents to the lines at P or P'. These tangents lie in the unique planes tangent to the sphere at P and P', and the angle at P has the same measure as the angle at P' (why?).

6. The area of the elliptic plane (surface of the sphere) is $4\pi r^2$.

7. Let A, B, C be three distinct, noncollinear points (not lying on the same great circle). Then no two of these three points are antipodal (why?). Thus A and B have a unique great circle going through them. Let $\overset{\frown}{AB}$ denote the shorter arc of this great circle with endpoints A and B. Similarly, let $\overset{\frown}{BC}$ and $\overset{\frown}{AC}$ denote the shorter arcs of the distinct great circles going through the corresponding pair of points. The *spherical triangle* $\triangle ABC$ consists of the three vertices A, B, C, and the three arcs $\overset{\frown}{AB}, \overset{\frown}{BC}$, and $\overset{\frown}{AC}$. This implies the measure of each angle in a spherical triangle is less than $180°$, and each triangle is confined to one hemisphere. Allowing "triangles" with angles greater than $180°$ or sides of length greater than πr is not very productive, as doing so leads to having obvious counterexamples to many basic results such as SSS, SAS, and ASA, etc. This will be explored is Now Try This 11.1.

8. All straight lines perpendicular to any given straight line meet in a pair of antipodal points. These points, called the *poles* of the line, have distance $\pi r/2$ from the line (why?).

9. The plane separation axiom holds in this model, since each line separates the elliptic plane into two half-planes, and any two distinct lines intersect in two antipodal points.

11.2 Models of Elliptic Geometry

> **Now Try This 11.1: SSS, SAS, ASA on the Whole Sphere**
>
> When defining triangles on the sphere, what happens if we allow triangles to have angles of measure greater than 180° or sides of length greater than πr?
>
> 1. Figure 11.12 shows that the Euclidean congruence condition SSS fails. To see this, consider $\triangle ABC$ (facing you) whose interior angles are α, β, and γ, and whose sides are the shorter arcs $\widehat{AB}$, $\widehat{BC}$, and $\widehat{AC}$. Can you see that the surface of the sphere minus the front-facing $\triangle ABC$ is a "triangle" covering more than half the sphere, and whose angles are the shaded angles at A, B, and C? Clearly these two "triangles" are not congruent despite the fact that they share the sides $\widehat{AB}$, $\widehat{BC}$, and $\widehat{AC}$.
>
> 2. Draw similar figures on the sphere to show that SAS and ASA also fail if we allow triangle sides of length more than πr.

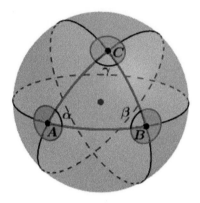

Figure 11.12

Single Elliptic Geometry

Requiring two lines in elliptic geometry to intersect in exactly one point leads to what is known as *single elliptic* geometry. We can try to use the surface of a sphere to model single elliptic geometry, but we need to resolve the issue of antipodal points having more than one line through them. This can be achieved by identifying pairs of antipodal points as single points, which results in the real projective plane. This plane is denoted by $\mathbb{RP}^2$ when formed starting with the unit sphere. One convenient way to model the real projective plane is to consider its points to be single points on the upper hemisphere and pairs of antipodal points on the equator. We will refer to this as the Klein half-sphere model (see below for more details). Projecting the points on the upper hemisphere vertically onto the points of the disc bounded by the equator, and identifying pairs of antipodal points on the equator as single points yields what will be referred to as the Klein disc model.

Next, we give more detailed descriptions of the Klein half-sphere and disc models in which any two lines intersect in exactly one point, and where the plane separation axiom does not hold. Showing that these models satisfy the first four Euclidean axioms (without the infinite extent property of lines) and the elliptic parallel postulate is the subject of Problem 18 at the end of this chapter.

The Klein Half-Sphere Model

We can visualize the surface of a sphere of radius r (with pairs of antipodal points identified as single points) as the upper hemisphere, including the equator, with pairs of antipodal points on the equator identified as single points. Any two points P and Q lying above the equator have a unique great semicircle going through them which intersects the equator in two antipodal points. Since these antipodal points are identified as a single point, the great semicircle through P and Q turns into a simple closed curve. The same is true if P lies on the equator and Q lies above it. If P and Q happen to be nonantipodal points on the equator, then the equator itself, with pairs of antipodal points on it identified as single points, represents a great semicircle through P and Q with its endpoints identified (see Figure 11.13).

Figure 11.13 schematically shows how identifying antipodal points on the equator can be visualized. First, hold the equator at C and C' and twist (rotate) the right-hand side by $180°$ so that A coincides with A'. Second, fold the resulting two small circles on top of each other so that B, C and D coincide with B', C' and D', respectively. This results in any two antipodal points being identified as a single point (the eight points in the figure are labeled merely to illustrate the required orientation).

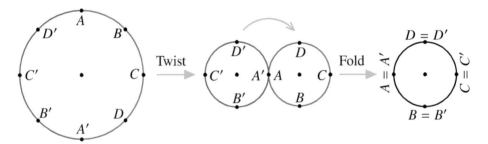

Figure 11.13

To summarize, points in the Klein half-sphere model are defined to be single points on the upper hemisphere (without the equator) together with pairs of antipodal points on the equator, and lines are great semicircles with their endpoint identified. A line being a simple closed curve ensures that a segment connecting two points can be extended continuously without bounds.

The Klein half-sphere model for single elliptic geometry satisfies the following [33]:

1. If P and Q are any two points, then exactly one elliptic line goes through them. You can think of such a line as a great semicircle with its endpoints identified.

2. If P and Q are distinct points, the elliptic line through them defines two arcs connecting P and Q. We define the distance between P and Q to be the length of the shorter arc. Recall that if r is the radius of the sphere and θ (measured in radians)

11.2 Models of Elliptic Geometry

is the central angle corresponding to the arc of the great circle going through P and Q, then the length of this arc is $r\theta$. For instance, let P and Q be on a great circle C with center O and radius r, and let P' and Q' be their respective antipodal points, as in Figure 11.14 (drawn in the plane). Then the distance $d(P,Q) = \min\{r\alpha, r\beta\}$. This is because $\{P, P'\}$ and $\{Q, Q'\}$ represent two points in this model and β may be smaller than α.

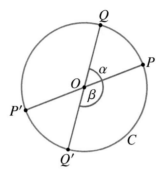

Figure 11.14

3. If r is the radius of the hemisphere, then the length of any elliptic line is πr, so the maximum distance between any two points $\pi r/2$. Hence the side length of a triangle in single elliptic geometry is at most $\pi r/2$.

4. The area of the elliptic plane (surface area of the hemisphere) is $2\pi r^2$.

5. The plane separation axiom does not hold in this model, as lines do not separate the elliptic plane into two disjoint half-planes. In Figure 11.15 the arc $\overparen{PAQ}$ does not intersect the line l (assume here that you are viewing the upper hemisphere from a point directly above the north pole).

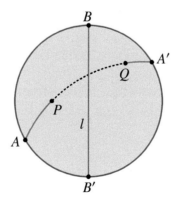

Figure 11.15

Figure 11.16 (see [33, page 257]) shows that the distance between P and Q on the hemisphere should be measured along the arc $\widehat{PAQ}$ not $\widehat{PEQ}$. It also shows two congruent vertical angles α and α'. Finally, it shows a number of triangles in the half-sphere model such as $\triangle ACD$. Note that $\triangle CRT$ is a triangle in this model while the three-sided figure $C'ETRDC'$ is not a triangle in this model (why?).

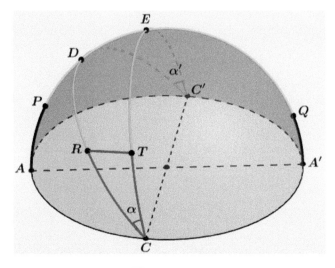

Figure 11.16

The Klein Disc Model

The projection of each point (x, y, z) lying on the upper hemisphere onto $(x, y, 0)$ in the xy-plane, takes the upper hemisphere to the disc bounded by the equator. Identifying pairs of opposite points (endpoints of diameters) on the boundary of the disc as single points leads to the Klein disc model for single elliptic geometry (don't confuse this model with the Klein disc model for hyperbolic geometry). In this model:

1. A point is either a point in the interior of the disc or a pair of endpoints of a diameter of the disc.
2. Lines are diameters of the disc or arcs of circles that intersect the boundary of the disc in two points lying on a diameter of the disc (see Figure 11.17) (why?).
3. Angles are measured as angles in the Euclidean plane.

This model can be viewed intuitively as taking the half-sphere model and flattening it onto a Euclidean plane, as in Figure 11.17.

11.2 Models of Elliptic Geometry

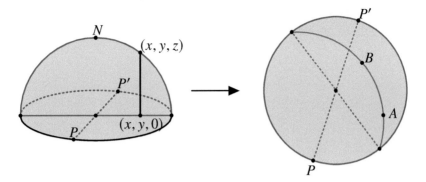

Figure 11.17

Given two points in the Klein disc model, how do we construct the elliptic line going through them? Recall that we used inversion to construct hyperbolic lines in the Poincaré disc, so can we use inversion again to construct elliptic lines in the Klein disc? Fortunately, we can, and it is very convenient to do so compared to carrying out the construction without the use of inversion.

As a way to motivate the construction, we will use stereographic projection instead of orthogonal projection to derive the Klein disc model. The construction below and its motivation were adapted from [49]. Figure 11.18 shows two antipodal points P and P' on the sphere of radius r centered at the origin O, and their respective images Q and Q' under stereographic projection from N (the north pole) onto the xy-plane halfway between the north and south poles. Note that this projection maps the lower hemisphere onto the disc bounded by the equator and the upper hemisphere onto the portion of the xy-plane lying outside the equator.

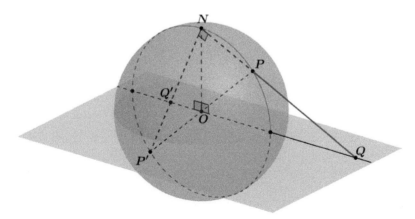

Figure 11.18

We now investigate how Q and Q' are related. The great circle going through N, P, and P' gets projected onto $\overleftrightarrow{OQ}$, a line going through O, the center of the sphere. Hence, O, Q, and Q' are collinear, and N, P, Q, O, and Q' lie in the same plane. Note that $\angle PNP'$ is a right angle (why?), and $\triangle ONQ'$, $\triangle OQN$, $\triangle Q'QN$ are right triangles. Hence we have

$(NQ')^2 = (OQ')^2 + r^2$, and $(NQ)^2 = (OQ)^2 + r^2$, where $r = ON$ is the radius of the sphere.

Substituting the values of $(NQ')^2$ and $(NQ)^2$ into the equation $(NQ')^2 + (NQ)^2 = (Q'Q)^2$, and using the fact that $Q'Q = OQ' + OQ$, we arrive at:

$$(OQ')^2 + 2r^2 + (OQ)^2 = (OQ')^2 + 2 \cdot OQ' \cdot OQ + (OQ)^2.$$

Therefore, $OQ \cdot OQ' = r^2$. The less careful reader might now be tempted to conclude that Q and Q' are inverse images of each other under inversion in the equator, but certainly that is not the case, since Q and Q' are on opposite sides of O.

However, the discussion above is very helpful, as it tells us that if Q' is a point on the disc bounded by the equator, P' is its preimage (under the stereographic projection defined above), and P is the antipodal point of P', then Q, the projection of P onto the xy-plane, satisfies $OQ \cdot OQ' = r^2$. It also shows that if Q and Q' are collinear with O in the xy-plane, have O between them, and satisfy $OQ \cdot OQ' = r^2$, then the preimages of Q and Q' on the sphere are antipodal points on a great circle passing through N (why?).

In general, given two points A and B in the Klein disc model, we now describe how to construct an elliptic line through them. We can think of the Klein disc as the disc bounded by the equator of a sphere, and consider the stereographic projection π from the north pole N onto the plane containing the equator. Let O be the center of the Klein disc.

1. If A, B, and O are distinct collinear points, then the preimages $\pi^{-1}(A)$ and $\pi^{-1}(B)$ lie on a great circle on the sphere going through N. This is true since the projection of a circle passing through N on the plane of the Klein disc is a line going through O (see Theorem 9.12 in Section 9.5). Hence the diameter of the Klein disc on which A and B lie is the elliptic line going through A and B.

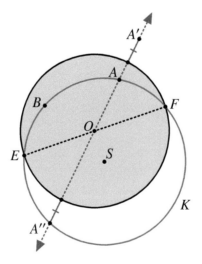

Figure 11.19

11.2 Models of Elliptic Geometry

2. If A, B and O are distinct noncollinear points, then WLOG let A' be the image of A under inversion in the boundary of the Klein disc (the equator). Let A'' be a point on $\overleftrightarrow{OA}$ such that $OA' = OA''$ and O is between A' and A'', as shown in Figure 11.19. Note that the preimages $\pi^{-1}(A)$ and $\pi^{-1}(A'')$ are antipodal points on the sphere (why?). Note also that the unique plane determined by $\pi^{-1}(A)$ and $\pi^{-1}(A'')$ and $\pi^{-1}(B)$ determines a great circle going through these three points on the sphere. This circle intersects the boundary of the Klein disc in two antipodal points E and F (why?). Since this circle does not go through N, its projection on the plane of the Klein disc is a circle going through A, B, A'', $\pi(E) = E$, and $\pi(F) = F$ (see Theorem 9.13 in Section 9.5). Hence the Euclidean circle K lying on the three points A, B and A'' determines the elliptic line going through A and B. Note that $\overline{EF}$ is a diameter of the Klein disc. The center S of the circle K is determined by the intersection of the perpendicular bisectors of the chords $\overline{AB}$ and $\overline{A''B}$.

Figure 11.20 shows an elliptic circle C centered at point S (in the Klein disc model) and two lines $\overleftrightarrow{PS}$ and $\overleftrightarrow{QS}$ that are orthogonal to the circle; hence their intersection with C determines the endpoints of two diameters of C. Note that E, the Euclidean center of C, is not necessarily the same as the elliptic center S (why?).

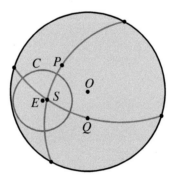

Figure 11.20

At first glance, Figure 11.21 appears as if to show an elliptic circle C centered at S consisting of two pieces (disconnected). However, the careful reader should quickly realize that this is not the case because U and V are the antipodal points of U' and V', respectively, and any two antipodal points are defined as a single point in the Klein disc model. Hence the circle C is connected and is not broken into two pieces. As mentioned before, the Euclidean center E of a circle C is not necessarily the same as the elliptic center S.

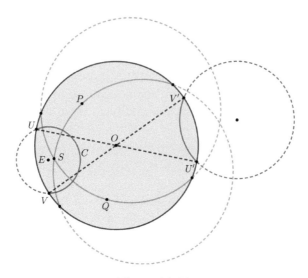

Figure 11.21

> **Now Try This 11.2: AAS Fails in Elliptic Geometry**
>
> 1. Figure 11.22, drawn in the Klein disc, shows that the Euclidean congruence condition AAS does not hold in elliptic geometry. Can you find two triangles in the figure that satisfy AAS but are not congruent? *Hint: γ is an elliptic circle centered at C.*
>
> 2. Draw a figure on the sphere showing that AAS does not hold in double elliptic geometry as well.
>
>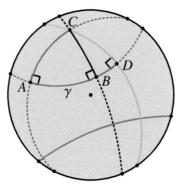
>
> Figure 11.22

Recall that we can think of the real projective plane as a sphere in $\mathbb{R}^3$ with pairs of antipodal points identified as single points. The following theorem, due to the French mathematician Albert Girard (1595–1632), demonstrates the fundamental result in elliptic geometry that the area of a triangle is determined by its angle sum. The proof of Girard's

11.2 Models of Elliptic Geometry

theorem for spherical triangles on the unit sphere is explored in the Problem 21 at the end of this chapter. Extending the proof to elliptic triangles in the real projective plane is the subject of Problem 22.

> **Theorem 11.10: Girard's Theorem**
>
> *The area of a triangle on the real projective plane is equal to its angle sum minus π. That is, the area of such a triangle is equal to its excess. (Note that we use radians here since we are working on the sphere.)*

The elliptic geometry models discussed earlier allow us to visualize how lines and figures behave in this geometry, at least on a portion of the elliptic plane. Results such as those in Theorems 11.2 and 11.6 no longer seem so strange, and one might even get an intuitive feel for how this geometry behaves. Applications of the results of elliptic geometry can be found in areas such as navigation (the shape of the earth is, after all, roughly spherical) and cosmology. Scientists believe that the shape of the universe may determine its future. If the universe is shaped so as to exhibit elliptic geometry, it is theorized that our universe at some point may cease to expand and begin to implode!

11.2.1 Problem Set

1. Describe how it is possible to have a triangle with three right angles in spherical geometry. What does this say about whether the Pythagorean theorem holds in spherical geometry or not?

2. Find an upper bound for the sum of the angles of a triangle in spherical geometry.

3. Can two lines in elliptic geometry have a common perpendicular? How about having more than one common perpendicular? Justify your answer.

4. The figure below shows a triangle $\triangle ABC$ in the elliptic Klein disc model and a line k going through the vertex C. What does the figure demonstrate?

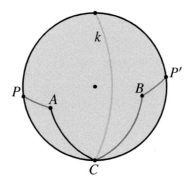

5. List the axioms and assumptions we used in our development of elliptic geometry.

6. Explain why Corollary 11.1 follows from Theorem 11.1.

7. Discussion of Theorem 11.8:
   a. Write a proof of Theorem 11.8.
   b. Consider the case in Theorem 11.8 where we partition the original triangle into two smaller triangles. Then cut the two smaller triangles into even smaller triangles. How do the excesses of the smaller triangles relate to the excess of the original triangle? In general, if we proceed in this manner of cutting up a triangle into smaller and smaller triangles, how do the excesses of the smaller triangles relate to the excess of the original triangle?

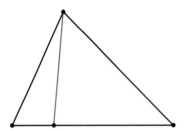 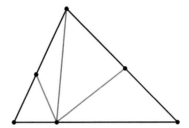

8. Does an upper bound for the excess of an elliptic triangle exist? If so, what is it?

9. It can be shown that the area of an elliptic triangle is proportional to its excess. The proof of this is outlined in Problem 21 for a triangle on a sphere of radius 1 (in double elliptic geometry). A similar proof can be given for a triangle on the upper hemisphere (in single elliptic geometry). If an elliptic triangle T_e of excess e has its area defined by $A(T_e) = k \cdot e$ for a positive constant k, what is an upper bound for the area of a triangle? You can think of k as R^2, where R is the radius of the sphere.

10. Given an elliptic triangle, construct a Saccheri quadrilateral of equal area by following the steps in (a) and (b) below. Use great circles of the sphere as lines.
    a. Let $\triangle ABC$ be a triangle in the elliptic plane so that $\overline{AB}$ is the longest side. Construct the midpoints D and E of $\overline{AC}$ and $\overline{BC}$, respectively, and draw the line $\overleftrightarrow{DE}$.
    b. Construct the perpendicular to $\overleftrightarrow{DE}$ through A, the perpendicular to $\overleftrightarrow{DE}$ through B, and the perpendicular to $\overleftrightarrow{DE}$ through C. Label the points of intersections of these perpendiculars with $\overleftrightarrow{DE}$ as F, G, and H, respectively.
    c. Prove that $ABGF$ is a Saccheri quadrilateral, and that the area of $ABGF$ is equal to the area of $\triangle ABC$

11. Given a Saccheri quadrilateral in the elliptic plane, describe a procedure for constructing an elliptic triangle of equal area.

12. Prove that in elliptic geometry an angle inscribed in a semicircle is always obtuse.

13. Use a spherical object such as a ball to represent the sphere. Using rubber bands to represent great circles, construct a Saccheri quadrilateral on the surface of the ball.

11.2 Models of Elliptic Geometry

Approximate the measure of the summit angles with a protractor. Are they acute, obtuse, or equal? How does the upper base compare to the lower base?

14. Prove that in the double elliptic geometry on the sphere, no two vertices of a triangle are antipodal points.

15. In the double elliptic geometry on the sphere, two distinct lines form a *lune*, which is defined to be the area bounded by two great semicircles meeting at antipodal points. Prove that the two angles of a lune are congruent.

16. In the model of elliptic geometry on the unit sphere, one way of ensuring the uniqueness of a line determined by two points is to identify antipodal points. Do the following:
  a. Draw a sphere and demonstrate three pairs of antipodal points on it.
  b. When we identify antipodal points, the resulting surface is called the real projective plane. Imagine what such a surface must look like. Try to draw a picture of how you imagine it.
  c. Explain why identifying antipodal points ensures the uniqueness of the line determined by two distinct points on the sphere.

17. Consider the real projective plane (formed by identifying antipodal points on the surface of the unit sphere) as a model of elliptic geometry.
  a. The length of a line on the unit sphere is 2π. What is the length of a line in the real projective plane?
  b. The area of the unit sphere is 4π. What is the area of the real projective plane?

18. Show that the Klein half-sphere and disc models for single elliptic geometry satisfy the first four Euclidean axioms (without the infinite extent property of lines) and the elliptic parallel postulate which is equivalent to the statement that any two lines intersect.

19. In elliptic geometry, prove that if O is the pole of l and O' is the pole of l', then the distance from O to l measured along a perpendicular is the same as the distance from O' to l'. This distance is called the *polar distance*.

20. Use the polar distance to show that all lines in elliptic geometry have the same finite length. What is this length?

21. Exploring the proof of Theorem 11.10 (Girard's theorem) on the unit sphere:
  a. Let $\triangle ABC$ be a triangle on the unit sphere determined by lines l, m, and k, as in the figure below.

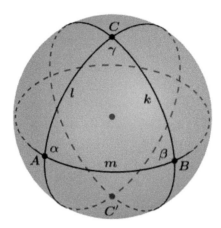

Then lines *l* and *k* meet at antipodal points *C* and *C'* and determine a lune. What is the area of the lune (see Problem 15) in terms of the angle γ? You may want to use the two hints given below.

- *Hint 1:* γ is the angle of the triangle at point *C*, but it is also the central angle formed by the great circle that divides the lune into two congruent parts, as shown in the figure above.
- *Hint 2:* Since $360° = 2\pi$ radians, the area of the lune times $2\pi/\gamma$ equals the area of the sphere. Convince yourself that this is true, and then use it to determine the area of the lune.

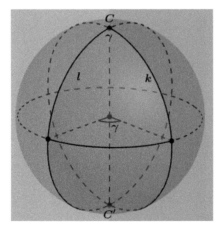

b. What is the area of the lune determined by lines *l* and *m*? By lines *m* and *k*?

c. Since great circles of the unit sphere meet at antipodal points and divide the sphere into two congruent hemispheres, for every triangle in one hemisphere there is a corresponding *antipodal triangle* in the opposite hemisphere. Show that the area of a triangle is equal to the area of its antipodal triangle.

11.2 Models of Elliptic Geometry

d. Imagine we could break apart and flatten the sphere and put it on a table to look at. Then it might look something like the figure below, where Δ'_i is the antipodal triangle of Δ_i, C' is the antipodal point of C, and so on. Then the area of Δ equals the area of Δ' (not shown), the area of Δ_1 equals the area of Δ'_1, and so on. Suppose that Δ is the same triangle as in part (a); what is the area of $\Delta + \Delta_1$, $\Delta + \Delta_2$, and $\Delta + \Delta_3$?

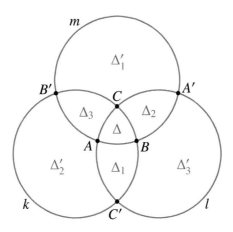

e. The area of the unit sphere is 4π, so $2\Delta + 2\Delta_1 + 2\Delta_2 + 2\Delta_3 = 4\pi$. Use this, and your results from part (d), to show that the area of Δ equals $\alpha + \beta + \gamma - \pi$ as required by Theorem 11.10.

22. Make the necessary adjustments to the proof outlined in Problem 21 to show that Girard's theorem holds for elliptic triangles on the real projective plane. You can visualize the elliptic triangles as triangles in the Klein half-sphere or disc model.

23. Using dynamic geometry software, construct a regular quadrilateral in the Klein disc model such that each of its angles is $2\pi/3$. What is the area of this quadrilateral?

24. Using dynamic geometry software, justify that Euclid's Proposition 16 does not hold in general in single or double elliptic geometry (see figure below). How is the validity of Proposition 16 related to the lengths of the medians of a triangle?

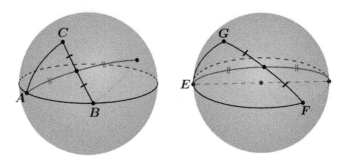

25. Prove that the Euclidean congruence condition AAS does not hold in elliptic geometry.

26. Given any two points A and B in the Klein disc model, use dynamic geometry software to construct the elliptic line $\overleftrightarrow{AB}$. How does the line change as the locations of A and B change?

27. Given points S and P in the Klein disc model, use dynamic geometry software to construct the elliptic circle centered at S and going through P. What happens when S is close to the center of the disc? How about when S is close to the boundary of the disc?

12 Projective Geometry

12.1 Introduction and Early Results

We saw in Chapters 10 and 11 that abandoning Euclid's fifth postulate led to the discovery of the hyperbolic and elliptic geometries. If, in addition to parallelism, other familiar notions such as congruence, distance, angle measure, circles, and betweenness are abandoned, what type of geometry can that lead to and is it worth studying? Surprisingly enough, such drastic changes lead to a practical and important geometry called *projective geometry*, which has important applications in many fields, especially in the area of computer graphics and computer vision. In computer graphics the primary objective is to convert a three-dimensional object into a photo realistic two-dimensional image on the computer screen. Conversely, identifying a three-dimensional object from a two-dimensional image of it is a typical problem in computer vision. Projective geometry is better suited than Euclidean geometry for establishing a mathematical correspondence between the points of a three-dimensional object and the points in its two-dimensional image.

Projective geometry has its roots in the visual arts, as artists and architects investigated ways to accurately and more realistically portray a three-dimensional scene on a two-dimensional plane in the form of a piece of paper or canvas. Those investigations led to the discovery of the *theory of perspective* reportedly by the Italian architect Bunelleschi in 1415, which was later documented by Leon Battista Alberti in 1435. This theory suggests that the eye of the painter is connected to points on the landscape by straight lines, the intersection of which with a two-dimensional plane generates a drawing of the landscape. Among the first artists to consider this problem of perspective drawing in a scientific way was Leonardo da Vinci (1452–1519), whose work provides rich examples of paintings where mathematical perspective is employed. To investigate projective geometry we will assume basic knowledge of some fundamental principles of three-dimensional geometry such as the fact that any three noncollinear points determine a unique plane, distinct planes that intersect do so in a line, a line connecting two points that lie in a plane lies entirely in that plane, and through a point not on a given plane exactly one plane passes that is parallel to the given plane. Some of these fundamental results are discussed in Chapter 15. Some of the material in this section was adapted from [93].

Central Projection

In projective geometry the concept of *central projection*, defined below, is of primary importance. The reader is encouraged to contrast this with *parallel projection*, which plays an important role in Euclidean geometry.

> **Definition: Central Projection (Central Perspectivity) in $\mathbb{R}^3$**
>
> Let H and H' be two-dimensional planes in three-dimensional space and E be a point that does not lie on either plane. *Central projection (central perspectivity)* from H to H' assigns to each point P in H the point P' in H' determined by the intersection of $\overleftrightarrow{EP}$ with H' (if it exists). The point E (for eye) is called the *center of projection* and we say that P' is the projection of P onto the plane H' from (or with respect to) the center E.

Figure 12.1 shows a square $ABCD$ lying in the plane H in three-dimensional space and its central projection $A'B'C'D'$ onto the plane H' from the point E, which does not lie in H or H'. For example, the image of the point $A \in H$ is the point $A' \in H'$, which is the intersection of $\overleftrightarrow{EA}$ with the plane H'. Note that each point in the plane H has a unique image in H', except for the points lying on h, the *vanishing line* in the plane H. This line is the intersection of H and the unique plane parallel to H' and going through E.

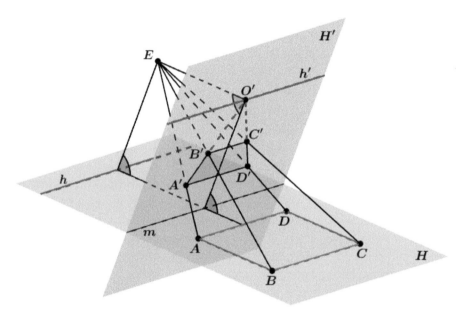

Figure 12.1

Conversely, each point in H' has a unique preimage in H, except for the points on h', the vanishing line in H'. The line h' is the intersection of the unique plane parallel to H and going through E. Therefore, central projection is not a one-to-one correspondence between the points of H and the points of H'. Families of parallel lines in H (that are not parallel to H') are projected onto families of lines in H' that intersect at points on h'. Hence h' is often referred to as the vanishing line or horizon of the plane H. Similarly, h is the vanishing line or horizon of H'.

12.1 Introduction and Early Results

Common examples of central projections are shadows cast by point light sources such as street lamps, and images cast on a screen by a video projector. Figure 12.2 shows the shadows cast by window frames from a point light source inside a room.

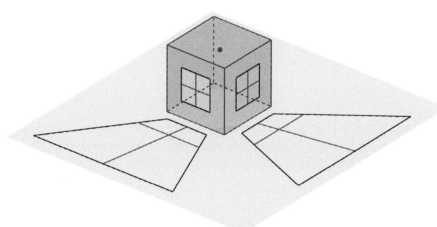

Figure 12.2

Careful examination of Figure 12.1 reveals several properties that central projections have. See if you can discover some of these before reading on. Next, we state a few of these properties and provide some justification for them.

1. Angle measure is not necessarily preserved under central projection, and the same is true for distance. Note that the image of the square $ABCD$ in Figure 12.1 is a trapezoid in H' whose angles and sides are not congruent to the corresponding angles and sides of $ABCD$.

2. Lines in H (other than h) that are parallel to m (the line of intersection of H and H') are projected onto lines in H' that are parallel to m. However, these are not the only lines in H that get projected onto parallel lines in H'; can you think of any other such lines?

3. Parallel lines in H are not necessarily projected onto parallel lines in H'. For instance, $\overleftrightarrow{AB}$ and $\overleftrightarrow{CD}$, which are parallel in H, are projected onto $\overleftrightarrow{A'B'}$ and $\overleftrightarrow{C'D'}$ which intersect at a *vanishing point* O' on the vanishing line h' in H'. In fact, the family of all lines parallel to $\overleftrightarrow{AB}$ in H are projected onto a family of lines in H' all of which intersect in the point O'. This should remind the reader of the image of railroad tracks going off into the distance appearing to converge at a distant vanishing point, as in Figure 12.3.

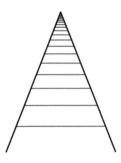

Figure 12.3

4. Central projection is a one-to-one correspondence between points in H (other than those on h) and points in H' (other than those on h'). We will deal with the vanishing lines h and h' a little later.

5. Central projection maps points onto points and lines onto lines. More importantly, it preserves the *incidence relations* between points and lines. That is, if a point P is on a given line in H, then its image P' is on the image of the given line in H'. Also if two lines intersect in a point Q in H not on the vanishing line in H, then their images intersect in the image of Q in H'.

6. As a consequence of Property 5, we can say the image of a triangle under central projection is a triangle, and the same is true for a quadrilateral and other polygons.

7. If the plane H is parallel to the plane H', then the central projection is a central similarity (also known as homothety or dilation).

8. Lines in H, intersecting in a point on the vanishing line h, are projected onto parallel lines in H' that are not parallel to m. Conversely, the preimages of parallel lines in H' that are not parallel to m are lines in H that meet at points on h.

9. Families of parallel lines in H that are not parallel to h are projected onto families of lines in H' that intersect in points on h'.

To show that a central projection maps lines in H onto lines H' (see Property 5), let $l \neq h$ be a line in H. Note that the family of lines joining the points on l to the center of projection E form a plane that is not parallel to H'. The intersection of this plane with H' is a line l', which is the image of l under central projection from E. Similarly, every line $l' \neq h'$ in H' is the image of a line $l \neq h$ in H. The lines l and l' are not shown in Figure 12.1.

To justify Property 8, let l_1 and l_2 be two lines in H intersecting in a point P on the vanishing line h in plane H, as in Figure 12.4. Note that the planes H_1 and H_2 determined by E and each of the lines l_1 and l_2, respectively, have $\overleftrightarrow{EP}$ in common. Since h is the intersection of H and the plane parallel to H' and going through E, we have $\overleftrightarrow{EP}$ is parallel to H'. Hence l'_1 and l'_2, the images of l_1 and l_2 under central projection from E, are parallel in H' (why?). A similar argument shows that the converse of Property 8 is also true.

12.1 Introduction and Early Results

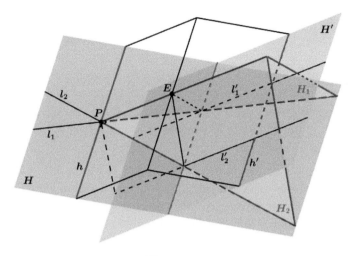

Figure 12.4

Now Try This 12.1

1. Justify Property 9 by showing that the plane H_1 determined by the center of projection E and the line l_1 and the plane H_2 determined by E and l_2 intersect in a line parallel to H, and have a point P' in common with the vanishing line h' in H' (see Figure 12.5). Hence l'_1 and l'_2 intersect in a point P' on h'.

2. How are lines in H that are parallel to h projected?

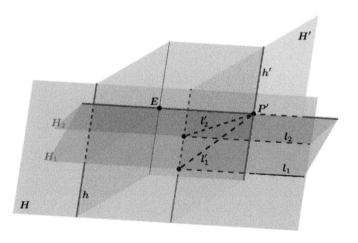

Figure 12.5

We now describe how to deal with h and h', the vanishing lines in the planes H and H', respectively. What line in H is h' the image of and what line in H' is h the preimage of? Clearly, ordinary lines in H and H' cannot be the answer, since these are already paired in a one-to-one fashion by the central projection. We could try to extend the planes H and H' by adding new points to each plane infinitely far from m (the line where the planes intersect) to correspond to points on h' and h, respectively. The new points are called *points at infinity* and each family of parallel lines in H or H' will be assigned a single point at infinity. The set of points at infinity appended to H' are then taken to be a *line at infinity* onto which the line h is projected. Similarly, the points at infinity appended to H are taken to be a line at infinity whose image is the vanishing line h'. This results in the central projection being a bijection between the extended planes. Extending the Euclidean plane by adding points at infinity to produce the real projective plane will be discussed in more detail in Section 12.3.

Despite the fact that central projection (as shown in Figure 12.1) does not preserve notions such as parallelism, angle measure, or distance, the square $ABCD$ and its image $A'B'C'D'$, however, continue to share some geometric properties. For instance, both are quadrilaterals and the incidence relations (whether a point lies on a line or whether two lines intersect) between their corresponding sides are preserved. What if the square $ABCD$ or its image $A'B'C'D'$ is projected from a different point E' (or onto a different plane) yielding a different image $A''B''C''D''$ (not shown in Figure 12.1); how are these images related? Clearly, the images will be different depending on the position of the plane of projection and the center of projection. But both images are projections of the same square, so surely they must share some common geometrical properties as we will see later. For now, we will just say that it's not always possible to find a central projection taking A' to A'', B' to B'', C' to C'', and D' to D''.

Figure 12.6 (drawn in a two-dimensional plane) shows the projection of line l onto line m with respect to the center E, and the projection of line m onto line n with respect to the center F. It also shows that the composition of the projection from E and the projection from F, which maps l onto n (taking A to A'', B to B'', and C to C''), is not a central projection, since $\overleftrightarrow{AA''}$, $\overleftrightarrow{BB''}$, and $\overleftrightarrow{CC''}$ are clearly not concurrent in a point X. This shows that the composition of two central projections is not necessarily a central projection.

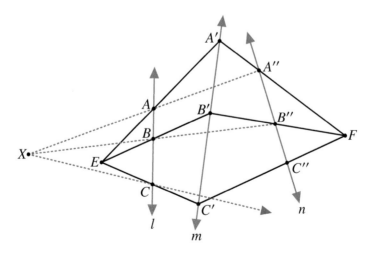

Figure 12.6

12.1 Introduction and Early Results

Desargues' Theorem

Projective geometry focuses on studying those properties of figures that remain unchanged under projection. The systematic study of projective geometry began with the French mathematician Girard Desargues (1591–1661) who investigated properties preserved under perspective mapping. Desargues concluded that *parallel lines have a common point at an infinite distance, and if none of the points on a line are at a finite distance, then the line itself is at an infinite distance.* Desargues discovered and proved a basic result in projective geometry dealing with perspective triangles, now known as Desargues' theorem. We next provide a proof of Desargues' theorem using central projection, which shows the power of this transformation as problem-solving tool. Before doing so, we will define what it means for triangles to be in perspective.

Definition: Perspective

Two triangles $\triangle ABC$ and $\triangle A_1 B_1 C_1$ are said to be *perspective* (or *in perspective*) from a point O if the lines $\overleftrightarrow{AA_1}$, $\overleftrightarrow{BB_1}$, and $\overleftrightarrow{CC_1}$ are concurrent in the point O (the *perspective center*).

Theorem 12.1: Desargues' Theorem

Let $\triangle ABC$ and $\triangle A_1 B_1 C_1$ be two triangles that are in perspective, as in Figure 12.7. If P, Q, and R are the points of intersection of the corresponding sides of these triangles, then these points are collinear.

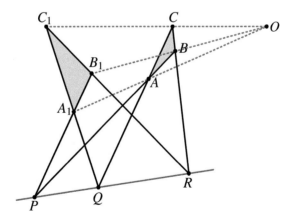

Figure 12.7

Proof. Consider two planes H and H' intersecting in a line m and a point E not on either plane. We have shown earlier (see Property 8) that under central projection from H to H' with center E, the preimages of a set of parallel lines in H' (that are not parallel to m) is a set of lines in H that meet at a point on h (the vanishing line in H). Thus, suppose we are able to project the plane H containing Figure 12.7 onto a plane H' such that the corresponding sides of the projected triangles $\triangle A'B'C'$ and $\triangle A'_1 B'_1 C'_1$ in H' are parallel, that is, $\overleftrightarrow{A'B'} \parallel \overleftrightarrow{A'_1 B'_1}$, $\overleftrightarrow{B'C'} \parallel \overleftrightarrow{B'_1 C'_1}$, and $\overleftrightarrow{A'C'} \parallel \overleftrightarrow{A'_1 C'_1}$. Then we can conclude that the corresponding sides of the triangles $\triangle ABC$ and $\triangle A_1 B_1 C_1$ in H, that is, $\overleftrightarrow{AB}$ and $\overleftrightarrow{A_1 B_1}$, $\overleftrightarrow{BC}$ and $\overleftrightarrow{B_1 C_1}$, $\overleftrightarrow{AC}$ and $\overleftrightarrow{A_1 C_1}$, intersect in points lying on the vanishing line h in the plane H [93].

We now describe how to determine the plane H' and the center of projection E so that $\overleftrightarrow{PQ}$ is the vanishing line in H. Let H' be a plane perpendicular to H and intersecting it in a line parallel to $\overleftrightarrow{PQ}$ (other than $\overleftrightarrow{PQ}$). Next, choose the center of projection E (not shown in Figure 12.8) to be a point on the intersection line of a convenient plane parallel to H and the plane perpendicular to H and intersecting it in $\overleftrightarrow{PQ}$. Figure 12.8 shows the result of projecting the plane H (containing Figure 12.7) onto the plane H' determined above. As desired, $\overleftrightarrow{PQ}$ is the vanishing line in H under this projection (it is the intersection of H and the plane parallel to H' and going through the center of projection).

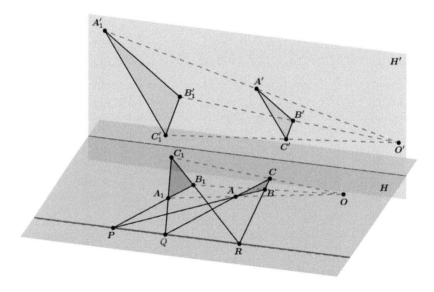

Figure 12.8

Note that since $\overleftrightarrow{AB}$ and $\overleftrightarrow{A_1 B_1}$ intersect at the point P on the vanishing line in H, their projections $\overleftrightarrow{A'B'}$ and $\overleftrightarrow{A'_1 B'_1}$ are parallel in H', and hence $\triangle O'A'B'$ is similar to $\triangle O'A'_1 B'_1$. Similarly, since $\overleftrightarrow{AC}$ and $\overleftrightarrow{A_1 C_1}$ intersect at the point Q on the vanishing line in H, then $\overleftrightarrow{A'C'}$ and $\overleftrightarrow{A'_1 C'_1}$ are parallel in H', and hence $\triangle O'A'C'$ is similar to $\triangle O'A'_1 C'_1$.

This implies that $\frac{O'B'}{O'B'_1} = \frac{O'A'}{O'A'_1}$ and $\frac{O'C'}{O'C'_1} = \frac{O'A'}{O'A'_1}$. Hence $\frac{O'B'}{O'B'_1} = \frac{O'C'}{O'C'_1}$, which means that $\triangle O'B'C'$ is similar to $\triangle O'B'_1 C'_1$, and hence $\overleftrightarrow{B'C'}$ is parallel to $\overleftrightarrow{B'_1 C'_1}$ (why?). Therefore, R, the point of intersection of $\overleftrightarrow{BC}$ and $\overleftrightarrow{B_1 C_1}$, also lies on the vanishing line in H. Hence the

12.1 Introduction and Early Results

points P, Q, and R are collinear. Note that the fact that H' was chosen to be perpendicular to H did not play any role in the proof, so this requirement was just for convenience and is not necessary. ■

The converse of Desargues' theorem states that *if the points of intersection of the corresponding sides or two coplanar triangles are collinear, then the triangles are in perspective*. The reader will be asked to prove this in Problem 1 at the end of this section.

What if $\triangle ABC$ and $\triangle A_1 B_1 C_1$ are in perspective, but a pair of their corresponding sides are parallel and hence do not intersect at a finite point in the plane containing the triangles, as in Figure 12.9? Does Desargues' theorem still hold? It does if we assume (as Desargues concluded) that lines parallel to the same line all have in common a point at infinity, and all these points at infinity form a line called the line at infinity. For instance, suppose $\overleftrightarrow{AB}$ and $\overleftrightarrow{A_1 B_1}$ are parallel in the Euclidean sense, as in Figure 12.9. Suppose also that $\overleftrightarrow{BC}$ and $\overleftrightarrow{B_1 C_1}$ intersect in R, and $\overleftrightarrow{AC}$ and $\overleftrightarrow{A_1 C_1}$ intersect in Q. Then $\overleftrightarrow{QR}$ is parallel to $\overleftrightarrow{AB}$ and $\overleftrightarrow{A_1 B_1}$ in the Euclidean sense (see Problem 5 at the end of this section), and hence $\overleftrightarrow{AB}$ and $\overleftrightarrow{A_1 B_1}$ meet $\overleftrightarrow{QR}$ in the point at infinity appended to $\overleftrightarrow{QR}$. Similarly, if two pairs of corresponding sides of the triangles are parallel, then the third pair is parallel (why?), and hence each pair of corresponding sides meets at a point at infinity, and all these points lie on the line at infinity. Therefore, Desargues' theorem holds in general in the real projective plane, which will formally be defined in Section 12.3.

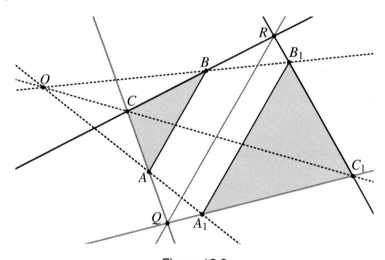

Figure 12.9

Cross Ratio

Desargues also discovered that the *cross ratio* of four collinear points is invariant under central projection. More specifically, if A, B, C, and D lie on a line l, and A', B', C', and D' are their respective projections on l' with respect to a point O, as in Figure 12.10, then

$$\frac{AC}{BC} \div \frac{AD}{BD} = \frac{A'C'}{B'C'} \div \frac{A'D'}{B'D'}.$$

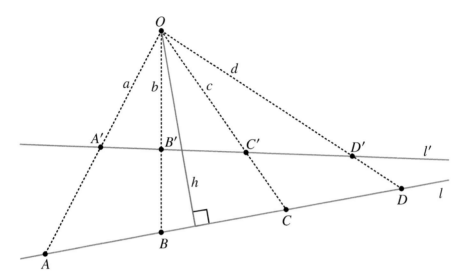

Figure 12.10

But where does this ratio come from and how do we know that it's invariant under central projection? To answer this, we will give an elementary motivation for the concept of cross ratio, which will lead to an elementary proof of its invariance under projection. Figure 12.10 shows four collinear points A, B, C, and D lying on line l (in that order) and their images A', B', C', and D' (lying on l') under projection from O. How are the distances between the points on l and the distances between their projections on l' related? For example, is the ratio $\frac{AB}{BC}$ always equal to the ratio $\frac{A'B'}{B'C'}$? You can easily see that these ratios are not equal by slightly rotating l about B which changes the ratio $\frac{AB}{BC}$ but does not change $\frac{A'B'}{B'C'}$. Let's now investigate other types of ratios to see how they are affected by projection.

For clarity purposes and ease of notation, let $a = OA$, $b = OB$, $c = OC$, and $d = OD$. Also, let h be the vertical distance from O to l. Recall the basic Euclidean result stating that the area of an arbitrary triangle XYZ equals $\frac{1}{2} xy \sin(\angle Z)$ (why?), where x and y are the lengths of the sides opposite $\angle X$ and $\angle Y$, respectively. Applying this result to $\triangle OAB$, $\triangle OBC$, $\triangle OCD$, $\triangle OAC$, $\triangle OAD$, and $\triangle OBD$ gives the following:

(1) $AB \cdot h = ab \sin(\angle AOB)$,   (4) $AC \cdot h = ac \sin(\angle AOC)$,

(2) $BC \cdot h = bc \sin(\angle BOC)$,   (5) $AD \cdot h = ad \sin(\angle AOD)$,

(3) $CD \cdot h = cd \sin(\angle COD)$,   (6) $BD \cdot h = bd \sin(\angle BOD)$.

Our goal is to find ratios among the lengths of the segments connecting pairs of the points A, B, C, and D that depend only on the angles at O and not on a, b, c, or d. Such ratios will be invariant under projection (why?).

For example, dividing Equations (4) by (2) and (5) by (6) gives

(7) $\dfrac{AC}{BC} = \dfrac{a \sin(\angle AOC)}{b \sin(\angle BOC)}$,   (8) $\dfrac{AD}{BD} = \dfrac{a \sin(\angle AOD)}{b \sin(\angle BOD)}$.

12.1 Introduction and Early Results

Note the common factor $\frac{a}{b}$ in (7) and (8). Dividing Equations (7) by (8) eliminates this factor and results in an invariant ratio of ratios that depends only on the angles at O:

$$\frac{AC}{BC} \div \frac{AD}{BD} = \frac{\sin(\angle AOC)}{\sin(\angle BOC)} \div \frac{\sin(\angle AOD)}{\sin(\angle BOD)}.$$

Repeating the steps above using the points A', B', C', and D' instead of A, B, C, and D gives

$$\frac{A'C'}{B'C'} \div \frac{A'D'}{B'D'} = \frac{\sin(\angle A'OC')}{\sin(\angle B'OC')} \div \frac{\sin(\angle A'OD')}{\sin(\angle B'OD')}.$$

But clearly, $\angle AOC \cong \angle A'O'C'$, $\angle BOC \cong \angle B'OC'$, etc. Hence

$$\frac{AC}{BC} \div \frac{AD}{BD} = \frac{A'C'}{B'C'} \div \frac{A'D'}{B'D'}.$$

This shows that the ratio $\frac{AC}{BC} \div \frac{AD}{BD}$, which is often denoted by $(A, B; C, D)$, is invariant under projection.

It should be mentioned here that cross ratios are generally computed using directed segments (for example, $AB = -BA$), but since we are only concerned with positive distances here (lengths of segments), we will neglect this aspect of the cross ratio.

> **Now Try This 12.2**
>
> In the discussion above, dividing Equations (4) by (2) and (5) by (6), and then taking the ratio of the outcomes (7) and (8) gave us the invariant cross ratio $\frac{AC}{BC} \div \frac{AD}{BD}$. Try to discover other nonequivalent invariant ratios by manipulating Equations (1)–(6) above in different ways. *Hint: You should be able to find two additional invariant ratios* $\frac{BD}{CD} \div \frac{AB}{AC}$ *and* $\frac{AB}{BC} \div \frac{AD}{CD}$. How are all these invariant ratios related to each other?

Pascal's Theorem

Blaise Pascal (1623–1662) built on Desargues' work related to properties invariant under perspectivity. He proposed and proved a result, now known as Pascal's theorem, which states that *if a hexagon is inscribed in a circle (see Figure 12.11), then the pairs of opposite sides intersect in three points that lie on a straight line.* Pascal further argued that since the theorem holds for the circle, it must also be true for all conics by the principle of *projection and section.* That is, if the plane of Figure 12.11 is projected onto a second plane from a point not on either plane, then the section on the second plane (the collection of points where the lines of projection intersect the second plane) will contain a conic and a hexagon inscribed in it. Intuitively, the projection lines connecting the center of projection to points on the circle form a cone and the intersection of the second plane with this cone will be a conic section (ellipse, hyperbola, or parabola). Unfortunately, there is no record of Pascal's original proof of this theorem.

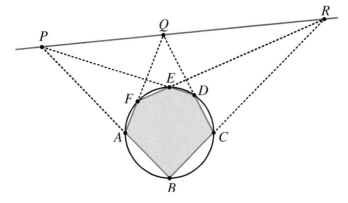

Figure 12.11

> **Remark 12.1**
>
> Pascal's theorem continues to hold if we degenerate one or two sides of the hexagon inscribed in the circle to single points. For example, for the degenerate hexagon $AABCCD$ in Figure 12.12, the sides $\overline{AA}$ and $\overline{CC}$ are the tangents to the circle at A and C. Think of $\overline{AA}$ in Figure 12.12 as the limit of the side $\overline{AF}$ in Figure 12.11 as F gets closer and closer to A.
>
>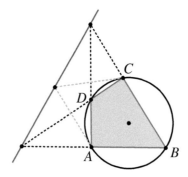
>
> Figure 12.12

The reader will be asked to prove Pascal's theorem for the case where the conic is a circle in Problem 2 at the end of this section. The proof strategy is adapted from [93], and relies on the ability to project a plane H containing a circle C and a line l (that does not intersect C) onto a plane H' such that the image of C is a circle C' in H', and l is the vanishing line in H under this projection (see Figure 12.13). To explain how this can be achieved, we enlist the help of stereographic projection (see Section 9.5).

Stereographic projection of the surface of the sphere from a point N (north pole) onto a plane tangent to the sphere at the point S (south pole) takes circles on the sphere not passing through N onto circles in the plane, and takes circles on the sphere passing through N onto lines in the plane. Each line or circle in the plane has a circle on the sphere as its preimage. Let Ω be a sphere passing through C, α be the plane containing l and tangent to the sphere

12.1 Introduction and Early Results

at a point labeled N, and H' be a plane parallel to α and tangent to the sphere at S, the point diametrically opposite N. Stereographic projection from N projects the circle C contained in H onto a circle C' contained in the plane H', as in Figure 12.13. This defines a central projection from H to H' with its center at N taking C to C' and having l as the vanishing line in H (the plane α through the center of projection N is parallel to H' so the intersection of α and H is the vanishing line in H) [93].

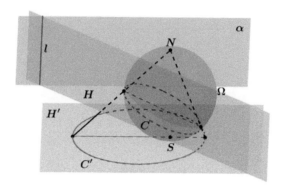

Figure 12.13

Now Try This 12.3 outlines how to construct Ω, N, and H' to achieve the desired central projection. Recall that, in Euclidean geometry, if C is a circle centered at O and P is a point outside C, then the points of intersection of C and the circle centered at the midpoint of $\overline{OP}$ and going through P determine where the tangents to C from P meet C. Think about how this idea can be extended to construct a plane tangent to a sphere through a line not meeting the sphere.

> **Now Try This 12.3**
>
> 1. Let C be a circle in a plane H and let m be the line that goes through the center of C and is perpendicular to H. Let O be any point on m. Prove that O is the center of a sphere Ω whose intersection with H is C.
>
> 2. Let Ω be a sphere that intersects H in the circle C. Let l be a line in H that is disjoint from C. Use the three-dimensional features of a dynamic geometry software such as GeoGebra to construct a plane tangent to Ω and going through the line l, as in Figure 12.13. Hint: (i) The sphere centered at the midpoint of the perpendicular segment from the center of Ω to the line l intersects Ω in a circle. (ii) The plane through the line connecting the centers of the spheres and the line m [defined in (1) above] intersects the circle common to the two spheres in two antipodal points N and S. (iii) The plane α through line l and one of these points is tangent to Ω.
>
> 3. The plane H' in Figure 12.13 should not be hard to construct as it is simply tangent to Ω at S and parallel to α. Describe precisely how this can be done.

Pappus' Hexagon Theorem

One of the earliest theorems in projective geometry is the so called *Pappus' hexagon theorem*, named after Pappus of Alexandria (290–350 C.E.), one of the last great geometers of antiquity. This theorem states that *if A, B, and C are points on a straight line and X, Y, and Z are points on another line, then the points of intersection of $\overleftrightarrow{AY}$ and $\overleftrightarrow{BX}$, $\overleftrightarrow{AZ}$ and $\overleftrightarrow{CX}$, and $\overleftrightarrow{BZ}$ and $\overleftrightarrow{CY}$ lie on a straight line* (see Figure 12.14).

Note that Pappus' theorem is a special case of Pascal's theorem. This can be seen when the radius of the circle in Pascal's theorem gets larger and larger.

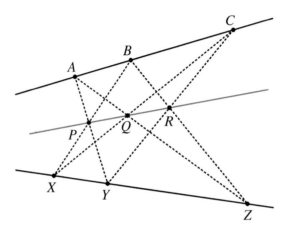

Figure 12.14

Proof. Pappus' theorem can be proved using techniques adapted from [93], and similar to those used in the proof of Desargues' theorem. Therefore, we will only give an outline of the proof and leave the details as an exercise for the reader.

First, we project H, the plane of Figure 12.14, onto a plane H' such that the vanishing line h in the plane H is the line $\overleftrightarrow{PR}$. This is possible if H' is parallel to the plane determined by $\overleftrightarrow{PR}$ and the center of projection (H' doesn't contain $\overleftrightarrow{PR}$). As a result, we have $\overleftrightarrow{A'Y'} \parallel \overleftrightarrow{B'X'}$ and $\overleftrightarrow{B'Z'} \parallel \overleftrightarrow{C'Y'}$ (why?).

Second, assume $\overleftrightarrow{A'C'}$ and $\overleftrightarrow{Z'X'}$ intersect in a point O' in H' (O' is not shown in Figure 12.15). This gives $\triangle O'A'Y' \sim \triangle O'B'X'$ and $\triangle O'B'Z' \sim \triangle O'C'Y'$ (why?). Third, working with these similar triangles, we can conclude that $\frac{O'A'}{O'C'} = \frac{O'Z'}{O'X'}$ (why?). This implies that $\overleftrightarrow{A'Z'} \parallel \overleftrightarrow{C'X'}$ because a line that divides two sides of a triangle proportionally is parallel to the third side (the converse of the *side splitting* theorem). Hence $\overleftrightarrow{AZ}$ and $\overleftrightarrow{CX}$ intersect at a point Q on $h = \overleftrightarrow{PR}$ (why?).

Proving Pappus' theorem when $\overleftrightarrow{A'C'}$ is parallel to $\overleftrightarrow{Z'X'}$ is the subject of Problem 4 at the end of this section.

12.1 Introduction and Early Results

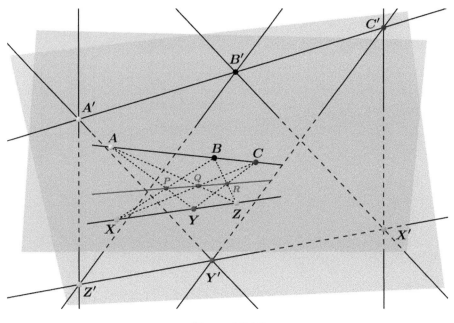

Figure 12.15

Interest in projective geometry faded away during the 17th and 18th centuries, but was revived in the 19th century by Jean-Victor Poncelet (1788–1867) who published *Traité des propriétés projectives des figures* in 1822, which was a further study of those properties which remain invariant under projection. Poncelet advanced the ideas of Desargues by postulating a common point at infinity for all lines parallel to a given line and a line at infinity lying on all these points. Also, his work with poles and polars associated with conic sections resulted in the principle of duality (see Section 12.5). Building on Poncelet's work, Karl von Staudt (1798–1867) was the first mathematician to advance the study of projective geometry independent of any geometric measurements such as distances and angles.

12.1.1 Problem Set

1. Prove the converse of Desargues' theorem. *Hint: Project the plane H of Figure 12.7 onto a plane H' such that the line joining the points of intersection P, Q, and R is the vanishing line in H.*

2. Prove Pascal's theorem following the steps outlined below:
   a. Take the plane of Figure 12.11 to be the plane H in the figure below and project it onto a plane H' such that the image of the circle circumscribing the hexagon in H is a circle in H' and the vanishing line in H has the points P and R on it, as shown below (O is the center of projection in the figure).

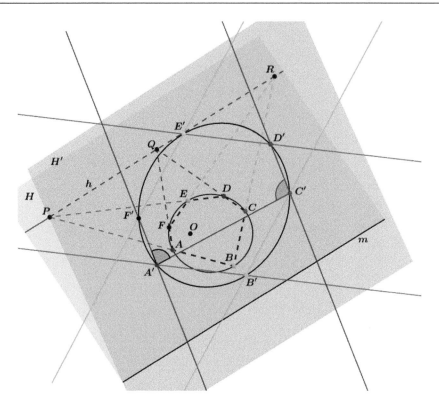

    b. Show that $\overleftrightarrow{A'B'}$ is parallel to $\overleftrightarrow{E'D'}$ and $\overleftrightarrow{E'F'}$ is parallel to $\overleftrightarrow{B'C'}$ in H'.

    c. Show that $\angle F'A'C'$ and $\angle D'C'A'$ are supplementary, and hence $\overleftrightarrow{A'F'}$ is parallel to $\overleftrightarrow{C'D'}$.

    d. Show that $\overleftrightarrow{AF}$ must intersect $\overleftrightarrow{CD}$ at a point Q on the vanishing line in H.

    e. Conclude that P, Q, and R are collinear.

3. Provide the missing details in the proof of Pappus' theorem for the case where $A'C'$ and $X'Z'$ intersect at a point O' (see Figure 12.15).

4. Prove Pappus' theorem in the case where $\overleftrightarrow{A'C'}$ and $\overleftrightarrow{X'Z'}$ do not intersect (see Figure 12.15).

5. Follow the steps outlined below to prove that $\overleftrightarrow{AB}$ is parallel to $\overleftrightarrow{QR}$ in Figure 12.9. *Hint: Use the Euclidean result stated below in your proof.*

    Menelaus' theorem: Consider $\triangle EFG$ and points X, Y, and Z on $\overleftrightarrow{FG}$, $\overleftrightarrow{EG}$, and $\overleftrightarrow{EF}$, respectively, as in the figure below. If X, Y, and Z are collinear, then $\frac{GY}{YE} \cdot \frac{EZ}{ZF} \cdot \frac{FX}{XG} = 1$. The line passing through X, Y, and Z is referred to as a transversal of $\triangle EFG$.

12.2 Projective Planes

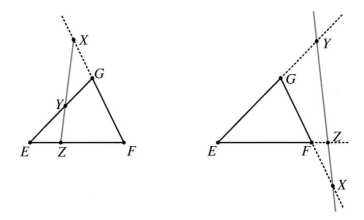

   a. Apply Menelaus' theorem to $\triangle OBC$ for the transversal $\overleftrightarrow{RC_1}$ in Figure 12.9 to arrive at $\frac{CC_1}{C_1O} \cdot \frac{OB_1}{B_1B} \cdot \frac{BR}{RC} = 1$.

   b. Apply Menelaus' theorem to $\triangle OAC$ for the transversal $\overleftrightarrow{QC_1}$ in Figure 12.9 to arrive at $\frac{CC_1}{C_1O} \cdot \frac{OA_1}{A_1A} \cdot \frac{AQ}{QC} = 1$.

   c. Use the fact that $\triangle OAB$ is similar to $\triangle OA_1B_1$ to conclude that $\frac{BR}{RC} = \frac{AQ}{QC}$ and hence $\overleftrightarrow{AB}$ is parallel to $\overleftrightarrow{QR}$.

12.2 Projective Planes

Projective geometry can be developed using different approaches. We will focus our attention here on the synthetic approach, and will briefly introduce homogeneous coordinates (an important tool in the algebraic approach) in the next section. In the synthetic approach, projective geometry can be achieved as an extension of Euclidean geometry by denying the existence of parallel lines. Material in this section was adapted from [86].

A *projective plane* consists of a set of objects called *points*, a set of objects called *lines*, and a relation called *incidence* between the points and lines satisfying the following axioms:

   Axiom 1: Any two distinct points have exactly one line incident with both of them.

   Axiom 2: Any two distinct lines have at least one point incident with both of them.

   Axiom 3: There exist four distinct points, no three of which are incident with the same line.

Note that Axiom 2 implies that there are no parallel lines, while Axiom 3 implies that the plane is nondegenerate (not a single point, not a single line, etc.). Henceforth, we will say that "a point lies on a line" and "a line goes through a point" instead of saying "the point and the line are incident." The assertions in the following theorem follow directly from the axioms above. We will prove (1) and (3) and leave the rest as exercises.

> **Theorem 12.2**
>
> 1. *Any two distinct lines have exactly one point lying on both of them.*
> 2. *Any two distinct points have at least on line lying on both of them.*
> 3. *There exist four distinct lines, no three of which are concurrent.*
> 4. *Every line has at least three points on it.*
> 5. *Every point has at least three lines passing through it.*
> 6. *Given any point P, there exists a line not passing through P.*

Proof. To prove (1), suppose l and m are distinct lines. By Axiom 2, l and m have at least one point P in common. Suppose l and m have another point Q in common. If P and Q are distinct, then Axiom 1 is contradicted, hence $P = Q$.

To prove (3), note that by Axiom 3, there are four distinct points P_1, P_2, P_3, P_4, no three of which are collinear. Hence the four lines lying on P_1 and P_2, P_2 and P_3, P_3 and P_4, and P_4 and P_1 are distinct (since no three of the four points are collinear), and no three of the lines are concurrent (since the four points are distinct). ∎

A finite projective geometry is a geometry with a finite number of points (hence a finite number of lines) satisfying Axioms 1–3 discussed earlier. Fano's geometry, pictured in Figure 12.16, is an example of such a geometry. Note that this geometry has seven points and seven lines. Also each point has three lines on it and each line has 3 points on it. The Fano plane is the smallest projective plane (why?).

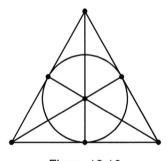

Figure 12.16

Duality

Projective geometry has a special elegance because of the strong symmetry it exhibits. For example, in the projective plane any two distinct points determine a unique line, and any two distinct lines intersect in a unique point (hence parallel lines do not exist). This leads to the *planar principle of duality*, which states that *exchanging the words "line" and "point" (and making the necessary grammar adjustments) in a valid statement leads to a valid statement.* For instance, the plane dual of the statement "*any two distinct points have a unique line on them*" is "*any two distinct lines have a unique point in common.*" Also, the plane dual of "*three collinear points*" is "*three concurrent lines*," etc. This striking feature of projective geometry allows us to transform points into lines and lines into points, which can produce new results from given ones or transform a given problem into one that is easier to solve.

Note that assertions (1), (2), and (3) in Theorem 12.2 are the duals of Axioms 1, 2, and 3, respectively. Hence the duals of the projective plane axioms are consequences of these axioms, which means that the dual of a valid statement in a projective plane is also a valid statement (why?).

12.2.1 Problem Set

1. Prove Theorem 12.2.
2. What is the plane dual of Fano's geometry? Justify your answer.
3. Prove that if one line in a finite projective plane has $n + 1$ points (n is referred to as the *order* of the plane), then
    a. Every line has $n + 1$ points on it.
    b. Every point has $n + 1$ lines on it.
    c. There are $n^2 + n + 1$ points in the plane.
    d. There are $n^2 + n + 1$ lines in the plane.
4. Construct a projective plane with exactly 13 points in it. Use the previous problem to determine the number of points on a line.

We next describe an important example of a projective plane called the *real projective plane* or the *extended Euclidean plane*.

12.3 The Real Projective Plane

In the Euclidean plane, given a line m and a point P not on m, each line through P intersects m at a single finite point except the line l that is parallel to m. If we want any two lines to intersect, how can we remove this exception? The lines l and m clearly do not have an ordinary (finite) point in common, but they have the same "slope" or "direction." So we can in fact say that any line through P has a point or a direction in common with m [16].

But terms like "slope" and "direction" are problematic in projective geometry (due to the absence of the concepts of distance and angle measure), so we need to avoid such terms. Fortunately, since points and lines are undefined terms, identifying the direction of a line as a point allows us to simply say that any line l through P has a point in common with m. The line l has been extended by appending to it a "nonordinary" or "nonfinite" point corresponding to its slope. This point is often referred to as the *point at infinity* (or *ideal point*) associated with l. This terminology makes sense since the point where a given line l through P meets m recedes farther and farther from any fixed point on m (e.g., the perpendicular projection of P onto m) as the given line approaches being parallel to m.

Having associated a point at infinity with each line l through P, how does the set of points at infinity behave? Since this set has exactly one point in common with each line, it is natural to consider the locus of points at infinity to be a line, the *line at infinity*. This stipulation allows us to say that any two points have a unique line through them and any two lines have a unique point in common regardless of whether the points or lines are ordinary or ideal.

An analogous approach can be used to justify associating a line at infinity with each plane in $\mathbb{R}^3$ and taking these lines at infinity to be a *plane at infinity*.

Having decided to introduce points at infinity where parallel lines would intersect, how exactly should this be done to ensure that not all of the familiar structure of the Euclidean plane is lost? For instance, we should still want any two distinct points to have a unique line through them and any two distinct lines to have a unique point in common.

Given an *ordinary point* (Euclidean point) P, we can think of the Euclidean plane as the set of all lines in a plane that contain P. For each line l containing P, let's agree to append a unique ideal point $\bar{L}$. Let us also agree that the set of all ideal points form a line l_∞ called the *ideal line*, as shown schematically in Figure 12.17. Note that line l has exactly one ideal point $\bar{L}$ on it, so traveling along l infinitely far from P in either direction we arrive at the same ideal point $\bar{L}$. Hence, l is intuitively like a circle and its ideal point should not be thought of as an endpoint.

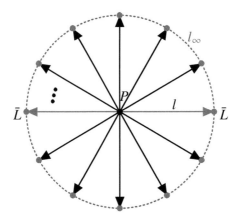

Figure 12.17

Now if m is a line not through P, then by Euclid's fifth postulate, there is a line l through P that does not meet m at an ordinary point, so it must meet it at $\bar{L}$, the ideal point appended to l. Now, if $k \neq m$ is another line not going through P and is parallel to m (in the Euclidean sense), then it must meet m at $\bar{L}$, otherwise, m would have more than one ideal point on it. This implies that all lines parallel to l (in the Euclidean sense) meet at $\bar{L}$. We are now ready to define the real projective plane.

12.3 The Real Projective Plane

> **Definition: The Real Projective Plane**
>
> Let P be an ordinary point in the Euclidean plane. Extend each Euclidean line l passing through P by appending exactly one ideal point $\bar{L}$ to it. Let the set of all ideal points form a line l_∞ called the ideal line.
>
> Assume the points (ordinary and ideal) and lines (extended lines through P and l_∞) satisfy the following conditions:
>
> 1. If m, a line not through P, has no ordinary points in common with line l through P ($m \parallel l$ in the Euclidean sense), then the intersection of m and l is the ideal point $\bar{L}$ of l. (This implies that each Euclidean line not passing through P has also been extended by appending an ideal point to it.)
>
> 2. If m, a line not through P, has only the ideal point $\bar{L}$ in common with line l through P, then either m is parallel to l (in the Euclidean sense) or m is the ideal line l_∞.
>
> The *real projective plane* consists of the ordinary points in the Euclidean plane and the ideal points appended to each line through a given ordinary point P as described above. Lines in the real projective plane are all extended lines (Euclidean lines with an ideal point appended to each line) together with the ideal line l_∞ consisting of all ideal points.

We now investigate which of the Euclidean postulates continue to hold in the real projective plane [90].

First, if A and B are any two points in this plane, do they lie on a unique line? To answer this, we need to consider the nature of the two points: are they ordinary or ideal?

1. If both points are ordinary points, then Euclid's first postulate guarantees a unique Euclidean line through A and B. This line, together with the ideal point appended to it, is the unique extended line through A and B.

2. If both A and B are ideal points, then we have l_∞ going through them, and this is the only line, because any other line has exactly one ideal point on it.

3. Without loss of generality, if A is an ideal point and B is an ordinary point, then they cannot have more than one extended line through them, since such lines would have to be parallel in the Euclidean sense (they have the ideal point A in common) and hence cannot have B in common.

Second, let l and m be any two lines; do they meet at exactly one point?

1. If both lines are extended lines, and if they have an ideal point in common, then they are parallel in the Euclidean sense, and hence cannot have an ordinary point in common. Also, they cannot meet at more than one ideal point because to each line we appended exactly one ideal point. So the ideal point that the lines have in common is unique.

2. If l and m are extended lines that meet at an ordinary point A, then by Euclid's first postulate, then cannot have another ordinary point in common. Furthermore, since the lines are not parallel, their ideal points are distinct, and hence their intersection point A is unique.

3. If one of the lines is the ideal line l_∞, it has exactly one ideal point in common with the other line, and since the ideal line contains no ordinary points, the two lines intersect at a unique ideal point.

The discussion above yields the following theorem:

> **Theorem 12.3**
>
> *In the real projective plane, the following are true:*
> 1. *Every ordinary line m has exactly one ideal point $\bar{M}$.*
> 2. *Any two distinct points determine a unique line.*
> 3. *Any two distinct lines have exactly one point in common.*

Starting with the Euclidean plane, we were able to construct the real projective plane by appending to each line through a given point P an ideal point and by taking the set of these ideal points to be an ideal line. We can use a similar approach (outlined below) to define the three-dimensional projective space.

1. Given a line l in $\mathbb{R}^3$, extend each plane H containing l using the same procedure as in the definition of the real projective plane.
2. Assume that planes parallel to the the same plane in $\mathbb{R}^3$ share the same ideal line.
3. Assume that if the planes H_1 and H_2 are not parallel, then their ideal lines are distinct.
4. Let the set of all ideal points and ideal lines form an ideal plane H_∞.

> **Definition: The Real Projective Space**
>
> The projective three-dimensional space is defined to be the set of all ordinary points in $\mathbb{R}^3$ together with the points of H_∞ (the ideal plane). Planes in the projective space are extended planes (planes along with their ideal lines) together with H_∞. Finally, lines are extended lines as well as ideal lines.

If l is a Euclidean line in $\mathbb{R}^3$, we can think of $\mathbb{R}^3$ as the set of all planes in $\mathbb{R}^3$ that contain the line l. Figure 12.18 schematically depicts a number of extended planes containing the line l and intersecting the ideal plane H_∞ in their ideal lines. It also depicts a plane H not parallel to l whose intersection with H_∞ is the ideal line h_∞ of H.

12.3 The Real Projective Plane

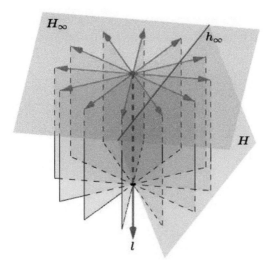

Figure 12.18

Having defined the real projective plane, you may wonder what kind of transformations can be applied in this plane. Since the notions of distance and betweenness do not apply in projective geometry (why?), familiar Euclidean transformations such as the various isometries and central similarity (homothety) do not apply. Instead, we can apply *central perspectivity*, defined below, which gives a bijection between the points of two lines in the projective plane based solely on incidence relations.

> **Definition: Central Perspectivity in the Real Projective Plane**
>
> Given two lines l and m in the real projective plane and a point E not on either line, a *central perspectivity* T_E from l to m centered at E takes a point A on l to the point B on m, where B is the intersection of $\overleftrightarrow{EA}$ with m, as in Figure 12.19. The point X on l for which $\overleftrightarrow{EX} \parallel m$ gets projected onto the ideal point of m. Similarly, the point Y on m for which $\overleftrightarrow{EY} \parallel l$ is the preimage of the ideal point of l.

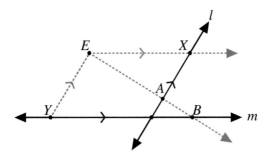

Figure 12.19

As we have seen before (see Figure 12.6), the composition of two central perspectivities is not necessarily a central perspectivity. However, this composition is still a transformation from one line to another which relies only on incidence relations between points and lines.

> **Definition: Projectivity in the Real Projective Plane**
>
> Given two lines l and m in the real projective plane, a *projectivity* from l to m is the composition of a finite number of central perspectivities.

The following definition should remind the reader of the definition of central projection in $\mathbb{R}^3$ given near the beginning of Section 12.1. Note, however, that the planes H and H' in the definition below are projective planes.

> **Definition: Central Perspectivity and Projectivity in the Real Projective Space**
>
> Given two planes H and H' in the real projective space and a point E not on either plane:
>
> 1. A *central perspectivity* from H to H' takes a point A on H to the point B on H' where B is the intersection of $\overleftrightarrow{EA}$ with H', as in Figure 12.20. The vanishing line h in H gets mapped onto the ideal line of H', and the ideal line of H gets mapped onto the vanishing line h' in H'.
> 2. A *projectivity* from plane H to plane H' in the real projective space is the composition of a finite number of central perspectivities.

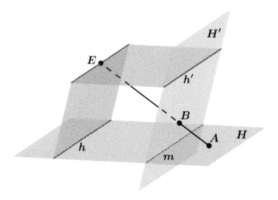

Figure 12.20

Models of the Real Projective Plane

We now give brief descriptions of several models of the real projective plane. All these models satisfy the projective plane Axioms 1–3 discussed earlier. In particular, in all of these models, any two straight lines always have a unique common point, and any two distinct points determine a unique line.

12.3 The Real Projective Plane 549

1. The real projective plane can be viewed as the extended two-dimensional Euclidean plane as described earlier. Here points are ordinary points or ideal points and lines are extended lines or the ideal line l_∞.

2. Similar to what we did in elliptic geometry (see Chapter 11), we can picture the real projective plane as a disc. The interior of the disc represents the Euclidean plane and the boundary circle represents the line at infinity. Any two antipodal (diametrically opposite) points on the boundary of the disc are identified as a single point. Lines are diameters of the disc or arcs of circles that intersect the boundary of the disc in points that determine a diameter. In this model, parallel lines in the Euclidean plane correspond to lines that meet in antipodal points, and the boundary of the disc represents l_∞.

3. We can view the real projective plane as the surface of a sphere with points being represented by pairs of antipodal points and lines by great circles.

4. We can use the surface of a semi-sphere to represent the real projective plane, with points being points on the surface of the semi-sphere and pairs of antipodal points on its boundary circle. Lines here would be great semicircles with their end points identified. What happens if this model is projected orthogonally on the plane going through the boundary of the semi-sphere?

5. Finally, we can think of the points of the real projective plane as lines in $\mathbb{R}^3$ going through the origin, and lines of the real projective plane as planes in $\mathbb{R}^3$ also going through the origin. So if (x_1, y_1, z_1) is a point in $\mathbb{R}^3$, the set of points (rx_1, ry_1, rz_1), for all real numbers r, represents a point in the projective plane, and a plane of the form $ax + by + cz = 0$ represents a line in the projective plane. Note that any two distinct planes through the origin intersect in a unique line through the origin, and any two distinct lines through the origin determine a unique plane through the origin (we are assuming basic knowledge of three-dimensional geometry here).

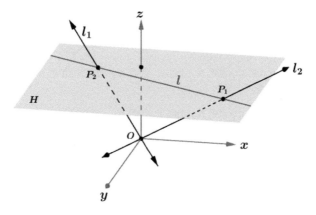

Figure 12.21

As a further nontechnical explanation of this model, let H be a plane parallel to the xy-plane, as in Figure 12.21. Any line through the origin, except for lines in the xy-plane, intersects H at exactly one point. So there is a one-to-one correspondence between the points in H and lines through the origin not lying in the xy-plane. We can think of lines through the origin lying in the xy-palne as intersecting H at infinity, so each line corresponds to a point at infinity appended to H. Two lines l_1 and l_2 going through the origin and not lying in the xy-plane determine a unique plane through the origin whose intersection with H is a line l. So a line in H corresponds to a plane going through the origin. We can think of the xy-plane as the line at infinity consisting of all the points at infinity appended to H.

> **Now Try This 12.4**
>
> To determine a one-to-one correspondence between points and lines in Model (5) and points and lines in Model (1) of the real projective plane, verify the following steps:
>
> 1. Let (a, b, c) be a point in $\mathbb{R}^3$.
>   a. If $c \neq 0$, then (a, b, c) is on the same line l through the origin as the point $(a/c, b/c, 1)$, which is the intersection of l with the plane $z = 1$. In Model (5), line l is regarded as a single point, and hence we can use any of its points, in particular $(a/c, b/c, 1)$, to identify it. Projecting $(a/c, b/c, 1)$ onto $(a/c, b/c)$ in the xy-plane establishes a one-to-one correspondence between points (a, b, c) in Model (5), where $c \neq 0$, and finite (ordinary) points in Model (1).
>   b. If $c = 0$, then point (a, b, c) is on a line lying in the xy-plane and passing through the origin. In fact, $(a, b, 0)$ is on the line $y = (b/a)x$ if $a \neq 0$ or the line $x = 0$ if $a = 0$ (why?). For each of these lines, using the point at infinity associated with the line to identify it, establishes a one-to-one correspondence between points $(a, b, 0)$ in Model (5) and points at infinity in Model (1).
>
> 2. If l is a line in the projective plane as viewed in Model (5), then l is a nondegenerate plane $ax + by + cz = 0$ through the origin. We now show how l corresponds to a unique line in Model (1).
>   a. If $b \neq 0$, then l corresponds to the line $y = -(a/b)x - c/b$ in the projective plane viewed as in Model (1) (why?); otherwise, it corresponds to the line $x = -c/a$.
>   b. If $a = b = 0$, then l (i.e., the plane $z = 0$) corresponds to the line at infinity in the projective plane viewed as in Model (1).

12.3.1 Problem Set

1. Use the axioms in Section 12.2 to prove that a central perspectivity in the real projective plane is a one-to-one mapping.

2. Explain the difference between a dilation (homothety) and a central perspectivity.
3. Prove Theorem 12.3.

12.4 Homogeneous Coordinates

Some of the material in this section was adapted from [10]. Homogeneous coordinates is an important tool in the algebraic approach toward the development of projective geometry. This tool was first introduced by Möbius in 1827, and later advanced and popularized by Plücker in 1834. Homogeneous coordinates, as described below, allow us to assign finite coordinates to objects at infinity, and lead to elegant algebraic representation of other geometric concepts including transformations. In this section we present a brief motivation and treatment of homogeneous coordinates to help the reader better understand objects at infinity, and to emphasize the principle of duality in projective geometry, which will be discussed in the following section. The name "homogeneous" comes from the fact that nonzero multiples of the coordinates of an object such as a point or a line represent the same object.

To motivate the concept of homogeneous coordinates, consider the intersections of the lines through the origin of $\mathbb{R}^3$ with the plane $z = 1$ (chosen for convenience). This establishes a one-to-one correspondence between points in the plane $z = 1$ and lines going through the origin other than lines in the xy-plane (why?). As discussed earlier, it is reasonable to stipulate that each of the lines through the origin and lying in the xy-plane meets the plane $z = 1$ in a point at infinity, which can be interpreted as the direction of that line (since points are undefined; see Section 12.3). Points in the plane $z = 1$ have three coordinates $(x, y, 1)$. Furthermore, since all nonzero multiples of $(x, y, 1)$ lie on the same line through the origin, we can consider $(\omega x, \omega y, \omega)$, where $\omega \neq 0$, as all representing the same point. If you find this a little bit concerning, remember that you encountered the same idea in elementary school when you considered the fractions $\frac{1}{3}, \frac{2}{6}, \frac{3}{9}$, etc., to all have the same value and took $\frac{1}{3}$ (fraction in lowest terms) to represent these equivalent fractions. We also encountered this idea in Model (5) of the projective plane when we interpreted points as lines through the origin of $\mathbb{R}^3$.

Since finite points in the plane $z = 1$ are represented by three coordinates, we should try to represent points at infinity using three coordinates as well. To do so, for each line in the xy-plane of the form $ax + by = 0$ (hence passing through the origin) choose one of its points, for instance $(-b, a, 0)$, to represent its point at infinity. Note that a and b determine the slope of the line, and these are present in the coordinates $(-b, a, 0)$ without actually including the value of the slope which can be infinite (if $b = 0$). Using ∞ as a coordinate is not a good option, as this will hinder calculations with coordinates. Moreover, it will make it necessary to continue to distinguish between ordinary points and points at infinity, which is something we should avoid, since there ought not to be such distinction in projective geometry. Since any nonzero multiple of $ax + by = 0$ represents the same line, we should consider all nonzero multiples of $(-b, a, 0)$ as representing the same point at infinity.

Having discussed how to view points in the projective plane as points in the $z = 1$ plane and points of the form $(-b, a, 0)$ (representing lines in the xy-plane and going through the origin), we now turn our attention to lines. Let P_1 and P_2 be two points in the plane $z = 1$ with coordinates $(x_1, y_1, 1)$ and $(x_2, y_2, 1)$, respectively, as in Figure 12.22.

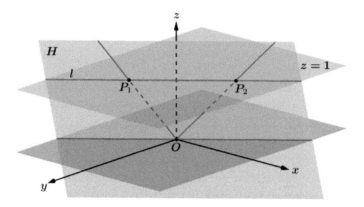

Figure 12.22

The line l through P_1 and P_2 has the equation $(y_1 - y_2)x + (x_2 - x_1)y + y_2x_1 - y_1x_2 = 0$ (why?). Also, the plane H through the origin and the points P_1, and P_2 has the equation $(y_1 - y_2)x + (x_2 - x_1)y + (y_2x_1 - y_1x_2)z = 0$ (why?). Note that l is the intersection of the plane H determined by $\overleftrightarrow{OP_1}$ and $\overleftrightarrow{OP_2}$ with the plane $z = 1$. Taking l as the line determined by P_1 and P_2 in the $z = 1$ plane establishes a one-to-one correspondence between lines in the $z = 1$ plane and planes through the origin other than the xy-plane, which is parallel to plane $z = 1$. It is reasonable to stipulate that the xy-plane meets the plane $z = 1$ at infinity; hence we can view the xy-plane (or the plane $z = 0$) as the line at infinity. This is consistent with our earlier stipulation that each line in the xy-plane of the form $ax + by = 0$ represents a single point at infinity, namely $(-b, a, 0)$.

Having motivated the concept of homogeneous coordinates using Cartesian coordinates in $\mathbb{R}^3$, we now get back to developing this concept for the real projective plane without reference to the z coordinate, since we are not technically in $\mathbb{R}^3$. We outline below how homogeneous coordinates can be defined for points and lines (both ordinary and ideal) in the projective plane.

1. Points in the xy-plane with old coordinates (x, y) are assigned the new homogeneous coordinates $(x, y, 1)$ or any nonzero multiple of that [i.e., $(\omega x, \omega y, \omega)$, where $\omega \neq 0$]. Conversely, points with homogeneous coordinates (x, y, ω) for $\omega \neq 0$ have old coordinates $(\frac{x}{\omega}, \frac{y}{\omega})$ obtained by dividing by ω. So points now have three coordinates: An x-coordinate, a y-coordinate, and a ω-coordinate, where $\omega = 1$ for all ordinary points. Again, we did not want to consider the third coordinate to be the z-coordinate since, technically, we are not in $\mathbb{R}^3$.

2. A line in the xy-plane having the old equation $ax + by + c = 0$, where not both a and b are zero, is assigned the new equation $ax + by + c\omega = 0$ and the coordinates $[a, b, c]$

12.4 Homogeneous Coordinates

(or any nonzero multiple of that). We use square brackets to distinguish coordinates of lines from coordinates of points. Unlike the slope-intercept form $y = mx + b$, the general form $ax + by + c = 0$ (where not both a and b are zero) represents any line in the xy-plane regardless of the slope of the line. Furthermore, the form $ax + by + c = 0$ allows us to think of a, b, and c as the coordinates for the line. For instance, we can use $[2, -3, 4]$ to represent the line $2x - 3y + 4 = 0$. Note that any nonzero multiple of $[2, -3, 4]$ represents the same line as $[2, -3, 4]$ (why?). We now can say that a point (x, y, ω) is on the line $ax + by + c\omega = 0$ if $[a, b, c] \cdot (x, y, \omega) = ax + by + c\omega = 0$. That is, if the *dot product* (the sum of the products of the corresponding entries) of the coordinates of the line and the point equals 0.

3. Points at infinity are assigned new (finite) homogeneous coordinates $(a, b, 0)$. To further explain this important concept, remember that points at infinity are the points where parallel lines meet. So given two distinct parallel lines $ax + by + c\omega = 0$ and $ax + by + d\omega = 0$, subtracting the equations gives $(c - d)\omega = 0$, and hence $\omega = 0$. Substituting this back in one of the equations gives $ax + by = 0$. Any multiple of $(-b, a, 0)$ satisfies this equation, and so if $b \neq 0$, then we can choose $(1, \frac{-a}{b}, 0)$ as the point of intersection of these parallel lines. Note that $-\frac{a}{b}$ is the slope of these lines. Consequently, lines parallel to $ax + by + c\omega = 0$ all have in common the point at infinity with homogeneous coordinates $(-b, a, 0)$ (or any nonzero multiple of that).

4. The line at infinity has the equation $\omega = 0$ and coordinates $[0, 0, 1]$ (or any nonzero multiple of that). To explain this, note that any ordinary line has a new equation of the form $ax + by + c\omega = 0$, where not both a and b are zero. Hence it is plausible to determine the line at infinity by taking $a = 0$ and $b = 0$ (why?). This gives $c\omega = 0$, and so $\omega = 0$. Therefore, the line at infinity has coordinates $[0, 0, c]$, which is the same as $[0, 0, 1]$, since $c \neq 0$ (why?). As further evidence of the appropriateness of our choice of the line at infinity, note that the dot product of $(a, b, 0)$ and $[0, 0, 1]$ is zero, so any point at infinity $(a, b, 0)$ lies on the line at infinity represented by $[0, 0, 1]$. Moreover, the intersection of any ordinary line $[a, b, c]$ with $[0, 0, 1]$ is $(b, -a, 0)$, which is the point at infinity associated with $[a, b, c]$. We know this since the point of intersection satisfies $ax + by + c\omega = 0$ and $\omega = 0$, and hence it satisfies $ax + by = 0$.

> **Now Try This 12.5**
>
> 1. Show that if a line has slope m, then it has one point at infinity with homogeneous coordinates $(1, m, 0)$ (or any nonzero multiple of that).
>
> 2. Show that if a line has slope $-1/m$, then it has one point at infinity with homogeneous coordinates $(-m, 1, 0)$ (or any nonzero multiple of that).
>
> 3. How are the lines in (1) and (2) related? What is the dot product of the homogeneous coordinates of their points at infinity?

Recall that if $\mathbf{u} = (a_1, b_1, c_1)$ and $\mathbf{v} = (a_2, b_2, c_2)$ are two vectors in $\mathbb{R}^3$, then their *cross product* is the vector $\mathbf{u} \times \mathbf{v} = (b_1 c_2 - b_2 c_1, a_2 c_1 - a_1 c_2, a_1 b_2 - a_2 b_1)$. We next use the cross product to give rules for finding the point of intersection of two given lines, and the line determined by two given points. The justification of the first rule is the subject of Now Prove This 12.1, and the justification of the second rule is the subject of Problem 11 at the end of this section.

1. To find the point of intersection of two lines, we simply compute the cross product of their homogeneous coordinates.

2. To find the line going through two points, we simply compute the cross product of their homogeneous coordinates.

> **Now Prove This 12.1**
>
> To prove (1) above, let l_1, l_2 be the lines $a_1 x + b_1 y + c_1 \omega = 0$ and $a_2 x + b_2 y + c_2 \omega = 0$, respectively.
>
> 1. Use basic algebra or linear algebra techniques to show that if l_1 and l_2 intersect in an ordinary point (x, y), then $x = \frac{b_1 c_2 - b_2 c_1}{a_1 b_2 - b_1 a_2}$ and $y = \frac{a_2 c_1 - a_1 c_2}{a_1 b_2 - b_1 a_2}$. Note that $a_1 b_2 - b_1 a_2 \neq 0$ since l_1 is not parallel to l_2.
>
> 2. Show that the homogeneous coordinates of the point of intersection are $(b_1 c_2 - b_2 c_1, a_2 c_1 - a_1 c_2, a_1 b_2 - a_2 b_1)$.
>
> 3. Compute the cross product of the coordinates of the lines l_1 and l_2, and verify that the corresponding coordinates of the cross product and the point of intersection of l_1 and l_2 are equal.

12.4.1 Problem Set

1. Use homogeneous coordinates to find the point of intersection of $x + y - 1 = 0$ and $-x + 2y + 5 = 0$.

2. Use homogeneous coordinates to find the equation of the line going through $(-1, 2)$ and $(3, -2)$.

3. Use homogeneous coordinates to find the equation of the line passing through $(2, -3)$ and having slope 2.

4. Use homogeneous coordinates to find the point of intersection of the line at infinity and $2x + y + 2 = 0$.

5. Use homogeneous coordinates to show that all lines parallel to the same line share the same point at infinity.

6. Use homogeneous coordinates to show that all points at infinity lie on the line at infinity.
   Note: *The following problems require basic knowledge of matrix algebra.*

7. Given three lines l_1, l_2, and l_3 given by $a_1x + b_1y + c_1 = 0$, $a_2x + b_2y + c_2 = 0$, and $a_3x + b_3y + c_3 = 0$, respectively, show that these lines are concurrent if and only if the determinant

$$\begin{vmatrix} a_1 & b_1 & c_1 \\ a_2 & b_2 & c_2 \\ a_3 & b_3 & c_3 \end{vmatrix} = 0.$$

What are the homogeneous coordinates of the point of concurrency?

8. Show that a translation by (h, k) in the xy-plane can be represented using homogeneous coordinates as multiplication by the matrix

$$\begin{bmatrix} 1 & 0 & h \\ 0 & 1 & k \\ 0 & 0 & 1 \end{bmatrix}.$$

Can a translation in the xy-plane be represented as multiplication by a 2×2 matrix?

9. Show that a scaling by a factor of λ in the xy-plane can be represented using homogeneous coordinates as multiplication by the matrix

$$\begin{bmatrix} \lambda & 0 & 0 \\ 0 & \lambda & 0 \\ 0 & 0 & 1 \end{bmatrix}.$$

10. Show that counterclockwise rotation about the origin by angle θ in the xy-plane can be represented using homogeneous coordinates as multiplication by the matrix

$$\begin{bmatrix} \cos\theta & -\sin\theta & 0 \\ \sin\theta & \cos\theta & 0 \\ 0 & 0 & 1 \end{bmatrix}.$$

11. Prove that if P_1 and P_2 are two points with homogeneous coordinates $(x_1, y_1, 1)$ and $(x_2, y_2, 1)$, respectively, then the cross product $P_1 \times P_2$ of the coordinates of the points determines the coordinates of the line connecting them. *Hint: First show that the line connecting P_1 and P_2 is $(y_1 - y_2)x + (x_2 - x_1)y + y_2x_1 - y_1x_2 = 0$, and then show that the line determined by $P_1 \times P_2$ has the same coordinates.*

12.5 Duality: Poles, Polars, and Reciprocation

As mentioned earlier, the *planar principle of duality* in projective geometry allows us to transform points into lines and lines into points, which can produce new results from given ones or transform a given problem into one that is easier to solve. None of the transformations discussed in previous chapters, including inversion (see Chapter 9), have this property. In contrast, a new transformation called *reciprocation* with respect to a fixed circle ω (formal definition to follow) takes any point to a line, and any line to a point. Furthermore, it takes a circle $\alpha \neq \omega$ to a conic (ellipse, parabola, or hyperbola) having the center of the circle as a focus. As implied by Problem 12 at the end of this section, this depends on the ratio of the distance between the centers of α and ω and the radius of α.

In order to define reciprocation we first define *poles* and *polars*. We will use inversion to define these terms (see Problem 10 for an alternative way to define poles and polars). As customary, capital letters and their corresponding small letters are used to denote non-ideal poles and their corresponding polars, respectively.

> **Definition: Poles and Polars**
>
> Let ω be a circle centered at a point O with radius k, as in Figure 12.23.
>
> 1. Let P be any nonideal point other than O, and let P' be the image of P under inversion in the circle ω. The *polar* of P is defined to be the line p that is perpendicular to $\overleftrightarrow{OP}$ at P'.
>
> 2. Conversely, if $p \neq l_\infty$ is a line not passing through O, let P' be the perpendicular projection of O onto p. The *pole* of p is defined to be the point P that is the image of P' under inversion in ω.
>
> 3. For O, the center of ω, and lines through O, define:
>   a. The polar of O to be l_∞ and the pole of l_∞ to be O, where l_∞ is the line at infinity associated with the plane containing ω.
>   b. The pole of a line l through O to be the ideal point $\bar{M}$ associated with the line m passing through O and perpendicular to l.
>   c. The polar of an ideal point $\bar{N}$ to be the line passing through O that is perpendicular to the line n with which $\bar{N}$ is associated.

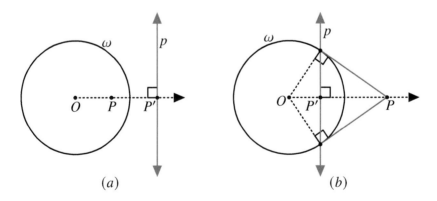

Figure 12.23

Figure 12.23 shows an ordinary P and its polar p for both P lying inside ω and P lying outside ω. Clearly, if P lies on ω then p would be the line tangent to ω at P (why?). Note

12.5 Duality: Poles, Polars, and Reciprocation

that if P lies outside ω as in Figure 12.23(b), then p would be the line joining the points of intersection of ω with the tangents to ω from P (why?).

From the definition of poles and polars, it should be clear that each point P in the real projective plane has a unique polar p, and each line p has a unique pole P (why?). Having defined poles and polars, we are now ready to define reciprocation.

> **Definition: Reciprocation**
>
> Let ω be a circle centered at O with radius k. Reciprocation with respect to ω is a transformation of the set of points and lines of the real projective plane which takes any point onto its polar, and any line onto its pole.

We next show that if point P is the pole of a line p not going through O, then the polar of any point Q on p goes through P. This establishes a one-to-one correspondence between points on the line p and lines through P other than $\overleftrightarrow{OP}$ (why?), and implies that P can be determined as the intersection of polars corresponding to any two points on p. This is stated formally in the following theorem.

> **Theorem 12.4: La Hire's Theorem**
>
> Let ω be a circle centered at O with radius k. If a point Q lies on the polar of a given point P (other than O), then the polar of Q goes through P. Furthermore, any line through P (other than $\overleftrightarrow{OP}$) is the polar of some point lying on the polar of P.

Proof. Figure 12.24 shows a point P, its polar p, an arbitrary finite (ordinary) point Q on p, and the line q which is the perpendicular line to $\overrightarrow{OQ}$ through P.

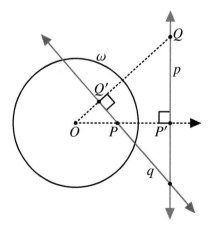

Figure 12.24

We now show that q is the polar of Q. Let Q' be the perpendicular projection of P on $\overrightarrow{OQ}$. To show that q is the polar of Q we only need to show that Q' is the image of Q

under inversion in ω (why?). This can be achieved by showing that $OQ \cdot OQ' = k^2$. To this end, note that $\triangle OPQ' \sim \triangle OQP'$ by AAA, and hence $\frac{OP}{OQ} = \frac{OQ'}{OP'}$. This implies that $OQ \cdot OQ' = OP \cdot OP' = k^2$, since P' is the image of P under inversion in ω. Therefore, Q' is the image of Q under inversion in ω, and hence q, which is identical to $\overleftrightarrow{PQ'}$, is the polar of Q. Conversely, any line q through P (other than $\overleftrightarrow{OP}$) has a line perpendicular to it and going through O that intersects p at a point Q, which turns out to be the pole of q. ∎

Theorem 12.4 implies that if p and q are the polars of P and Q, respectively, then Q is on line p if and only if P is on line q. It also implies that the polars of a set of collinear points are a set of concurrent lines.

Does Theorem 12.4 still hold if the point P is allowed to be equal to O? From the definition of poles and polars, we know that if $P = O$, then its polar is $p = l_\infty$. Furthermore, if $\bar{Q} \in l_\infty$ is the ideal point associated with the ordinary line q going through $P = O$, then its polar is the line through $P = O$ that is perpendicular to q. Conversely, if m is any line through $P = O$, then m is the polar of the ideal point associated with the line through $P = O$ that is perpendicular to m. This ideal point clearly lies on l_∞, the polar of $P = O$.

Therefore, polars of all points on any line l go through the pole of l, and each line through the pole of l has its pole lying on l, regardless of the nature of points or lines. Hence, Theorem 12.4 holds in general in the real projective plane.

Figure 12.25 schematically shows lines l, m, and k and their respective poles $\bar{M}$, $\bar{L}$, and K (poles and polars are color coded). Note that lines m and k meet at an ideal point on the ideal line l_∞. In fact, all the polars of points on l meet at this ideal point (why?). Per our discussion in Section 12.3, the ideal points labeled $\bar{L}$ represent a single point in the projective plane, and so do the ideal points labeled $\bar{M}$.

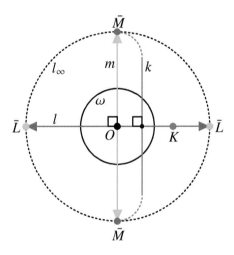

Figure 12.25

Points such as P and Q in Theorem 12.4 are called *conjugate points* (each lies on the polar of the other) and their polars p and q are called *conjugate lines* (each goes through the

12.5 Duality: Poles, Polars, and Reciprocation

pole of the other). It follows that the polar of a point P is the locus of points conjugate to P, and the pole of p is the intersection point of lines conjugate to p.

If a point P lies on its polar p, then P is called *self-conjugate*. Similarly, if a line p goes through its pole P, then p is called *self-conjugate*.

> **Now Try This 12.6**
>
> 1. Which points and which lines are self-conjugate with respect to a circle ω?
> 2. Can a line joining two self-conjugate points be self-conjugate? Justify your answer.
> 3. Can a line contain more than two self-conjugate points? Justify your answer.
> 4. If P is a self-conjugate point, how many self-conjugate points does each line through P have?

Theorem 12.5 [19, Theorem 6.12] summarizes some results that can be used to construct the pole of a line and the polar of a point depending on where they lie relative to ω. The reader will be asked to prove these results in Problem 1 at the end of this section.

> **Theorem 12.5**
>
> Let ω be a circle centered at a point O of radius k.
> 1. The pole of any secant $\overleftrightarrow{AB}$ not passing through O is the point of intersection of the tangents to ω at A and B.
> 2. The polar of any point lying outside ω is the line joining the points of contact of the two tangents to ω from this point.
> 3. The pole of any line p not passing through O is the point of intersection of the polars of two points on p.
> 4. The polar of any point P other than O is the line joining the poles of two secants through P.

Reciprocation of poles and their polars leads to the *principle of planar duality* which allows us to define the *plane dual* of any configuration of points and lines. For example, in Figure 12.26, the dual of the *quadrangle ABCD*, consisting of the four points A, B, C, D (no three of them are collinear) and the six lines joining them, is the *quadrilateral abcd*, consisting of the four lines a, b, c, d (no three of them are concurrent) and their six intersection points. Note that in $ABCD$ any point has three lines on it and each line has two points on it, while in $abcd$ each line has three points on it and each point has two lines on it.

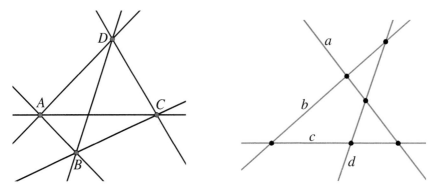

Figure 12.26

Reciprocation with respect to a circle ω takes a line p to its pole P, and a point P to its polar p, but how does it transform curves, circles in particular? To understand this we need to think of the curves as being approximated by their tangents. Formally, the *envelope* of a family of curves in the plane is defined to be the curve that is tangent to each curve in the family at some point; these points of tangency are the points of the envelope. Hence we can think of a circle as the locus of points equidistant from a center or as the envelope of its tangents. Now given a point P on the circle ω, the image of P under reciprocation with respect to ω is the line tangent to ω at P. Hence reciprocation takes ω viewed as a locus of points onto ω viewed as the envelope of its tangents and vice versa.

Next, we investigate the image (under reciprocation with respect to a unit circle ω centered at O) of an arbitrary circle γ of radius r and centered at a point $B \neq O$. Consider an arbitrary point Q on γ and let a denote the tangent to γ at that point (here we are thinking of γ as the envelope of its tangents). Let A be the pole of a and b be the polar of B with respect to ω, as in Figure 12.27.

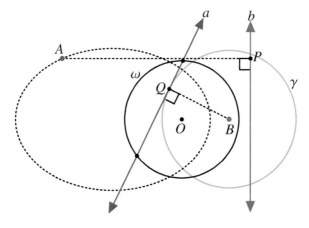

Figure 12.27

12.5 Duality: Poles, Polars, and Reciprocation

By Problem 11 (at the end of this section), we know that $\frac{OA}{AP} = \frac{OB}{BQ} = \frac{OB}{r}$, where AP and BQ are the distances from A to b and from B to a, respectively. Hence any tangent a to γ gets mapped to a point A such that the ratio $\frac{OA}{AP}$ is equal to $\frac{OB}{r}$, which is constant for all a (why?). That is, the ratio of the distance between A and the fixed point O and the distance between A and the fixed line b is constant, and is equal to $\frac{OB}{r}$.

This should remind the reader of the well-known definition of a conic section as the locus of points whose distances to a point (the focus) and a line (the directrix) are in a constant ratio. This ratio is called the eccentricity of the conic and its value determines the type of conic. Therefore, the image of a circle γ centered at B of radius r under reciprocation in ω centered at O of radius 1 is

1. an ellipse if $\frac{OB}{r} < 1$ (or O lies inside γ); see Figure 12.27;
2. a parabola if $\frac{OB}{r} = 1$ (or O lies on γ),
3. a hyperbola if $\frac{OB}{r} > 1$ (or O lies outside γ).

In Problem 12 at the end of this section, the reader will be asked to use dynamic geometry software to verify the conclusions regarding reciprocation of circles listed above.

12.5.1 Problem Set

1. Prove Theorem 12.5.
2. Prove that three points are collinear if and only if their polars are concurrent.
3. Determine the plane duals of Desargues' theorem and Pascal's theorem.
4. Let ω be a circle centered at a point O of radius k, and let $\alpha \neq \omega$ be a circle centered at O of radius r. What is the image of α under reciprocation with respect to ω? Justify your answer.
5. Let ω be a circle centered at a point O of radius k. Show that reciprocation with respect to ω takes a rectangle whose diagonals intersect at O onto a rhombus.
6. Prove that under reciprocation with respect to a circle ω having center O, the polar of any point A other than O can be constructed as the radical axis of the circle ω and the circle having $\overline{OA}$ as a diameter. *Hint: The radical axis of two nonconcentric circles centered at O_1, O_2 with radii r_1, r_2, respectively, is the set of points P satisfying $(PO_1)^2 - r_1^2 = (PO_2)^2 - r_2^2$. Note that this implies that the radical axis of two circles intersecting in two points is the common secant line.*
7. In the figure below, line a is the polar of the point A and line b is the polar of the point B with respect to the circle ω. Determine how $\angle AOB$ is related to the angles between a and b and prove your answer.

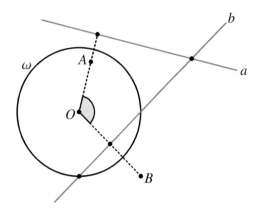

8. In the figure below, suppose that the circle ω centered at a point O is tangent to the lines a, b, and c.
   a. If $a \parallel b$, determine the measure of the angle α. *Hint: Transform the figure using reciprocation with respect to ω.*
   b. Solve this problem using regular Euclidean geometry; what do you notice?

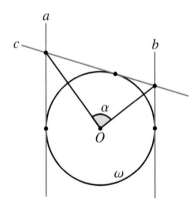

9. A *complete quadrilateral* consists of four lines a, b, c, and d, no three of them concurrent, and their six points of intersection. The four lines represent the sides and the six points represent the vertices of the complete quadrilateral. The pairs of opposite vertices (not connected by the any of the four lines) determine three additional lines called diagonals. Thus a complete quadrilateral has four sides, six vertices and three diagonals. Prove that the three diagonals of a complete quadrilateral are never concurrent ($\overleftrightarrow{AB}$, $\overleftrightarrow{EF}$, and $\overleftrightarrow{CD}$ are the three diagonals of the complete quadrilateral in the figure below).

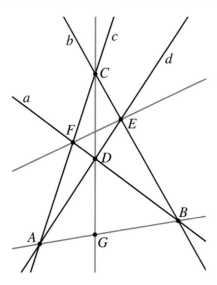

10. Let $ABCD$ be a cyclic quadrilateral inscribed in a circle ω centered at a point O. Let $\overleftrightarrow{AB}$ and $\overleftrightarrow{CD}$ meet in Q, and $\overleftrightarrow{AD}$ and $\overleftrightarrow{BC}$ meet in R. Prove that $\overleftrightarrow{RQ}$ is the polar of P, the point of intersection of the diagonals of $ABCD$. Note that this gives an alternative way for determining the polar of a point P. This procedure works for points both inside and outside ω.

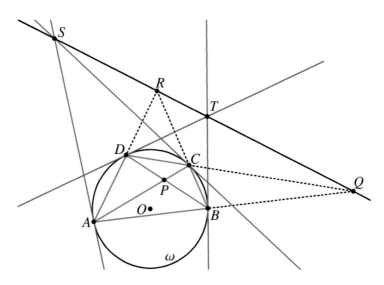

Hint: Let the tangents at A and C meet at S and the tangents at B and D meet at T. Now apply Pascal's theorem to the degenerate hexagon AABCCD to conclude that S, R, and Q are collinear (see Remark 12.1). Next apply Pascal's theorem to the degenerate hexagon ABBCDD to conclude that R, T, and Q are collinear. Hence S, R, T, and Q are collinear. But S is the pole of $\overleftrightarrow{AC}$ and T is the pole of $\overleftrightarrow{BD}$, and hence $\overleftrightarrow{RQ}$ is the polar of P, by Theorem 12.4.

11. Let A and B be two points, and lines a and b be their polars with respect to a circle ω centered at O [93, Problem 52]. Let AP and BQ be the distances from A to b and from B to a, respectively. Prove that $\frac{OA}{AP} = \frac{OB}{BQ}$.

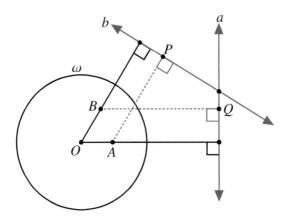

12. Given a circle ω centered at O and a circle α centered at A of radius r, let $\epsilon = \frac{OA}{r}$. Use dynamic geometry software to show the following:

    a. If $\epsilon < 1$, then the image of α under reciprocation in ω is an ellipse.

    b. If $\epsilon = 1$, then the image of α under reciprocation in ω is a parabola.

    c. If $\epsilon > 1$, then the image of α under reciprocation in ω is a hyperbola.

13. Do the conclusions of the previous problem still hold if α is a parabola with its focus at A. What if α is hyperbola with center at A?

12.6 Polar Circles and Self-Polar Triangles

Material in this section was adapted from [19]. A triangle is said to be *self-polar* with respect to a circle ω if each of its sides (considered as lines) is the polar of the opposite vertex with respect to ω. Note that this implies that any two vertices and any two sides are are conjugate (why?). The existence of self-polar triangles follows from Theorem 12.4. In fact, $\triangle PRQ$ in Figure 12.28 is self-polar with respect to ω (why?). Note that p is the polar of P and q is the polar of Q; hence the polar of their point of intersection R is the line passing through P and Q. Therefore, $\overleftrightarrow{OR}$, $\overleftrightarrow{OP}$, and $\overleftrightarrow{OQ}$ are the altitudes to $\overleftrightarrow{PQ}$, $\overleftrightarrow{QR}$, and $\overleftrightarrow{PR}$, respectively. This implies that O, the center of ω, is the orthocenter of $\triangle PRQ$. The circle ω is called the *polar circle* of $\triangle PRQ$. Since P' is the image of P under inversion in ω, we have $OP \cdot OP' = k^2$. Hence the radius k of ω is equal to $\sqrt{OP \cdot OP'}$.

12.6 Polar Circles and Self-Polar Triangles

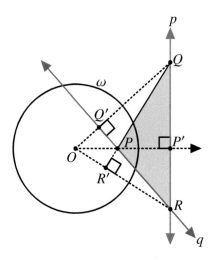

Figure 12.28

> **Now Try This 12.7**
>
> 1. Consider all typical possible locations of P and Q relative to ω in Figure 12.28, that is, P inside and Q outside ω, P outside and Q inside ω, and P and Q outside ω. What can you conclude about $\triangle PRQ$?
> 2. Show that a right triangle or an acute triangle is not self-polar with respect to any circle.
> 3. Can you picture a self-polar triangle having two of its vertices on l_∞? What does it look like?

The discussion above shows how to construct the polar circle ω of any obtuse $\triangle PRQ$. We simply take O, the orthocenter of the triangle, to be the center of ω and take its radius k to be $\sqrt{OP \cdot OP'}$. It turns out that the polar circle is unique (see Problem 3 at the end of this section).

There is nothing special about choosing P and P', the foot of the altitude opposite P; we could have chosen any vertex of $\triangle PRQ$ and the foot of the altitude opposite of it to determine the radius k. This is due to the fact that $\triangle OPQ' \sim \triangle OQP'$ and $\triangle OPR' \sim \triangle ORP'$ (why?), where Q' and R' are the feet of the altitudes opposite Q and R, respectively. Hence, $\sqrt{OP \cdot OP'} = \sqrt{OQ \cdot OQ'} = \sqrt{OR \cdot OR'}$.

Inversion in this polar circle takes the vertices of $\triangle PRQ$ onto the feet of its altitudes. Now, since the circumcircle of $\triangle PRQ$ does not go through its orthocenter (why?), inversion in ω takes the *circumcircle* to the unique circle passing through the feet of the altitudes which is the *nine-point circle* (see Section 9.4). What can you say about the centers of the polar circle, nine-point circle, and circumcircle? This gives us Theorem 12.6 [19, Theorem 6.12]. How does this theorem relate to Theorem 9.10?

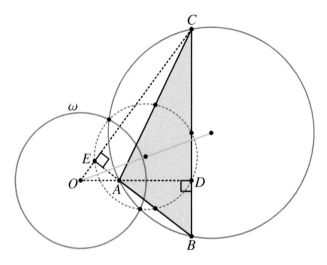

Figure 12.29

> **Theorem 12.6**
>
> *For any obtuse triangle, the circumcircle and the nine-point circle are inverted onto each other under inversion in the polar circle of the triangle.*

12.6.1 Problem Set

1. Prove that if $ABCD$ is a cyclic quadrilateral, as in the figure below, then
   a. $\overleftrightarrow{RT}$ is the polar of S.
   b. $\triangle RST$ is a self-polar triangle with respect to the circle circumscribing $ABCD$.

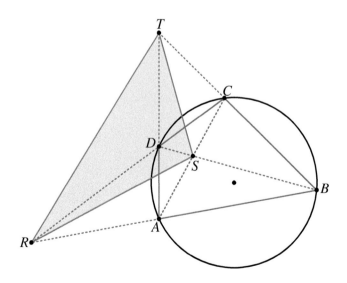

12.6 Polar Circles and Self-Polar Triangles

2. Given a circle ω centered at O with radius k, show that it is possible to draw an infinite number of self-polar triangles with respect to ω.

3. Show that the polar circle of an obtuse triangle ABC is unique.

4. Prove that if a triangle is inscribed in a circle, any line conjugate to one side meets the other two sides in conjugate points.

5. Prove that three points are collinear if and only if their polars with respect to a circle ω centered at O with radius k are concurrent.

6. Prove that if α and β are two orthogonal circles, then the endpoints of a diameter of α are conjugate with respect to β and vice versa.

7. Let $\overline{AB}$ be a diameter of a circle ω, and let P and Q be points on ω such that P is closer to A than Q [73]. If $\overline{PA}$ and $\overline{QB}$ intersect at S outside the circle, and if the tangents at P and Q meet at R, prove that $\overleftrightarrow{RS}$ is perpendicular to $\overleftrightarrow{AB}$.

8. A circle ω is inscribed in a quadrilateral $ABCD$ such that it's tangent to the sides $\overline{AB}, \overline{BC}, \overline{CD}, \overline{DA}$ at the points E, F, G, H, respectively.
   a. Show that $\overleftrightarrow{AC}, \overleftrightarrow{EF}$, and $\overleftrightarrow{GH}$ are concurrent in the pole of $\overleftrightarrow{BD}$ with respect to ω.
   b. What can you say about $\overleftrightarrow{AC}, \overleftrightarrow{BD}, \overleftrightarrow{EG}$, and $\overleftrightarrow{FH}$? Justify your answer.

13 Taxicab Geometry

13.1 Introduction

Studies have shown that most middle and high school students possess a limited understanding of the Euclidean geometric concepts taught at that level. For students to truly understand and appreciate some of these concepts, they need to examine what happens when one or more of the building blocks of Euclidean geometry (points, lines, axioms, etc.) are altered, thus creating a non-Euclidean geometry. For instance, altering the parallel postulate to say that a line has more than one parallel line through a point not on it results in hyperbolic geometry (see Chapter 10), while altering it to say that a line has no parallel lines results in elliptic geometry (see Chapter 11). Unfortunately, trying to introduce these geometries in a meaningful way into a typical high school geometry classroom is somewhat unfeasible. Nonetheless, how can Euclidean geometry be altered to create a non-Euclidean geometry suitable for studying at the high school level? Such a non-Euclidean geometry should not require students to seriously change familiar notions they have about basic building blocks such as points and lines. One convenient way to do this is by introducing taxicab geometry, which only requires a different metric, that is, a different way of measuring the distance between points in the plane.

As we will see later in this chapter, using the taxicab metric (defined below) alters many of the notions we have about familiar geometric objects, thus producing some surprising and interesting results. For instance, students learn that a circle is the set of all points in the plane that are equidistant from a fixed point called the center. Figure 13.1 shows a Euclidean circle C and a taxicab circle K, both centered at A and of radius $\overline{AB}$.

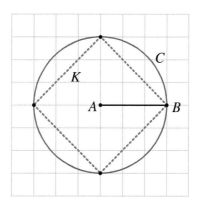

Figure 13.1

Notice that K is a square whose diagonals are parallel to the x- and y-axes, which is very different from what students normally think of as a circle. Such a drastic change in the shape of a familiar figure requires students to pay more attention to mathematical definitions,

© The Author(s), under exclusive license to Springer Nature Switzerland AG 2024
S. Libeskind, I. S. Jubran, *Euclidean, Non-Euclidean, and Transformational Geometry*,
https://doi.org/10.1007/978-3-031-74153-1_13

and helps illustrate their important role in the study of mathematics. Furthermore, such surprising results should intrigue students and raise their level of curiosity regarding how other familiar objects might behave in this geometry.

An excellent introduction to taxicab geometry is Krause's *Taxicab Geometry: An Adventure in Non-Euclidean Geometry* [52]. In this book, Krause presents the basics of taxicab geometry and poses conceptual problems that allow the reader to explore taxicab geometry in greater depth. He also presents many practical applications of taxicab geometry to real-world problems such as urban planning. The book emphasizes original mathematical thought and the ability to apply theoretical ideas to help solve practical problems. Some of the material in this chapter was adapted from [44].

The so-called taxicab metric or Manhattan metric was first introduced by Minkowski in the late 19th century and is defined as follows:

> **Definition: Taxicab Metric**
>
> Let $P_1(x_1, y_1)$ and $P_2(x_2, y_2)$ be two points in the coordinate plane. The taxicab distance d_T between P_1 and P_2 is defined to be the sum of the horizontal and vertical distances between them. That is, $d_T(P_1, P_2) = |x_1 - x_2| + |y_1 - y_2|$.

Formally proving that d_T is a metric on $\mathbb{R}^2$ is the subject of Problem 9 at the end of this chapter. While this method of measuring distance may sound strange to the average person, it makes perfect sense for taxicab drivers who operate their cabs in the downtown of a hypothetical city arranged as a uniform rectangular grid, where blocks are congruent and streets run only north–south and east–west. Using the usual Euclidean distance between two locations to determine how much to charge the passenger is not practical here. Figure 13.2 shows two points P_1 and P_2 whose Euclidean distance $d_E = \sqrt{(x_2 - x_1)^2 + (y_2 - y_1)^2}$ is 5, while their taxicab distance is 7. Note that there are many paths of length 7 that lead from P_1 to P_2. A path in *discrete taxicab geometry* (where points are only lattice points) means a set of horizontal moves $\rightarrow$ and vertical moves $\uparrow$ along the grid lines. Simple counting techniques allow us to conclude that there are $\binom{7}{3} = \frac{7 \cdot 6 \cdot 5}{3!} = 35$ such paths (why?).

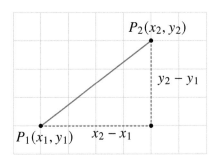

Figure 13.2

In this chapter, we will be dealing with *continuous taxicab geometry* where points are all points in the plane not just lattice points. Hence the line segment connecting P_1 and P_2

13.2 Taxicab Versus Euclidean

in taxicab geometry is the same as the Euclidean segment $\overline{P_1 P_2}$, which is not necessarily a path along grid lines.

13.2 Taxicab Versus Euclidean

In this section we explore some basic ideas from Euclidean geometry and see whether these continue to hold in taxicab geometry or not. It is important to remember that taxicab geometry satisfies all the Euclidean postulates, and points, lines, and the relationship "on" are the same in both geometries. That being said, some important basic Euclidean results cease to hold in taxicab geometry due to the use of a different metric. The concepts of π, construction of equilateral triangles, angle measurement, and congruence of triangles will be considered next.

Taxicab π

The number π is defined to be the ratio of the circumference of a circle to its diameter. The value of π in Euclidean geometry is approximately 3.14159, and we know that the circumference and area of a circle satisfy $C = 2\pi r$ and $A = \pi r^2$, respectively. What is the status of these concepts in taxicab geometry? The reader is asked to investigate this in Now Try This 13.1. The notation π_T is used to represent π in taxicab geometry.

> **Now Try This 13.1**
>
> The reader is encouraged to investigate (1) and (2) below before reading the discussion that follows.
>
> 1. Use the taxicab circle in Figure 13.3 to compute π_T, the ratio of the circumference of a circle to its diameter in taxicab geometry. Does this value of π_T hold for all taxicab circles? Explain.
> 2. Do the circumference and area formulas $C = 2\pi_T \cdot r$ and $A = \pi_T \cdot r^2$ hold for taxicab circles? Explain.
>
>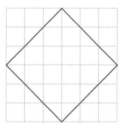
>
> Figure 13.3

Now Try This 13.1 should have led the reader to conclude that $\pi_T = 4$, and that $C = 2\pi_T \cdot r$ holds for all taxicab circles. To justify this, think of the taxicab circle as being centered at the origin and going through the point $(a, 0)$. Then each side of the circle has taxicab length $2a$ and the circumference has taxicab length $8a$. Hence $\pi_T = 4$, and $C = 2\pi_T \cdot r$ holds for all taxicab circles regardless of where they are centered (why?).

On the other hand, the area formula $A = \pi_T \cdot r^2$ does not hold for all taxicab circles. To explain this, remember that the notion of area in Euclidean geometry is based on the area of a rectangle, which in turn is defined to be the number of unit squares that can fit in the rectangle. Using this notion of area, the taxicab circle in Figure 13.3 has an area of 18 square units (this circle is a Euclidean square of side length $\sqrt{18}$). However, the taxicab length of the side of this square is 6 units, so squaring the side would give an area of 36 square units. Furthermore, what happens to the area of the square if it is rotated 45° so that its sides are parallel to the coordinate axes? For this rotated square, the taxicab length of the side is $\sqrt{18}$ units, and hence squaring the side would give an area of 18 square units. So it appears that rotating the square has changed its area! Now what happens if we apply the area formula $A = \pi_T \cdot r^2$ to the circle in Figure 13.3? The radius of this circle is 3 units, and using $\pi_T = 4$ we conclude that the area is 36 square units. To arrive at an area of 18 square units would mean that $\pi_T = 2$, but this violates the notion that π_T is the ratio of the circle's circumference to its diameter. Needless to say, trying to develop a taxicab area notion different from the Euclidean notion is ambiguous. Therefore, we will agree to use the Euclidean area notion in taxicab geometry, which implies that many of the well-known area formulas for familiar objects such as triangles, rectangles, circles, etc., will cease to hold. Additionally, we will choose $\pi_T = 4$, which satisfies the circumference formula $C = 2\pi_T \cdot r$ for all taxicab circles.

Equilateral Triangles

We now consider the concept of constructing an equilateral triangle on a given segment $\overline{AB}$. This is achieved in Euclidean geometry by constructing two circles centered at A and B, each having $\overline{AB}$ as radius. This construction seems to work in taxicab geometry if the two circles intersect in two points, as in Figure 13.4. Looking ahead, we will see later in this section that $\triangle ABC$ in Figure 13.4 is an example of an equilateral triangle that is not equiangular.

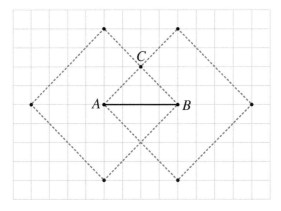

Figure 13.4

But what if the taxicab circles centered at A and B and having $\overline{AB}$ as radius intersect in more than two points as in Figure 13.5? Here the circles C_1 and C_2 intersect in the segments $\overline{EF}$ and $\overline{E'F'}$, so $\triangle ABC$ is an equilateral triangle for any point C on $\overline{EF}$ or $\overline{E'F'}$. Therefore, unlike in Euclidean geometry, this construction can produce infinitely many equilateral triangles each having $\overline{AB}$ as a side. This is due to the fact that contrary to Euclidean circles, two distinct taxicab circles can intersect in infinitely many points.

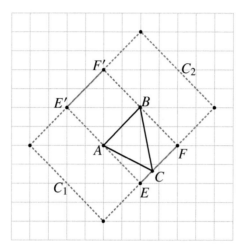

Figure 13.5

For a related problem, recall that in Euclidean geometry, it is not possible to construct an equilateral triangle with vertices lying on lattice points of a rectangular grid (cannot construct an equilateral triangle on a geoboard or geoboard paper). Is constructing such a triangle possible in taxicab geometry? Figure 13.6 shows three taxicab equilateral triangles with vertices lying on lattice points. Notice that the first is a right triangle, the second has no equal angles, and the third has two equal angles (angles measured in the Euclidean sense). The reader is encouraged to use graph paper to draw their own triangles with vertices lying on lattice points.

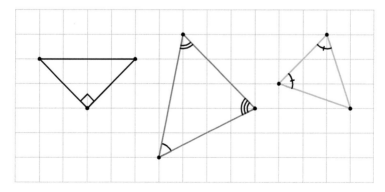

Figure 13.6

Angle Measurement

In "pure" taxicab geometry (as in [52]), the standard Euclidean definition of angle measure is used. In the discussion that follows we will define and use a new taxicab angle measure analogous to the radian measure for Euclidean angles. This measure will allow us to analogously define taxicab versions of the trigonometric functions sine and cosine. We will continue to use "taxicab geometry" to refer to taxicab geometry with this new angle measure.

In Euclidean geometry there is a natural way to define the measure of an angle in terms of the geometry of the circle. If C is a Euclidean circle and θ is a central angle relative to C, the *radian* measure of θ is defined to be the ratio of the arc length of C subtended by θ to the length of the radius of C. This is well defined, since this ratio is the same for all circles (by similarity). Similarly, if C is a taxicab circle and θ is a central angle relative to C, we define the *t-radian* (taxicab radian) measure of θ to be the ratio of the taxicab arc length of C subtended by θ to the taxicab length of the radius of C.

Figure 13.7 shows a taxicab unit circle and an angle θ whose measure is 1 t-radian. This means that the Euclidean angle $\frac{\pi}{4}$ in the standard position (its vertex is at the origin and its initial side is along the positive x-axis) has a taxicab measure of 1 t-radian. Similarly, the Euclidean angles $\frac{\pi}{2}$ and π in the standard position have measures 2 and 4 t-radians, respectively. Hence the circumference of the taxicab unit circle subtends an angle of $8 = 2\pi_T$ t-radians. A word of caution is needed here: The t-radian measure of a Euclidean angle in nonstandard position is not necessarily equal to the t-radian measure of the same Euclidean angle in standard position. This is due to the fact that taxicab circles are squares whose diagonals are parallel to the coordinate axes. Hence rotating an angle whose vertex is at the center of a given circle might change the taxicab length of the arc subtended by the angle. The reader will be asked to verify this in Problem 14 at the end of this chapter. However, all right angles are congruent and their measure is 2 t-radians, and the measure of any straight angle is 4 t-radians (why?).

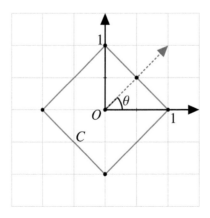

Figure 13.7

13.2 Taxicab Versus Euclidean

To summarize, we have defined the *t*-radian measure of an angle in taxicab geometry to be the length of the arc it subtends (as a central angle with respect to some circle) divided by the radius of that circle. For example, the measure of ∠AOB in Figure 13.8 is $(1 + 1 + 4 + 4 + 1 + 1)/4 = 3$ *t*-radians. Note that the measure of ∠AOB is independent of which circle centered at O is used to determine this measure (why?).

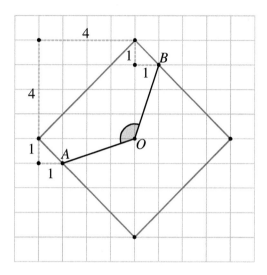

Figure 13.8

Figure 13.9 shows an equilateral triangle of taxicab side length 4 and two circles K_A and K_B each of radius 4 and centered at A and B, respectively.

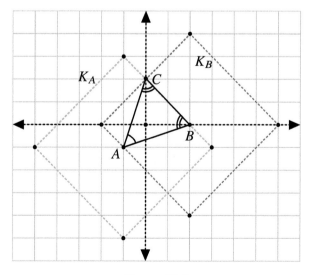

Figure 13.9

Note that the taxicab length of the arc of K_A subtended by $\angle A$ is 4, and since the radius of K_A is 4, we conclude that the measure of $\angle A$ is $\frac{4}{4} = 1$ t-radian. Similarly, the taxicab length of the arc of K_B subtended by $\angle B$ is 6, and since the radius of K_B is 4, we conclude that the measure of $\angle B$ is $\frac{6}{4} = 1.5$ t-radians. A similar argument can be used to show that the taxicab measure of $\angle C$ is 1.5 t-radians as well. This shows that the equilateral triangle $\triangle ABC$ is not equiangular (its angles are not congruent). That being said, note that the angle sum of $\triangle ABC$ is 4 t-radians. A proof of the fact that the sum of the angles in any taxicab triangle is 4 t-radians can be found in [88].

Let C be the taxicab unit circle centered at the origin and θ be an angle in the standard position relative to C. As in Euclidean geometry, let's define the sine and cosine of a taxicab angle θ to be the x- and y-coordinates of P, respectively, where P is the point of intersection of the terminal side of the angle and the circle C. The reader should keep in mind that sin and cos are defined here using the taxicab metric not the Euclidean metric. So despite using the same notation, these are not quite the same as the trigonometric functions normally taught in high school.

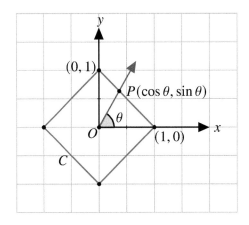

Figure 13.10

Then for an angle θ (used to denote both the angle and its measure) in the first quadrant as in Figure 13.10 we have

1. $\cos\theta + \sin\theta = 1$, since P is on the taxicab unit circle C.

2. $(1 - \cos\theta) + \sin\theta = \theta$, since the length of the arc subtended by the angle θ is equal to the measure of θ (why?).

3. $\cos\theta + (1 - \sin\theta) = 2 - \theta$, since the length of the taxicab arc of C between $\overrightarrow{OP}$ and the positive y-axis is $2 - \theta$.

Combining (1) and (3) gives $\cos\theta = 1 - \frac{\theta}{2}$ and combining (1) and (2) gives $\sin\theta = \frac{1}{2}\theta$. Similarly, the reader should be able to verify that

$$\cos\theta = \begin{cases} 1 - \frac{1}{2}\theta & \text{if } 0 \leq \theta < 4, \\ -3 + \frac{1}{2}\theta & \text{if } 4 \leq \theta < 8, \end{cases}$$

13.2 Taxicab Versus Euclidean

$$\sin\theta = \begin{cases} \frac{1}{2}\theta & \text{if } 0 \leq \theta < 2, \\ 2 - \frac{1}{2}\theta & \text{if } 2 \leq \theta < 6, \\ -4 + \frac{1}{2}\theta & \text{if } 6 \leq \theta < 8. \end{cases}$$

Now Try This 13.2

1. Graph the taxicab functions $y = \cos(x)$ and $y = \sin(x)$ defined above in the same coordinate system. How are these graphs related to each other and how do they compare to the graphs of the Euclidean sine and cosine functions?

2. Use the graphs in (1) to determine the missing entries in the table below (where k is an integer). How do these taxicab trigonometric identities compare to the corresponding Euclidean identities?

$\cos(-\theta) = $ _____	$\cos(\theta - 2) = $ _____
$\sin(-\theta) = $ _____	$\sin(\theta - 2) = $ _____
$\cos(\theta - 4) = $ _____	$\cos(\theta + 8k) = $ _____
$\sin(\theta - 4) = $ _____	$\cos(\theta + 8k) = $ _____

To learn more about taxicab trigonometry, the interested reader can refer to [88].

Congruence of Triangles

We next explore the concept of congruence of triangles. Recall that two triangles are congruent if and only if there is a one-to-one correspondence between their vertices such that corresponding sides are equal in length and corresponding angles have the same measure. Unfortunately, since the taxicab length of a segment and the t-radian measure of an angle are rotation dependent, Euclidean congruence conditions such as SSS, SAS, ASA, etc., do not hold in taxicab geometry.

Theorem 13.1

The triangle congruence conditions SAS, ASA, SAA, and SS do not hold in taxicab geometry.

Proof. This theorem can be proved by simply providing visual counterexamples for each of the conditions. Consider $\triangle ABC$ and $\triangle A'B'C'$ shown in Figure 13.11. Clearly $AB = A'B' = 4$, $BC = B'C' = 4$, $AC = 8$, and $A'C' = 4$. Furthermore, $m(\angle A) = m(\angle A') = 1$ t-radian, $m(\angle C) = m(\angle C') = 1$ t-radian, and $m(\angle B) = m(\angle B') = 2$ t-radians (why?). Hence, $\triangle ABC$ and $\triangle A'B'C'$ satisfy SAS, ASA, and SAA, but they are not congruent since $AC \neq A'C'$.

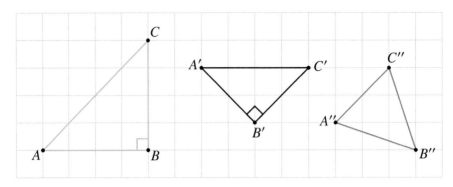

Figure 13.11

To show that SSS does not hold in taxicab geometry, consider $\triangle A'B'C'$ and $\triangle A''B''C''$ shown in Figure 13.11. Clearly these are equilateral triangles of side length 4, but they are not congruent since $m(\angle A'B'C') = 2$ t-radians, whereas $m(\angle A'') = m(\angle C'') = 1.5$ t-radians and $m(\angle B'') = 1$ t-radian (why?). ∎

It turns out that for two taxicab triangles to be congruent they have to satisfy the very strict condition SASAS, and no weaker condition will suffice. Can you verify this claim?

We next investigate the status of the Pythagorean theorem in taxicab geometry.

> **Now Try This 13.3**
>
> 1. Use the taxicab right triangles in Figure 13.12 to determine whether the Pythagorean theorem holds in taxicab geometry or not. If it does not hold, investigate any other general relationship between the length of the hypotenuse and the lengths of the other sides of a right triangle in taxicab geometry.
>
> 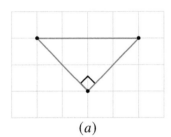　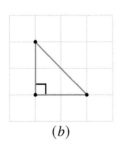　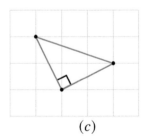
>
> (a)　　　　　　　　(b)　　　　　　　　(c)
>
> Figure 13.12
>
> 2. Does the converse of the Pythagorean theorem hold in taxicab geometry? *Hint: Consider the triangle with vertices at $(2,0)$, $(1,3)$, and $(0,-1)$.*

13.2.1 Problem Set

1. State whether the following statements are true or false. Justify your answers.

a. If $d_T(A, B) = d_T(C, D)$, then $d_E(A, B) = d_E(C, D)$.
b. If $d_E(A, B) = d_E(C, D)$, then $d_T(A, B) = d_T(C, D)$.
c. $d_E(A, B) = d_T(A, B)$ only if both points lie on a horizontal or vertical line.
d. $d_E(A, B) \leq d_T(A, B)$ for any points A and B.

2. Given $A = (3, 5)$ and $B = (-3, 1)$, use graph paper to do the following:
   a. Plot all points P such that $d_T(P, A) = 3$.
   b. Plot all points P such that $d_T(P, A) = 3$ and $d_T(P, B) = 4$. What do you notice?
   c. Plot all points P such that $d_T(P, A) = 6$ and $d_T(P, B) = 4$. What do you notice?
   d. Plot all points P such that $d_T(P, A) = d_T(P, B)$. What do you notice?

3. In Euclidean geometry, given two points A and B, the set of points P such that $PA + PB$ is minimal is the segment $\overline{AB}$. Does the same hold in taxicab geometry? If not, what is the set of points P such that $PA + PB$ is minimal?

4. In Euclidean geometry, given three points A, B, C, the set of points P such that $PA + PB + PC$ is minimal is a single point (called the Fermat–Torricelli point of $\triangle ABC$). Does the same hold in taxicab geometry? Explain how to determine this set in taxicab geometry. *Hint: WLOG, consider A, B, C as in the figure below, which shows a horizontal line through A and a vertical line through B, and argue that their intersection point P is the unique point in the plane such that PA + PB + PC is minimal. Note that the x-coordinate of B is between the x-coordinates of A and C and the y-coordinate of A is between the y-coordinates of B and C* [52].

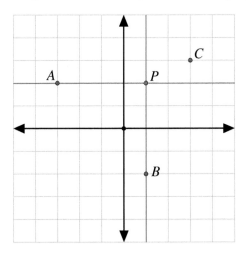

5. In taxicab geometry, given four points $A = (x_1, y_1)$, $B = (x_2, y_2)$, $C = (x_3, y_3)$, $D = (x_4, y_4)$, what is the set of points P such that $PA + PB + PC + PD$ is minimal? *Hint: First, arrange the x-coordinates of the points in increasing order, and assume WLOG that $x_1 \leq x_2 \leq x_3 \leq x_4$. Then the middle two x-coordinates are x_2 and x_3. Second,*

arrange the y-coordinates of A, B, C, D in increasing order, and assume WLOG that $y_1 \leq y_2 \leq y_3 \leq y_4$. Then the middle two y-coordinates are y_2 and y_3. Show that the cross product of the intervals $[x_2, x_3] \times [y_2, y_3]$ is the required set. What possible forms can this set take?

6. In taxicab geometry, given five points A, B, C, D, E, what is the set of points P such that $PA + PB + PC + PD + PE$ is minimal? *Hint:* Since 5 is odd, the middle x-coordinate of the 5 points crossed by the middle y-coordinate gives a single point as the minimal set.

7. In taxicab geometry, given n points in the plane, what can you say in general about the set of points P such that the sum of distances from P to each of the n points is minimal?

8. In taxicab geometry, for each of the following sets of points, find the set of points P such that the sum of distances from P to the points in the set is minimal.
   a. $A = (-2, 1), B = (3, 2)$.
   b. $A = (-3, 4), B = (4, 3), C = (0, -2)$.
   c. $A = (-4, 0), B = (-1, 3), C = (3, -1), = D(1, -3)$.
   d. $A = (0, 1), B = (1, 2), C = (2, 0), D(4, -2), E = (1, -1)$.

9. Prove that the taxicab distance d_T is a metric on $\mathbb{R}^2$. That is, prove that d_T satisfies the following conditions:
   a. For any P_1, P_2 in $\mathbb{R}^2$, $d_T(P_1, P_2) \geq 0$ and $d_T(P_1, P_2) = 0$ if and only if $P_1 = P_2$.
   b. For any P_1, P_2 in $\mathbb{R}^2$, $d_T(P_1, P_2) = d_T(P_2, P_1)$.
   c. For any P_1, P_2, P_3 in $\mathbb{R}^2$, $d_T(P_1, P_3) \leq d_T(P_1, P_2) + d_T(P_2, P_3)$.

10. In Euclidean geometry, given three points A, B, and C, we always have $d_E(A, B) + d_E(B, C) \geq d_E(A, C)$. This is the famous triangle inequality. Does this inequality hold in taxicab geometry? If so, when is $d_T(A, B) + d_T(B, C) = d_T(A, C)$?

11. In taxicab geometry, for each of the following draw pictures distinct from ones you have seen earlier in the chapter.
    a. Give a counterexample for each of the following congruence conditions: ASA, SAA, and SSS.
    b. Draw an isosceles triangle with noncongruent base angles.
    c. Draw a triangle with two congruent angles which is not isosceles.
    d. Draw a right triangle which is equilateral.

12. Find the taxicab measure of each of the angles of the equilateral triangle $\triangle ABC$ in the figure below.

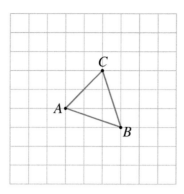

13. Investigate the area concept in taxicab geometry. In particular, do the formulas

   Area of a triangle $= \frac{1}{2}(base \times height)$ and   Area of a square $= (side\ length)^2$

   hold in this geometry? If these formulas do not hold in general, for which triangles/squares do they hold?

14. Use the figure below to verify that the taxicab measure of a Euclidean angle in nonstandard position is not necessarily equal to the taxicab measure of the same Euclidean angle in standard position. We showed earlier that the taxicab measure of a 45° angle in standard position is 1 t-radian. Now consider the two 45° angles in nonstandard position shown below.

   a. What is the taxicab measure of the $\angle AOB$?

   b. What is the taxicab measure of the $\angle COD$?

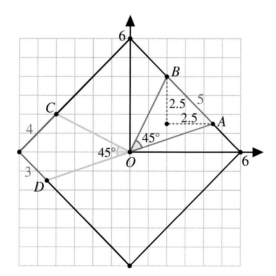

13.3 Distance from a Point to a Line

Having defined earlier how to find the distance between two points in Taxicab geometry, we now turn our attention to defining the distance between a point and a line. In Euclidean geometry this distance is defined to be the length of the perpendicular segment from the point to the line, which is the minimum distance between the given point and any point on the line.

Let P be a point and l be a line as in Figure 13.13. Notice that $d_T(P, Q_1) = 3$, $d_T(P, Q_2) = 4$, and $d_T(P, Q_3) = 5$, etc., so how should we define the distance between P and l? Naturally, it makes sense to define $d_T(P, l)$ to be the minimum of the distances $d_T(P, Q)$, where Q is a point on l. Therefore, $d_T(P, l) = 3$ for P and l in Figure 13.13.

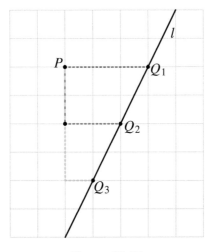

Figure 13.13

Examining this issue further, the reader should come to the conclusion that the way to measure the distance between a point and a line depends on the slope of the line as stated in Theorem 13.2.

> **Theorem 13.2**
>
> Let l be a line in the Euclidean coordinate plane with slope m. Let P be a point not on l.
> 1. If $|m| \leq 1$, then $d_T(P, l)$ is the vertical distance from P to l.
> 2. If $|m| > 1$, then $d_T(P, l)$ is the horizontal distance from P to l.

The following should serve as motivation and intuitive justification for Theorem 13.2. We can find the distance between a point P and a line l by considering a taxicab unit circle centered at P. If the circle does not intersect the line l, imagine increasing its radius gradually until the resulting circle is tangent to l. The distance between P and the first point of intersection between the enlarged circle and l is the minimum distance between

13.3 Distance from a Point to a Line

P and any point on the line. If the taxicab unit circle centered at P intersects l, the same argument applies but now we imagine its radius decreasing gradually until the resulting circle is tangent to l.

Figure 13.14 shows a point P and three lines l_1, l_2, and l_3 of slopes $m > 1$, $m = 1$, and $m < 1$, respectively. It also shows a series of taxicab circles approaching these lines.

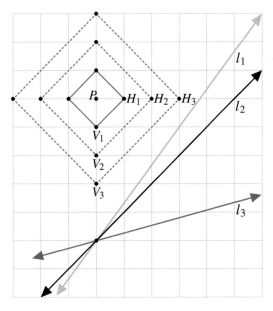

Figure 13.14

Notice that as the radii of the circles centered at P increase, the "horizontal" vertices H_i reach l_1 before the "vertical" vertices V_i. This is not surprising as the vertical distance from P to l_1 divided by the horizontal distance from P to l_1 equals the slope of l_1 (why?), which is greater than 1. Hence $d_T(P, l_1)$ is the horizontal distance from P to l_1. On the other hand, the vertical vertices V_i reach l_3 before the horizontal vertices H_i, since the vertical distance from P to l_3 divided by the horizontal distance from P to l_3 equals the slope of l_3, which is less than 1. Hence $d_T(P, l_3)$ is the vertical distance from P to l_3. Finally, the vertices H_i and V_i reach l_2 at the same time because $\overleftrightarrow{V_i H_i}$ is parallel to l_2. In this case, $d_T(P, l_2)$ can be taken to be either the vertical or horizontal distance from P to l_2. This allows us to conclude that how the distance between a point and a line is measured in taxicab geometry is determined by the slope of the line as stated in Theorem 13.2.

> **Now Prove This 13.1**
>
> Given a point $P(x_1, y_1)$ and a line l defined by $ax + by + c = 0$. Follow the steps outlined below to derive a general formula for the distance between P and l in taxicab geometry.
>
> 1. Show that if $b \neq 0$, then the vertical distance between P and l is $\frac{1}{|b|}|ax_1 + by_1 + c|$.
> 2. Show that if $a \neq 0$, then the horizontal distance between P and l is $\frac{1}{|a|}|ax_1 + by_1 + c|$.
> 3. Use (1) and (2) to conclude that the distance between P and l is
> $$d(P, l) = \frac{1}{\max\{|a|, |b|\}}|ax_1 + by_1 + c|.$$

13.4 Taxicab Midsets

We will first address the concept of the *midset* of two points. The reader should be familiar with the fact that in Euclidean geometry, the set of points equidistant from two points P_1 and P_2 (their *midset*) is the perpendicular bisector of the segment $\overline{P_1 P_2}$. Does the same also hold in taxicab geometry? It does only if P_1 and P_2 lie on a horizontal line or a vertical line (why?), that is, when the slope of the line $\overleftrightarrow{P_1 P_2}$ is zero or undefined. Figure 13.15 shows the midset $M_{P_1 P_2}$ of the points P_1 and P_2 when they lie on a horizontal line. Note that the two intersection points of the taxicab circles C_{12} and C_{22}, each of radius 2 and centered at P_1 and P_2, respectively, lie on the midset. The same is true for the intersection points of any taxicab circles C_{1r} and C_{2r} of radius r, where $1 < r < \infty$, and centered at P_1 and P_2, respectively. Therefore, $M_{P_1 P_2}$ is the perpendicular bisector of $\overline{P_1 P_2}$. The reader should investigate the midset of P_1 and P_2 when they lie on a vertical line.

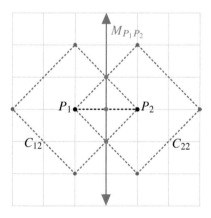

Figure 13.15

13.4 Taxicab Midsets

Figure 13.16 shows an example of the midset of points P_1 and P_2, where $\overleftrightarrow{P_1P_2}$ is neither horizontal or vertical and also does not have a slope of 1 or -1. Note that $M_{P_1P_2}$ is not a straight line. Before reading on, can you tell from the figure how this midset was constructed?

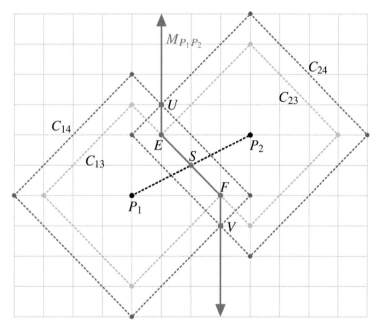

Figure 13.16

Figure 13.16 shows two taxicab circles C_{13} and C_{14} of radii 3 and 4, respectively, and centered at P_1. It also shows two taxicab circles C_{23} and C_{24} of radii 3 and 4, respectively, and centered at P_2. The intersection of C_{13} and C_{23} is $\overline{EF}$, which is part of $M_{P_1P_2}$. The intersection of C_{14} and C_{24} consists of the points U and V; hence, these are also part of $M_{P_1P_2}$. In general, for $3 < r < \infty$, any two circles C_{1r} and C_{2r} of radius r and centered at P_1 and P_2, respectively, intersect in two points: one lying on $\overrightarrow{EU}$ and the other lying on $\overrightarrow{FV}$ (why?). Therefore, as shown in Figure 13.16, $M_{P_1P_2}$ consists of $\overline{EF}$, $\overrightarrow{EU}$, and $\overrightarrow{FV}$.

Figure 13.17 shows the midset of two points P_1 and P_2 where the slope of $\overleftrightarrow{P_1P_2}$ is equal to -1. Note that this midset consists of two "quadrants" joined by a line segment. The taxicab circles C_{12}, C_{13} and C_{14} are centered at P_1 of radii 2, 3, and 4, respectively. Similarly, C_{22}, C_{23} and C_{24} are circles centered at P_2 of radii 2, 3, and 4, respectively. The intersection of C_{12} and C_{22} is $\overline{EF}$; hence it's part of the midset. The intersection of the circles C_{13} and C_{23} is the segments x and x', and the intersection of C_{14} and C_{24} is the segments y and y'. In general, for $2 < r < \infty$, any two circles C_{1r} and C_{2r} of radius r and centered at P_1 and P_2, respectively, intersect two segments of slope -1, one lying in each shaded quadrant. Therefore, for $2 < r < \infty$, the union of the intersections of C_{1r} and C_{2r} is the two shaded quadrants. The reader should verify that the same approach can be applied to

P_1 and P_2 when the slope of $\overleftrightarrow{P_1P_2}$ is 1 to show that their midset consists of two "quadrants" joined by a line segment.

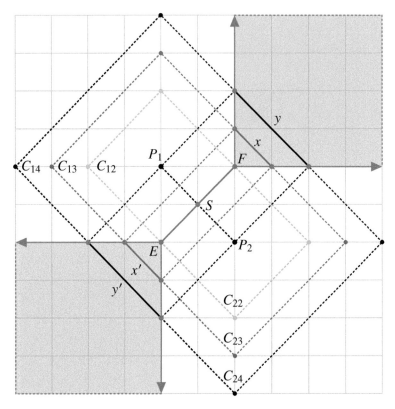

Figure 13.17

As an application of the concept of the taxicab midset between two points, consider the following situation: There are three elementary schools in the area. Assume School A is located at $(-3, -1)$, School B at $(0, -4)$, and School C at $(3, 3)$. Which school should students in the community attend so that they have the least distance to travel according to where they live?

Figure 13.18, where $O = (0, 0)$, shows where the three schools are located, and the three midsets M_{AB}, M_{BC}, and M_{AC}. Note that M_{AB} is a "quadrant" midset, while M_{BC} (orange) and M_{AC} (blue) are linear. Examining Figure 13.18 closely, we arrive at the following conclusions:

(1) Students living in the yellow region should attend School C.

(2) Students living in the green region should attend School A.

(3) Students living in the gray region should attend School B.

(4) Students living in the the pink regions can attend either School A or School B.

(5) Students living on the segment $\overline{EF}$ can attend any of the three schools as points on $\overline{EF}$ are equidistant from the three schools.

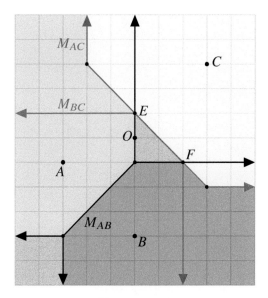

Figure 13.18

We next investigate the concept of the midset of two lines. In Euclidean geometry, the midset of two intersecting lines consists of the two perpendicular lines that bisect the angles formed by the intersecting lines (why?). Figure 13.19 shows two lines L_1 and L_2 (intersecting in point O) and their taxicab midset $M_{L_1 L_2}$ consisting of the lines m_1 and m_2. Are the lines m_1 and m_2 perpendicular in taxicab geometry?

The steps below outline how to determine the midset of two intersecting lines [52].

(1) Determine two pairs of distinct lines j_1, j_2 and k_1, k_2 that are at a convenient distance d_1 from L_1 and L_2, respectively (see Figure 13.19). These lines determine four points of intersection A_1, B_1, C_1, and D_1 that are distance d_1 from both L_1 and L_2. Note that these four points are the vertices of a Euclidean parallelogram centered at O.

(2) Determine two pairs of distinct lines j_3, j_4 and k_3, k_4 that are at a convenient distance $d_2 \neq d_1$ from L_1 and L_2, respectively. These lines determine four points of intersection A_2, B_2, C_2, and D_2 that are at distance d_2 from both L_1 and L_2. Note that these four points are the vertices of a second Euclidean parallelogram centered at O.

(3) The points O, B_1, B_2, D_1, and D_2 are collinear. To justify this you can show that the slopes of $\overleftrightarrow{B_1 D_1}$ and $\overleftrightarrow{B_2 D_2}$ are equal, and hence these lines are the same since they share the point O (why?). Similarly, the points O, A_1, A_2, C_1, and C_2 are collinear.

Hence, the lines m_1 and m_2 determined by the diagonals of the parallelogram in (1) or (2) are the midset $M_{L_1L_2}$ of the lines L_1 and L_2.

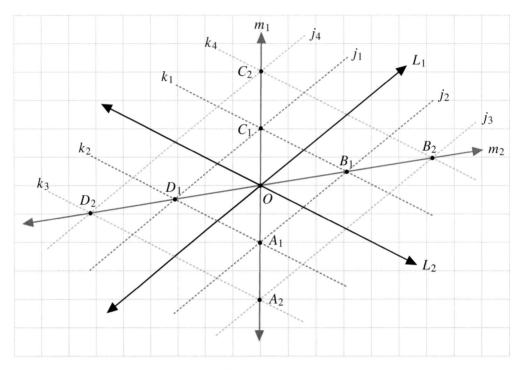

Figure 13.19

Note that there isn't anything mysterious about the distances d_1 and d_2 used in steps (1) and (2) above. In Figure 13.19 we chose $d_1 = 2$ and $d_2 = 4$, but any other convenient choices would work as well. Also, having shown that $M_{L_1L_2}$ consists of the lines $\overleftrightarrow{B_1D_1}$ and $\overleftrightarrow{A_1C_1}$ determined by the four points of intersection in (1) renders step (2) above unnecessary for the actual construction of $M_{L_1L_2}$. Better yet, since O is on m_1 and m_2, we can simply use the points O, A_1, and B_1 to determine $M_{L_1L_2}$.

To close this section, recall that in Euclidean geometry a point is on the angle bisector if and only if it is equidistant from the sides of the angle. Therefore, it is appropriate to consider the midset of two intersecting lines in taxicab geometry as the equivalent of the angle bisector in Euclidean geometry. We will make use of this idea later in this chapter when determining the incenter and excenters of a triangle.

13.4.1 Problem Set

1. For which lines in the coordinate plane does the taxicab distance from a point to a line equal the Euclidean distance from a point to a line?

2. Let l be the line through the points $(1, 2)$ and $(-4, 3)$.

13.5 Circle(s) Through Three Points   589

    a. Graph the set of all points P such that $d_E(P, l) = 3$.

    b. Graph the set of all points P such that $d_T(P, l) = 3$.

3. Let l be the line through the points $(0,-6)$ and $(3,0)$.

    a. Graph the set of all points P such that $d_T(P, l) = 2$.

    b. Graph the set of all points P such that $d_T(P, l) = d_T(P, F) = 2$, where $F = (0, 0)$.

4. There are three schools located as follows: S_1 at $(-4, 3)$, S_2 at $(1, 2)$, and S_3 at $(-1, -4)$. Draw school district boundaries such that every student in the community is going to the closest school.

5. [52] Do the following in taxicab geometry:

    a. There are three high schools in Ideal City: Fillmore at $(-4, 3)$, Grant at $(2, 1)$ and Harding at $(-1, -6)$, as in the figure below. Draw school district boundary lines so that each student in Ideal City attends the high school nearest their home.

    b. If Burger Baron wants to open a hamburger stand equally distant from each of the three high schools, where should it be located?

    c. A fourth high school, Polk High, has just been built at $(2, 5)$. Redraw the school district boundary lines.

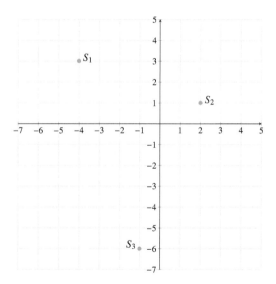

13.5 Circle(s) Through Three Points

The midset concept plays an important role in determining the number of taxicab circles passing through any three distinct points P_1, P_2, and P_3. In Euclidean geometry, any three distinct noncollinear points determine a unique circle. However, the situation is much different in taxicab geometry as three distinct noncollinear points can determine zero, exactly one, or infinitely many taxicab circles. Furthermore, any three distinct collinear points can determine zero or infinitely many taxicab circles (why?). It turns out that the number of taxicab circles determined by P_1, P_2, P_3 is equal to the number of points of intersection of the three midsets $M_{P_1P_2}$, $M_{P_2P_3}$, and $M_{P_1P_3}$ of the pairs of points $\{P_1, P_2\}$, $\{P_2, P_3\}$, and $\{P_1, P_3\}$, respectively [13].

It is worth mentioning at this point that determining a taxicab circle C passing through three noncollinear points P_1, P_2, and P_3 can be tricky sometimes, since a taxicab circle is a square whose diagonals are parallel to the coordinate axes. To simplify this process, we can first rotate the given three points by 45° about some convenient point, the origin for example, which results in three new distinct noncollinear points P'_1, P'_2, and P'_3 as in Figure 13.20. Then, when possible, we can easily determine a square S with sides parallel to the axes lying on these three new points. Finally, rotating this square about the center of rotation used earlier by −45° yields the desired taxicab circle C passing through the original three points.

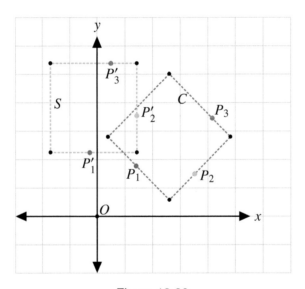

Figure 13.20

Figure 13.21 shows three distinct noncollinear points P_1, P_2, and P_3, which pairwise determine three lines (not shown), each with slope $|m| > 1$.

13.5 Circle(s) Through Three Points

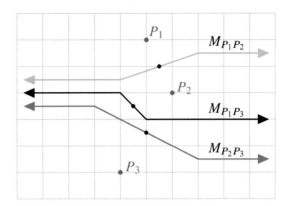

Figure 13.21

Note that the midsets $M_{P_1P_2}$, $M_{P_2P_3}$, and $M_{P_1P_3}$ do not intersect; hence there are no points in the plane that are equidistant from the three points. Therefore, there is no taxicab circle passing through these three points.

Figure 13.22 shows three distinct noncollinear points P_1, P_2, and P_3 such that $\overleftrightarrow{P_1P_2}$ has slope $|m| > 1$, $\overleftrightarrow{P_2P_3}$ has slope $|m| < 1$, and $\overleftrightarrow{P_1P_3}$ has slope $|m| > 1$. Note that the midsets $M_{P_1P_2}$, $M_{P_2P_3}$, and $M_{P_1P_3}$ are linear and have a single point in common. Therefore, there is a unique taxicab circle passing through P_1, P_2, and P_3, as shown in Figure 13.22.

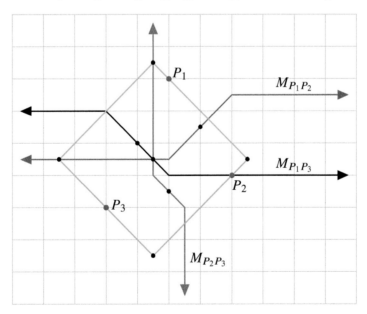

Figure 13.22

Figure 13.23 shows three distinct noncollinear points P_1, P_2, and P_3 such that $\overleftrightarrow{P_2P_3}$ has slope $|m| = 1$, and $\overleftrightarrow{P_1P_2}$ and $\overleftrightarrow{P_1P_3}$ have slope $|m| > 1$. Note that the midsets $M_{P_1P_2}$ and $M_{P_1P_3}$ are linear, while the midset of $M_{P_2P_3}$ is a "quadrant" midset. The intersection of the three midsets is the ray $\overrightarrow{AB}$, so there are infinitely many points in the plane that are equidistant from P_1, P_2, and P_3. Any of the points on $\overrightarrow{AB}$ can serve as the center of a possible taxicab circle going through P_1, P_2, and P_3. Therefore there are infinitely many taxicab circles through these three points; two of these circles are shown in Figure 13.23.

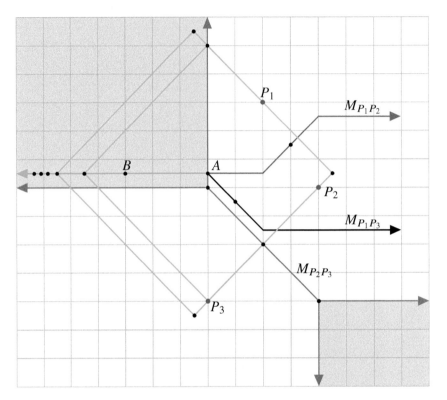

Figure 13.23

Çolakoğlu and Kaya in [13] used a synthetic approach to investigate the problem of the number of taxicab circles determined by three distinct points. Their conclusions are listed below without proof. The reader is encouraged to draw pictures verifying these conclusions.

1. No taxicab circle can be constructed through the three points if:
    a. The three points lie on a line with slope $|m| < 1$ or a line with slope $|m| > 1$.
    b. The three points are noncollinear and determine three lines all with slopes $|m| < 1$ or all with slopes $|m| > 1$.

2. Infinitely many taxicab circles can be constructed through the three points if:
    a. The three points lie on a line with slope $|m| = 1$.
    b. The three points are noncollinear and they determine one line with slope $|m| = 1$ and two lines with slopes $|m| < 1$ or two lines with slopes $|m| > 1$.
    c. The three points are noncollinear and they determine two lines with slopes $|m| = 1$ and one line with slope $|m| < 1$ or one line with slope $|m| > 1$.

d. The three points are noncollinear and they determine one line with slope $|m| = 1$, one line with slope $|m| < 1$, and one line with slope $|m| > 1$. Additionally, the point not on the line with slope $|m| = 1$ lies outside the two circles having the other two points as vertices (see Figure 13.24).

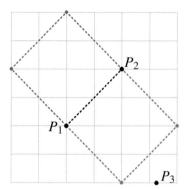

Figure 13.24

3. A unique taxicab circle can be constructed through the three points if:
   a. The three points are noncollinear and they determine one line with slope $|m| < 1$ and two lines with slopes $|m| > 1$.
   b. The three points are noncollinear and they determine two lines with slopes $|m| < 1$ and one line with slope $|m| > 1$.
   c. The three points determine lines as in (2)-d. above, but now together with the condition that the point which is not on the line with slope $|m| = 1$ does not lie outside the two circles having the remaining two points as vertices.

The discussion above yields the following theorem:

> **Theorem 13.3**
>
> *Given any three distinct points in the plane, they can determine zero, one and only one, or infinitely many taxicab circles.*

13.6 Conics in Taxicab Geometry

Having defined the distance between a point and a line and the concept of the midset of two points, we are now ready to explore objects such as taxicab ellipses, hyperbolas, and parabolas. Algebraic equations for these objects in taxicab geometry are discussed first, and then their geometric shapes are explored.

An *ellipse* is defined as the set of points $P(x, y)$ in the plane such that the sum of the distances from $P(x, y)$ to two fixed points $F_1(x_1, y_1)$ and $F_2(x_2, y_2)$, the foci of the ellipse, is a constant $d \geq 0$. Using the taxicab metric, the equation of an ellipse is given by

$$|x - x_1| + |y - y_1| + |x - x_2| + |y - y_2| = d.$$

Figure 13.25 shows two ellipses and their foci in taxicab geometry; think about how these were constructed. Note that if the line segment joining the foci $F_1(x_1, y_1)$ and $F_2(x_2, y_2)$ is horizontal or vertical then the ellipse is a six-sided polygon, while if the segment has slope m such that $0 < |m| < \infty$ then the ellipse is an eight-sided polygon (why?). Note also that when the sum of the distances from the foci approaches the distance between the foci, the ellipse in taxicab geometry approaches being a rectangular region whose dimensions are $|x_2 - x_1|$ and $|y_2 - y_1|$ and having the foci as two diagonally opposite vertices (why?). When the same happens in Euclidean geometry, the ellipse degenerates into the line segment joining the foci.

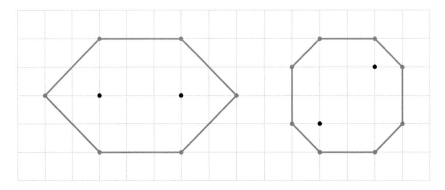

Figure 13.25

A *hyperbola* is defined to be the set of points $P(x, y)$ in the plane such that the absolute value of the difference of the distances between $P(x, y)$ and two fixed points $F_1(x_1, y_1)$ and $F_2(x_2, y_2)$, the foci of the hyperbola, is a constant $d \geq 0$. Using the taxicab metric, the equation of this hyperbola is given by

$$\left|(|x - x_1| + |y - y_1|) - (|x - x_2| + |y - y_2|)\right| = d.$$

Hyperbolas in taxicab geometry will be investigated in Problems 2 and 4 at the end of this chapter. The shape of a hyperbola will depend on the slope of the line determined by its foci. The use of dynamic geometry software is strongly encouraged here.

Finally, a *parabola* is defined as the set of points $P(x, y)$ in the plane that are equidistant from a fixed point $F_1(x_1, y_1)$ (the focus) and a fixed line $ax + by + c = 0$ (the directrix). Unfortunately, since the distance from a point to a line depends on the slope of the line, an exact, compact equation for a parabola in taxicab geometry is not as easily determined as the equation of an ellipse or a hyperbola.

Figure 13.26 shows a taxicab parabola with focus F and directrix l. Each point on the parabola is equidistant from F and l. For example, the point P is 5 units from F and 5 units from l.

13.7 Taxicab Incircles, Circumcircles, Excircles, and Apollonius' Circle

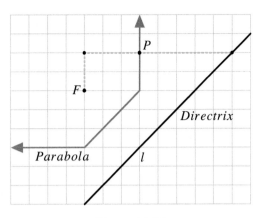

Figure 13.26

13.7 Taxicab Incircles, Circumcircles, Excircles, and Apollonius' Circle

In Euclidean geometry, the *incircle* and *circumcircle* are uniquely determined for any triangle. The *incenter* of a Euclidean triangle is the point of concurrency of the angle bisectors and the *circumcenter* is the point of concurrency of the perpendicular bisectors of the sides. Knowing that the existence of a taxicab circle through three points is determined by the slopes of the lines determined by these points, it is not a big surprise to realize that the existence of incircles and circumcircles is not always guaranteed in taxicab geometry.

In Euclidean geometry, if a line l is tangent to a circle C, then l intersects C in exactly one point and the circle lies entirely on one side of the line. Since taxicab circles are squares with sides having slope ± 1, such circles can intersect a line in one or more points and still lie entirely on one side of the line. In the discussion that follows, we will consider a taxicab circle to be tangent to a line if it intersects the line in one or more points and it lies entirely on one side of the line. This allows us to consider a line intersecting a taxicab circle along one of its sides to be tangent to the circle, as shown in Figure 13.27.

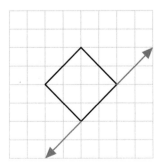

Figure 13.27

Suppose that at least one of the sides of a triangle in taxicab geometry has a slope of ±1. Does such a triangle have a taxicab incircle and circumcircle? Figure 13.28 shows two such triangles $\triangle ABC$ and $\triangle A'B'C'$ together with their incircles and circumcircles (interiors of circles are shaded). Convince yourself that incircles and circumcircles *always* exist for taxicab triangles with at least one side having a slope of ±1.

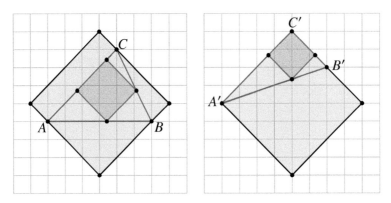

Figure 13.28

Specific conditions for the existence of the incircle and circumcircle of a taxicab triangle having no sides of slope ±1 are given without proof in Theorem 13.4.

> **Theorem 13.4**
>
> Let $\triangle ABC$ be a triangle having no sides of slope ±1. Then $\triangle ABC$ has a taxicab circumcircle and incircle if and only if two sides of the triangle have slope $|m| < 1$ (respectively $|m| > 1$) and the remaining side has slope $|m| > 1$ (respectively $|m| < 1$) [28, Theorem 1].

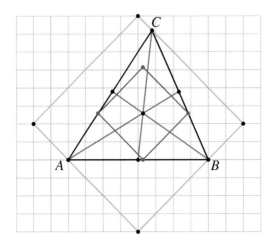

Figure 13.29

13.7 Taxicab Incircles, Circumcircles, Excircles, and Apollonius' Circle

Figure 13.29 shows a taxicab triangle $\triangle ABC$ with its incircle and circumcircle. Note the slopes of the sides of the triangle.

> **Now Try This 13.4**
>
> Use dynamic geometry software to justify the claims in Theorem 13.4. Notice how the slopes of the sides of the triangle (hence the measure and position of the angles) determine whether the circumcircle and incircle exist or not.

The excircles of any triangle are uniquely determined in Euclidean geometry. The center of the excircle opposite a certain vertex is the point of intersection of the angle bisectors of the two exterior angles at the remaining vertices. As mentioned earlier, we will use the Euclidean property that an angle bisector is equidistant from the sides of the angle to construct the equivalent of an angle bisector in taxicab geometry. Hence the center of a taxicab excircle (if it exists) is a point of intersection of the midsets of the sides of the exterior angles adjacent to the excircle under consideration.

Figure 13.30 shows one of the taxicab excircles of $\triangle ABC$, namely, the excircle opposite vertex A.

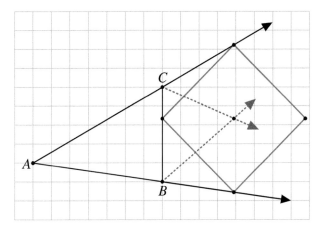

Figure 13.30

> **Now Try This 13.5**
>
> Use dynamic geometry software to investigate the existence of the excircles for various taxicab triangles. How is this related to the slopes of the sides of the triangle?

In Euclidean geometry, it is well known that each triangle has a circle that is internally tangent to each of its three excircles. This circle is called the *Apollonius circle* of the triangle and its construction was discussed in Section 9.3.2. The situation is much different in taxicab geometry since the existence of the Apollonius circle, as expected, depends on the slopes of the sides of the triangle. Conditions necessary for a triangle in taxicab geometry to *always* have an Apollonius circle are given in Theorem 13.5 below. Triangles not satisfying these conditions may or may not have Apollonius circles. For more on this topic, the interested reader can refer to [27].

Figure 13.31 shows the three excircles (with shaded interiors) of a triangle $\triangle ABC$ together with its Apollonius circle. Note that $\overline{AB}$, $\overline{BC}$, and $\overline{AC}$ have slopes less than 1, greater than 1, and equal to 1, respectively.

> **Theorem 13.5**
>
> Let $\triangle ABC$ be a triangle in taxicab geometry. Then there is always an Apollonius taxicab circle if and only if the three sides of the triangle have slopes $|m_1| = 1$, $|m_2| < 1$, and $|m_3| > 1$ [27, Theorem 2.3].

Proof. Let $\triangle ABC$ be a triangle with exactly one side whose slope is ± 1, as in Figure 13.31. Since taxicab circles are squares with sides having slopes of ± 1, each of the excircles of the triangle has a side lying on the side (or its extension) of the triangle whose slope is ± 1. A taxicab circle of side length equal to the sum of the lengths of the sides of the three excircles can always be drawn that is internally tangent to all three excircles.

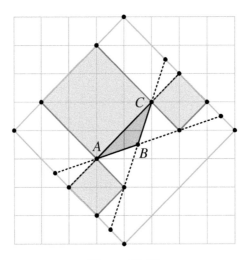

Figure 13.31

Figure 13.32(a) shows a triangle ABC that violates the conditions in the statement of the theorem (why?). This triangle does not have an excircle opposite the vertex C (why?), and hence does not have an Apollonius circle internally tangent to all three excircles. This complete the proof of Theorem 13.5. Figure 13.32(b) shows a triangle DEF that violates the conditions in the statement of Theorem 13.5, but still has an Apollonius circle.

13.8 Inversion in Taxicab Geometry

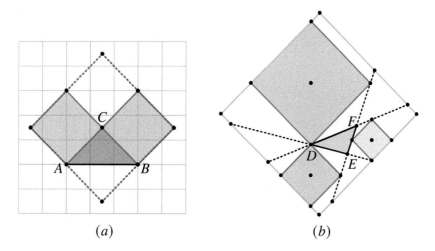

Figure 13.32

13.8 Inversion in Taxicab Geometry

In this section we investigate how the transformation inversion in a circle works in the taxicab geometry. Recall that the image of a point P, under inversion in a circle C centered at point O with radius r, is the point P' on the ray $\overrightarrow{OP}$ satisfying $OP \cdot OP' = r^2$. As expected, using the taxicab metric in the definition of inversion is bound to produce some unexpected and interesting results.

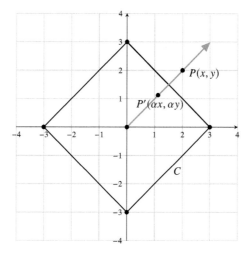

Figure 13.33

In the Cartesian plane, inversion in a taxicab circle C can be described as follows:

1. If C is centered at the origin $O = (0,0)$ with radius r, as in Figure 13.33, then a point $P(x, y)$ inverts onto the point $P'(\alpha x, \alpha y)$ on $\overrightarrow{OP}$, where $\alpha = \frac{r^2}{(|x|+|y|)^2}$. This can be easily justified using the taxicab metric in the equation $OP \cdot OP' = r^2$.

2. If C is centered at an arbitrary point $O = (a, b)$, then since translation is an isometry in taxicab geometry, we can shift (a, b) to the point $(0, 0)$, apply the inversion formula discussed in (1), and then shift $(0, 0)$ back to (a, b), see Now Prove This 13.2.

Now Prove This 13.2

Given a taxicab circle C centered at $O = (a, b)$ with radius r. Prove that the image of a point $P(x, y)$ under inversion in C is the point $P'(x', y')$, where

$$x' = \frac{r^2(x-a)}{(|x-a|+|y-b|)^2} + a \quad \text{and} \quad y' = \frac{r^2(y-b)}{(|x-a|+|y-b|)^2} + b.$$

Figure 13.34 shows how lines not going through the center of inversion and having slopes 0, -1, and $\frac{1}{3}$ are inverted in the taxicab circle centered at the origin with radius 3. Investigating how vertical lines and lines going through the center of the taxicab inversion circle are transformed is left as an exercise.

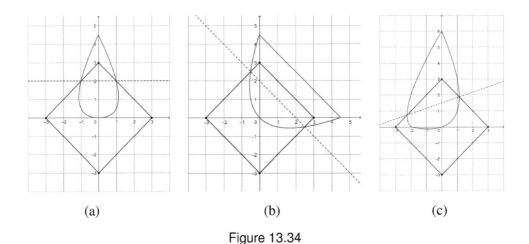

(a)   (b)   (c)

Figure 13.34

Figure 13.35 shows the images of various circles under inversion in the black circle centered at the origin with radius 3. To help identify how the various parts of the circle are transformed, the same color is used for each image and its preimage with preimages being styled as dashed lines.

13.8 Inversion in Taxicab Geometry

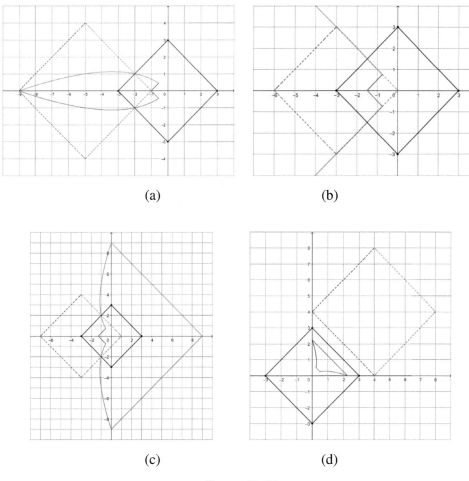

(a) (b)

(c) (d)

Figure 13.35

Given a taxicab circle C centered at O with radius r, Figure 13.35 leads to the following conclusions regarding inversion in C:

1. The image of a taxicab circle centered at O is a taxicab circle centered at O (why?).
2. The image of a taxicab circle not centered at O is not a taxicab circle.
3. The image of a taxicab circle going through O is not a line going through O.
4. A taxicab circle orthogonal to C does not invert onto itself.

13.8.1 Problem Set

1. On the figure below, draw the parabola with focus F and directrix l.

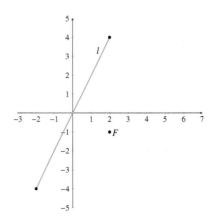

2. Draw the hyperbola with foci $F_1 = (-2, -2)$, $F_2 = (1, 3)$ and distance difference constant 1.

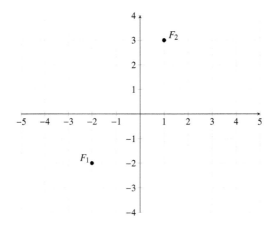

3. Use graph paper to sketch the graphs of the following taxicab parabolas:
   a. The parabola with focus $F = (4, 2)$ and directrix $y = -x$.
   b. The parabola with focus $F = (4, -4)$ and directrix $y = -x$.
   c. The parabola with focus $F = (4, 0)$ and directrix $y = -3x$.

4. A taxicab hyperbola can be described as $\{P : |d_T(P, F_1) - d_T(P, F_2)| = c\}$, where F_1 and F_2 are the foci, and c is the distance difference constant. Use graph paper to sketch the graphs of the taxicab hyperbolas satisfying:
   a. $F_1 = (0, 0)$, $F_2 = (4, 4)$, and $c = 4$.
   b. $F_1 = (0, 0)$, $F_2 = (4, -6)$, and $c = 2$.
   c. $F_1 = (0, 0)$, $F_2 = (4, 0)$, and $c = 2$.

5. A taxicab ellipse can be described as $\{P : d_T(P, F_1) + d_T(P, F_2) = c\}$, where F_1 and F_2 are the foci, and c is the sum of the distances from the foci. Use graph paper to sketch the graphs of the taxicab ellipses satisfying:

13.8 Inversion in Taxicab Geometry

    a. $F_1 = (0,0)$, $F_2 = (-4, 4)$, and $c = 12$.
    b. $F_1 = (0,0)$, $F_2 = (4, 0)$, and $c = 8$.
    c. $F_1 = (0,0)$, $F_2 = (4, -2)$, and $c = 10$.

6. Let $A = (-3, 1)$, $B = (0, 2)$, and $C = (5, 6)$. Try circumscribing a circle about $\triangle ABC$. What goes wrong? Can you find another triangle about which it is impossible to circumscribe a circle?

7. Let $A = (3, -3)$, $B = (-3, 6)$, and $C = (5, 4)$. Recall from Euclidean geometry that the center of the inscribed circle of a triangle (its incircle) is the intersection of the angle bisectors of the triangle, which corresponds to the intersection of the midsets of the edges of the triangle in taxicab geometry.
    a. Inscribe a Euclidean circle in $\triangle ABC$.
    b. Think of a procedure for finding the center of the taxicab circle inscribed in a triangle.
    c. Inscribe a taxicab circle in $\triangle ABC$.

8. Do you think it is always possible to inscribe a taxicab circle in a triangle? Do you think the inscribed taxicab circle is unique? If not, give an example of a triangle that has no incircle and an example of a triangle that has more than one incircle.

9. Let $A = (3, -1)$, $B = (0, 7)$, and $C = (-7, -3)$.
    a. Construct a Euclidean circle that circumscribes $\triangle ABC$.
    b. Construct a taxicab circle that circumscribes $\triangle ABC$.

10. Given two points A and B, how do you construct a circle in taxicab geometry going through A and B? Is such a circle unique? *Hint: Consider the slope of the line segment $\overline{AB}$.*

11. Devise a construction in taxicab geometry for dividing a given line segment into n equal parts.

12. Investigate the existence of a centroid, circumcenter, incenter, and orthocenter for various taxicab triangles [24].

13. Given two points $A = (x_1, y_1, z_1)$ and $B = (x_2, y_2, z_2)$ in three-dimensional space, define the taxicab distance between A and B by $d_T(A, B) = |x_1 - x_2| + |y_1 - y_2| + |z_1 - z_2|$. The unit sphere centered at the origin is the set of all points satisfying the equation $|x| + |y| + |z| = 1$. Determine the nature of the unit sphere in taxicab geometry. Justify your answer. *Hint: Consider the regular octahedron with vertices at $(1, 0, 0), (-1, 0, 0), (0, 1, 0), (0, -1, 0), (0, 0, 1), and (0, 0, -1)$.*

14. In Euclidean geometry, a three-dimensional unit sphere centered at the origin can be constructed by rotating the unit circle $x^2 + y^2 = 1$ about any of its diameters. Can this idea (rotation in a Euclidean manner) be extended to construct the unit sphere in taxicab geometry [43]? *Hint: The taxicab unit circle has three types of diameters: (1) diameters connecting two opposite vertices (vertical or horizontal diameters),*

(2) diameters connecting midpoints of opposite sides (diameters having slope 1 or −1), and (3) diameters having finite slope not equal to 0, 1, or −1. The figure below shows a taxicab unit circle and the surface produced by rotating the circle about the diameter lying on line k.

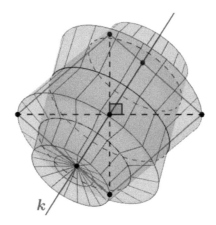

15. Show that a taxicab unit sphere can be obtained by rotating a taxicab unit circle (lying in the xy-plane) about the x-axis in a taxicab manner (such that the cross sections parallel to the yz-plane are taxicab circles [89].

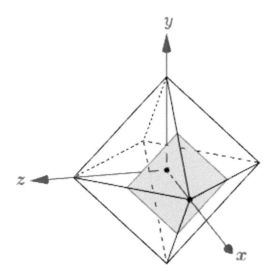

16. Draw a taxicab cylinder of height 8 units, whose base has radius 2, and whose axis lies on the x-axis. *Hint: Cross sections parallel to the yz-plane are taxicab circles of radius 2.*

17. Plot the images of the four solid lines in the figure below under inversion in the given circle.

13.8 Inversion in Taxicab Geometry

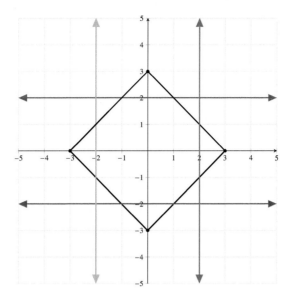

18. Plot the images of the four solid lines in the figure below under inversion in the given circle.

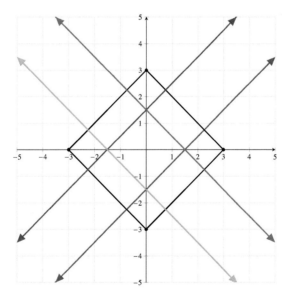

14 Fractal Geometry

14.1 Introduction

A great deal of the credit for the development of fractal geometry should go, and deservedly so, to B. Mandelbrot, who in 1980 discovered the Mandelbrot set and in 1982 published the influential book titled *The Fractal Geometry of Nature* [59]. That being said, we should not neglect contributions by other mathematicians who came before him and who laid the foundation for his work. Late 19th- and early 20th-century mathematicians such as Cantor, Hausdorff, Julia, Levy and many others made valuable contributions to the field now known as fractal geometry, notwithstanding the disadvantage of not having any computer technology in their days.

Many well-known mathematical objects that were studied decades before the discovery of fractal geometry are examples of fractals. The Cantor set (see Figure 14.1) and space filling curves such as the Hilbert curve (see Figure 14.2) are two important examples of such objects.

Figure 14.1 shows the first four stages of the construction of the most common Cantor set, known as the Cantor middle-thirds set, which has fractal dimension $\frac{\ln 2}{\ln 3} \approx 0.631$. The Cantor set was first introduced by the German mathematician George Cantor in 1883. It has many unintuitive properties such as being uncountable, having zero length, and being totally disconnected (containing no intervals). To construct the Cantor middle-thirds set, begin with the interval $K_0 = [0, 1]$ and remove the open middle third interval $(\frac{1}{3}, \frac{2}{3})$. Let $K_1 = [0, \frac{1}{3}] \cup [\frac{2}{3}, 1]$ be the set of the remaining points. Next, remove the open middle third intervals $(\frac{1}{9}, \frac{2}{9})$ and $(\frac{7}{9}, \frac{8}{9})$ from K_1. This gives $K_2 = [0, \frac{1}{9}] \cup [\frac{2}{9}, \frac{3}{9}] \cup [\frac{6}{9}, \frac{7}{9}] \cup [\frac{8}{9}, 1]$. Repeating this process indefinitely gives the Cantor middle-thirds set $K_\infty = \bigcap_{i=0}^{\infty} K_i$.

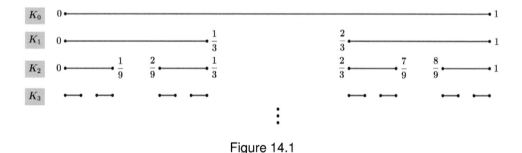

Figure 14.1

Figure 14.2 shows the first five stages of the construction of the Hilbert curve [40], which has fractal dimension 2 and represents an example of a curve that is continuous everywhere, but differentiable nowhere (why?).

© The Author(s), under exclusive license to Springer Nature Switzerland AG 2024
S. Libeskind, I. S. Jubran, *Euclidean, Non-Euclidean, and Transformational Geometry*,
https://doi.org/10.1007/978-3-031-74153-1_14

608   Chapter 14. Fractal Geometry

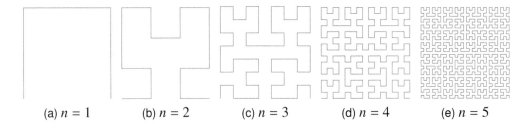

(a) $n = 1$ (b) $n = 2$ (c) $n = 3$ (d) $n = 4$ (e) $n = 5$

Figure 14.2

> **Now Try This 14.1**
>
> Figure 14.3 shows the first and second stages and part of the third stage of the Hilbert curve drawn on a 16×16 grid. Can you give specific instructions to a friend on how to draw the first, second, and third stages of the curve? Furthermore, given stage n of the curve, how do you construct stage $n + 1$? Give this some serious thought and try to figure it out on your own before reading the discussion that follows. Refer to Figure 14.2 if necessary.
>
>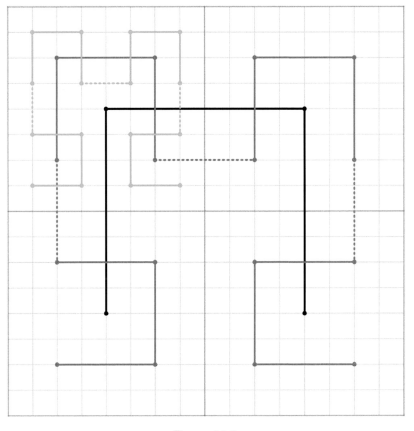
>
> Figure 14.3

14.1 Introduction

Now Try This 14.1 should have led the reader to conclude the following:

1. Think of the 16×16 square grid as the union of four 8×8 square grids. Then the first stage H_1 of the Hilbert curve consists of the three segments connecting the centers of the four 8×8 grids in an upside-down U shape.

2. Shrink the curve H_1 by a factor of $\frac{1}{2}$ and place a copy of it in each of the four 8×8 square grids such that the copies in the lower left and lower right 8×8 grids are rotated 90° clockwise and 90° counterclockwise, respectively. Then the second stage H_2 of the Hilbert curve consists of these four copies of the first stage and the three dashed segments connecting them.

3. Shrink the curve H_2 by a factor of $\frac{1}{2}$ and place copies of it in each of the 8×8 square grids such that the copies in the lower left and lower right 8×8 square grids are rotated 90° clockwise and 90° counterclockwise, respectively. Then the third stage H_3 of the curve consists of these four copies of the second stage and the three segments connecting them. Only the copy in the upper left 8×8 grid is shown in Figure 14.3.

4. Repeating this process indefinitely results in the Hilbert curve $H = \lim_{n \to \infty} H_n$.

Informally, *fractals* are defined as objects having fractional (noninteger) dimension, or as objects that are *self-similar*; that is, when any small part of the fractal is magnified the result resembles the original fractal. A more precise definition uses a concept called *Hausdorff dimension* which was introduced by Felix Hausdorff in 1918. Mandelbrot (1924–2010) coined the term "fractal" and defined a fractal as a set having Hausdorff dimension strictly greater than its topological (geometric) dimension. The Sierpinski gasket (also known as the Sierpinski triangle), Koch curve, Sierpinski carpet, and Apollonian gasket described below are well-known examples of fractals. How to determine the fractal dimension of each of these objects will be discussed in Section 14.2.

Many fractals are constructed using an iterative process. For example, begin with a filled-in triangle and iterate the following process: Divide the triangle into four congruent triangles by connecting the midpoints of the sides, and then remove the middle triangle. Repeat this process on the three filled-in triangles left, and so on (see Figure 14.4). Iterating this process ad infinitum produces, in the limit, the Sierpinski gasket, which has fractal dimension $\frac{\ln 3}{\ln 2} \approx 1.585$.

Figure 14.4

In the case of the Koch curve, start with a line segment and at each stage remove the middle third of all line segments and replace it with two congruent scaled-down line segments (each having the same length as the removed segment) meeting at an angle of $60°$, as shown in Figure 14.5. Iterating this process ad infinitum produces, in the limit, the Koch curve, which has fractal dimension $\frac{\ln 4}{\ln 3} \approx 1.26$.

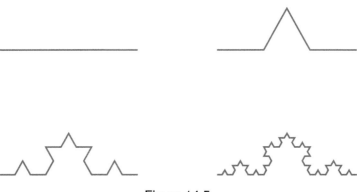

Figure 14.5

For the Sierpinski carpet (see Figure 14.6), start with a filled-in square and iterate the following process: At each stage, divide each filled-in square into nine congruent squares and remove the middle square. Iterating this process ad infinitum produces the Sierpinski carpet, which has fractal dimension $\frac{\ln 8}{\ln 3} \approx 1.893$.

Figure 14.6

Finally, for the Apollonian gasket shown in Figure 14.7, start with three mutually tangent circles, and construct their inner *Soddy circle* (a circle tangent to all three circles). Then construct the inner Soddy circles of this circle and each pair of the original three circles, and repeat this process ad infinitum. The points which are never inside any of the circles form the Apollonian gasket, which has fractal dimension ≈ 1.306.

14.2 Fractal Dimension

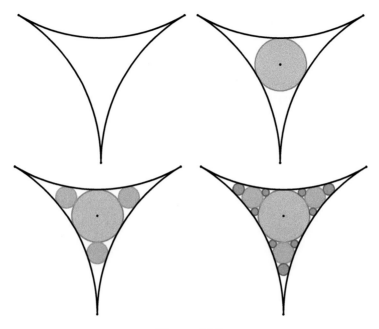

Figure 14.7

14.2 Fractal Dimension

In fractal geometry we study objects that are geometrically intricate, and the concept of fractal dimension is used to quantitatively measure the complexity of these objects. Using the concept of length to measure fractals such as the Koch curve or fractal-like objects such as the coastline of an island produces different results depending on the length of the measuring stick used. The shorter the measuring stick is, the more details and variations we are able to capture. Consequently, as the length of the measuring stick goes to zero, infinitely many sticks will be needed to capture the infinitesimal details of the fractal object being measured. Clearly, this approach as well as the traditional topological (geometric) dimension approach are not suitable for distinguishing fractals based on their complexity, hence the need for the concept of fractal dimension.

Before attempting to define fractal dimension, a very brief discussion of *topological dimension* is needed to see how these are related. We will not try to give a technical definition of the topological dimension of an arbitrary set here since we are mainly dealing with Euclidean spaces. Instead we can, roughly speaking, define the dimension of an object to be the number of coordinates needed to specify a point in the object. The reader should be familiar with the fact that the Euclidean n-space $\mathbb{R}^n$ has dimension n. Hence a point is zero-dimensional, a line is one-dimensional, a plane is two-dimensional, etc.

To compute the fractal dimension of an object we can use the concepts of *similarity dimension*, *box-counting dimension*, or *Hausdorff dimension*. Similarity dimension works well primarily for self-similar objects and hence is limited in its application. Box-counting dimension is more general and can therefore be used for real world objects such as the coastline of an island, however it can be hard to compute. Hausdorff dimension is most general, however it requires mathematical skills beyond the scope of this book, and therefore a technical treatment of it is not appropriate here.

Fractal dimension has many surprising and beneficial real-world applications. For example, in medicine doctors have studied the fractal dimensions of the blood vessels in the retina and bronchial trees as potential biomarkers for the detection of diseases such as diabetes and cancer, respectively [53]. It also has applications in computer graphics, chemistry, finance, cosmology, and many other unexpected fields where fractal-like objects exist and need to be studied.

Similarity Dimension

This approach to defining fractal dimension works very well for self-similar fractals, hence the name. To motivate the definition, we start by calculating the dimension of some simple sets in a way that can be generalized to more complicated sets. To this end, consider the unit interval, the filled-in unit square, and the unit cube shown in Figure 14.8. These clearly have topological dimension d equal to 1, 2, and 3, respectively.

We can view the unit interval $[0, 1]$ as $[0, \frac{1}{2}] \cup [\frac{1}{2}, 1]$, that is, the union of two subinterval each of which is a scaled-down copy of $[0, 1]$ by a factor of $\frac{1}{2}$. It can also be viewed as $[0, \frac{1}{3}] \cup [\frac{1}{3}, \frac{2}{3}] \cup [\frac{2}{3}, 1]$, that is, the union of three subintervals each of which is a scaled-down copy of $[0, 1]$ by a factor of $\frac{1}{3}$. Similarly, the unit square can be viewed as the union of four copies of itself where the side length of each copy is scaled down by a factor of $\frac{1}{2}$, or as a union of nine copies of itself where the side length of each copy is scaled down by a factor of $\frac{1}{3}$. Finally, the unit cube can be viewed as the union of eight copies of itself where the side length of each copy is scaled down by a factor of $\frac{1}{2}$, or as the union of 27 copies of itself where the side length of each copy is scaled by a factor of $\frac{1}{3}$.

For each of the three objects, how is their topological dimension d related to the scaling factor s and the corresponding number N of scaled-down copies that make up the object? For example, for the filled-in unit square, how are the values $d = 2$, $s = \frac{1}{2}$, and $N = 2^2$ related, and does the same relationship hold when $s = \frac{1}{3}$ and $N = 3^2$? Furthermore, does this relationship hold for the unit interval and the unit cube as well?

The discussion above should have led the reader to conclude that for each of the objects shown in Figure 14.8, the equation $N = \left(\frac{1}{s}\right)^d$ is satisfied, where d, s, N are as defined earlier. Solving this equation for d results in $d = \frac{\ln N}{\ln \frac{1}{s}}$. Since for regular objects we expect the fractal and topological dimensions to be the same, it makes sense to use the formula we just discovered to define the fractal dimension of self-similar fractals.

Definition: Similarity Dimension

If $X \subset \mathbb{R}^n$ can be divided into N congruent copies of itself each scaled by a factor s, then the similarity dimension of X is given by $dim(X) = \dfrac{\ln N}{\ln \frac{1}{s}}$.

14.2 Fractal Dimension

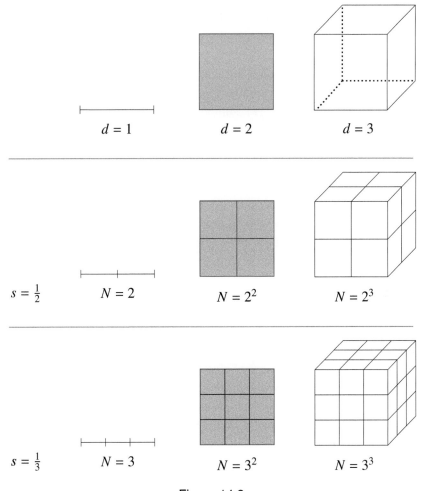

Figure 14.8

To illustrate how to compute the similarity dimension of a fractal object, consider the first two stages in the construction of the Koch curve shown in Figure 14.5. The first stage is a line segment of length 1 unit and the second stage consists of 4 line segments each of length $\frac{1}{3}$ units. Hence $N = 4$, and $s = \frac{1}{3}$, and therefore $d = \frac{\ln 4}{\ln 3}$. Notice that the third stage of the Koch curve can be viewed as the union of $N = 16$ copies of the first stage, each of which is scaled by a factor $s = \frac{1}{9}$. Using these values of N and s gives $d = \frac{\ln 4}{\ln 3}$ (why?). Furthermore, using the second and third stages to compute d gives the same value, since the third stage can be viewed as the union of $N = 4$ congruent copies of the second stage, each scaled by a factor $s = \frac{1}{3}$. Therefore, as anticipated, the similarity dimension does not depend on the stages used to determine N and s.

Box Dimension

Box dimension (or box-counting dimension) is another way to calculate fractal dimension and it can be applied to fractal objects that are not necessarily self-similar such as the coastline of some island. The formal definition of box dimension given below may look somewhat intimidating at first glance, but the motivation and example given below should make it clear.

To motivate the definition, consider the unit interval, unit square, and unit cube shown in Figure 14.8. What is the minimum number of intervals of length ϵ needed to cover the unit interval? Obviously, the answer is $1/\epsilon$. Similarly, what is the minimum number of squares (resp. cubes) of side length ϵ needed to cover the unit square (resp. unit cube)? The answer is $1/\epsilon^2$ squares and $1/\epsilon^3$ cubes, respectively. Note that for each object the exponent of ϵ is the topological dimension of the object.

This can be extended to define the box dimension of more complicated sets $X \subset \mathbb{R}^n$ as follows: For any $\epsilon > 0$, let $N(X, \epsilon)$ denote the minimum number of n-dimensional cubes of side length ϵ needed to cover X. If there is a number D such that $N(X, \epsilon)$ approaches $1/\epsilon^D$ as ϵ approaches 0, then we say the box dimension of X is D.

More precisely, if $\lim_{\epsilon \to 0} \frac{N(X,\epsilon)}{1/\epsilon^D} = k$, where $k > 0$, then taking the natural logarithm of both sides gives

$$\lim_{\epsilon \to 0} \left(\ln(N(X, \epsilon)) - D \ln(1/\epsilon) \right) = \ln(k).$$

Dividing both sides by $\ln(1/\epsilon)$ gives

$$\lim_{\epsilon \to 0} \frac{\ln(N(X, \epsilon))}{\ln(1/\epsilon)} - D = \lim_{\epsilon \to 0} \frac{\ln(k)}{\ln(1/\epsilon)}.$$

Since k is finite, so is $\ln(k)$; and as $\epsilon \to 0$, $\ln(\frac{1}{\epsilon}) \to \infty$. Hence, $D = \lim_{\epsilon \to 0} \frac{\ln(N(X,\epsilon))}{\ln(\frac{1}{\epsilon})}$.

We now give the formal definition of box dimension and use it to compute the dimension of the Sierpinski gasket.

> **Definition: Box Dimension**
>
> Let X be a bounded subset of $\mathbb{R}^n$, and $\epsilon > 0$. Let $N(X, \epsilon)$ be the minimal number of boxes (n-dimensional cubes) of side length ϵ required to cover X. We say that X has box dimension D if $\lim_{\epsilon \to 0} \frac{\ln(N(X, \epsilon))}{\ln(\frac{1}{\epsilon})}$ exists and has value D.

To illustrate how to compute the box dimension, consider the Sierpinski gasket shown in Figure 14.9. Use rectangular grids of side length $\epsilon = 1, 1/2, 1/4, \ldots, 1/2^n$ to cover the gasket and count $N(X, \epsilon)$, the minimal number of boxes needed to cover the gasket. These results are summarized in the table below.

14.2 Fractal Dimension

ϵ	$N(X, \epsilon)$
1	1
1/2	3
1/4	9
1/8	27
$\vdots$	$\vdots$
$1/2^n$	3^n

Table 14.1

Taking the limit as $\epsilon \to 0$ (or equivalently as $n \to \infty$), we are able to conclude that
$$D = \lim_{n \to \infty} \frac{\ln(3^n)}{\ln(2^n)} = \frac{\ln 3}{\ln 2}.$$

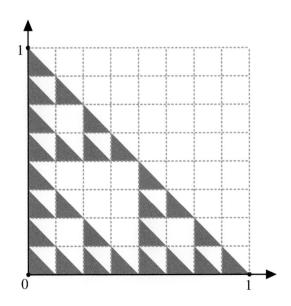

Figure 14.9

Another way to think about D is as follows: If you plot the results in the table above on a graph with $\ln(N(X, \epsilon))$ on the vertical axis and $\ln(\frac{1}{\epsilon})$ on the horizontal axis, then the slope of the line that best fits the data will be a good approximation of the box dimension of the fractal. The reader is encouraged to verify this using the data in Table 14.1.

14.2.1 Problem Set

1. Let T_0 be a filled-in equilateral triangle (see Figure 14.4). Let T_1 be obtained from T_0 by removing the triangle formed by the midpoints of the sides of T_0. Next, let T_2

be obtained from T_1 by removing the triangles formed by the midpoints of each of the smaller three filled-in triangles making up T_1. Continue this process indefinitely and let $T_\infty = \lim_{n \to \infty} T_n$. The fractal T_∞ is called the Sierpinski triangle or gasket.

   a. If the initial stage T_0 has perimeter P_0, what is the perimeter of the nth stage T_n? What is the perimeter of T_∞?

   b. If the initial stage T_0 has area A_0, what is the area of the nth stage T_n? What is the area of T_∞?

   c. Compute the fractal dimension of T_∞.

2. Let C_0 be a filled-in square whose side is 1 unit long (see Figure 14.6). Let C_1 be obtained from C_0 by dividing C_0 into 9 congruent $\frac{1}{3} \times \frac{1}{3}$ squares and removing the middle square. Let C_2 be obtained from C_1 by dividing each filled-in square in C_1 into 9 congruent $\frac{1}{9} \times \frac{1}{9}$ squares and removing each middle square. Continue this process indefinitely and let $C_\infty = \lim_{n \to \infty} C_n$. The fractal C_∞ is called the Sierpinski carpet.

   a. What are the area and perimeter of C_n?

   b. What are the area and perimeter of C_∞?

   c. Compute the fractal dimension of C_∞.

3. A $1 \times 1 \times 1$ cube M_0 is composed of twenty seven $\frac{1}{3} \times \frac{1}{3} \times \frac{1}{3}$ cubes. First, remove the six $\frac{1}{3} \times \frac{1}{3} \times \frac{1}{3}$ cubes having a face in the center of one of the faces of M_0, as well as the $\frac{1}{3} \times \frac{1}{3} \times \frac{1}{3}$ cube at the center of M_0; this gives M_1 shown in the figure below. Second, apply the previous construction to each of the 20 remaining $\frac{1}{3} \times \frac{1}{3} \times \frac{1}{3}$ cubes; this gives M_2. Continue this process indefinitely and let $M_\infty = \lim_{n \to \infty} M_n$, this fractal is called the Menger Sponge.

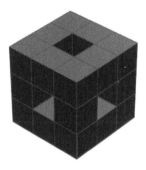

   a. What are the surface area and volume of M_n?

   b. What are the surface area and volume of the M_∞?

   c. Compute the fractal dimension of M_∞.

4. Let S_0 be an equilateral triangle with side length 1 unit. Divide each side of the triangle into three congruent parts each of length $\frac{1}{3}$ unit. Replace the middle segment by two segments each of length $\frac{1}{3}$ unit and meeting at an angle of $60°$; this gives S_1

14.2 Fractal Dimension

(see the figure below). Continue this process indefinitely and let $S_\infty = \lim_{n \to \infty} S_n$. The fractal S_∞ is called the Koch snowflake.

a. If the initial stage S_0 has perimeter P_0, what is the perimeter of the nth stage S_n? What is the perimeter of S_∞?

b. If the region enclosed by the initial stage S_0 has area A_0, what is the area of the region enclosed by S_n? What is the area of the region enclosed by S_∞?

c. Compute the fractal dimension of S_∞.

5. Draw the first four stages of two fractals, one with similarity dimension $\frac{\ln 6}{\ln 4}$, and the other with similarity dimension $\frac{3}{2}$.

6. Compute the fractal dimension of the Hilbert curve discussed earlier (see Figure 14.2).

7. The figure below shows the first three stages in the construction of a certain fractal curve. Compute the fractal dimension of this curve.

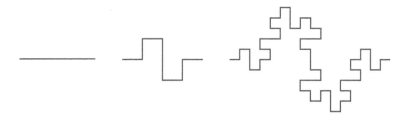

8. The figure below shows the first three stages in the construction of a certain fractal curve. Compute the fractal dimension of this curve.

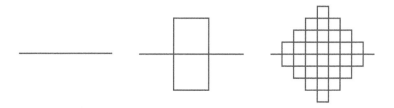

9. Consider the Cantor middle-thirds set shown in Figure 14.1.

a. Find a compact expression for the total length of the middle-thirds intervals missing from $[0, 1]$ at the nth stage K_n of the construction of the Cantor set.

b. Find the total length of all the middle-thirds intervals removed from $[0, 1]$ in the construction of the Cantor set.

c. Compute the fractal dimension of the Cantor set.

d. The endpoints of the open middle-thirds intervals removed during the construction of the Cantor set belong to the Cantor set (why?). Are these the only points in the Cantor set? If not, find a point in the Cantor set that is not an endpoint of one of the removed intervals.

10. It can been shown that the fractal dimension of broccoli is ≈ 2.7 [45]. Use the concept of box dimension to justify this claim. To do this, obtain a large bunch of broccoli and measure and record its size. Break up the broccoli into smaller pieces that are roughly the same size and record the size and number of pieces. Repeat the previous step 3 or 4 times. Since the pieces of broccoli at each stage are not exactly uniform in size, you will probably need to consider a convenient size range at each stage. For the data collected at each stage, plot a point whose x-coordinate is the natural logarithm of the average of the size range at that stage and whose y-coordinate is the natural logarithm of the number of pieces at that stage. The slope of the line that best fits the data points is roughly the fractal dimension of the broccoli. Note that since things are not necessarily uniform here, one or more outlying results might have to be discarded.

11. There are many resources online for constructing fractals using paper folding and as purely origami figures (see, for example, [31]). Investigate how this folding is done. Also, if you have access to a 3D printer, investigate designing and printing your own fractals or use code available online to print some 3D fractals.

14.3 Affine Transformations

To generate all but the simplest fractals, we need to understand the geometry of plane transformations. Of particular interest are *affine transformations*, which are geometric transformations that preserve lines and parallelism but not necessarily distances or angles. Examples of such transformations are scalings, reflections, rotations, translations, dilations, shears, and any compositions of these mappings.

The coordinate representation of an affine transformation $T : \mathbb{R}^2 \to \mathbb{R}^2$ is given by $T((x, y)) = (x', y') = (ax + by + e, cx + dy + f)$, where $a, b, c, d, e,$ and f are constants. For computing purposes, it's easier and more convenient to work with the matrix representation of an affine transformation which is given by

$$T\left(\begin{pmatrix} x \\ y \end{pmatrix}\right) = \begin{pmatrix} a & b \\ c & d \end{pmatrix} \cdot \begin{pmatrix} x \\ y \end{pmatrix} + \begin{pmatrix} e \\ f \end{pmatrix}.$$

Furthermore, if r and s are the scaling factors in the x- and y-directions, respectively; θ and ϕ are the angles of rotation of horizontal and vertical lines, respectively; and e and f measure the horizontal and vertical translations, respectively, then we have

14.4 Iterated Function Systems

$$\begin{pmatrix} a & b \\ c & d \end{pmatrix} = \begin{pmatrix} r\cos(\theta) & -s\sin(\phi) \\ r\sin(\theta) & s\cos(\phi) \end{pmatrix}.$$

It can be easily shown that $r^2 = a^2 + c^2$, $s^2 = b^2 + d^2$, $\theta = \arctan(c/a)$, and $\phi = \arctan(-b/d)$. The reader is encouraged to justify these claims.

For example, in Figure 14.10, $A'B'C'D'$ is the image of the unit square $ABCD$ under a transformation where $r = 2$, $s = 1.5$, $\theta = 30°$, $\phi = 60°$, $e = -0.5$, and $f = 0.5$. For more on affine transformations and 2×2 matrices, see Section 14.6.

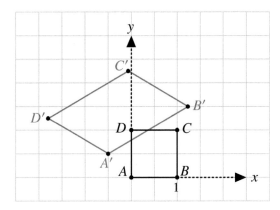

Figure 14.10

14.4 Iterated Function Systems

Being able to determine the matrix representation of plane transformations allows us to generate fractals using a method called iterated function systems (IFS). An *iterated function system* in $\mathbb{R}^2$ is a collection $\{F_1, F_2, \ldots, F_N\}$ of contraction mappings from $\mathbb{R}^2$ to $\mathbb{R}^2$. The transformation F_i is a *contraction mapping* if there is a constant $0 \leq s < 1$ such that for any x and y in $\mathbb{R}^2$, the distance between $F_i(x)$ and $F_i(y)$ is less than or equal to s times the distance between x and y, that is, $d(F_i(x), F_i(y)) \leq s\, d(x, y)$.

For example, the Sierpinski gasket (see Figure 14.9) can be generated using the IFS $\{T_1, T_2, T_3\}$, where the transformation T_1 scales the original triangle by a factor of $\frac{1}{2}$ and places the right-angle corner of the scaled triangle at (0,0). Similarly, T_2 and T_3 scale the original triangle by a factor of $\frac{1}{2}$ and place the right-angle corners of the scaled copies at $(\frac{1}{2}, 0)$, and $(0, \frac{1}{2})$, respectively.

In matrix form, the IFS for this Sierpinski gasket is $\{T_1, T_2, T_3\}$, where

$$T_1\begin{pmatrix} x \\ y \end{pmatrix} = \begin{pmatrix} .5 & 0 \\ 0 & .5 \end{pmatrix} \cdot \begin{pmatrix} x \\ y \end{pmatrix} + \begin{pmatrix} 0 \\ 0 \end{pmatrix},$$

$$T_2\begin{pmatrix} x \\ y \end{pmatrix} = \begin{pmatrix} .5 & 0 \\ 0 & .5 \end{pmatrix} \cdot \begin{pmatrix} x \\ y \end{pmatrix} + \begin{pmatrix} .5 \\ 0 \end{pmatrix},$$

$$T_3\begin{pmatrix} x \\ y \end{pmatrix} = \begin{pmatrix} .5 & 0 \\ 0 & .5 \end{pmatrix} \cdot \begin{pmatrix} x \\ y \end{pmatrix} + \begin{pmatrix} 0 \\ .5 \end{pmatrix}.$$

To explain how this IFS can be used to generate the Sierpinski gasket, consider the simple filled-in unit square S_0 shown in Figure 14.11(a). By Theorem 14.1, we don't have to start with a filled-in right triangle; we can start with any compact (closed and bounded) set $S_0 \subset \mathbb{R}^2$. Applying the IFS $\{T_1, T_2, T_3\}$ to this initial set S_0 produces the set $S_1 = T_1(S_0) \cup T_2(S_0) \cup T_3(S_0)$ shown in Figure 14.11(b).

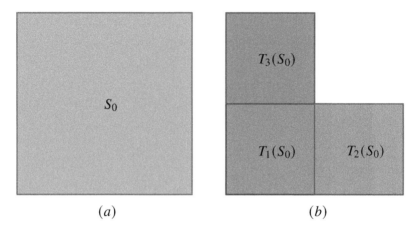

Figure 14.11

Applying the IFS to S_1 gives $S_2 = T_1(S_1) \cup T_2(S_1) \cup T_3(S_1)$ shown in Figure 14.12.

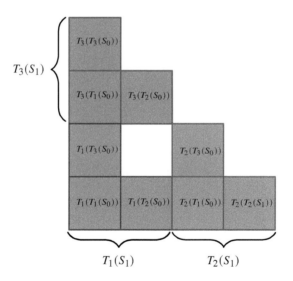

Figure 14.12

Continuing this process gives $S_{k+1} = T_1(S_k) \cup T_2(S_k) \cup T_3(S_k)$, and as k goes to infinity S_k limits to an invariant set S, which in this case is the Sierpinski gasket. The set S satisfies $S = T_1(S) \cup T_2(S) \cup T_3(S)$ and is called the *attracting set* or *attractor* of the IFS $\{T_1, T_2, T_3\}$.

14.4 Iterated Function Systems

It's worth mentioning at this point that applying the IFS $\{T_1, T_2, T_3\}$ to any compact subset S_0 of $\mathbb{R}^2$, no matter what the subset is, will yield the same Sierpinski gasket. The reader is encouraged to use fractal-generating software such as the *IFS Construction Kit by Larry Riddle* [76] to verify this claim. The fact that the attractor of an IFS exists and does not depend on the the initial set S_0 is guaranteed by the *collage theorem* stated below. The proof of this theorem requires mathematical skills beyond the scope of this book (a technical proof can be found in [4]).

Theorem 14.1: Collage Theorem

Given an IFS $\{F_1, F_2, \ldots, F_N\}$ on $\mathbb{R}^2$, there is a unique nonempty compact subset $A \subset \mathbb{R}^2$ called the attractor such that $A = F_1(A) \cup F_2(A) \cup \ldots \cup F_N(A)$.

Now Try This 14.2

Find an iterated function system that can be used to generate the Sierpinski carpet. The first four stages of the construction of this fractal are shown in Figure 14.6. *Hint: The IFS will have eight transformations.*

Given an IFS $\{T_1, T_2, \ldots, T_N\}$, there are two algorithms for determining its attractor on the computer: the *deterministic algorithm* and the more efficient *random iteration algorithm*. When applying the deterministic algorithm, we start with a compact set $S_0 \subset \mathbb{R}^2$, and for $n = 1, 2, 3, \ldots$, successively compute the sets $S_n = \bigcup_{i=1}^{N} T_i(S_{n-1})$. Then by Theorem 14.1, the sets S_n "converge to the attractor" of the IFS as n goes to infinity. On the other hand, when applying the random iteration algorithm, probabilities $p_i > 0$, satisfying $\sum_{i=1}^{N} p_i = 1$, are assigned to each transformation T_i, and a point $x_0 \in \mathbb{R}^2$ is chosen at random. Then for $n = 1, 2, 3, \ldots$, the point x_n is determined by choosing it from $\{T_1(x_{n-1}), T_2(x_{n-1}), \ldots, T_N(x_{n-1})\}$, where the probability of choosing $T_i(x_{n-1})$ is p_i. The sequence of points x_n "converges to the attractor" of the IFS. That is, the points x_n come arbitrarily close to every point in the attractor of the IFS as n goes to infinity. A precise discussion of convergence here would require mathematical skills beyond the scope of this book. The interested reader can refer to [4] for more on convergence, and on the role of the probabilities p_i and how they can be determined. For the fractals in this section applying equal probabilities to each of the transformations in the IFS works well.

Note that for the deterministic algorithm each of the transformations is applied to the whole set S_{n-1} at the nth stage and the union of these images is taken to be S_n while for the random iteration algorithm each of the transformations is evaluated at a single point x_{n-1} at the nth stage and only one of these points is chosen as x_n. Therefore, the random iteration algorithm is a more efficient algorithm for determining the attractor of an IFS, whereas the deterministic algorithm reveals more of the geometry involved in determining the attractor.

To truly understand and appreciate fractal geometry, it is important for students to generate their own fractal images, not just study static images generated by others. The excellent, easy-to-use, multifaceted, and freely available fractal software tool *IFS Construction Kit* [76] was used to generate most of the fractal images in this section. Using fractal-generating software such as the *IFS Construction Kit* is key to exploring the material discussed henceforth. Advanced mathematical software such as *Mathematica* can also be used here.

In the problem set at the end of this section, the reader is presented with fractal images and asked to determine an IFS that can be used to generate each image. We have already seen how this can be done for the Sierpinski gasket and carpet. We next discuss an example where the fractal, on the surface, looks more complicated and less mathematical than fractal objects we have seen so far.

Figure 14.13

Careful examination of the fractal image in Figure 14.13 reveals that it can be viewed as the union of three congruent copies of the whole image, each shrunk by a factor of 1/2. The blue copy placed in the upper left quadrant is oriented the same way as the whole image, the green copy placed in the lower right quadrant is a 90° clockwise rotation of the blue copy, and the red copy placed in the lower left quadrant is a vertical reflection of the green copy. Therefore, to generate this fractal we can start with a square, shrink it by 1/2 and place three shrunken copies in the original square, as shown Figure 14.14. To help us determine orientation and the position of the lower left corner of the original square, a large "L" is placed there.

14.4 Iterated Function Systems

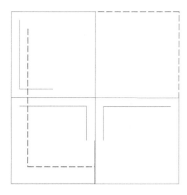

Figure 14.14

This geometric representation of the IFS appears in the "design window" in the *IFS Construction Kit*, while the corresponding algebraic representation appears instantly in the "IFS window." Being able to geometrically design the IFS is a great way for students to avoid messy calculations and focus on developing their geometric intuition.

The algebraic representation of the IFS depicted geometrically in Figure 14.14 is given below. Some basic knowledge of the algebra of 2×2 matrices is helpful here. In matrix form, the IFS for the fractal shown in Figure 14.13 is the IFS $\{F_1, F_2, F_3\}$ defined below.

$$F_1\left(\begin{pmatrix}x\\y\end{pmatrix}\right) = \begin{pmatrix}.5 & 0\\ 0 & .5\end{pmatrix}\cdot\begin{pmatrix}x\\y\end{pmatrix} + \begin{pmatrix}0\\.5\end{pmatrix},$$

$$F_2\left(\begin{pmatrix}x\\y\end{pmatrix}\right) = \begin{pmatrix}0 & .5\\ -.5 & 0\end{pmatrix}\cdot\begin{pmatrix}x\\y\end{pmatrix} + \begin{pmatrix}.5\\.5\end{pmatrix},$$

$$F_3\left(\begin{pmatrix}x\\y\end{pmatrix}\right) = \begin{pmatrix}0 & -.5\\ -.5 & 0\end{pmatrix}\cdot\begin{pmatrix}x\\y\end{pmatrix} + \begin{pmatrix}.5\\.5\end{pmatrix}.$$

The careful reader should recognize $\begin{pmatrix}0 & .5\\-.5 & 0\end{pmatrix}$ in F_2 as the product $\begin{pmatrix}0 & 1\\-1 & 0\end{pmatrix}\cdot\begin{pmatrix}.5 & 0\\0 & .5\end{pmatrix}$, which is a shrinking by 1/2 followed by a 90° clockwise rotation about the origin. Similarly, $\begin{pmatrix}0 & -.5\\-.5 & 0\end{pmatrix}$ in F_3 is the product $\begin{pmatrix}-1 & 0\\0 & 1\end{pmatrix}\cdot\begin{pmatrix}0 & .5\\-.5 & .0\end{pmatrix}$, which is $\begin{pmatrix}0 & .5\\-.5 & 0\end{pmatrix}$ from F_2 followed by reflection about the y-axis. As mentioned earlier, the *IFS Construction Kit* allows us to determine the IFS geometrically, which eliminates the need for many unnecessary calculations. This tool was used to generate the fractal images and design pictures in the exercises that follow.

14.4.1 Problem Set

1. For each of the fractals below: Sierpinski gasket, Koch curve, Levy dragon, and Barnsley's fern, the design picture next to it shows a geometric representation of

an IFS that can be used to generate the fractal. Use fractal-generating software such as the *IFS Construction Kit* to verify that indeed each IFS generates the fractal associated with it.

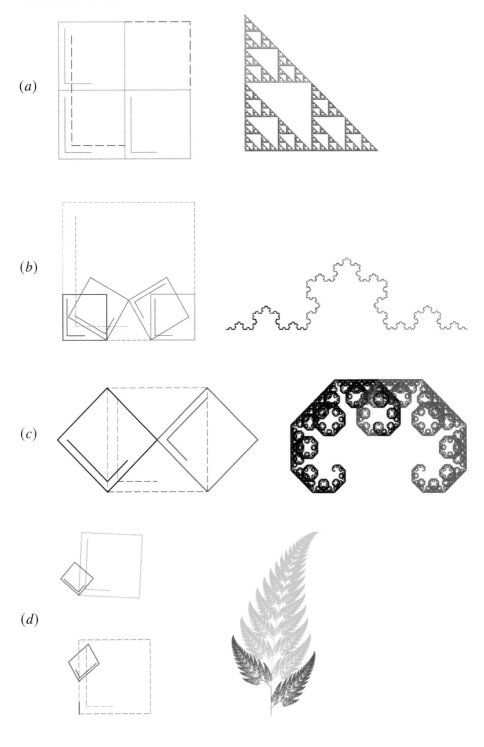

(a)

(b)

(c)

(d)

2. Determine an IFS that can be used to generate each of the fractals below. An IFS in the form of a design picture is sufficient here; that is, draw the initial polygon and geometrically determine the set of transformations needed to generate the fractal. Some of the images below appear on Yale University's Fractal Geometry website [32].

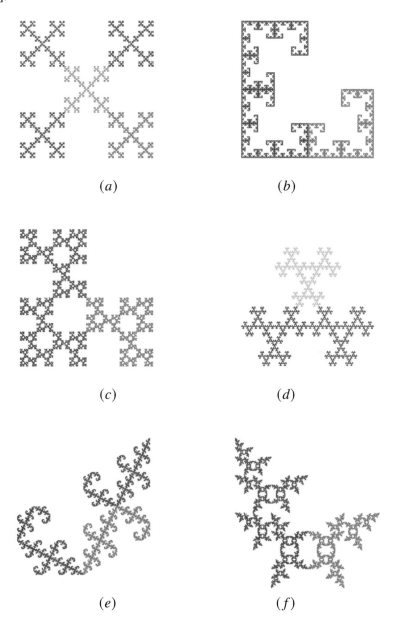

(a)

(b)

(c)

(d)

(e)

(f)

3. The following fractals are somewhat more challenging. Determine an IFS that can be used to generate each of them. An IFS in the form of a design picture is sufficient here.

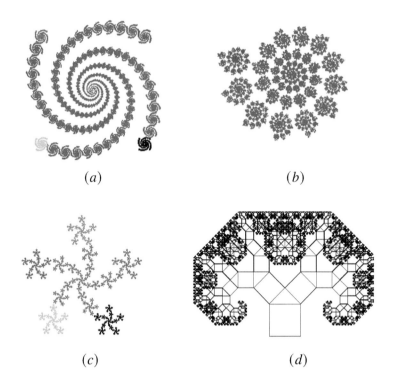

(a)   (b)

(c)   (d)

14.5 The Julia and Mandelbrot Sets

The Julia and Mandelbrot sets are fractal sets that are generated using iteration of complex functions. Such complex functions don't need to be complicated; even simple functions such as quadratic polynomials can result in fascinating, complex fractal sets.

Consider the simple quadratic polynomial $f : \mathbb{C} \to \mathbb{C}$ defined by $f(z) = z^2 + c$, where c is a complex constant and $\mathbb{C}$ is the complex plane. To motivate the definition of the Julia and Mandelbrot sets, we first consider the simple case when $c = 0$. If we represent $z \in \mathbb{C}$ as $z = re^{i\theta}$, then the nth iterate of z under f is

$$z_n = f^n(z) = \underbrace{f \circ f \circ \cdots \circ f}_{n \text{ times}}(z) = (r)^{2^n} e^{i(2^n \theta)}.$$

Now let's investigate the dynamics of f on $\mathbb{C}$, that is, see what happens to points $z_0 \in \mathbb{C}$ under iteration of $f(z) = z^2$. To do this, pick a point $z_0 = r_0 e^{i\theta_0}$ and consider the behavior of $z_1 = z_0^2$, $z_2 = z_1^2$, $z_3 = z_2^2$, ... These are called the iterates of z_0 under f. Clearly, if $r_0 < 1$, then $z_n = (r_0)^{2^n} e^{i(2^n \theta_0)}$ gets closer and closer to the origin as n gets larger and larger (why?). In fact, since the angle is doubled after each iteration, the points z_n spiral in on $(0, 0)$ as n goes to infinity. If $r_0 = 1$, then z_n lies on the unit circle for all n. Finally, if $r_0 > 1$, then z_n gets farther and farther from the origin as n gets larger and larger. In fact, the points z_n spiral out toward infinity as n goes to infinity.

14.5 The Julia and Mandelbrot Sets

Denote the modulus of z_n [equivalently the distance between z_n and $(0,0)$] by $|z_n|$. Next, let E_0 be the set of all points $z_0 \in \mathbb{C}$ such that $|z_n| \to \infty$ as $n \to \infty$, and let P_0 be the set of points $z_0 \in \mathbb{C}$ such that $|z_n| \not\to \infty$ as $n \to \infty$. The *Julia set* corresponding to $c = 0$, denoted by J_0, is defined to be the boundary between the E_0 and P_0. Hence the Julia set for $f(z) = z^2$ is the unit circle, which is not very interesting. The *filled-in Julia* set K_c is defined to be the set of points $z_0 \in \mathbb{C}$ such that $|z_n| \not\to \infty$ as $n \to \infty$. Hence the filled-in Julia set for $f(z) = z^2$ is the unit disc.

To generate more intricate and interesting Julia sets we need to consider $f(z) = z^2 + c$ for $c \neq 0$. Figures 14.15(a), 14.15(b), 14.15(c) represent J_c for $c = 0.5 + 0.5i$, $c = -1.168 - 0.01i$, and $c = -0.27 - 0.88i$, respectively.

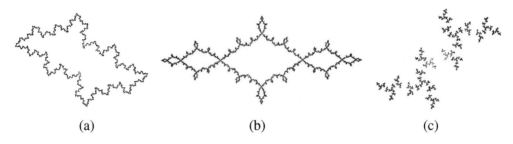

(a) (b) (c)

Figure 14.15

Julia sets can be much more intricate as evident in Figure 14.16, which shows the Julia set for $c = -0.52 + 0.52i$. This image was generated using *The Online Fractal Generator v1.06* [91]. Online software tools for investigating the Julia and Mandelbrot sets can also be found at [14] and [63].

Figure 14.16

It was proved in 1919 independently by G. Julia and P. Fatou that the filled-in Julia set K_c is either connected (one piece) or totally disconnected (like the Cantor set). Furthermore, Mandelbrot proved in 1966 that K_c for $f(z) = z^2 + c$ is connected if and only if the the *orbit* of the point 0 under f is bounded. That is, if the points $0, f(0), f^2(0), f^3(0), \ldots$ lie within a circle of finite radius centered at the origin. This is often referred to as the *Mandelbrot criterion*.

The *Mandelbrot set* associated with $f(z) = z^2 + c$ is denoted by M and is defined to be the set of $c \in \mathbb{C}$ for which K_c is connected. Equivalently, using the Mandelbrot criterion, M is the set of $c \in \mathbb{C}$ for which the orbit of 0 under f is bounded. Figure 14.17 shows a picture of M [63]. For example, to determine whether the point $c = 1$ is in the Mandelbrot set or not, consider the orbit of $z = 0$ under $f(z) = z^2 + 1$. Clearly, $f(0) = 1$, $f^2(0) = 2$, $f^3(0) = 5$, $f^4(0) = 26$, $f^5(0) = 677, \ldots$. Since the orbit of 0 is not bounded, $c = 1$ is not in the Mandelbrot set. The reader should verify that $c = i$ is in the Mandelbrot set, since the orbit of $z = 0$ under $f(z) = z^2 + i$, namely, $i, -1+i, -i, -1+i, -i, \ldots$ is bounded (contained in the circle $|z| = 2$, for example).

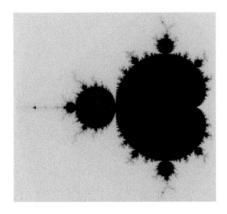

Figure 14.17

14.5.1 Problem Set

1. Consider $f : \mathbb{C} \to \mathbb{C}$ defined by $f(z) = z^2 + c$.
    a. For $c = -1, -i, 1+i, -0.2+0.7i, -0.54+.054i$ determine if the filled-in Julia set is connected (all one piece). *Hint: You can do this by starting with the point 0 and determining if the orbit of 0 is bounded or not.*
    b. Use fractal-generating software to determine the Julia sets for $c = -1.313$, -0.739, $-0.125+0.752i$, $-0.137 - 0.743i$, $0.289 + 0.528i$, $0.432 - 0.025i$, $-0.824 + 0.554i$, $0.187 - 0.613i$, $-0.756 + 0.231i$.
    c. Determine which values of c in part (b) are in the Mandelbrot set corresponding to f.

2. Let $f : \mathbb{C} \to \mathbb{C}$ be defined by $f(z) = z^2 + c$. Use the triangle inequality to prove that if $|c| > 2$ then the orbit of 0 is unbounded. Use this to conclude that if $|c| > 2$, then c is not in the Mandelbrot set corresponding to f.

3. For $f(z) = z^2 + c$, determine the Julia sets J_c for $c = -2$ and for c satisfying $|c| > 2$.

14.6 Linear Transformations and 2 × 2 Matrices: A Brief Summary

$T : \mathbb{R}^2 \to \mathbb{R}^2$ is a linear transformation if it satisfies the conditions:

1. $T(\mathbf{u} + \mathbf{v}) = T(\mathbf{u}) + T(\mathbf{v})$ for all $\mathbf{u}, \mathbf{v}$ in $\mathbb{R}^2$.
2. $T(c\mathbf{u}) = cT(\mathbf{u})$ for all $\mathbf{u}$ in $\mathbb{R}^2$ and c in $\mathbb{R}$.

Any 2×2 matrix $M = \begin{pmatrix} a & b \\ c & d \end{pmatrix}$ naturally defines a linear plane transformation $T_M : \mathbb{R}^2 \to \mathbb{R}^2$ given by

$$T_M(\begin{pmatrix} x \\ y \end{pmatrix}) = \begin{pmatrix} a & b \\ c & d \end{pmatrix} \cdot \begin{pmatrix} x \\ y \end{pmatrix}.$$

Not all plane transformations can be represented by 2×2 matrices; take translation for example. However, linear transformations of the plane such as scalings, rotations, shears, etc., can be conveniently represented as 2×2 matrices.

For example, the matrix $\begin{pmatrix} a & 0 \\ 0 & d \end{pmatrix}$ represents a *scaling* by a factor of a in the x direction and by a factor of d in the y direction, whereas the matrix $\begin{pmatrix} \cos(\theta) & -\sin(\theta) \\ \sin(\theta) & \cos(\theta) \end{pmatrix}$ represents a *counterclockwise rotation* about the origin $(0,0)$ by an angle θ. Also, the matrix $\begin{pmatrix} 1 & b \\ c & 1 \end{pmatrix}$, represents a *shear* by a factor of b in the x direction and a factor of c in the y direction.

In each of Figures 14.18(a)-(c), the quadrilateral $A'B'C'D'$ is the image of the square $ABCD$ under one of the transformations $\begin{pmatrix} \frac{3}{4} & 0 \\ 0 & \frac{1}{2} \end{pmatrix}, \begin{pmatrix} \frac{1}{\sqrt{2}} & -\frac{1}{\sqrt{2}} \\ \frac{1}{\sqrt{2}} & \frac{1}{\sqrt{2}} \end{pmatrix}, \begin{pmatrix} 1 & 0 \\ 2 & 1 \end{pmatrix}$, respectively. Note that in these figures $A = (0,0)$ is fixed so $A = A'$, and the angle of rotation in Figure 14.18(b) is $45°$.

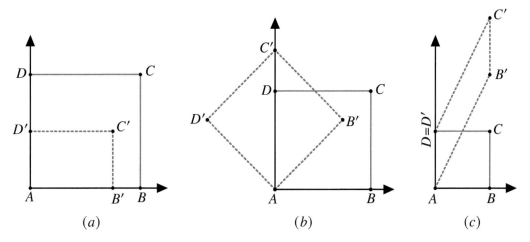

Figure 14.18

15 Solid Geometry

15.1 Objectives

The world around us is three-dimensional, and yet it is often more difficult for people to perceive geometric concepts in three dimensions than in two dimensions. Being in three-dimensional space allows us when working with two-dimensional objects to step outside the plane containing them, and thus be able to see their detailed structure. Unfortunately, we cannot do the same when studying three or higher dimensional objects. Despite that, we are often able to study three-dimensional objects by studying their two-dimensional components. In this chapter we aim to do the following:

1. Briefly explore the fundamental concepts of three-dimensional space such as parallelism and orthogonality of lines and planes. Furthermore, the concepts of angles between lines and planes and angles between planes will be discussed.

2. Define polyhedra and investigate some of their basic properties. Unfortunately, the word *polyhedron* (plural *polyhedra*) has been used to describe different objects over the years. It has been used to refer to a solid region in space like a solid cube, the boundary of such a region (the set of faces, edges, and vertices), or the frame of such a region (the set of vertices and edges) (see Figure 15.1). In this chapter, a polyhedron is understood to be a solid satisfying certain conditions, which will be discussed in Section 15.3.

Figure 15.1

3. Investigate *Descartes' lost theorem*, which states that *the sum of the deficiencies of the solid angles in any polyhedron is eight right angles (4π)*. Here the deficiency of a solid angle at a given vertex means 2π minus the sum of the face angles at that vertex. This can be visualized by cutting along an edge at the given vertex, and then flattening out the faces containing that vertex into the plane as in Figure 15.2, which shows the deficiencies at typical vertices of a cube, a dodecahedron, and an octahedron.

For example, a typical vertex of a dodecahedron has three congruent regular pentagons meeting at it. Hence the deficiency at that vertex is $2\pi - 3(3\pi/5) = \pi/5$, where $3\pi/5$ is the measure of one of the face angles at the vertex, which is the angle of a regular pentagon. The cube, dodecahedron, and octahedron have 8, 20, and 6 vertices, respectively, and hence the sum of deficiencies of the solid angles of each polyhedron is $8(\pi/2)$, $20(\pi/5)$, and $6(2\pi/3)$, respectively, which equals 4π.

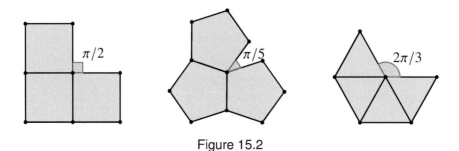

Figure 15.2

4. Prove *Euler's formula* $V + F - E = 2$ and discuss some of its consequences. Here V, F, and E are the number of vertices, faces, and edges of a polyhedron, respectively. This formula holds true for all convex polyhedra and in general for all simply connected polyhedra (ones with no holes).

5. Prove that there are only five *convex regular polyhedra*: tetrahedron, cube, octahedron, dodecahedron, and icosahedron, as shown in Figure 15.3.

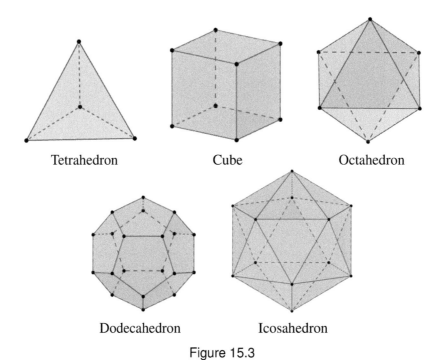

Figure 15.3

15.2 Fundamental Concepts

In this section we briefly explore the concepts of parallelism and orthogonality in three-dimensional space. We also discuss how angles between lines and planes or planes and planes are defined. A number of proofs will be provided to give the reader a taste for how proofs in three-dimensional space are constructed. This is by no means a comprehensive treatment of the subject; for such a treatment the reader can refer to [50] or [83].

We begin our investigation of solid geometry – the geometry of three-dimensional objects – by listing axioms dealing with planes.

Plane Axioms

1. If two distinct points of a line lie in a plane, then the entire line lies in the plane.
2. If two planes intersect at a point, then they intersect in a line containing that point.
3. There exists a unique plane containing any three noncollinear points.
4. Every plane contains at least three noncollinear points.
5. There exist at least four noncoplanar points in space.

The following theorem lists three statements that are equivalent to Axiom 3 above. The proof of the theorem is straightforward and hence is left as an exercise.

> **Theorem 15.1**
>
> *The following statements are equivalent to plane Axiom 3.*
> 1. *Given a line and a point not on it, there exists a unique plane containing the point and the line.*
> 2. *Given two distinct, intersecting lines, there exists a unique plane that contains both lines.*
> 3. *Given two distinct, intersecting lines and a third line intersecting both lines in distinct points, then the triangle formed lies in the same plane as the lines.*

Parallel and Perpendicular Planes and Lines

Two lines l and m are *parallel* if they lie in the same plane and they do not intersect. If l and m do not intersect in three-dimensional space and no plane contains both of them, then they are called *skew* lines.

> **Now Prove This 15.1**
>
> If l and m are two lines in three-dimensional space, then exactly one of the following must hold:
>
> **1.** l and m are parallel.
>
> **2.** l and m intersect in a single point.
>
> **3.** l and m coincide.
>
> **4.** l and m are skew.

Whether lines and planes or planes and planes are parallel or not is determined in the usual manner. That is, a line is *parallel* to a plane if the two never intersect, and two planes are *parallel* if they have no point in common.

> **Now Try This 15.1**
>
> **1.** Explain why the plane containing any two parallel lines in three-dimensional space is unique.
>
> **2.** Explain how (1) implies that lines in space that are parallel to the same line are parallel to each other.

What does it mean for a line l to be perpendicular to a plane H? This must have something to do with right angles, but we have not yet defined the angle between a line and a plane. To do this, suppose l intersects H at a point P. We know how to measure the angles between l and any line m in H that goes through P. Hence we can say that a line l is *perpendicular* to a plane H in three-dimensional space if every line m in H passing through P forms a right angle with l (in the plane containing l and m). If k is a line in H not through P, then k is perpendicular to l if l is perpendicular to the line k' that is the unique parallel to k through P and lying in H.

> **Now Try This 15.2**
>
> **1.** Given a line l and a point P not on it in three-dimensional space, does there exist a unique line parallel to l through P? Explain.
>
> **2.** Given a line l and a point P on it in three-dimensional space, does there exist a unique line perpendicular to l through P? Explain.

Theorem 15.2 shows that a plane containing a line l that is parallel to a given plane intersects the given plane in a line parallel to l.

> **Theorem 15.2**
>
> Let l be a line parallel to a plane H. If H' is any plane containing l that intersects H, then l is parallel to the intersection line m of H and H'.

15.2 Fundamental Concepts

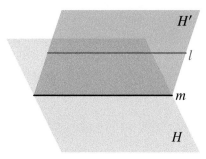

Figure 15.4

Proof. By way of contradiction, suppose the plane H' intersects H in a line m that is not parallel to l. Then, since l and m lie in the same plane H', they must intersect at some point P. The point P lies on line l but is also in plane H, which contradicts the fact that l and H are parallel. ∎

Theorem 15.3 shows that through any point P not on a given plane there exists at least one line that is parallel to the given plane. Theorem 15.4 shows that the parallel lines, guaranteed by Theorem 15.3, determine a unique plane through the point P that is parallel to the given plane.

Theorem 15.3

Let H be a plane, l be a line in H, and P be a point not on H. Then H is parallel to l', where l' is the unique line through P and parallel to l.

Proof. Since P is not on l, part (1) of Now Try This 15.2 establishes the existence of a unique line l' though P that is parallel to l. By way of contradiction, suppose l' intersects H in a point Q. Since l is parallel to l', the point Q is not on l, and hence there exists a unique plane containing Q and l by part (1) of Theorem 15.1. This plane must be H, since H contains both Q and l. However, two parallel lines determine a unique plane (why?), hence there exists a unique plane H' containing the parallel lines l and l', and this plane is not H because it contains the point P. But since Q is on l', there exist two planes, namely H and H', containing l and Q which is absurd. Therefore l' is parallel to H. ∎

Theorem 15.4

Given a plane in space and a point not on the plane, there exists a unique plane through the point that is parallel to the given plane.

Proof. Let H be the given plane, P be a point not on H, and l and m be any two intersecting lines in H (Axiom 4 assures us that such lines exist). Let L and M be the unique planes through l and P and m and P, respectively, that exist by Theorem 15.1. Let l' and m' be the unique lines in L and M passing through P and parallel to l and m, respectively, this follows from the Euclidean parallel postulate. Finally, by Theorem 15.1, let H' be the unique plane containing the lines l' and m'. Theorem 15.3 tells us that lines l' and m' are parallel to H; hence H' is parallel to H. Otherwise, if H and H' intersect in a line k, then l' and m' are parallel to k (why?), which is absurd since l' and m' are intersecting lines. Hence H' is the desired plane parallel to H and containing P. ∎

The following theorem indicates how to construct a line perpendicular to a given plane at a point on the plane.

Theorem 15.5

If a line is perpendicular to two distinct intersecting lines, then it is perpendicular to the plane determined by these lines.

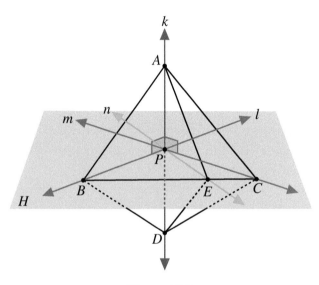

Figure 15.5

Proof. Let l and m be two distinct lines intersecting at a point P. Let H be the plane determined by l and m. Let k be a line perpendicular to l and m, as shown in Figure 15.5. Without loss of generality, assume that k goes through the point P. Let n be an arbitrary line lying in plane H going through P. If we can show that k is perpendicular to n, then we have that k is perpendicular to H (why?). To do this, choose two points A and D on k such that $\overline{AP} \cong \overline{DP}$, and show that line n lies on the perpendicular bisector of $\overline{AD}$. Accordingly, let B and C be points on l and m, respectively, and let E be the intersection point of line n and $\overline{BC}$. Then show that E is equidistant from A and D. Note that $\overline{AB} \cong \overline{DB}$, since $\triangle APB \cong \triangle DPB$ by SAS. Similarly, $\overline{AC} \cong \overline{DC}$. Hence $\triangle ABC \cong \triangle DBC$ by SSS, which means that $\angle ABE \cong \angle DBE$. Thus, $\triangle ABE \cong \triangle DBE$ by SAS, which means that $\overline{AE} \cong \overline{DE}$. Therefore, $\overleftrightarrow{PE}$ is the perpendicular bisector of $\overline{AD}$, and k is perpendicular to H. ∎

15.2 Fundamental Concepts

Let C be any point on a given line $\overleftrightarrow{AB}$. Consider two distinct planes H and H' containing $\overleftrightarrow{AB}$ and in each plane construct a perpendicular to $\overleftrightarrow{AB}$ through C, as in Figure 15.6. The plane determined by these two perpendiculars, call them l and l', is perpendicular to $\overleftrightarrow{AB}$ (why?). Now given any plane perpendicular to $\overleftrightarrow{AB}$ at C, it must intersect H and H' in l and l' (why?). Hence the plane perpendicular to $\overleftrightarrow{AB}$ at C is unique. This proves the following Theorem 15.6.

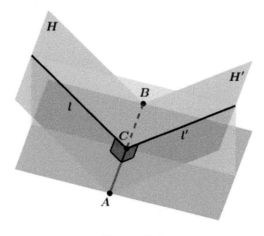

Figure 15.6

Theorem 15.6

Given a line and a point on the line, there exists a unique plane through the point that is perpendicular to the line.

The following theorem shows that: (1) lines perpendicular to a given plane are parallel to each other, (2) if one line is perpendicular to a given plane then all lines parallel to this line are perpendicular to the given plane, and (3) a line perpendicular to a given plane is perpendicular to all planes parallel to this plane. The proofs of (1) and (3) are left as exercises.

Theorem 15.7

Let l and m be lines, and H and H' be planes.
1. *If $l \perp H$ and $m \perp H$, then $l \parallel m$.*
2. *If $l \perp H$ and $l \parallel m$, then $m \perp H$.*
3. *If $H \parallel H'$, and $l \perp H$, then $l \perp H'$.*

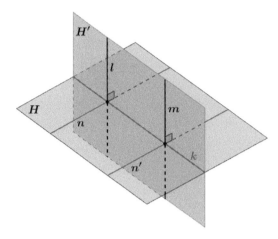

Figure 15.7

Proof. To prove (2), let k be the line of intersection of plane H with the plane through l and m. Then $l \perp k$, and since l, m, and k lie in the same plane H', this means that $m \perp k$. Let n be the perpendicular to k in the plane H through the foot of l, and construct n' in H parallel to n and going through the foot of m. Since l and n meet at right angles, so do m and n' (why?). Hence m is perpendicular to two intersecting lines in H, which means that m is perpendicular to H. ∎

The following theorem shows that through any point, exactly one line can be drawn that is perpendicular to a given plane.

> **Theorem 15.8**
>
> 1. *Given a plane and a point on it, there exists a unique line perpendicular to the plane through the point.*
> 2. *Given a plane and a point not on it, there is a unique line perpendicular to the plane through the point.*

Proof. To prove (1), consider a plane H and a point C on it, as shown in Figure 15.8. Construct two lines l and m in H and going through C. By Theorem 15.6 there exist unique planes H' and H'' through C that are perpendicular to l and m, respectively. The intersection line k of the planes H' and H'' contains C and is perpendicular to both l and m (why?); hence it is perpendicular to H. To prove uniqueness, any line n that is perpendicular to H at point C is perpendicular to l and m at C, and hence it lies in H' and H''. But k is the intersection line of H' and H'', and therefore the lines n and k coincide.

15.2 Fundamental Concepts

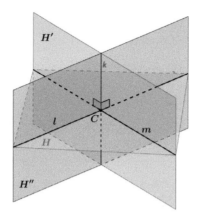

Figure 15.8

To prove (2), consider Figure 15.9, which shows the construction of a line perpendicular to a plane H from a point A not on H.

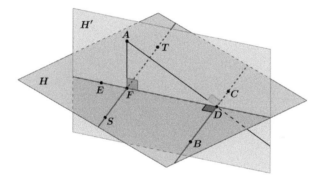

Figure 15.9

First, let B and C be two points on the plane H, and A be a point not on H. The points A, B, and C determine a unique plane (not shown in Figure 15.9). In this plane, use the usual Euclidean construction to draw the perpendicular $\overleftrightarrow{AD}$ from A to $\overleftrightarrow{BC}$. If $\overleftrightarrow{AD}$ is perpendicular to H, then we have the desired perpendicular to H through A. Second, if $\overleftrightarrow{AD}$ is not perpendicular to H, use the usual Euclidean construction in plane H to draw the perpendicular $\overleftrightarrow{DE}$ to $\overleftrightarrow{BC}$. Let H' be the plane determined by A, D, and E. Construct $\overleftrightarrow{AF}$, the perpendicular from A to $\overleftrightarrow{DE}$. Third, in the plane H construct $\overleftrightarrow{ST}$ through F parallel to $\overleftrightarrow{BC}$. Now $\overleftrightarrow{BC} \perp \overleftrightarrow{AD}$ and $\overleftrightarrow{BC} \perp \overleftrightarrow{DE}$; hence by Theorem 15.5 $\overleftrightarrow{BC}$ is perpendicular to H', the plane determined by $\overleftrightarrow{AD}$ and $\overleftrightarrow{DE}$. Fourth, Theorem 15.7 implies that $\overleftrightarrow{ST}$ is perpendicular to H', since $\overleftrightarrow{ST} \parallel \overleftrightarrow{BC}$. Therefore $\overleftrightarrow{AF} \perp \overleftrightarrow{ST}$ and $\overleftrightarrow{AF} \perp \overleftrightarrow{DE}$, which means that $\overleftrightarrow{AF}$ is perpendicular to the plane H. To show that $\overleftrightarrow{AF}$ is unique, let $\overleftrightarrow{AG}$, where G (not shown) is a point on H, be another perpendicular from A to H. Then in the plane determined by $\overleftrightarrow{AF}$

and $\overleftrightarrow{AG}$, we have two perpendiculars from A to the line $\overleftrightarrow{FG}$, which contradicts the exterior angle theorem (Proposition 16 of Euclid's *Elements*).

Angles Between Planes and Lines

> **Theorem 15.9**
>
> *Two angles (not necessarily in the same plane) whose corresponding sides are parallel are either congruent or supplementary.*

Proof. If two angles have corresponding parallel sides, then their sides lie on two sets of parallel lines. Let l, m and l', m' be two sets of intersecting lines such that $l \parallel l'$ and $m \parallel m'$, and let O and O' be the respective vertices of the angles formed. Connect $\overline{OO'}$, and choose points A on l and B on m such that $\angle AOB$ is one of the given angles, as in Figure 15.10.

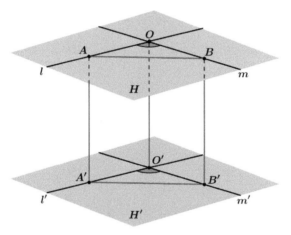

Figure 15.10

Let A' be the intersection of l' with the line parallel to $\overline{OO'}$ through A. Similarly, construct B' on m' such that $\overline{BB'} \parallel \overline{OO'}$. Let H be the plane determined by the lines l and m, and let H' be the plane determined by the lines l' and m'. The lines l and m are parallel to H' and hence H is parallel to H' (see Problems 6 and 7 at the end of this section). Then $OAA'O'$ and $OBB'O'$ are parallelograms. Since $\overline{AA'} \cong \overline{BB'}$ and $\overline{AA'} \parallel \overline{BB'}$, we have $AA'B'B$ is a parallelogram (why?). This means that $AB = A'B'$, and hence by SSS we have $\triangle AOB \cong \triangle A'O'B'$, which gives $\angle AOB \cong \angle A'O'B'$. We have shown that the $\angle AOB$ is congruent to one of the angles formed by the intersection of l' and m'. The other angles formed by these lines are the vertical and supplementary angles to $\angle A'O'B'$. In the former case, the angle is congruent to $\angle AOB$, and in the latter it is supplementary to it.

Skew lines do not lie in the same plane; hence we cannot define the angle between them in the normal fashion. So how else can we determine the angle between two skew lines? Ponder this question before reading on. Given two skew lines l and m as in the Figure 15.11,

15.2 Fundamental Concepts

choose a point P in space, construct l' the unique line parallel to l through P, and then construct m', the unique line parallel to m through P. Let the angle between l and m be defined as the smaller of the two angles formed by l' and m' in the unique plane containing these lines. The uniqueness of l' and m' and the plane containing them guarantees that this definition of the angle between skew lines is well defined for the point P. But does the angle between l and m depend on our choice of the point P? The answer is no, since if $P' \neq P$ is chosen and l'' and m'' are lines through P' that are parallel to l and m, respectively, then the smaller angle between l'' and m'' has the same measure as the smaller angle between l' and m'. This is because these two angles have parallel corresponding sides.

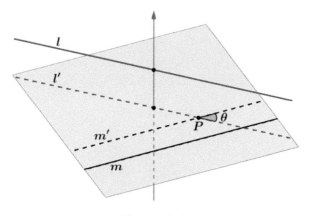

Figure 15.11

Given a plane H and a line $\overleftrightarrow{AB}$ intersecting H at B, as in Figure 15.12, the angle between $\overleftrightarrow{AB}$ and H is $\angle ABC$, where C is the foot of the perpendicular from A to H.

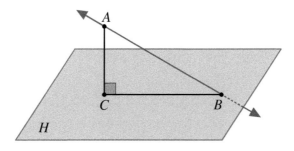

Figure 15.12

Given two half-planes H and H' intersecting in $\overleftrightarrow{AB}$ (see Figure 15.13), how can the angle between the half-planes be defined? The angle bounded by H, H', and $\overleftrightarrow{AB}$ is called the *dihedral* angle formed by H and H'. The two half-planes bounding the angle are called the *faces* of the dihedral angle and the common edge is called the *edge* of the dihedral angle. Let C be an arbitrary point on $\overleftrightarrow{AB}$, and construct $\overleftrightarrow{CD}$ and $\overleftrightarrow{CE}$ perpendicular to $\overleftrightarrow{AB}$ in H and H', respectively. The angle $\angle DCE$ is called the *plane angle* of the *dihedral angle* formed by H and H'. The measure of the dihedral angle is defined to be the measure of its plane angle. Also, it turns out that all plane angles corresponding to the same dihedral angle are congruent (why?); hence it does not matter which point C on $\overleftrightarrow{AB}$ is used to determine the plane angle.

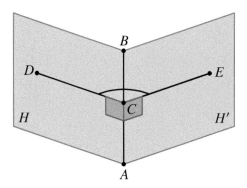

Figure 15.13

A view as in Figure 15.14 shows the two half-planes H and H' as edges and it shows the line of intersection $\overleftrightarrow{AB}$ as a point. The angle between the two edge views of the planes is the plane angle of the dihedral angle. Note that this view shows intuitively that all plane angles of a given dihedral angle are congruent.

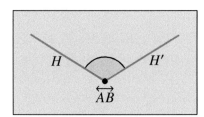

Figure 15.14

15.2.1 Problem Set

1. Explain why removing one of the legs of a wobbly four-legged chair, while making it unsafe for sitting, would at least fix the wobble.

2. In three-dimensional space, answer the following and justify your answers.
    a. What is the locus of points equidistant from two given points?
    b. What is the locus of points equidistant from two given parallel lines?
    c. What is the locus of points equidistant from three given noncollinear points?
    d. What is the locus of points equidistant from a circle?

3. Complete the proof of Theorem 15.7.
4. Complete the proof of Theorem 15.8.

15.2 Fundamental Concepts

5. Can a line be perpendicular to two intersecting planes? Explain.

6. Given two intersecting lines l and m, prove that if a plane is parallel to both l and m, then it is parallel to the plane determined by these lines.

7. Given two parallel lines l and m, if H is a plane containing l and not m, then m is parallel to H.

8. [83] Given a plane H and a point P not in H, find the locus of the midpoints of all segments connecting P to points in H. Hint: A line parallel to the base of a triangle and bisecting one side bisects the other side as well.

9. [50] Let $\overleftrightarrow{AB}$ and $\overleftrightarrow{CD}$ be skew lines as in the figure below. Prove that the midpoints of the segments $\overline{AC}, \overline{AD}, \overline{BC}$, and $\overline{BD}$ are vertices of a parallelogram, and that its plane is parallel to the lines $\overleftrightarrow{AB}$ and $\overleftrightarrow{CD}$.

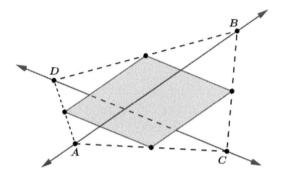

10. If P is a point not in plane H, and if the foot of the perpendicular from P to H is C, prove the following:

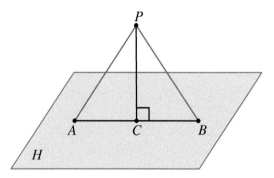

   a. The distance from P to C is smaller than the distance from P to any other point in H.

   b. $PA = PB$ if and only if $AC = BC$.

   c. Describe how you can use this to determine how to erect a flagpole perpendicular to a level yard.

15.3 Polyhedra

In order to avoid any confusion, we will start by defining some basic terms needed to formally define a polyhedron. To this end, we define a *polygon* to be a simple closed curve in the plane made up of three or more line segments where the endpoints of any two adjacent segments are noncollinear. The line segments are called the *sides* or *edges* of the polygon and the points where the sides meet are called the *vertices* of the polygon. A *regular polygon* is defined to be a polygon with congruent sides and congruent angles. Lastly, a *polygonal region* is defined to be a polygon together with the two-dimensional region it bounds.

> **Definition: Polyhedron**
>
> We define a *polyhedron* to be a solid (three-dimensional object) bounded by a finite number (at least four) of plane *faces* (polygonal regions) joined together such that (1) any two faces can have at most one side or one vertex in common, (2) each side of each face is the common side of exactly two faces, (3) no two adjacent faces (that have a common edge) lie in the same plane, and (4) if a number of faces meet at a vertex v, it is possible to get from any face to any other face without having to go through v. The sides where two faces meet are called *edges* of the polyhedron, and the points where two or more edges meet are called *vertices* [20].

Note that this definition rules out three-dimensional objects such as those in Figure 15.15. The first is made up of two solid pyramids meeting at a vertex, the second is made up of two solid pyramids having one edge in common, and the third is made up of two noncongruent solid cubes one sitting on top of the other. Can you tell why each of these violates the definition of a polyhedron given above?

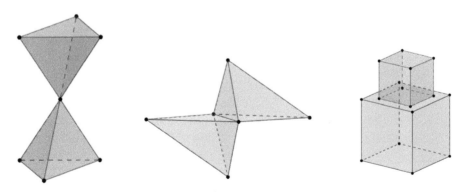

Figure 15.15

15.3 Polyhedra

> **Definition**
>
> 1. A *regular polyhedron* is defined to be a polyhedron whose faces are all bounded by congruent regular polygons such that the same number of regular polygons meet at each vertex.
> 2. A polyhedron is *convex* if any two points in the polyhedron can be joined by a line segment contained completely in the polyhedron.

The following lemma can be easily proved using the fact that any polygon can be divided into a finite number of nonoverlapping triangles each of which has angle sum 180° (by Euclid's Proposition 32).

> **Lemma 15.1**
>
> If P is a polygon with n sides, then the following are true.
> 1. The sum of the interior angles of P is $(n-2)180°$.
> 2. The sum of the exterior angles of P is $360°$.
> 3. If P is a regular polygon, then the measure of each of its interior angles is $(n-2)180°/n$.

The results in Lemma 15.2 follow directly from the definition of a polyhedron.

> **Lemma 15.2**
>
> Let $\mathcal{P}$ be any polyhedron, then the following are true.
> 1. Every edge in $\mathcal{P}$ is common to precisely two faces.
> 2. Every vertex in $\mathcal{P}$ is common to at least three edges.
> 3. Every face in $\mathcal{P}$ has at least three edges.

Letting V, E, and F be the number of vertices, edges, and faces of a polyhedron $\mathcal{P}$, Lemma 15.3 gives lower bounds for V, E, and F, and follows directly from Lemma 15.2.

> **Lemma 15.3**
>
> For any polyhedron $\mathcal{P}$, we have $V \geq 4$, $E \geq 6$, and $F \geq 4$.

The proof of Lemma 15.4 is straightforward and is the subject of Problem 3 at the end of this section.

> **Lemma 15.4**
>
> Let $\mathcal{P}$ be any polyhedron. For any integer $n \geq 3$, let F_n denote the number of faces having n edges each, and let V_n denote the number of vertices common to n edges each. Then the following are true [7]:
>
> 1. $F = F_3 + F_4 + F_5 + \ldots$
> 2. $V = V_3 + V_4 + V_5 + \ldots$
> 3. $2E = 3F_3 + 4F_4 + 5F_5 + \ldots$
> 4. $2E = 3V_3 + 4V_4 + 5V_5 + \ldots$
> 5. $2E \geq 3F$ and $2E \geq 3V$.

15.4 Descartes' Lost Theorem

Descartes' lost theorem (see Theorem 15.10) was part of a manuscript titled *Progymnasmata de solidorum elementis* (Exercises on the Elements of Solids) written around 1623 [20, Chapter 5]. Descartes (1596–1650), a French philosopher and mathematician, died during a visit to Sweden and was buried there. While his belongings were being transported back to France, a box containing his manuscripts fell into the river in Paris. Most of the manuscripts were rescued, and some were later published, while others were made available to scholars for study. In 1676, Leibniz made copies of several of the manuscripts, including the one dealing with polyhedra. Descartes' original copy of this manuscript was lost, and Leibniz's copy remained undiscovered until 1860. After Descartes' death, the general study of polyhedra laid dormant until the Swiss mathematician Euler (1707–1783) came on the scene. Euler made substantial contributions in this area and many others, including the discovery of what is now known as Euler's formula (see Section 15.5).

Before proving Descartes' lost theorem, we need to investigate a few results dealing with angles in a polyhedron.

> **Lemma 15.5**
>
> Given any convex polyhedron $\mathcal{P}$, the sum of the face angles at each vertex v is less than 2π.

Proof. Suppose n faces $F_1, F_2, \ldots, F_n$ meet at the vertex v. Consider a plane intersecting only those edges meeting at v. Each of the faces at v meets this plane in a line segment. Slicing off the corner of the polyhedron $\mathcal{P}$ at v along this plane gives a pyramid whose base is n-sided. Figure 15.16 illustrates this process when the vertex v is surrounded by three faces, and hence three face angles $\alpha_1, \alpha_2, \alpha_3$.

15.4 Descartes' Lost Theorem

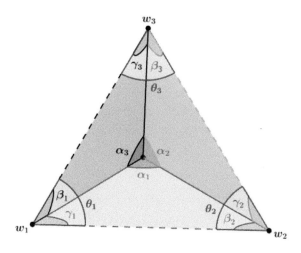

Figure 15.16

Denote the vertices of the base of this pyramid by $w_1, w_2, \ldots, w_n$. Note that these are not vertices of the polyhedron $\mathcal{P}$. Let T_i be the boundary of the triangular face of this pyramid that came from the face F_i, and use α_i to denote the face angle in T_i whose vertex is v. Let θ_i denote the face angle in the dashed base of the pyramid (facing the reader) whose vertex is at w_i. Finally, let the face angles adjacent to θ_i be denoted by β_i and γ_i.

Let the sum of the face angles at v be S; then $S = \alpha_1 + \alpha_2 + \cdots + \alpha_n$. By Euclid's Proposition 32, the sum of the face angles of the sliced pyramid $(\alpha_1 + \alpha_2 + \cdots + \alpha_n) + (\beta_1 + \beta_2 + \cdots + \beta_n) + (\gamma_1 + \gamma_2 + \cdots + \gamma_n)$ is equal to $n\pi$. Hence $S + (\beta_1 + \gamma_1) + (\beta_2 + \gamma_2) + \cdots + (\beta_n + \gamma_n) = n\pi$. Clearly, $\beta_i + \gamma_i$ is greater than θ_i (why?). Also, the sum $\theta_1 + \theta_2 + \cdots + \theta_n$ of the angles of the base of the pyramid is equal to $(n-2)\pi$ (why?). Therefore, $n\pi = S + (\beta_1 + \gamma_1) + (\beta_2 + \gamma_2) + \cdots + (\beta_n + \gamma_n) > S + (\theta_1 + \theta_2 + \cdots + \theta_n) = S + (n-2)\pi$. This leads to the conclusion that $S < 2\pi$. ∎

There are several types of angles in a polyhedron including *face angles*, *dihedral angles*, and *solid angles*. Precise descriptions of these angles are given below.

1. Angles in a face determined by two sides are called *face angles*.

2. Angles between two adjacent faces are called *dihedral angles*. If the two faces have the side $\overline{AB}$ in common, we denote the dihedral angle between them by $\angle \overline{AB}$. As mentioned in Section 15.2, the measure of the dihedral angle is the same as the measure of its *plane angle* (see Figure 15.13).

3. Given a vertex v, the *solid angle* at v is the angle bounded by the faces meeting at v. More specifically, consider a sphere of radius r centered at v. Then the *measure of the solid angle at vertex v is the ratio of the area of the surface of the sphere bounded by the faces meeting at v to the square of the radius of the sphere.* This is analogous to defining the measure of an angle in the Euclidean plane as the ratio of the length of the arc subtended by the angle to the radius of the circle having the vertex of the angle at its center.

Solid angles are typically measured in units called *steradians*. A *steradian* is defined to be the measure of a solid angle having its vertex at the center of a sphere and subtending a spherical surface area equal to the square of the radius of the sphere. Therefore, a whole sphere corresponds to a solid angle of 4π steradians. This is analogous to the radian's being defined as the measure of a central angle subtended by an arc equal in length to the radius of the circle, and consequently a whole circle corresponds to an angle of 2π radians.

> **Definition: Angular Excess**
>
> Given a spherical triangle ABC, its *angular excess* (also known as *spherical excess*) is the sum of the radian measures of its angles minus π. Similarly, the angular excess of an n-sided spherical polygon is the sum of the radian measures of its angles minus $(n-2)\pi$ (why?).

Girard's theorem (see Theorem 11.10) states that *the area of a spherical triangle is equal to its angular excess (in radians) times the square of the radius of the sphere*. This means that the area of any spherical polygon is its angular excess (in radians) times the square of the radius of the sphere (why?). Hence, if the sum of the angles of the spherical n-polygon subtended by a solid angle γ (surrounded by n faces) is S, then its angular excess is $(S-(n-2)\pi)$ radians, and we can say that the measure of γ is $\frac{(S-(n-2)\pi)r^2}{r^2} = (S-(n-2)\pi)$ steradians. Equivalently, since the whole sphere corresponds to a solid angle whose measure is 4π steradians, we can say that the measure of γ is $\frac{(S-(n-2)\pi)r^2}{4\pi r^2} \cdot 4\pi = (S-(n-2)\pi)$ steradians. For example, the measure of the solid angle at the vertex of a cube (see Figure 15.17) is $(\frac{\pi}{2} + \frac{\pi}{2} + \frac{\pi}{2} - \pi) = \frac{\pi}{2}$ steradians. Equivalently, since the solid angle at the vertex of a cube subtends a spherical area equal to $\frac{1}{8}$ of the surface area of the sphere, the measure of this solid angle is $\frac{1}{8} \cdot 4\pi = \frac{\pi}{2}$ steradians.

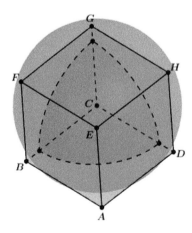

Figure 15.17

15.4 Descartes' Lost Theorem

Next, we show how to find the measure of the solid angle at the vertex of a regular tetrahedron. Consider the regular tetrahedron $ABCD$ shown in Figure 15.18 with its vertex C at the center of a sphere of radius r. The measure of the solid angle at vertex C is the ratio of the area of the spherical triangle $A'B'D'$ to the square of the radius of the sphere, or equivalently, $\frac{(S-\pi)r^2}{r^2} = (S - \pi)$ steradians, where S is the angle sum of the triangle $A'B'D'$.

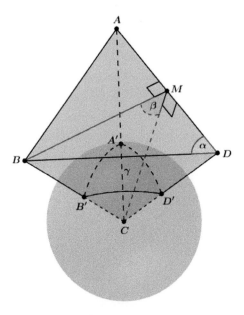

Figure 15.18

We now show how to find S, the angle sum of the spherical triangle $A'B'D'$. A face angle of the regular tetrahedron $ABCD$, such as α in the face ABD, is an angle in an equilateral triangle, and hence $\alpha = 60°$. The measure of the dihedral angle bounded by the faces ABD and ACD is the measure of the plane angle β between $\overline{BM}$ (which lies in the face ABD and is perpendicular to $\overline{AD}$) and $\overline{CM}$ (which lies in the face ACD and is perpendicular to $\overline{AD}$). Figure 15.19 shows $\triangle CMB$, which can be used to determine the measure of β.

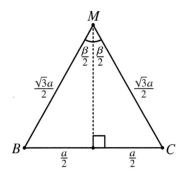

Figure 15.19

Note that if $\overline{BC}$ has length a, then $\overline{BM}$ and $\overline{MC}$ each has length $\frac{\sqrt{3}a}{2}$, since these are altitudes of equilateral triangles of side length a. Hence $\sin(\frac{\beta}{2}) = \frac{1}{\sqrt{3}}$, which results in a β value of approximately 70.529° or 1.231 radians.

The measure of the dihedral angle between the faces ACD and ACB is equal to the measure of the angle between the great circles determined by the arcs $A'B'$ and $A'D'$. Also, the lengths of the arcs $A'B'$, $A'D'$, and $B'D'$ divided by the radius of the sphere give the measures of the face angles at C (why?). By symmetry of the regular tetrahedron, all the dihedral angles are congruent, so all the angles of the spherical triangle $\triangle A'B'D'$ have the same measure as the dihedral angle β. Therefore, the angle sum of $A'B'D'$ is 3β, which is $\approx 3(1.231) = 3.693$ radians, and this means that the measure of the solid angle γ at C is $\approx (3.693 - \pi) \approx 0.551$ steradians.

Let ψ denote the solid angle in Figure 15.20 surrounded by the face angles α, β, and θ. Construct three planes such that each plane is perpendicular to a pair of adjacent faces around the solid angle ψ. The point of intersection of these planes is the vertex of χ, the solid angle surrounded by the face angles x, y, and z (the labels ψ and χ are not included in the figure). The solid angle ψ is sometimes referred to as an interior solid angle and χ as its corresponding exterior solid angle. How is χ related to ψ?

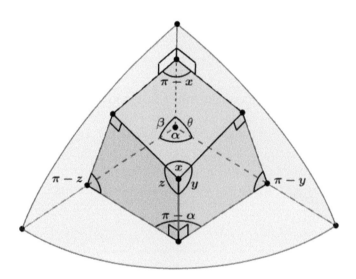

Figure 15.20

The careful reader should realize that the construction above is analogous to the construction in the plane shown in Figure 15.21, where perpendiculars to the sides of $\angle B$ are constructed from a point D in the interior of $\angle B$. How is $\angle D$ related to $\angle B$?

15.4 Descartes' Lost Theorem

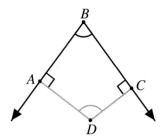

Figure 15.21

Clearly, ∠B and ∠D in Figure 15.21 are supplementary. So, it is plausible to expect the solid angles χ (surrounded by angles x, y, z) and ψ (surrounded by angles α, β, θ) in Figure 15.20 to be supplementary [20]. Recall that the steradian measure of a solid angle is the angular excess of the spherical polygon subtended by the solid angle. Also, the interior angles of this polygon are the dihedral angles between the planes surrounding the solid angle. Hence the measure of a solid angle is the sum of the dihedral angles between the faces around that angle minus $(n-2)\pi$ (why?), where n is the number of faces around the solid angle.

Therefore, in Figure 15.20, the measure of ψ is $(\pi-x)+(\pi-y)+(\pi-z)-\pi = 2\pi-(x+y+z)$, and the measure of χ is $(\pi-\alpha)+(\pi-\beta)+(\pi-\theta)-\pi = 2\pi-(\alpha+\beta+\theta)$. But $\alpha = \pi - x$, $\beta = \pi - y$, and $\theta = \pi - z$ (why?). Hence χ and ψ are supplementary (since their measures add up to π steradians).

Define the *deficiency* of a solid angle to be 2π minus the sum of the face angles around the solid angle. This means that the deficiency of ψ in Figure 15.20, which is $2\pi - (\alpha+\beta+\theta)$, is equal to its supplementary angle χ and vice versa. The same result holds if ψ is surrounded by n plane angles for $n > 3$, as the following lemma shows.

Lemma 15.6

The deficiency of a solid angle ψ is equal to its supplementary solid angle χ [20].

Proof. Suppose the solid angle ψ is surrounded by n face angles $\alpha_1, \alpha_2, \ldots, \alpha_n$. The measure of its supplementary solid angle χ is the angular excess of the spherical region subtended by χ. The angles of the spherical polygon bounding this region are the dihedral angles of the n faces surrounding χ, namely $(\pi - \alpha_1), (\pi - \alpha_2), \ldots, (\pi - \alpha_n)$. This n-sided spherical polygon can be divided into $n-2$ spherical triangles with nonoverlapping interiors; hence its angular excess is $(\pi - \alpha_1) + (\pi - \alpha_2) + \ldots + (\pi - \alpha_n) - (n-2)\pi$. Therefore, the measure of χ is $2\pi - (\alpha_1 + \alpha_2 + \cdots + \alpha_n)$, which is the deficiency of ψ. ∎

The supplementary solid angles can be thought of as the exterior angles of the polyhedron, which when arranged around a point form a whole sphere (with no gaps or overlap). This is analogous to the exterior angles of a polygon forming a circle when arranged around a point and hence having angle sum 2π (see Figure 15.22, where the polygon is the pentagon ABCDE).

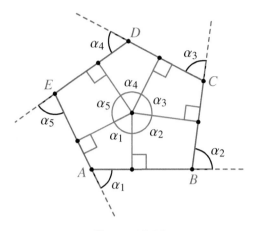

Figure 15.22

Lemma 15.7

In any polyhedron $\mathscr{P}$, the sum of the measures of all the supplementary solid angles is 4π.

Proof. The following is an adaptation of a proof given by E. Prouhet [30]. Let P be a point in the interior of the polyhedron $\mathscr{P}$, and drop perpendicular planes from P to each face of $\mathscr{P}$. The perpendicular planes to the faces surrounding a given solid angle ψ form the supplementary solid angle χ corresponding to ψ (see Figure 15.20). Each of the solid angles of $\mathscr{P}$ has a corresponding supplementary solid angle with vertex at P and vice versa. The supplementary solid angles, with their common vertex at P, fill the space around P without gaps or overlapping. Hence the sum of their measures is equal to the total area of a sphere centered at P divided by the square of its radius, that is, $\frac{4\pi r^2}{r^2} = 4\pi$. ∎

In one of the theorems in *Progymnasmata de solidorum elementis*, Descartes stated the following [20]: *As in a plane figure all the exterior angles, taken together, equal four right angles, so in a solid body all the exterior solid angles, taken together, equal eight solid right angles.* A "plane figure" here is interpreted as a polygon and a "solid body" is interpreted as a polyhedron. Descartes did not provide a complete proof of this important theorem. It is worth mentioning that this result holds not only for convex polyhedra, but also for *simply connected* nonconvex polyhedra (nonconvex polyhedra with no holes).

Theorem 15.10: Descartes' Lost Theorem

The sum of deficiencies of the solid angles in any polyhedron $\mathscr{P}$ is eight right angles.

Proof. By Lemma 15.6, the measure of the supplementary solid angle of a given solid angle is 2π minus the sum of the face angles surrounding the given solid angle. Hence, the sum of the measures of all the supplementary solid angles is equal to 2π times the number of solid angles, minus the sum of all the face angles in $\mathscr{P}$. By Lemma 15.7, the sum of the measures of all the solid supplementary angles in any polyhedron is 4π. This implies that $(2V-4)\pi$ is equal to the sum of the face angles of $\mathcal{P}$ (why?). Hence the sum of the deficiencies of the the solid angles of $\mathcal{P}$ is $2\pi V - (2V-4)\pi = 4\pi$, or equivalently, eight solid right angles. ∎

Next, we present a proof of Theorem 15.10 that uses Euler's formula, which is the subject of Section 15.5.

Proof. Suppose the polyhedron $\mathcal{P}$ has n faces $F_1, F_2, \ldots, F_n$. Suppose the face F_i has E_i edges, and hence E_i vertices. Since each edge is common to exactly two faces, the total number of edges E satisfies $2E = E_1 + E_2 + \cdots + E_n$. By Euler's formula, the number of vertices V of $\mathcal{P}$ satisfies $V = E - F + 2 = (E_1 + E_2 + \cdots + E_n)/2 - n + 2$. Hence $2V = (E_1 + E_2 + \cdots + E_n) - 2n + 4$.

Since $\mathcal{P}$ has V vertices, the sum of deficiencies of the solid angles of $\mathcal{P}$ is $2\pi V$ minus the sum of the angles at all the vertices (equivalently, the sum of the angles in all the faces). Since the sum of the interior angles of a polygon with m sides is $(m-2)\pi$, the sum of deficiencies of the solid angles of $\mathcal{P}$ is

$$= 2\pi V - \big((E_1-2)\pi + (E_2-2)\pi + \cdots + (E_n-2)\pi\big)$$
$$= 2\pi V - (E_1 + E_2 + \cdots + E_n)\pi + 2n\pi$$
$$= \pi\big(2V - (E_1 + E_2 + \cdots + E_n) + 2n\big) = 4\pi. \qquad \blacksquare$$

15.5 Euler's Formula and Its Consequences

Given a convex polyhedron, Euler's formula states that $V + F - E = 2$, where V, F, and E are the number of vertices, faces, and edges, respectively. This formula was announced by Euler in 1750. Some claim that Descartes (1596–1650) discovered Euler's formula nearly a century before Euler. However, many are of the opinion that although Descartes did indeed discover facts about three-dimensional polyhedra that could have led him to deduce Euler's formula, he did not actually arrive at the formula itself. More specifically, in 1630 Descartes concluded that (1) the number of face angles equals $2F + 2V - 4$, and (2) the number of face angles is equal to $2E$, which clearly leads to $V + F - E = 2$, but he did not conclude the final form of the formula.

> **Theorem 15.11: Euler's Formula**
>
> *For any convex polyhedron, the number of vertices plus the number of faces minus the number of edges is equal to 2, that is, $V + F - E = 2$.*

Proof. We will illustrate the proof using a simple convex polyhedron, namely the cube (see [21]). The techniques used in the proof can be generalized to any convex polyhedron. Let $ABCDA_1B_1C_1D_1$ be a cube. "Project" the cube from a point O above it onto a plane below it to obtain the "planar graph" $A'B'C'D'A_1'B_1'C_1'D_1'$ shown in Figure 15.23.

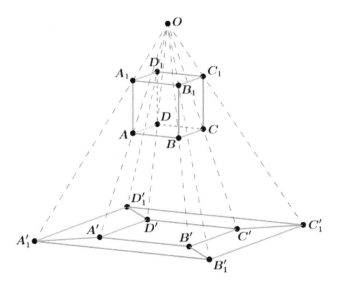

Figure 15.23

Note that each edge in the cube corresponds to an edge in the projected image, and the same is true for each vertex. Also, the faces of the cube except for the one closest to O correspond to polygonal regions with disjoint interiors in the projected image. To remedy this exception, let this face, namely $A_1B_1C_1D_1$, correspond to the infinite region in the projection plane bounded by $A_1'B_1'C_1'D_1'$. This allows us to conclude that the number of vertices, edges, and faces of the cube is the same as the number of vertices, edges, and regions in the projection plane. More importantly, the incidence relationships between the vertices, edges, and, faces in the cube are preserved in the projected image. That is, how these objects lie in relationship to each other is the same in both the cube and its projection. Let R denote the number of regions in the projection plane which correspond to the faces of the cube. If we can prove that $V + R - E = 2$ in the projection plane, then that would mean the formula $V + F - E = 2$ holds for the cube as well.

We now examine how the value of the quantity $V + R - E$ is affected when edges of the projected image are removed one at a time. Working from the outside, if we remove an edge such that $A_1'B_1'$ (an edge that bounds a finite region), then the number of regions and the number of edges are each decreased by 1. Hence $V + R - E$ becomes $V + (R - 1) - (E - 1) = V + R - E$, which means it's left unchanged. Similarly, removing the edges $B_1'C_1'$, $C_1'D_1'$, and $A_1'D_1'$ does not change the value of $V + R - E$.

This leaves us with the region $A'B'C'D'$ and the edges $A'A_1'$, $B'B_1'$, $C'C_1'$, and $D'D_1'$. Note that these edges no longer bound finite regions. How would removing these edges one at a time affect $V + R - E$? If any of these edges is removed, then the number of edges and the number of vertices are each decreased by 1. Hence $V + R - E$ becomes $(V - 1) + R - (E - 1) = V + R - E$, which means it's left unchanged. Having removed the edges $A'A_1'$, $B'B_1'$, $C'C_1'$, and $D'D_1'$ leaves us with the region $A'B'C'D'$.

15.5 Euler's Formula and Its Consequences

We can next remove the edge $A'B'$, and then the edges $A'D'$ and $B'C'$ as before, while keeping $V + R - E$ unchanged. This leaves us with a single edge $C'D'$, two vertices C' and D', and one region (the infinite one). At this final stage $V + R - E = 2 + 1 - 1 = 2$, and since this quantity remained unchanged throughout the process, it must have always been equal to 2. Therefore, $V + F - E = 2$ for any convex polyhedron. Convexity is important in the proof as it ensures that the projection of the polyhedron does not have overlapping edges. ∎

It is worth mentioning that Euler's formula also holds for *simply connected* nonconvex polyhedra. Informally, this means nonconvex polyhedra with no holes in them. It turns out that a polyhedron with n holes satisfies the more general formula $V + F - E = 2 - 2n$, known as the Poincaré formula. We will not try to prove this formula here.

> **Now Try This 15.3**
>
> The polyhedron in Figure 15.24 is not simply connected (note the hole in the shape of a rectangular prism). Find V, F, and E for this polyhedron and compute $V + F - E$. What do you notice?
>
>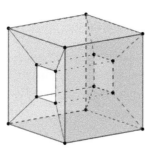
>
> Figure 15.24

In Theorem 15.12 we use Euler's formula to prove that there are only five convex regular polyhedra. These polyhedra, also called *Platonic solids*, have been known since antiquity. It is believed that Pythagoras discovered the dodecahedron, and Theaetetus discovered the octahedron and icosahedron. Theaetetus gave a mathematical description of all five polyhedra and may have been responsible for the first known proof that there are no other convex regular polyhedra. Plato associated Earth with the cube, air with the octahedron, water with the icosahedron, and fire with the tetrahedron. In Book XIII of the *Elements*, Euclid gave a complete mathematical description of the Platonic solids, and in Proposition 18 he demonstrated that there are only five convex regular polyhedra.

> **Theorem 15.12**
>
> *There are only five convex regular polyhedra: tetrahedron, cube, octahedron, dodecahedron, and icosahedron.*

Proof. This proof was adapted from [21]. Given a regular polyhedron, its faces are bounded by congruent regular polygons such that the same number of faces meet at each vertex. Let p denote the number of edges in each face, and let q denote the number of edges (hence faces) that meet at each vertex. Note that each edge is common to two faces, so if we simply multiply the number of faces F by p, we get twice the number of edges. Hence $2E = pF$, or equivalently $\frac{2}{p} = \frac{F}{E}$. Similarly, each edge connects two vertices, so if we simply multiply the number of vertices V by q, we get twice the number of edges. Hence $2E = qV$, or equivalently $\frac{2}{q} = \frac{V}{E}$.

Dividing Euler's formula $V + F - E = 2$ by E gives

$$\frac{V}{E} + \frac{F}{E} - 1 = \frac{2}{E}.$$

Next, substituting $\frac{2}{q}$ for $\frac{V}{E}$ and $\frac{2}{p}$ for $\frac{F}{E}$ yields

$$\frac{2}{q} + \frac{2}{p} - 1 = \frac{2}{E}, \text{ or equivalently } \frac{1}{q} + \frac{1}{p} = \frac{1}{E} + \frac{1}{2}.$$

Hence,

$$\frac{1}{q} + \frac{1}{p} > \frac{1}{2}, \text{ or equivalently } \frac{p+q}{pq} > \frac{1}{2}.$$

Manipulating this inequality using simple algebra gives $(p-2)(q-2) < 4$ (how?). Consequently, the only $\{p, q\}$ values that satisfy this inequality are $\{3, 3\}$, $\{4, 3\}$, $\{3, 4\}$, $\{5, 3\}$, and $\{3, 5\}$. These solutions correspond to the tetrahedron, cube, octahedron, dodecahedron, and icosahedron, respectively. ∎

The following corollaries give some additional consequences of Euler's formula. The proof of Corollary 15.2 is left as an exercise.

Corollary 15.1

There is no convex polyhedron with seven edges.

Proof. By way of contradiction, suppose there exists a convex polyhedron with seven edges. Part (5) of Lemma 15.4 tells us that $2E \geq 3F$, hence $3F \leq 14$. Furthermore, since the number of faces of a polyhedron is greater than 3, we conclude that $F = 4$. Similarly, since $2E \geq 3V$, we have $3V \leq 14$, which leads to $V = 4$. Now using Euler's formula we have $4 + 4 - 7 = 2$, which is absurd. ∎

Corollary 15.2

For any convex polyhedron $\mathcal{P}$, the number of vertices and number of faces satisfy the following:
1. $V \geq \frac{F}{2} + 2$.
2. $V \leq 2F - 4$.

15.5 Euler's Formula and Its Consequences

So what are the possible values of V and F that polyhedra can have? To determine this we solve the two inequalities in Corollary 15.2. Figure 15.25 shows the graphs of these inequalities. The lattice points outside the shaded region lying between lines l (representing $V = \frac{F}{2} + 2$) and k (representing $V = 2F - 4$) constitute values of V and F that polyhedra cannot have. Some of the material that follows was adapted from [20].

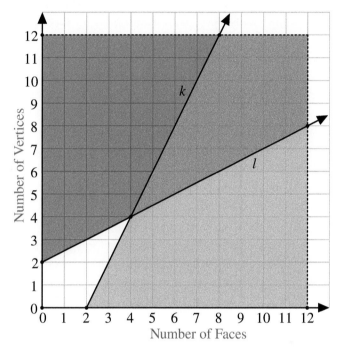

Figure 15.25

Let (n, m) be any lattice point in the shaded region between the lines l and k: is there a polyhedron with n faces and m vertices corresponding to this point? Considering the three cases: $n = m$, $n < m$, and $n > m$, the discussion below reveals that a polyhedron does indeed exist for any (n, m) in the shaded region between the lines l and k.

1. If $n = m$, consider a pyramid whose base is bounded by a regular polygon with $n - 1$ sides. Such a pyramid exists for $n \geq 4$ and has n faces, n vertices, and $2n - 2$ edges. Therefore, there exists a polyhedron corresponding to any lattice point (n, n) in the shaded region between l and k.

2. If $n < m$, consider a pyramid whose base is bounded by a regular polygon with s sides. We next show how to choose s (in terms of n and m) so that this pyramid can be transformed into a polyhedron with n faces and m vertices satisfying $n < m$. Note that each vertex of the base of the pyramid is surrounded by three faces. *Truncating* the corner at one of these vertices–cutting off a tetrahedron, as shown in Figure 15.26(a)– causes the number of faces to increase by 1, the number of edges to increase by 3, and the number of vertices to increase by 2. Note also that each of the new vertices

produced by the truncation process is surrounded by three faces; hence the corners at these new vertices can be truncated as well. Therefore, the truncation process can be applied indefinitely, which allows us to construct the desired polyhedron satisfying $n < m$.

More specifically, a pyramid whose base is bounded by a regular polygon with s sides corresponds to the lattice point $(s+1, s+1)$. Applying the truncation process h times to this pyramid yields a polyhedron with $h + s + 1$ faces and $2h + s + 1$ vertices (why?). We now solve for h and s, resulting in $n = h + s + 1$ and $m = 2h + s + 1$. Subtracting the first equation from the second gives $h = m - n$. Multiplying the first equation by 2 and subtracting that from the second equation gives $s = 2n - m - 1$. Thus to construct a polyhedron with n faces and m vertices for any (n, m) in the shaded region between l and k satisfying $n < m$, we can do the following:

- **a.** Start with a pyramid whose base is bounded by a regular polygon with $s = 2n - m - 1$ sides. Note that by Corollary 15.2, $2n - m \geq 4$ (why?), and hence such a pyramid exists and has $2n - m$ faces and $2n - m$ vertices.
- **b.** Apply the truncation process $m - n$ times to this pyramid to produce the desired polyhedron with n faces and m vertices.

For example, the point $(n, m) = (8, 10)$ is in the shaded region satisfying $m > n$. Starting with a pyramid whose base is a regular polygon with $s = 2n - m - 1 = 2(8) - 10 - 1 = 5$ sides, we have a polyhedron with 6 faces and 6 vertices. Applying the truncation process to this polyhedron h times, where $h = m - n = 10 - 8 = 2$, gives a polyhedron with 8 faces and 10 vertices, as desired.

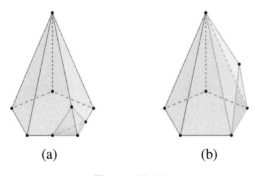

(a) (b)

Figure 15.26

3. If $n > m$, consider a pyramid whose base is bounded by a regular polygon with t sides. We next show how to choose t (in terms of n and m) so that this pyramid can be transformed into a polyhedron with n faces and m vertices satisfying $n > m$. *Augment* one of the faces of the pyramid as shown in Figure 15.26(b), that is replace that face by a pyramid that bulges to the outside such that the resulting polyhedron is still convex. This causes the number of faces to increase by 2, the number of edges to increase by 3, and the number of vertices to increase by 1. Note that each of the new faces produced by augmentation can be augmented as well. Therefore,

15.5 Euler's Formula and Its Consequences

the augmentation process can be applied indefinitely, which allows us construct the desired polyhedron satisfying $n > m$.

More specifically, for any (n, m) in the shaded region between l and k satisfying $n > m$, start with a pyramid whose base is bounded by a regular polygon with $2m - n - 1$ sides (why?). Such a pyramid exists because $2m - n \geq 4$, and it has $2m - n$ faces and $2m - n$ vertices. Applying the augmentation process $n - m$ times to this pyramid produces the desired polyhedron with n faces and m vertices.

The following two corollaries give some further consequences of Euler's formula. The proofs of these corollaries are straightforward and hence are left as exercises.

Corollary 15.3

For any convex polyhedron $\mathcal{P}$:
1. $3F - E \geq 6$ and $3V - E \geq 6$.
2. $2F \geq V + 4$ and $2V \geq F + 4$.
3. $\frac{2E}{F} < 6$.

Remark 15.1

The ratio $\frac{2E}{F}$ can be thought of as the average number of edges in a face. Furthermore, Corollary 15.3 says that $\frac{2E}{F} < 6$. Hence we can conclude that it is impossible to construct a polyhedron whose faces are all hexagons.

Corollary 15.4

1. *Any convex polyhedron $\mathcal{P}$ must have at least one three-sided, four-sided, or five-sided face.*
2. *Any convex polyhedron $\mathcal{P}$ must have at least one vertex where three, four, or five edges meet.*

Note that the previous two corollaries hold for all polyhedra satisfying $V - E + F = 2$, not just convex polyhedra.

15.5.1 Problem Set

1. Prove Lemma 15.3.
2. Let $\mathcal{P}$ be a polyhedron. Prove the following:
   a. If all faces of $\mathcal{P}$ are bounded by n sided polygons, then $nF = 2E$.
   b. If all faces of $\mathcal{P}$ are triangular, then F is even, and E is a multiple of 3.
   c. If q edges meet at each vertex of $\mathcal{P}$, then $qV = 2E$.

3. Prove Lemma 15.4.

4. Prove Corollary 15.2.

5. Prove Corollary 15.3.

6. Prove Corollary 15.4.

7. Is there a convex polyhedron with 5 vertices, 10 edges, and 6 faces? Explain.

8. Is there a convex polyhedron with 22 vertices, 32 edges, and 12 faces? Explain.

9. Is there a convex polyhedron having four triangles and five pentagons for faces? How about one having three triangles, four pentagons, and five heptagons? Explain.

10. Show that any convex polyhedron whose faces consist of squares or regular hexagons must have exactly six squares [21]. Do you have enough information to determine the number of hexagons? Explain.

11. Suppose the faces of a convex polyhedron consist of squares and regular hexagons such that each square is surrounded by hexagons and each hexagon is surrounded by three squares and three hexagons. How many hexagons are there?

12. Show that a convex polyhedron whose faces are pentagons or hexagons must have exactly 12 pentagons.

13. Let $\mathcal{P}$ be a convex polyhedron made up of m squares and n regular pentagons [55]. Assume exactly three edges meet at each vertex.
    a. Show that $3V = 2E$, $4m + 5n = 2E$, and $m + n = F$.
    b. Use Euler's formula to show that $\mathcal{P}$ satisfies $2m + n = 12$.
    c. How many such polyhedra are there? Use a geometric construction kit such as *Polydron* to build examples of these polyhedra.

14. Show that any convex polyhedron whose faces are congruent equilateral triangles, where the number of faces meeting at each vertex is 5 or 6, has exactly 12 vertices each common to 5 edges.

15. In general, are solid angles with the same measure congruent? Justify your answer.

16. Show that the measure of the solid angle at any of the vertices of a regular octahedron is ≈ 1.359 steradians. Recall that a regular octahedron has eight congruent equilateral triangular faces, four of which meet at every vertex.

17. *Legendre's Proof of Euler's Formula*: Legendre [75, Chapter 10] provided an ingenious proof of Euler's formula by using *radial projection* onto a sphere. Given a convex polyhedron, imagine this polyhedron is placed inside a sphere such that the center of the sphere is inside the polyhedron. Projecting the vertices of the polyhedron onto the surface of the sphere projects the edges of the polyhedron onto arcs of great circles on the surface of the sphere. These arcs divide the surface of the sphere into spherical polygons that are related to each other exactly as their corresponding plane polygons on the surface of the polyhedron. The figure below shows the radial projections of a regular tetrahedron and a cube.

15.5 Euler's Formula and Its Consequences

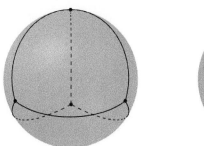
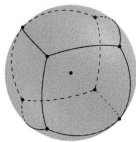

Start with a convex polyhedron having n faces, $F_1, F_2, F_3, \ldots, F_n$ such that F_i has m_i sides. Its radial projection onto a unit sphere will have n spherical polygons P_i each having angle sum S_i and m_i sides. The surface area of the sphere is 4π, which is equal to the sum of the areas of the spherical polygons corresponding to the faces of the polyhedron. Recall that the area of a spherical triangle is defined to be its *excess* (its angle sum minus π).

    a. Show that the area of each spherical polygon P_i is $S_i - (m_i - 2)\pi$.

    b. Show that the sum of the areas of P_i is $(S_1 + S_2 + \cdots + S_n) - (m_1 + m_2 + \cdots + m_n)\pi + 2n\pi$.

    c. Conclude that $4\pi = 2\pi V - 2\pi E + 2\pi F$ and hence $2 = V - E + F$, where V, E, F are the numbers of vertices, edges, and faces respectively. *Hint: The sum of the angles at a given vertex V_i on the sphere is equal to the full angle at V_i in the plane tangent to sphere at V_i which is 2π. Hence $2\pi V = (S_1 + S_2 + \cdots + S_n)$.*

18. The *Archimedean solids* are a set of 13 polyhedra each of which is a convex polyhedron whose faces are regular polygons of two or more types that meet in the same pattern around each vertex and all have equal side lengths. Refer to [20] for images of these polyhedra or find them online. They were described by Pappus of Alexandria around 340 C.E., who attributed them to the Greek mathematician Archimedes (287–212 B.C.E.). The figure below shows the *nets* for the truncated cube, truncated octahedron, and truncated icosahedron. A *net* of a polyhedron is an arrangement of nonoverlapping, edge-joined solid polygons in the plane, which when folded along the edges, form the faces bounding the polyhedron. A *solid polygon* here is the union of the polygon and its interior.

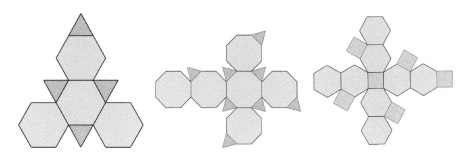

a. Use a geometric construction kit such as *Polydron* or dynamic geometry software such as GeoGebra to construct some of these solids.

b. Show that the sum of the deficiencies at the vertices of each of the polyhedra whose nets are shown above is 4π.

c. Traditionally, a soccer ball is constructed by stitching together 32 pieces of leather: 12 regular pentagons and 20 regular hexagon all having the same side length such that no two pentagons are adjacent (so each pentagon is surrounded by hexagons). Therefore, the surface of a soccer ball can be thought of as the radial projection (see Problem 17) of the surface of a truncated icosahedron. How many vertices, edges, and faces does a truncated icosahedron have? Justify your answers.

19. We can construct the *dual* of certain convex polyhedra using the procedure outlined below [7]. First, given a polyhedron $\mathcal{P}$, take the center of each face of $\mathcal{P}$ to represent a vertex of its dual $\mathcal{P}'$. Second, an edge common to two faces in $\mathcal{P}$ would correspond to an edge connecting the centers of these two faces in $\mathcal{P}'$. Finally, a vertex v of $\mathcal{P}$ would correspond to the face in $\mathcal{P}'$ determined by the centers of the faces containing v in $\mathcal{P}$ (if such a face exists). We restrict our consideration here to polyhedra where these centers of the faces containing v lie in a plane and hence define a face of the dual polyhedron $\mathcal{P}'$. This certainly holds for regular polyhedra, as shown in the figure below, which shows a tetrahedron and a cube and their dual polyhedra.

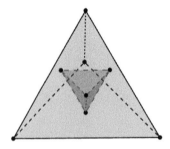 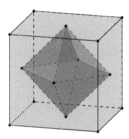

a. Construct the dual of the dodecahedron; what do you notice?

b. How is this related to the duality principle in projective geometry?

c. What can you say in general about the duals of the regular polyhedra?

d. A polyhedron $\mathcal{P}$ is *self-dual* if it is similar to its dual polyhedron; for example, the tetrahedron is self-dual. If E, F, and V are respectively the numbers of edges, faces, and vertices of the polyhedron, how are E and F related to V for a self-dual polyhedron?

Bibliography

[1] "Adrien-Marie Legendre, biography".
https://mathshistory.st-andrews.ac.uk/Biographies/Legendre/.

[2] Nicola Arcozzi. In: *Mathematicians in Bologna 1861-1960*. Ed. by S. Coen Editor. http://www.dm.unibo.it/~arcozzi/beltrami_sent1.pdf. Birkhäuser, 2012, 1–30.

[3] M. E. Barnes. "Morley's Triangle". In: *Pi Mu Epsilon Journal* 11.6 (2002), pp. 293–298.

[4] Michael F. Barnsley. *Fractals Everywhere*. Academic Press, 1988.

[5] "Bay Area Mathematical Olympiad 2008".
https://www.bamo.org/archives/problems_and_solutions/. 2008.

[6] George David Birkhoff and Ralph Beatley. *Basic Geometry*. Chelsea, 1940.

[7] Ethan D Bloch. *Polygons, Polyhedra, Patterns & Beyond*.
https://faculty.bard.edu/bloch/math107_notes.pdf. 2015.

[8] Benjamin Bold. *Famous Problems of Geometry and How to Solve Them*. New York: Dover, 1969.

[9] Roberto Bonola. *Non-Euclidean Geometry - A Critical and Historical Study of Its Development*. New York: Dover, 1956.

[10] David Butler. "The Line at Infinity: Coordinates".
https://blogs.adelaide.edu.au/maths-learning/2016/07/12/the-line-at-infinity-coordinates/. 2016.

[11] Debra S Carney and John Gilbert. *Chapter 5 Inversion - University of Texas at Austin*. https://web.ma.utexas.edu/users/rodin/333L/chp5.pdf. 2004.

[12] J. Castellanos, J. Austin, and E. Darnell. "NonEuclid: Interactive Javascript Software for Creating Straightedge and Collapsible Compass Constructions in the Poincaré Disk Model of Hyperbolic Geometry".
https://www.cs.unm.edu/~joel/NonEuclid/. 1994-2016.

[13] Baris H. Çolakŏglu and Rüstem Kaya. "A Synthetic Approach to the Taxicab Circles". In: *Applied Sciences* 9 (2007), pp. 67–77.

[14] Malin Christersson. *Animate Julia Sets*. http://www.malinc.se/m/JuliaSets.php.

[15] Malin Christersson. "Non-Euclidean Geometry".
http://www.malinc.se/noneuclidean/en/poincaretiling.php. 2018.

[16] N. A. Court. "Elements at Infinity in Projective Geometry". In: *National Mathematics Magazine* 19.3 (1944), pp. 141–146.

[17] H. S. M. Coxeter. *Introduction to Geometry*. Wiley, 1961.

[18] H. S. M. Coxeter. "Inversive Geometry". In: *Educational Studies in Mathematics* 3.3/4 (1971), pp. 310–321.

[19] H. S. M. Coxeter and S. L. Greitzer. *Geometry Revisited*. Mathematical Association of America, 1967.

[20] Peter R. Cromwell. *Polyhedra*. Cambridge University Press, 2004.

[21] Donald M. Davis. *The Nature and Power of Mathematics*. Princeton University Press, 1993.

[22] Tom Davis. "Inversion in a Circle". http://www.geometer.org/mathcircles/inversion.pdf. 2011.

[23] George Derise. *A Grand Tour of Physics: General Relativity - W&M*. https://www.wm.edu/offices/auxiliary/osher/course-info/classnotes/derise5.pdf. 2019.

[24] Keith M. Dreiling. "Triangle Construction in Taxicab Geometry". In: *Mathematics Teacher* 105.6 (2012), pp. 474–478.

[25] William Dunham. *Journey Through Genius: The Great Theorems of Mathematics*. New York: Penguin Books, 1990.

[26] Surajit Dutta. "A Simple Property of Isosceles Triangles with Applications". In: *Forum Geometricorum* 14 (2014), pp. 237–240.

[27] Temel Ermiş, Özcan Gelişgen, and Aybuke Ekici. "A Taxicab Version of a Triangle's Apollonius Circle". In: *Journal of Mahani Mathematical Research Center* 7.1 (2018), pp. 25–36.

[28] Temel Ermiş, Özcan Gelişgen, and Rüstem Kaya. "On Taxicab Incircle and Circumcircle of a Triangle". In: *KoG* 16 (2012), pp. 3–12.

[29] Euclid and Thomas Heath. *The Thirteen Books of Euclid's Elements*. Dover Publications, Inc., 1956.

[30] Pasquale J. Federico. *Descartes on Polyhedra - A Study of the De Solidorum Elementis*. Springer, 1982.

[31] *Fractal Folds*. https://users.math.yale.edu/public_html/People/frame/Fractals/Labs/PaperFoldingLab/PaperFoldingLab.html.

[32] Michael Frame, Benoit Mandelbrot, and Nial Neger. *Fractal Geometry*. https://users.math.yale.edu/public_html/People/frame/Fractals/. 2021.

[33] David Gans. *An Introduction to Non-Euclidean Geometry*. New York - London: Academic Press, 1973.

[34] Chaim Goodman-Strauss. "Compass and Straightedge in the Poincaré Disk." In: *The American Mathematical Monthly* 108.1 (2001), pp. 38–49.

[35] Marvin J. Greenberg. *Euclidean and Non-Euclidean Geometries: Development and History*. San Francisco: W. H. Freeman, 1980.

[36] George B. Halsted. *Girolamo Saccheri's Euclides Vindicatus*. Chicago - London: The Open Court Publishing Company, 1920.

[37] Sir Thomas L. Heath. *The Thirteen Books of Euclid's Elements - Volume I*. New York: Dover, 1956.

[38] David W. Henderson and Daina Taimiṇa. *Experiencing Geometry: Euclidean and Non-Euclidean with History*. Upper Saddle River, N. J.: Pearson Prentice Hall, 2005.

[39] D. Hilbert and S. Cohn-Vossen. *Geometry and the Imagination*. Chelsea Publishing Company, 1952.

[40] *Hilbert Curve*. https://mathworld.wolfram.com/HilbertCurve.html.

[41] Michael Hvidsten. *Exploring Geometry*. CRC Press, Taylor & Francis Group, 2017.

[42] "IMO Shortlist From 2003-2013". https://highschoolcam.files.wordpress.com/2014/07/imo-shortlist-2003-to-2013-full.pdf. 2014.

[43] Christina Janssen. *Taxicab Geometry: Not the Shortest Ride Across Town*. Dissertation (unpublished), Iowa State University, 2007. URL: https://web.archive.org/web/20111216052147/www.math.iastate.edu/thesisarchive/MSM/JanssenMSMSS07.pdf.

[44] Melissa Johnson and Shlomo Libeskind. "Non-Euclidean Geometry Topics to Accompany Euclidean and Transformational Geometry". Unpublished set of notes. 2009.

[45] A. Kapelner et al. *Fractal Dimension of Broccoli*. Ed. by Glenn Elert. https://hypertextbook.com/facts/2002/broccoli.shtml. 2002.

[46] David C. Kay. *College Geometry: A Discovery Approach with the Geometer's Sketchpad*. Addison-Wesley, 2001.

[47] Nikolaos L. Kechris. *Archimedes (Book of Lemmas)*. https://archive.org/details/enbibliolimmaton/page/n9/mode/2up. Athenes, 2018.

[48] James R. King. *Geometry Transformed: Euclidean Plane Geometry Based on Rigid Motions*. American Mathematical Society, 2021.

[49] L. Christine Kinsey, Teresa E. Moore, and Stratos Prassidis. *Geometry & Symmetry*. Wiley, 2011.

[50] A. Kiselev and Alexander Givental. *Kiselev's Geometry - Book II - Stereometry*. Sumizdat, 2009.

[51] Kenji Kozai and Shlomo Libeskind. "Circle Inversions and Applications to Euclidean Geometry". Unpublished set of notes. 2009.

[52] E. F. Krause. *Taxicab Geometry: An Adventure in Non-Euclidean Geometry*. Dover Books on Mathematics Series. Dover Publications, 1986.

[53] Lennon et al. "Lung Cancer—A Fractal Viewpoint". In: *Nature Reviews Clinical Oncology* 12.11 (2015), 664–675. DOI: 10.1038/nrclinonc.2015.108.

[54] I. Ed. Leonard et al. *Classical Geometry: Euclidean, Transformational, Inversive, and Projective*. Wiley, 2014.

[55] Martin Liebeck. *A Concise Introduction to Pure Mathematics, Third Edition*. CRC Press, 2011.

[56] G.A. Losa et al. *Fractals in Biology and Medicine*. Mathematics and Biosciences in Interaction v. 4. https://books.google.com/books?id=t9l9GdAt95gC. Birkhäuser Basel, 2005. ISBN: 9783764371722.

[57] Joseph MacDonnell. "Theorems of Girolamo Saccheri, S. J. (1667-1733) and his Hyperbolic Geometry". http://www.faculty.fairfield.edu/jmac/sj/sacflaw/sacther.htm.

[58] Benoit B. Mandelbrot. *How Fractals Can Explain What's Wrong with Wall Street*. https://www.scientificamerican.com/article/multifractals-explain-wall-street/. 2008.

[59] Benoit B. Mandelbrot. *The Fractal Geometry of Nature*. Freeman, 1982.

[60] Elena A. Marchisotto. "Mario Pieri and His Contributions to Geometry and Foundations of Mathematics". In: *Historia Mathematica* 20 (1993), pp. 285–303.

[61] Elena A. Marchisotto and James T. Smith. *The Legacy of Mario Pieri in Geometry and Arithmetic*. Boston · Basel · Berlin: Birkhḿyaccauser, 2007.

[62] *Math & the Art of MC Escher*. https://mathstat.slu.edu/escher/index.php/Hyperbolic_Geometry.

[63] Mark McClure. *Javascript Julia Set Generator*. https://www.marksmath.org/visualization/julia_sets/.

[64] Michael McDaniel. *Geometry by Construction: Object Creation and Problem-solving in Euclidean and Non-Euclidean Geometries*. United States: Universal Publishers, 2015.

[65] Edwin E. Moise. *Elementary Geometry from an Advanced Standpoint*. Addison-Wesley, 1990.

[66] C. Stanley Ogilvy. *Excursions in Geometry*. Oxford-England: Oxford University Press, 1969.

[67] AoPS Online. "American Invitational Mathematics Examination 1991". https://artofproblemsolving.com/wiki/index.php/1993_USAMO_Problems/Problem_2#Problem_2. 1993.

[68] AoPS Online. "American Invitational Mathematics Examination 1991". https://artofproblemsolving.com/wiki/index.php/1991_AIME. 2014.

[69] Jeremy Orloff. *Complex Variables with Applications (Orloff)*. https://math.libretexts.org/Bookshelves/Analysis/Complex_Variables_with_Applications_(Orloff). 2021.

[70] Boyd C. Patterson. "The Origins of the Geometric Principle of Inversion". In: *Isis* 19.1 (1933), pp. 154–180.

[71] Alfred S. Posamentier. *Advanced Euclidean Geometry: Excursions for Secondary Teachers and Students*. Key College Pub., 2002.

[72] Alfred S. Posamentier and Charles T. Salkind. *Challenging Problems in Geometry*. Dover Publications, Inc., 1996.

[73] Potla. *Poles and Polars - Another Useful Tool!* https://artofproblemsolving.com/community/c1642h1027724. 2012.

[74] Sanjay Ramassamy. "Miquel Dynamics for Circle Patterns". https://arxiv.org/pdf/1709.05509.pdf. 2008.

[75] David S. Richeson. *Euler's Gem: The Polyhedron Formula and the Birth of Topology*. Princeton University Press, 2019.

[76] Larry Riddle. *IFS Construction Kit*. https://larryriddle.agnesscott.org/ifskit/. 2004-2021.

[77] Joel W Robbin. *Coordinate Geometry*. https://people.math.wisc.edu/~jwrobbin/461dir/coordinateGeometry.pdf. 2005.

[78] "Romanian Mathematical Olympiad 1997". https://kheavan.files.wordpress.com/2011/10/mathematical-olympiads-1997-1998-problems-solutions-from-around-the-world-maa-problem-book-225p-b002kypabi.pdf. 1997.

[79] B. A. Rosenfeld and N. D. Sergeeva. *Stereographic Projection*. Mir Publ, 1986.

[80] Boris A. Rosenfeld and Adolf P. Youschkevitch. "Geometry". In: *Encyclopedia of the History of Arabic Science, Roshdi Rahed, ed.* 2 (1996), 447–494 [569].

[81] Skyler W. Ross. "Non-Euclidean Geometry". https://digitalcommons.library.umaine.edu/cgi/viewcontent.cgi?article=1365&context=etd. 2000.

[82] David C. Royster. "Non-Euclidean Geometry". https://www.ms.uky.edu/~droyster/courses/spring08/math6118/index.html. 2008.

[83] H. E. Slaught and N. J. Lennes. *Solid Geometry, with Problems and Applications*. Allyn and Bacon, 1919.

[84] Saul Stahl. *A Gateway to Modern Geometry: The Poincaré Half-Plane*. Jones and Bartlett, 2008.

[85] John Stillwell. *Sources of Hyperbolic Geometry*. Vol. 10. American Mathematical Society, 1996. Chap. Introduction and Translation to Klein's On the So-called Noneuclidean Geometry, pp. 63–111.

[86] R. Street. *Synthetic Plane Projective Geometry*. http://science.mq.edu.au/~street/ProjGeom.pdf. 1971.

[87] Richard Teukolsky. "An exciting enrichment topic, in Learning and Teaching Geometry, K-12". In: *1987 Yearbook-National Council of Teachers of Mathematics* (1987). Ed. by M. M. Lindquist, pp. 155–174.

[88] Kevin Thompson and Tevian Dray. "Taxicab Angles and Trigonometry". In: *Pi Mu Epsilon Journal* 11.2 (2000), pp. 87–96.

[89] Kevin P. Thompson. "The Nature of Length, Area, and Volume in Taxicab Geometry". In: *arXiv: Metric Geometry* (2011). https://api.semanticscholar.org/CorpusID:118789543.

[90] Edward C. Wallace and Stephen F. West. *Roads to Geometry*. Prentice Hall, 1992.

[91] Christopher Williams. *Online Fractal Generator*. http://usefuljs.net/fractals/about.html. 2014.

[92] James W. Wilson. "Jim Wilson's Home Page". http://jwilson.coe.uga.edu/. 2018.

[93] I. M. Yaglom. *Geometric Transformations III*. Ed. by Anneli Lax. Trans. by A. Shenitzer. Mathematical Association of America, 1973.

[94] Yufei Zhao. "Cyclic Quadrilaterals - the Big Picture". https://sites.google.com/site/imocanada/2009-winter-camp. 2009.

Index

AAA Congruence Condition (Elliptic), 507
AAA Congruence Condition (Hyperbolic), 453
Absolute value, 12, 344
Affine Transformations, 618
Altitude
    definition of, 47
    foot of, 94
Angle(s)
    adjacent, 24
    alternate exterior, 53
    alternate interior, 53
    bisector, 24
    central, 98
    complementary, 24
    definition of, 14
    dihedral, 641, 647
    exterior, 43
    face, 647
    formed by a tangent and a secant, 101
    inscribed, 102
    interior, 43
    interior of, 17
    linear pair, 24
    measure, 18
    measurement axiom, 19
    obtuse, 24
    plane, 641, 647
    remote exterior, 42
    remote interior, 43
    right, 24
    sides of, 102
    solid, 647, 648, 650
    straight, 14
    supplementary, 24
    vertical, 24
Angle of Parallelism, 450, 451, 457, 474
Angular Excess, 648
Apollonius, 203
    Circle of, 204
Apollonius' Problem, 393
Arc(s)
    adjacent, 98
    congruent, 98
    major, 98
    measure of, 98
    minor, 98
Archimedean Solids, 661
Archimedes, 233
    Archimedes approximation of π, 234
Area(s)
    axiomatic approach to, 145
    of a circle, 233
    of a parallelogram, 146
    ratio of areas of similar polygons, 219
    of similar figures, 217
    of a trapezoid, 153
    of a triangle, 148
Arbelos, 389
    Archimedes' Book of Lemmas, 389
    Pappus' *Synagoge*, 390
Asymptotic/Limiting/Horo Parallels, 434
Axiom, 10
    angle measurement, 19
    plane separation, 16

B.C.E., 2
Base
    of a parallelogram, 146
    of a triangle, 31

Beltrami, Eugenio, 436, 458
Betweenness of points, 12
Betweenness of rays, 18
Bisector of a segment, 26
    perpendicular, 26
Bolyai, Johann, 435, 436
Bolyai-Lobachevsky Theorem, 474

C.E., 58
CPCT, 43
Central angle
    intercepts, 98
    subtends, 98
Central Perspectivity, 547, 548
Central Projection, 525, 526
Central similitude, 290
Centrally similar, 290
Circle(s)
    of Apollonius, 204
    area of, 233
    center, 97
    chord, 97
    circumference of, 233
    circumscribed, 56
    congruent, 97
    diameter, 97
    incircle, 56
    inscribed in triangle, 56
    inscribed in polygon, 85
    Nine-Point Circle, 335
    radius, 97
    and similarity, 206
    similarity of, 291
Circumference
    of a circle, 233
Collage Theorem, 621
Common Perpendicular, 433, 445-449
Complete Quadrilateral, 562
Complex
    number, 342
    plane, 343
Complex numbers, 342
    argument of, 353
    geometric interpretation of, 343

    multiplication of, 348
    polar or trigonometric representation
        of, 353
Composition of reflections
    in three concurrent lines, 310
    in three lines, neither concurrent nor
parallel, 311
    in two intersecting lines, 304
    in two parallel lines, 309
Composition of rotations, 319
    half-turns, 323
Cone
    oblique circular, 422
    right circular, 422
Congruence
    of two figures, 267
Congruence conditions for triangles
    Angle, Side, Angle (ASA), 32
    Hypotenuse-Acute Angle, 46
    Hypotenuse-Leg (H-L), 45
    Side, Angle, Side (SAS), 31
Conjugate, 343
Construction(s)
    bisector of an angle, 41
    common tangents to two circles, 137
    equal distances, 86
    equilateral triangle, 39
    Euclidean, 36
    extending a line beyond
        an obstruction, 87
    division of a segment into congruent
        parts, 72
    of geometric mean, 205
    of golden rectangle, 525
    inscribed triangle with minimum
        perimeter, 278
    invalid, 38
    line through a point parallel to a line,
        65
    perpendicular bisector of a segment,
        39
    perpendicular to a line, 40
    redrawing a border, 150
    of regular pentagon, 226

INDEX

of regular pentagon with a given side
of regular polygons, 228
shortest highway, 271
shortest network, 285
steps in, 130
triangle, given two sides and an angle opposite one of the sides, 91
using dilations, 293
Convex set
    definition of, 15
Convex Regular Polyhedra, 632, 655
Coordinate system, 11
    origin, 11
    unit, 11
Cross Ratio, 534

Defect of a Triangle, 456, 458
Deficiency of a Solid Angle, 651
Degenerate parallelogram, 250
De Moivre, Abraham, 354
Desargues' Theorem, 531-533
Descartes, René, 646
Descartes' Lost Theorem, 631, 646, 652
Deterministic algorithm, 621
Dilation, 290
Directed segment, 250
Distance
    formula, 145-147
        from a point to a line, 47
Duality, 542, 555, 559

Elliptic Geometries, 499
Elliptic Parallel Postulate, 500
Euclid, 23
Euclid's *Elements*, 21
Euclid's 5th Postulate, 431
    statements equivalent to, 431
Euler, Leonhard, 293
Euler's formula, 355, 632, 653
    consequences of, 655-659
Excess of a Triangle, 507, 508, 519, 520, 648
Extended Line, 374, 543, 545

Fano's Geometry, 542
Fermat, 233
Fermat's point, 283
Fermat–Torricelli point, 579
Feuerbach's Theorem, 411, 414
Fibonacci Sequence, 231
Fixed point, 234
Foot of an altitude, 94
Fractal dimension, 611
    box-counting dimension, 613, 614
    Hausdorff dimension, 611
    similarity dimension, 612
Fractal geometry, 607
Fractals,
    Apollonian gasket, 610
    Cantor set, 607
    definition of, 609
        Hilbert curve, 608
        Koch curve, 610
        Koch snowflake, 617
        Menger sponge, 616
        paper folding, 618
        self-similar, 609
        Sierpinski carpet, 610
        Sierpinski gasket (triangle), 609

Gauss, Carl Friedrich, 229, 435
Gauss-Wantzel Theorem, 486
GeoGebra, ii
Geometric
    mean, 205
    sequence, 205
Geometry
    elliptic, 50, 499
    Euclidean, 50
        fractal, 607
    hyperbolic, 431, 437
    Non-Euclidean, 50
        projective, 525
        solid, 631
        taxicab, 569
Generalization
    of Pythagorean Theorem, 174, 220, 223 (in Problem 7)

of Treasure Island Problem, 323
  (Now Solve This 7.6)
Girard's Theorem, 519, 521, 523, 648
Glide reflections, 257-258
Golden
  ratio, 223
  rectangle, definition of, 223
Groups, transformation, 333-334,
  360-361

Harmonic
  division, 202
  mean, 191
Hart's Linkage, 408
Height
  of a parallelogram, 146, 150
  of a triangle, 146, 150
Hiker's path, 93, 50
Homogeneous Coordinates, 551
Homothety, 290
Hyper/ultra/super/divergent Parallels,
  433
Hyperbolic Metric, 465
Hyperbolic Parallel Postulate, 442
Hyperbolic Tessellations, 494
  regular Tessellations, 494
    Schläfli Symbol, 494
    dual of, 495
    M. C. Escher, 496

Identity transformation, 333
IFS Construction Kit, 621, 622
Imaginary
  axis, 343
  number, 343
  part, 343
Inverse, 247
Inversion, 363
  about a Euclidean circle, 364
  about a hyperbolic line, 471
  about a Taxicab circle, 599
  in the complex plane, 372
  of polar curves, 370
Involution, 363

Isometry(ies), 240
  first fundamental theorem of, 301
  second fundamental theorem of, 303
Isosceles triangle theorem, 31
Iterated Function System (IFS), 619
  attractor, 620
  contraction mapping, 619

Julia set, 626-628

Kite, 40
Klein, Felix, 298, 458
Krause's *Taxicab Geometry: An
  Adventure in Non-Euclidean
  Geometry*, 570

La Hire's Theorem, 557
Lambert Quadrilateral, 445, 500
Legendre, Adrien-Marie, 439, 441, 660
Line at Infinity (Ideal Line), 530, 543
Lobachevsky, Nikolai, 435
Locus, 105

Mandelbrot, Benoit, 437, 607
Mandelbrot set, 626-628
Mean
  arithmetic, 191
  geometric, 205
  harmonic, 191
Median(s)
  of a triangle, definition, 33
  property of, 75
Midpoint, definition of, 13
Minimal triangle, 279
Minkowski, Hermann 436, 570
Miquel's Theorem, 398, 399
Möbius, August Ferdinand, 245
Möbius Transformations, 470
Models of Elliptic Geometry, 508
  Double Elliptic Geometry, 509
    Beltrami-Riemann Sphere Model,
      509
  Single Elliptic Geometry, 511

The Klein Half-Sphere Model, 512
The Klein Disc Model (Elliptic), 514
Lines in the Klein Disc Model, 514-517
Circles in the Klein Disc Model, 517-518
Models of Hyperbolic Geometry, 458
Klein Disc Model (Hyperbolic), 463-465
Poincaré Half-Plane Model, 458-461
Poincaré Disc Model, 461-463
Constructions in, 476
Regular Hyperbolic Polygons, 486
Regular Hyperbolic Polygons with Right Angles, 487
Hypercycles, 490
Morley, Frank, 6, 338
Morley's Theorem, 6
proof of, 338

Nasir al-Din al-Tusi, 432
Nine-Point Circle, 335
NonEuclid, 481

Omar Khayyam, 432

Pappus, 174
Pappus' Hexagon Theorem, 538
Parallel lines
properties of, 52
Parallel and Perpendicular Planes and Lines, 633-639
Parallel postulate
Euclidean, 22
Hyperbolic, 22
Parallelogram
definition of, 63
degenerate, 250
height of, 146
Pascal's Theorem, 535
Pasch's Axiom, 499
Peaucellier-Lipkin Linkage, 394

Pedal triangle, 280
Pentagon
construction of regular, 223
property of regular, 226
Perspective, 531
π (pi)
Archimedes' approximation of, 208-211
modern evaluations of, 239
Plane
half, 15
separation axiom, 16
Plane Axioms, 633
Playfair, John, 58
Playfair's Axiom (see Parallel postulate), 58
Poincaré, Henri, 437, 458
Points
collinear, 7
coplanar, 7
equichordal, 15
Point at Infinity (Ideal Point), 364, 530, 543, 551
Polar Circles, 564
Polars (relative to a circle), 555, 556
conjugate lines, 558
self-Conjugate lines, 559
Poles (relative to a circle), 555, 556
conjugate points, 558
Polygon(s)
circumscribable, 115
circumscribes a circle, 115
inscribed in a circle, 107
ratio of areas of similar polygons, 219
Polyhedron, 631
augmentation of, 658
convex, 645
definition of, 644
dual of, 662
net of, 661
regular, 645
self-dual, 662
truncation of, 657

Poncelet, Jean-Victor, 539
Postulate, 8
    angle addition postulate, 19
    angle construction postulate, 19
    ruler, 12
Projection
    parallel, 69
    of a point, 69
    perpendicular, 47
    stereographic, 419, 515, 536
Projective Geometry, 525
Projective Planes, 541
    definition of, 545
    axioms of, 541
Projectivity, 548
Prouhet, E., 652
Ptolemy's Theorem, 396, 397
Pythagorean Theorem, 2, 159
    converse of, 164
    Euclid's proof, 164
    generalization of, 174, 220
    proof of, 160
    proof via similarity, 204

Quadrangle, 559
Quadrilateral(s)
    inscribed in a circle, 107
    Saccheri, 54

Random iteration algorithm, 621
Ratio and proportion
    properties of, 177
Ratio
    of areas of similar polygons, 219
    golden, 223
Ray
    definition of, 13
    definition of betweenness of, 12
Real Projective Plane, 543
    definition of, 545
Real Projective Space, 546
Rectangle, 54
    definition of a golden, 223
Reciprocation, 555

definition of, 557
Reflection in a line, 51
Rhombus, 41
Riemann, Bernhard, 436
Right triangle, 43
    hypotenuse, 43
    legs, 43
Ruler, 37
Russell, Bertrand, 21

Saccheri, Girolamo; 432, 434
    Hypothesis of Acute Angle, 433, 434, 499
    Hypothesis of Obtuse Angle, 432, 499
    Hypothesis of Right Angle, 432
    Quadrilateral, 432, 438, 500
    Saccheri - Legendre Theorem, 439
Scale factor, 179
Second fundamental theorem
    of isometries, 303
Segment
    definition of, 13
    directed, 250
Self-Polar Triangles, 564
Separation, 509
Sequence
    Fibonacci, 231
    geometric, 250
Shoemaker's Knife Problem, 389
Shortest highway, 271
Shortest network, 285
Similar, centrally, 290
Similar figures
    AA condition for triangles, 185
    definition of, 177, 290
    polygons, 179
    SAS similarity condition for triangles, 188
    SSS similarity condition for triangles, 187
Similarity of Circles, 291
Similitude, center of, 235
Similitude, central, 290

INDEX

Slope(s), 200 (problem 19)
   of perpendicular lines, 200
Solid geometry, 631
Square(s)
   definition of, 63
   on sides of a quadrilateral, 326
Steiner, Jacob, 278, 286
Steiner Chains, 402
Steiner's minimum distance problem, 4
Steiners networks, 285
Steradians, 648
Stereographic Projection, 419, 515, 536
Straight edge, 36
Symmetry(ies)
   of a figure, 254
   of a rectangle, 255

Tangent, 38, 100
Taxicab
   circle(s) through 3 points, 589
      Çolakoğlu and Kaya, 592
   circle, 569
   conic sections (ellipse, parabola, hyperbola), 593
   cylinder, 604
   distance from a point to a line, 582
   incircles, circumcircles, excircles and Apollonius circle, 595-599
   metric, 570
   midsets, 584
      applications of, 586, 587
   π, 571
   SAS, ASA, SAA, SSS, 577
   t-radian, 574
   trigonometric identities, 576, 577
   unit sphere, 603, 604
Theorem
   Bolyai-Lobachevsky, 474
   Circle of Apollonius, 204
   Collage, 621
   converse of side-spliiting, 183
   de Moivre's, 354
      Desargues', 531-533
      Descartes' Lost, 631, 646, 652

Euler line, 292
exterior angle, 43
Feuerbach's, 411, 414
first fundamental theorem of isometries, 301
Girard's, 519, 521, 523, 648
inscribed angle, 102
isosceles triangle, 31
La Hire's, 557
midsegment, 73
midsegment for trapezoids, 78
Miquel's, 398, 399
Morley's, 6, 338
Napoleon's, 328
Nine-Point Circle, 5, 335, 409, 566
Pappus' Hexagon, 538
   Pascal's, 535
Ptolemy's, 396, 397
   Pythagorean, 2, 159
SAS similarity condition for triangles, 188
SSS similarity condition for triangles, 187
second fundamental theorem of isometries, 303
   side-splitting, 182
   triangle inequality, 49
   Marion Walter's, 343 (Problem 8)
Transformation(s)
   composition of, 297
   identity, 247
   invariant properties of, 253-254
   inverse, 247
   orientation, 256
   similarity of the plane, 290
Transformation groups, 333-334, 360-361
   associative property, 333
   closure, 333
      in complex plane, 360
      identity, 333
      normal subgroup, 334
      subgroup, 333
Translations, 250

Trapezoid
    definition, 63
Treasure Island Problem, 4, 6, 80,
        85, 264
    solved using complex numbers, 351
    solved using composition of
        rotations, 321-323
Triangle(s)
    AA similarity conditions, 185
    acute, 31
    base, 31
    congruence, 29
    created by medians, 152
    cyclic, 107
    equilateral, 31
        height of, 47
        inequality, 40
        isosceles, 31
        minimal, 279
        obtuse, 31
        pedal, 280
    right, 43
    SAS similarity condition, 188
    SSS similarity condition, 187
    vertices of, 14
Triangle Inequality, 580

Vanishing Line (Horizon), 526, 529
Vanishing Point, 527
Vector(s), 250
    addition of, 346
    additive inverse, 346
    head, 251
    multiplication by a real number, 346
    position, 344
    subtraction of, 346
    tail, 251

Wantzel, Pierre, 85, 229

y-intercept, 149